AFRICAN MINING '91

Papers presented at the 'African Mining '91' conference, organized by the Institution of Mining and Metallurgy in association with the Zimbabwe Section of the Institution, the Geological Society of Zimbabwe and the Zimbabwe Institute of Engineers, and held in Harare, Zimbabwe, 10–12 June 1991.

ORGANIZING COMMITTEE

E. L. Dempster (Co-chairman) United Kingdom
Dr K. A. Viewing (Co-chairman) Zimbabwe

B. Basu Zimbabwe
B. Baverstock United Kingdom
W. M. F. Box United Kingdom
M. J. Brooks United Kingdom
C. Castelin Zimbabwe
B. Coghill Zimbabwe
M. E. Crowe United Kingdom
G. A. Ferguson United Kingdom
D. Ferreira Zimbabwe
Dr R. P. Foster United Kingdom
W. P. Furusa Zimbabwe
Dr J. D. G. Groom Zimbabwe
Dr F. P. Gudyanga Zimbabwe

D. I. Hodson United Kingdom
Dr F. D. Horscroft United Kingdom
R. M. Kadenhe Zimbabwe
P. Markham Zimbabwe
E. R. Morrison Zimbabwe
Dr F. Munezvenyu Zimbabwe
J. L. Nixon Zimbabwe
M. R. Richardson Zimbabwe
Dr A. E. Roberts Zimbabwe
T. J. A. Smith United Kingdom
J. R. Taylor Zimbabwe
W. D. Wallace Zimbabwe
A. D. Wilson Zimbabwe

AFRICAN MINING '91

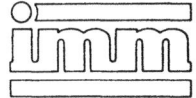 Published for the
INSTITUTION OF MINING AND METALLURGY
by
ELSEVIER APPLIED SCIENCE
LONDON and NEW YORK

ELSEVIER SCIENCE PUBLISHERS LTD
Crown House, Linton Road, Barking, Essex IG11 8JU, England

Sole Distributor in the USA and Canada
ELSEVIER SCIENCE PUBLISHING CO., INC.
655 Avenue of the Americas, New York, NY 10010, USA

WITH 81 TABLES AND 218 ILLUSTRATIONS

© 1991 INSTITUTION OF MINING AND METALLURGY
© 1991 SHABANIE AND MASHABA MINES (PVT) LTD—pp. 279–290
© J. STOCKS 1991—pp. 361–364

British Library Cataloguing in Publication Data

African Mining. Conference. *1991: Harare, Zimbabwe*
African mining '91
1. Africa. Mining
I. Title   II. Institution of Mining and Metallurgy
622.096

ISBN 1-85166-654-0

Library of Congress CIP data applied for

Foreword

The second 'African Mining' conference is planned for June 1991, and follows the first, very successful, event held in May 1987. That full four-year period was characterized by substantial changes in the political and economic climate of many countries in both hemispheres.

Copper prices were relatively firm, and the advance and steady demand for nickel and ferrochromium stabilized important sectors of the mineral industry, certainly in Zimbabwe. The promise for gold remained unfulfilled, but the smaller, relatively flexible, mines survived and only the large, deep and low-value mines seem seriously at risk. None of this has affected the hungry, and intensive exploitations from surface to the water-table have revealed many targets of promise to those willing to take the risks.

The pattern in Southern Africa was extraordinarily stable among the turmoil, with independence for Namibia, adjustments in South Africa and a gradual shift to market economies in the region. The pace of exploration has increased to recover some part of the progress that was lost in the Independence struggle, and at the end of the first decade in Zimbabwe, for example, oil is being sought in the Zambesi Rift, following the investigation of the Luangwa in Zambia, and there are exciting exploration projects for methane released from coal, deep in its basins.

The importance of these sources of inexpensive energy for developing countries is emphasized by the ransom in the wake of the Gulf War, so that much can be expected from those countries which have no oil but which do have coal.

These four years have seen also a dramatic increase in the use of lead-free petrol in the developed countries and the concomitant surge in the production of platinum group metals. The effect has been to stimulate the search for exploitable deposits; old prospects are re-evaluated, old shafts are re-opened, and those involved in the feasibility wrestle with metal values linked to inflationary trends, and to rates of exchange.

The results of many of these changing facets of our industry are described in the papers in *African Mining '91*, and those which are not are certain to be discussed either formally or informally at the conference. Many of the papers deal with advances in technology, which is the main reason for the meeting, and this conference is supported, as was the first, by the Mining Equipment Exhibition MINEX-91. This is proof of the substantial support to the mining industry that is available within the region for really appropriate technology. All of this can be seen in action during the visits to the mining and metallurgical operations, which are an integral part of the conference.

We know that 'African Mining '87' resulted in a surge of interest, and we have no doubt that the 1991 conference will extend the level of confidence in investment and industry in the region. That is the main motivation for the stimulus and organization led by the IMM, London, but with the support of the Minister and Ministry of Mines of Zimbabwe, and the mining industry in that country. Many are involved in the planning and implementation of the technical programme and of the mine and plant visits, and to all of these most generous sources the Chairmen offer their grateful thanks.

Keith Viewing and **Eric Dempster**
Co-Chairmen

March 1991

Contents

Metallurgy 2

Mining 2

Metallurgy 3

Mining 3

Metallurgy 4

Geology

General 2

General 1

Strategy for increasing gold production: the case of A.G.C. (Ghana) Ltd

S. E. Jonah
Lonrho Plc and Ashanti Goldfields Corporation, Accra, Ghana
J. A. Cox
Ashanti Goldfields Corporation, Ghana

ABSTRACT

This paper deals with the strategy adopted for expanding production at Ashanti Goldfields Corporation (Ghana) Limited. The history of the performance of the mine in recent times is traced. This includes how the mine survived during an extremely difficult period of the country's economic development. The mine went through phases of decline, consolidation, recovery and subsequently into a phased expansion programme.

Escalating mining costs, an uncertain gold price and the need to expand production in order to assist with Ghana's balance of payments problems (exacerbated by depressed cocoa prices), dictated a major change in direction for Ashanti.

The expansion programme includes changing the operations from traditional labour intensive mining of higher grade reserves to more mechanised and modern processes for dealing with all the ores on the mine including substantial tonnages of low grade material. The expansion which commenced in 1985 is expected to reach fruition in the mid to late 1990s when gold production will reach a multiple of the current 400,000 ounces.

The paper describes the major elements of the strategy to implement this expansion programme. Comments are also made on funding arrangements and on the significant parts of the operations, including retreatment of tailings, treatment of near surface oxides, open pit mining and expansion of the underground mine.

INTRODUCTION

The Ashanti Goldfields Corporation (Ghana) Limited (AGC) operates one of the mining world's best known gold mines. Its operations are based at Obuasi, in the Ashanti Region, some 100 miles north west of Accra, the capital city of the Republic of Ghana. Mining on a significant scale with substantial capital outlays commenced in 1897, at a site which had previously had a long history of indigenous small-scale mining activity. The venture is unique for its combination of longevity of existing operations coupled with its impressive prospects for future mining and diversification. In recent years AGC has attracted a great deal of attention for the success of its investment programmes to rehabilitate and expand its operations. The strategy of turnaround of AGC from an aged mining venture in decline to a mine with new challenges and achievements provides an example which may be of interest to other mining operations which require rejuvenation and growth.

HISTORICAL BACKGROUND

Gold production from the several reefs outcropping on surface at Obuasi commenced several centuries ago. In 1897, Ashanti Goldfields Corporation was incorporated as a public company in England and its shares floated on the London Stock Exchange. It acquired the Obuasi mine site and began underground operations.

The company's shareholding structure changed significantly in 1969 when it was acquired by Lonrho Plc. In return for agreeing to extend the company's lease over the mine site, the Government of Ghana acquired 20% of the shares of the company. Lonrho had the other 80% of the shares.

3

In 1972, pursuant to the then Government's policy of "capturing the commanding heights of the economy", a renegotiation occurred between the shareholders of the company. As a result, a new company, namely Ashanti Goldfields Corporation (Ghana) Limited was incorporated in Ghana under Ghanaian Law, in which the Ghana Government had 55% and Lonrho the remaining 45% of the shares. The technical management of the mine was entrusted to Lonrho under a management agreement between the company and Lonrho.

Ashanti Goldfields has been described as the single richest gold mine in the world. This is because from the earliest days of AGC's operations till the very early seventies, each ton of rock produced at the mine contained an average of about an ounce of gold. At various stages, the need to satisfy the dividend expectations of the shareholders of an independent company, the company's apprehension of what would happen to it after independence, and its responses to the mining tax structure in the country contributed to the focus on high grade operations.

In the 1970s, the company began to experience difficulties in obtaining foreign currency. In accordance with exchange control laws of the country, all of the company's foreign currency earnings had to be surrendered to the Bank of Ghana, the central bank, in exchange for the local currency, the cedi. On the other hand, to obtain foreign currency for any item that could not be obtained in Ghana with Ghanaian currency, the company had to obtain an import licence from the Ministry of Trade, and thereafter apply to the central bank for foreign currency to cover the transaction. At the same time, the official exchange rate of the cedi began to deviate from its rate in unofficial transactions. Foreign currency became scarce, and the discretion as to its allocation was in practice exercised in ways which were often quite irrational, discriminating against those in production. It thus became harder for the company to replace plant and equipment and also meet shareholders' expectations of dividend payments. This situation forced managers of the company to engage in practices which were expedient in the circumstance but which were clearly detrimental to the long term financial health of the company. For example, if and when foreign currency became available, managers simply multiplied their orders in excess of their requirements as an insurance against future shortages.

The cost of excess stock held was not viewed as a critical loss of financial resources that might be better employed elsewhere.

As the country's economy declined, Government tended to focus on a few sources for its revenue generation. The mining sector was one of the targeted sources. A higher burden of taxation was imposed on the mining sector. This had the effect, however, of encouraging concentration on the mine's high grade reserves. There was consistent high grading till 1972, culminating in the highest gold output in that year of 530,000 ounces. Fig.1. Thereafter, the mine entered a period of decline. The known high grade reserves had been depleted. An increasing proportion of the ore extracted from the mine contained sulphide material, from which there was a lower rate of gold recovery under the existing metallurgical treatment process. Thus overall gold extraction reduced from one ounce per ton milled to approximately one third of that level in a period of five years.

Between 1972 and 1977, there was a continuous increase in the volume of ore mined and treated. Fig.2. However, gold production declined significantly, to 360,000 ounces in 1977. The decline continued until 1987 when, despite the increase in ore production, gold output dropped to 246,000 ounces. The existing metallurgical process was not sufficient to ensure higher rates of gold recovery from the sulphide material. Besides, the mine infrastructure and related equipment imposed a limit on how much ore could be extracted. The basic mine plant and equipment had been designed for a small mine.

To some extent, the deterioration at the mine was cushioned for a while by the rise in gold prices which resulted from the dismantling of the Bretton Woods system of fixed exchange rate parities and a fixed price for gold. However, the overall impact of the country's economic policy and circumstance on the mining industry, and in particular on AGC was that little exploration activity occurred, essential redevelopment was not carried out, and only minimal replacement capital was invested in the company's operations throughout the 1970s and the first few years of the 1980s.

CONSOLIDATION AND REHABILITATION PROGRAMME

In 1983, the Ghana Government launched its Economic Recovery Programme. The new exchange rate policy it introduced, together with other financial arrangements outlined in the Minerals and Mining Law, 1986 enabled the company to earn sufficient cedis to cover its production costs and also gave ready access to foreign exchange for operations and re-investment. These developments enhanced the credit rating of the country and the company, and made it possible for the company to attract loan finance to undertake much needed rehabilitation and expansion.

The implementation of a US$160 million Rehabilitation and Expansion Programme was began in 1986, funded partly from an an IFC syndicated loan, with AGC contributing almost half of the cost of the programme from internally generated resources. The total package was used to replace plant and equipment in both the mining and metallurgical areas, in the construction and commissioning of a tailings treatment plant, and in the sinking of replacement shafts.

A major portion of the rehabilitation programme consisted of the development of the infrastructure of the underground mine to exploit the ore reserves in its central and southern section. These areas could not be easily accessed from the existing northern shaft systems. The programmes included continuing the development of a major shaft in the southern section and the start up of a new shaft in the central section, each with their attendant haulage furnishings and equipment, ventilation, compressed air and water handling facilities. This infrastructure investment allowed the rational development of AGC's Obuasi Mine southwards along strike.

Under the Re-Equipment Component of the investment package, funds were allocated for modernising underground production and development practices. The aim was to improve productivity and efficiency and hence to offset rising costs. Scrapers, loaders, conveyors and raiseborers were purchased. In comparison to many modern mines of today the re-equipment exercise was not mechanisation in the accepted sense but rather an exercise in replacing old equipment and bringing in the first stages of mechanisation.

At the treatment plant several key areas of the existing plant were upgraded and modified. These were again selected to provide improvements in operating efficiencies. The three principal areas of investment within the existing plant included the replacing of flotation cells with larger, more efficient units, the re-design and re-building of the

concentrate roasting facilities, and the erection of new leach tanks with draft tube agitators in the leach section. In addition, a new process facility was designed and brought on stream to re-process the tailings of previous AGC operations. This comprised a new mill, leach facilities and a carbon-in-pulp plant capable of processing 2000 tons per day of reclaimed tailings.

EXPANSION PROGRAMME

The confidence gained from the successful implementation of the Rehabilitation Project, and the assurance that under the prevailing economic climate good projects could be assured of funding, provided an impetus to transform the company into a dynamic, modern and efficient one. This could be achieved through:
(a) increasingly ambitious exploration and capital programmes;
(b) improvements in our cost competitiveness; and
(c) maintaining a diversified production base from underground and surface reserves, comprising a combination of heap leach and conventional processes, as well as the treatment of tailings, in order to ensure that Ashanti becomes capable of exploiting its considerable lower grade reserves.

The level of investment in exploration was increased and the targets redirected. Fig.3. The mine has traditionally concentrated its exploration activities in the underground. From 1988, AGC began an active exploration programme on the surface. From 605 metres in 1988, surface drilling increased to 53,000 metres in 1990. In order to avoid the risk of a decline in the asset base, it is expected that the current level of expenditure on exploration will be maintained. This aggressive exploration programme has resulted in the delineation of substantial surface reserves. Fig.4. (Total surface reserves currently stand at some 11.0 million tons, averaging 2.6 dwts per ton.)

The successes obtained from the exploration effort facilitated the preparation of a portfolio of projects which will form the core of the expansion programme. The first step is the implementation of the Sansu Project. This consists of the development of several open pits to feed a heap leach plant, and a 2.4 million tons per annum oxide plant. It also includes the exploration underground of ore bodies to the south of the Sansu shaft. The package also embraces the expansion of the tailings treatment plant to double production within a 12-month period from 60,000 to 120,000 tons per month, and permit finer grinding for improved recovery. An additional US$60 million IFC-syndicated loan agreement for the project was concluded in March 1990. The company itself contributed $33 million from internal cash generation.

At the time of writing, the heap leach and the tailings treatment plants have been commissioned. The addition of the heap leach plant has facilitated the treatment of very low grade ore and also provided a certain flexibility in the processing operations: it is now possible to cross-transfer ore feed between different treatment plants from different sources, thus allowing an optimisation of gold production. The result of all this has been an increase in gold output from 242,366 ounces in 1985 to 400,750 ounces in 1990. The original rehabilitation project anticipated a target of 400,000 ounces by 1992. The rapid transfer from rehabilitation into actual expansion has therefore taken place within the original rehabilitation project time. The present production target for 1992 is 620,000 ounces of gold. Fig. 5

EXPANSION STRATEGY

A major element of the expansion strategy involves a comprehensive review of all aspects of our current operations: the primary objective is to attain the optimum production level consistent with the resource base through the application of the most appropriate technology and systems. This will necessarily lead to the lowering of the mine's cut-off grade, thus leading to the profitable exploitation of the enormous but previously untapped low grade resources.

The review has covered the following key areas of the company's operation:
- the variety and potential of its reserves - oxide and sulphide;
- the range of production methods and possibilities:
  - open pit oxides
  - open pit sulphide
  - current underground
  - underground expansion for lower and high grade reserves;
- hoisting and haulage logistics for current operations and for future expansion;
- treatment processes:
  - options for sulphide ores;
  - oxide plant production;
- throughput levels;

- overall handling logistics for ore, waste, men and materials;
- supporting infrastructure; and
- change process from labour to capital intensive operations.

The assessment of the ore reserve potential of the oxides and sulphides has given current indications of some 25 million tons averaging 5 dwts per ton. An efficient and a more profitable exploitation of this would depend very much on the selection and correct sequencing of the mining and treatment options under consideration.

The open pits under the Sansu Project have been evaluated for both oxide and sulphide production. Investment to date has centred on the processing of the oxide ore fractions which overlay the deeper sulphide and quartz ore bodies. In the next phase the open pits will be systematically developed from the original oxide mining operations into deeper hard rock sulphide mining to the economic limit of each pit. The future mining and metallurgical processing of the sulphide ores so exposed is being incorporated into the wider scenario of the expansion of the lower grade underground mineralisations. A rationalisation of all of AGC's future sulphide handling capabilities is thus being incorporated into a single project, termed the Ashanti Mine Expansion Project.

It will be necessary to change radically the mining practices currently employed at AGC by introducing capital intensive mechanisation. New skills, habits and attitudes to work will have to be inculcated. This will have to be achieved without interfering with the current operations. Problems may arise from the coexistence of the manual and the more mechanised operations.

The change, which will take place over a number of years, will therefore require a comprehensive implementation strategy in order to ensure success. It will affect not only all personnel in the mining and engineering departments but also all the other service departments and sections. Unless the process is properly planned and implemented, much of the significant advantages will be dissipated.

The implementation strategy will include the conditioning and training of all levels of manpower from senior management down. This will involve visits and training periods for key personnel to other operations employing similar systems and the development and implementation of standards and procedures for all mining, engineering, and administrative functions. Work is already in progress in structuring the Manpower Development and Training Sections to cater for change by appointing senior managerial personnel to the department and creating the infrastructure for carrying out the tasks. This will include constructing training centres on surface and in the underground environment. Training equipment will be significantly more than that common in mature operations. There will also be provided specific mechanised equipment purchased strictly for training purposes similar to that that will be used in production.

The launching of the Sansu Project and its timely development into full-scale mining operations transformed AGC from being a single underground mining venture into a corporation running two distinct mines. Up to now, mining activity has been concentrated in a very small area of the concession. The company is now expanding into wider and more dispersed areas.

Besides, a typical mining project now involves a complex of operations embracing a variety of professions and skills. The highly centralised management which has been a feature of Ashanti's operations for most of the company's existence has thus become increasingly inadequate and we have responded with changes in organisational structure. In October 1990, separate management teams primarily responsible for production and treatment were created for the Obuasi and Sansu mines. The service functions of the organisation were similarly devolved into a Finance Group to cater for materials management, finance, accounting and information systems, together with a Corporate Services Group with responsibility for projects, exploration, manpower development and for all other aspects of administrative and technical support for the two mines. Each of these teams is now headed by a General Manager. It is intended that future mines will be developed with their own production management, and the centrally controlled service functions will extend their services to the new mines, thus reducing overhead expenditures by removing the necessity to duplicate service functions at each mine location. Lines of responsibility have been clarified and there is now a closer alignment between authority, accountability and responsibility.

HUMAN RESOURCE MANAGEMENT

One important area to which considerable attention is being devoted is the management of the human resource.

At the beginning of the Rehabilitation Programme in 1986, the total number of employees of the company stood at 14,700. In the 1970s and in the early 1980s, the state of the economy was such that the average worker's income from his regular job was not enough to cater for the basic needs of himself and his family. As a result, many workers sought to earn extra income from other activities, and this often involved them in taking time off from their regular employment. In order to deal with the persistent absenteeism of that period, managers overstated their labour requirements and there was therefore considerable overmanning in significant areas of our operations. Furthermore, in reaction to the collapse of the existing infrastructure necessary to service the mining industry the mine had to extend itself to make up for the deficiencies of the external environment by providing its own services.

The company was convinced that the plans for rehabilitation and expansion could only be effected with a loyal, committed and dedicated workforce, which saw its interests being advanced by the company, and felt part of the team through a comprehensive system consultation and communication. Labour also had to be rewarded better. But this could only be done with a leaner and more productive workforce.

One of the ways in which this could be achieved was by identifying the areas where fewer men could carry out the same tasks with greater efficiency and devising a scheme by which to effect this. In theory at any rate, the options for reducing labour were through a process of compulsory lay-offs, by attrition or by re-directing existing labour into new areas of activity. For two reasons, the first option was rejected. It would have been difficult and cumbersome to obtain government approval. Secondly, and far more importantly, it would have been inconsistent with the desire to obtain loyalty and commitment because in that case even those who remained would have been left with a feeling of insecurity. The company therefore opted for a combination of not replacing men who left the organisation and actively involving labour in the identification of the areas of slack and in the internal redeployment of excess personnel.

Previously, there had not been a clearly defined retirement policy, and there was a noticeable reluctance of men to retire. This was largely attributed to the unsatisfactory end of service benefits, which had the effect that employees preferred to die on the job than go on retirement. To address the specific issue of these benefits, a team of actuaries was invited to review the existing scheme and make proposals for their enhancement. The outcome was the introduction of what was generally acknowledged to be a more attractive package. This had the effect of encouraging more people to take voluntary retirement.

In relation to underground labour, different people had traditionally been involved in blasting, installation of support, drilling and mucking. We instituted a training programme to get workers to learn as many of these skills as they could to create 'composite crews'. A mine in Canada was identified to which a core team of twelve workers and two supervisors was sent for extended training and familiarisation in modern methods. The purpose was to demonstrate more effectively than was possible simply by talking, how a mine with a smaller multi-skilled workforce could be more productive. On their return, the team was deployed in stopes which had been selected for the introduction of a new system employing increasing mechanisation.

A scheme of remuneration was proposed in specified areas of underground activity which involved setting the price to be paid for a ton of material moved, and this price was payable to the men who did the job, no matter how many or how few they were. Following its successful implementation, the scheme, now popularly called "the contract bonus scheme", has been extended to other areas of activity, both underground and on surface. In the first eight months of its introduction, underground labour alone was reduced by 30%, and productivity in stopes increased by 44%. The increase in productivity in development work was even more impressive; it shot up by 200%.

In addition, a Christmas gift scheme calculated on the basis of an extra month's salary was replaced by a scheme based directly on gold output depending on the percentage of the year's production target attained. For management staff, the scheme was refined further to take account of the cost of production. Also, for management staff in charge of cost centres, a personal performance bonus scheme has been introduced, where the reward is based directly on how efficiently the budget in a particular centre is managed.

In setting targets, a lot of consultation was done with the workforce and management. This ensures the formulation of realistic targets, and also convinces those who can help achieve them that they are fair and based on relevant considerations. Unless this is done, the predominant responses that the targets would invoke are system circumvention, game playing and other practices which divert energy away from the job.

A major area of concern to our employees has been limited availability of adequate housing facilities at the mine or in its surrounding settlements. To address this, the company has in the past three years, made substantial investments in the provision of housing for various categories of employees. It has also made significant improvements in other social and recreational facilities. Our efforts at fostering greater community within the company have also involved the publication of a monthly newsletter, as well as the sponsoring of a Goldfields football team which now participates in the national football league.

We hope that in this presentation we have succeeded in demonstrating the variety of areas of challenge - technical, financial, organisation and human resource management - which have confronted us as well as our responses to those challenges in our efforts at rejuvenating Ashanti Goldfields and expanding its operations.

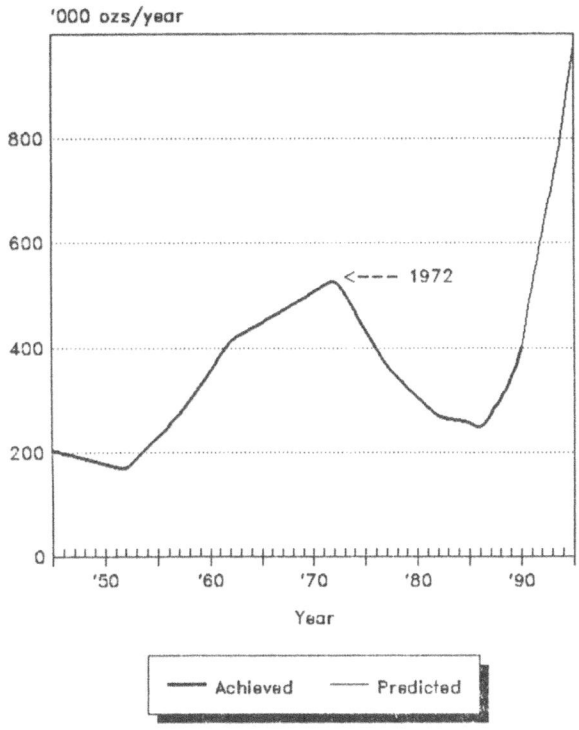

Fig 1. Gold Production over 50 years

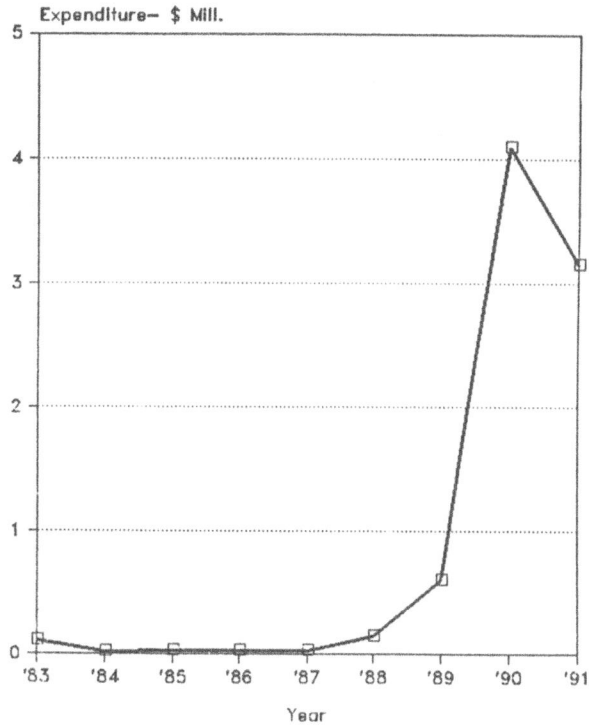

Fig 3. Exploration expenditure profile

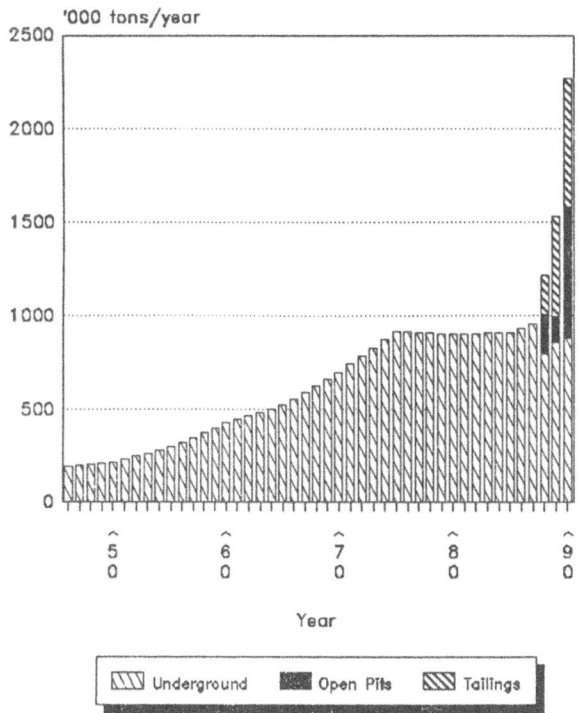

Fig 2. Ore production over 50 years

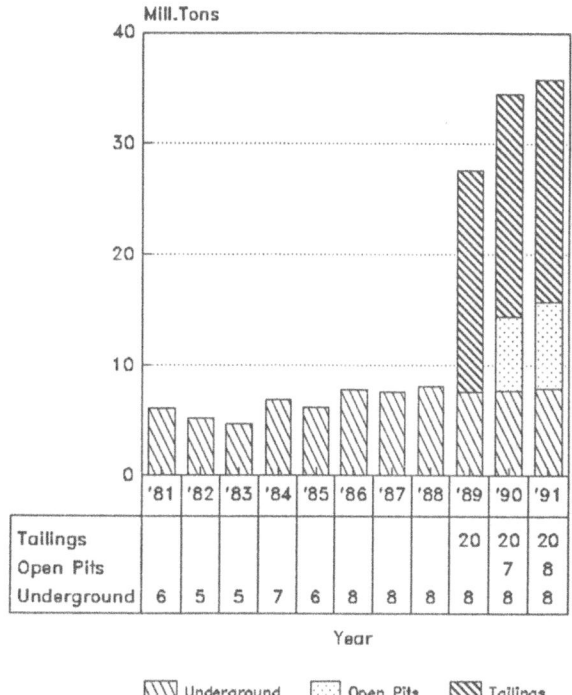

Fig 4. Available payable reserves

	'81	'82	'83	'84	'85	'86	'87	'88	'89	'90	'91
Tailings									20	20	20
Open Pits										7	8
Underground	6	5	5	7	6	8	8	8	8	8	8

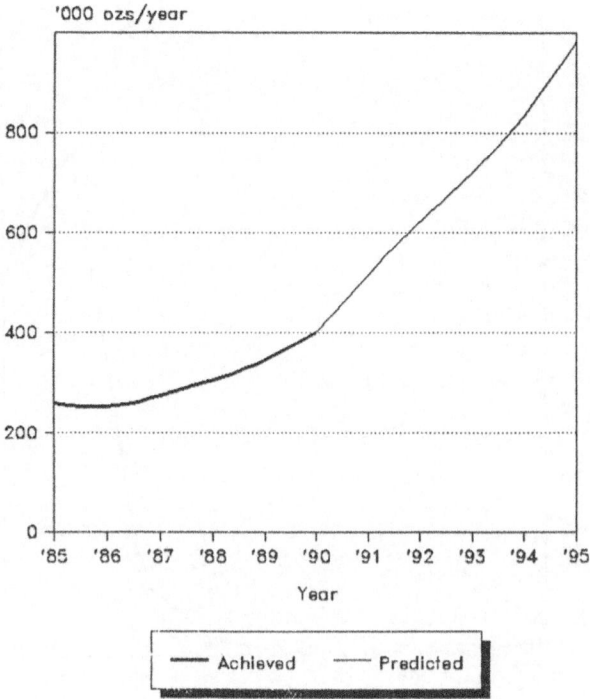

Fig 5. Recent gold production levels and projections

Debt funding for African mining projects: issues, options and sources

S. E. Jonah
Lonrho Plc and Ashanti Goldfields Corporation, Accra, Ghana
Kofi Ansah
Minerals Commission, Accra, Ghana

ABSTRACT

Securing adequate debt financing for mining projects is a difficult task in the best of environments. In Africa, the task is even more difficult on account of debt burdens, poor infrastructure, foreign exchange shortages and perceived political risk. These factors serve to limit the interest of international commercial banks, the main source of mining finance, in making long-term credit available for mining projects in the region. This paper examines issues related to the raising of debt funding for mining projects (particularly gold) in Africa. Issues of concern to potential lenders, such as completion risk, security arrangements, foreign exchange availability and political risk are discussed. Measures which tend to minimise lenders risks, such as escrow accounts, debt conversion, gold loans, hedging programmes, and political risk insurance, are examined. The government's role in enhancing project financing arrangements is also discussed. A survey of available sources of debt financing for African-based gold mining projects is provided. Emphasis is placed on specialized financial institutions which are active in the African lending market. Examples include governmental and inter-governmental credit agencies, export credit agencies, and suppliers' credits. Finally, the paper describes recent experiences in the financing of new and expanding gold mining operations in Ghana.

I. INTRODUCTION

In the last few years a number of gold mining projects have been developed in Ghana. The debt funding arrangements for these projects have had some interesting features and raised important issues. Since these funding arrangements are likely to become common for mining projects in other countries in Africa with similar economic conditions the issues that emerge from them merit thorough discussion.

Generally the development phase of a mining project may be debt-funded in two main ways:

a) Normal Corporate Borrowing: where the sponsors of the project borrow money from commercial banks on the strength of their balance sheets. The sponsors in this case are usually large multinational companies in strong financial positions.

b) Project Financing: this is debt funding where repayment is expected to come solely out of the cash flow from the project itself rather than from other cash sources of the project sponsors as is the case in conventional borrowing; the project assets and the cash expected to be generated provide the basic security for the lenders.

Of the two, project financing is generally the method of financing preferred by sponsors of mining projects and the discussions that follow relate solely to the issues that may arise from project financing in many countries in Africa.

II. ISSUES FOR CONSIDERATION BY THE PRINCIPAL PARTIES INVOLVED IN PROJECT FINANCING ARRANGEMENTS

A. Sponsor Considerations

(a) Reasons for Preferring Project Financing:-

The main reasons why sponsors of mining projects have a preference for project financing include the following:

(i) The sponsors may be foreign mining companies whose balance sheets cannot support conventional commercial borrowing.

(ii) The sponsor may have taken a loan which has covenants restricting further borrowing.

(iii) Even where there is no restriction on further borrowing a sponsor may wish to have borrowing for the project off the company's balance sheet so as to avoid a negative effect on key financial ratios which may in turn worsen the company's credit rating. Accounting practice in the sponsor's jurisdiction may, however, not allow this avoidance of consolidation of debts.

(iv) A key characteristic of project finance is that recourse to the sponsoring company is limited, and the lenders share in the risk.

(b) Choice of Project Entity

Tax and accounting considerations as well as investment laws in the host country are some of the factors that the project sponsors will take into account in choosing the form of entity under which the project will be operated. The forms of project entity that may be chosen by the sponsors include corporations, unincorporated joint ventures and partnerships. Sponsors of mining projects in most developing countries may have no choise but to operate their projects under incorporated joint ventures or companies since the investment laws of these countries often require incorporation to facilitate government shareholding.

(c) Accounting and Tax Considerations

If a key objective of project financing for the sponsors is to keep loans for the project off their balance sheets, it is important to structure the project so as to avoid consolidation of project financial statements with those of the sponsors' parent companies, though this objective may conflict with a sponsor's desire to control the project. Likewise tax considerations can determine the form of project financing that is used.

B. Concerns of Lenders

Only a small proportion of the cost of a mining project is funded from sponsors' equity; the bulk of the cost is funded through debt. Since in project financing, security for the loans is provided by the cash flow and the other assets of the project, the lenders' main concern relates to the factors that can affect the cash flow and the value of the assets. Because of their perception of exceptionally high risk associated with projects generally in Africa, lenders who provide project financing for mining projects in Africa tend to be doubly cautious and put in place complex devices to ensure repayment.

In what follows, we discuss some of the main risks associated with a mining project which lenders would assess before providing project financing, and the devices used to protect against them.

(a) Level of Gearing

The more shareholders' funds in the project the greater the debt service coverage for the senior lenders. Lenders therefore have a great desire to see a high level of shareholders' equity and quasi-equity in the financing structure for the project. A high level of sponsors' own funds in a project also serves to assure lenders that the sponsors (who would be directly responsible for running the project) will run the project efficiently and thereby ensure high cash flow generation. The appropriate level of gearing is usually determined by analysis of a financial model prepared for the project. Different risks associated with the project such as reserve risk, operating risks, gold price fluctuations, etc., together with their probabilities of occurrence are assessed and used in the model to produce a "banker's case".

That case will form an integral part of the lenders' evaluation of the debt capacity of the project. For the typical gold mining project, 60% of the project cost is financed by senior debt and the balance (40%) is financed by sponsors' funds in the form of equity and quasi-equity (e.g. subordinated shareholder loans).

The insistence on certain minimum levels of equity contribution from the sponsors could often have an impact on the eventual ownership structure for the project that is worth mentioning here. Exploration in a number of developing countries is initiated by junior mining companies who often are unable to raise monies to satisfy minimum requirements of equity contribution imposed by lenders when loan funding is sought during the development phase of the project. This results in the entry of a major company into the project with consequent dilution of the shareholding by the juniors to insignificant levels. A very good example of this is the Canadian Bogosu Resources gold mining project in Ghana, where Sikaman, a Canadian junior company that initiated the project, had its share in the project significantly reduced by the entry of Billiton, a major company, because Sikaman could not meet its share of the minimum equity requirement.

(b) Mineable Reserves

Lenders are understandably very concerned with the extent and the quality of mineable reserves presented in a bankable study. Most banks and institutions that provide long-term, non-recourse finance for mining projects in Africa require a minimum reserve life of 7-9 years. A project's mineable reserves must as a minimum be sufficient to cover the scheduled life of the long-term debt. Short mining reserve life poses difficulties for lenders as that reduces the lenders' safety margin. Most lenders require that reserves be independently certified by a reputable geological consulting firm. Similarly, they also usually require that the assay work be undertaken at an interna-tionally recognized laboratory. In evaluating the reserve risk of a project, lenders assess the potential for the mine to continue

to develop new reserves. Good future reserve potential can offset a shorter than desired existing mineable reserve life.

(c) Completion and Cost Overrun Risks

The main concern of lenders in respect of completion is the risk that the project's physical facilities get delayed in completion or that the project, when completed, operates below its design capacity. Completion risk is felt by lenders to be particularly high for mining projects located in Africa mainly on account of poor and unreliable infrastructural links to the project, frequent shortages of construction materials and remoteness of many African countries from centres of equipment manufacturing and technical services. Physical completion will not satisfy lenders; certain operating tests will have to be met. These include throughput targets and minimum recovery levels.

Delays in project completion usually cause cost overruns; but cost overruns can also occur even when construction is on schedule. The lenders are concerned that if cost overruns occur there will still be money beyond that which was originally borrowed to complete the project. Some of the measures commonly taken to protect against this risk are:
(i) the sponsors agree to provide all the excess funds or part of it, through additional direct equity contributions or quasi-equity contributions such as shareholder loans; (ii) the sponsors arrange for a stand-by line of credit from commercial banks to cover all or part of the overruns; (iii) the lenders themselves may agree to provide up to a certain proportion of the overruns; (iv) a host government that believes the project will contribute significantly to the country's economy when completed may agree to cover part of the overruns by on-lending to the project grant or soft money that it has obtained from bilateral or multilateral sources. The host government may also assist in the completion of the project in some other ways. For instance the government may agree to convert into foreign exchange, local currency which the project sponsors may be able to mobilize locally towards funding the foreign exchange portion of the cost overruns.

In Ghana the loan arrangements for the Ashanti Goldfields Corporation (AGC) expansion programme in 1985 and the Canadian Bogosu Resources (CBR) development programme in 1989 contained some of the above measures. In the AGC case it was agreed that in the event of cost overruns, the sponsors (Lonrho of U.K., and the Ghana Government) would provide up to a total of US$10 million (about 6% of the estimated project cost of about US$160 million); and the principal lender, the International Finance Corporation (IFC), agreed to provide up to US$10 million in addition, if it became necessary. The CBR arrangement whose principal sponsor was Billiton had similar provisions.

(d) Operating Risks

The cost competitiveness of a mine is of vital importance for its survival and profitability. Lenders expend much effort in trying to assess the competitive- ness of a potential gold mining project. The basic premise is that low cost producers will generally be viable regardless of what happens to product prices. Flexibility in mining operations, however, can sometimes help cope with a temporarily difficult cash flow situation. For example, the configuration of some mines allows 'sweet ore' to be exploited or the stripping ratio to be reduced without harming the long-term development of the mine when additional cash is needed. Some mines may be able to lay off workers to reduce costs when prices are low.

(e) Price Risk

Market prices are a key factor influencing the financial performance of a mining project. It is well known that gold prices fluctuate widely and that price projections are subject to a high degree of uncertainty. Lenders are indirectly exposed to this price risk and prefer that the mine manages its market risk through the use of hedging programmes. Sometimes lenders insist that a hedging programme be entered into prior to mine start-up and as a pre-condition for disbursement of the loan. At a minimum, the objectives of any hedging programme should be to lock in fixed prices for a portion of gold sales

sufficient to allow for operating costs, senior debt repayment, taxes and royalties to be covered in a depressed market. A carefully designed hedging programme can meet these objectives while at the same time allowing flexibility on delivery commitments and leaving some upside in the event of high gold prices. Floor price schemes, coupled with spot deferred commitments, are designed to achieve these objectives.

It must be ensured that hose governments approve of these hedging programmes since the tax authorities must first agree to tax the project on the basis of revenues actually made rather than those imputed on the basis of spot prices, and secondly accept the cost of participation for income tax purposes. Both of the Ghana projects mentioned earlier do have hedging programmes. The CBR programme is more elaborate and is frequently reviewed to ensure the achievement of the objectives of hedging. In this case the Bank of Ghana is the government agency that approves every programme.

(f) Foreign Exchange Risk

Adequate access to foreign exchange is necessary for the servicing of foreign exchange debts, the remittance of dividends to foreign shareholders and for purchasing spare parts and operating supplies. Foreign exchange availability for a project, therefore, has a great bearing on lenders' assessment of project risk and is especially relevant in the context of Africa. Chronic shortages of foreign exchange in most developing countries have led to the institution of exchange control measures which would normally restrict the availability of foreign exchange to mining projects located in these countries. Lenders will therefore like to see arrangements which effectively side-step these exchange control measures and guarantee free access to foreign exchange for operations and payment of debts. Typically, lenders prefer sales proceeds to be denominated in hard currency and paid to an account located in a jurisdiction where they believe expropriation is unlikely. Usually a specified proportion of sales is kept in the account (often referred to as the 'Retention Account') for the payment of all foreign exchange obligations including debt service.

It is common for the lenders to insist on trust arrangements for the operation of the retention account, whereby the trustee has instructions on how to disburse monies in the account. Debt service is given a high priority under the instructions to the trustee.

For added security regarding the availability of foreign exchange for debt service lenders would also try and get an agreement that in certain circumstances the host government would permit local currency held by the sponsor to be converted into hard currency for payment into the retention account. A typical such circumstance would be when the price of the product falls significantly, thus reducing the monies in the retention account to lower than expected levels.

(g) Political Risk

The perception of high political risk is a major consideration for lenders to projects located in African countries. It is believed that this is the principal reason why many international commercial banks are reluctant to extend long-term credit for mining projects in Africa.

To a large degree political risk may be mitigated through the use of political risk insurance. Project sponsors can take out political risk insurance policies and assign benefits to lenders. Insurance agencies such as MIGA (World Bank), OPIC (USA) and EDC (Canada) specialize in political risk insurance and provide coverage for obvious risks such as expropriation, war and civil unrest, (and even currency inconvertibility).

Lenders who co-lend in association with international organizations such as the IFC often perceive that association as offering protection against certain political risks such as expropriation because of the likely good relationship between the host country and the IFC.

(h) Security

At a minimum, the lenders' security package would consist of a mortgage over the project fixed assts and a floating charge over

the project's non-fixed assets (bank accounts, inventories etc.). But the package often covers many more of the sponsors' interests connected with the project. These include the sponsors' title to mine the concession (i.e. the mining lease), interest in insurance policies and other contracts related to the project. Some of these security arrangements give rise to issues of concern to host governments and are discussed later.

Because it is felt that some of the elements of the security package may be difficult to enforce in some developing countries in the event of default, some lenders attempt to get an overall guarantee from the project sponsors, especially where the host government is a major sponsor.

(i) Environmental Concern

In recent times the environmental impact of mining projects in developing countries has become of some concern to lenders, especially the bilateral and multilateral lending agencies. These lenders require that a full environmental impact study be included in the loan documentation, and that stringent environmental standards be maintained during operation of the project. The recent IFC-led loan to Ashanti Goldfields Corporation in Ghana required that, in addition to all this, an environmental expert be engaged full-time on the project.

C. Host Government Considerations

The matters that host governments consider and assess in any financing arrangement relate mainly to the form of entity under which the project will be run; the lenders' security arrangements; debt to equity gearing; and measures to guarantee foreign exchange to the project.

(a) Form of Project Entity

The host government would prefer that the mining project runs under a single entity incorporated in the country. This affords the government better control over the running of the project and also makes it easier for the country's tax authorities to tax profits from the project effectively. If it were to deal with the sponsors individually (as it might have to if the entity were an unincorporated joint venture, for example) the government could have difficulties getting access to adequate information for many purposes including taxation. It is also administratively easier for the government to arrange title to the concession for an incorporated single entity.

(b) Security for Lenders

An important point of consideration for the host government is the manner in which lenders would assess the risk associated with a mining project located in the country and the resultant level of security that would be required. It is the suspicion of most developing countries that risk assessment by most foreign lenders is based on popular sentiments regarding their countries generally, rather than on the basis of the likelihood of occurrence of adverse events based on actual historical facts about the country. For instance, where as a matter of historical fact expropriation of foreign-owned projects has rarely or never occurred, the host government would like the risk assessment model to have this low probability as an input, rather than the usually high expropriation probability factor based on general perceptions.

Another element of the lenders' security package that causes some concern to host governments relates to the sponsors' title to the concession. As part of their security, lenders invariably require that the government undertakes not to revoke sponsors' title to the concession even if they breach provisions that would normally call for revocation. Some of these provisions may be important to government; they may, for instance, serve to ensure prompt payment of royalties and rents to the government. Rendering them ineffective for the sake of lenders' security is therefore of considerable concern to the government. In the AGC and CBR projects in Ghana, the government very reluctantly agreed to eliminate provisions of this nature from its standard mining lease.

Other security measures such as the right of lenders to appoint successors to the sponsors to run the project in the event of default in loan repayment do not, generally speaking, conflict with the

interest of the host government. The implementation of the measure could also ensure that the government gets royalties, taxes and other obligations due it.

In addition to all these security arrangements some lenders would insist on sovereign guarantee where the host government is a major sponsor. This is vigorously resisted by host governments. But where it is compelled to give this guarantee, the government in turn requires the other sponsors to indemnify it to the extent of their ownership interest in the project.

(c) Debt/Equity Gearing

Usually the main interest of government here is that investors must put in substantial equity into the project. This ensures that interest payments on loans would not be inordinately high to reduce taxes. However, where the government wishes to participate in the project, it may propose a lower level of equity so as to make its contribution affordable, especially in the case where the government on its own is able to attract concessionary financing for the project.

(d) Availability of Foreign Exchange to Project

Most of the measures that project lenders and sponsors insist on to ensure foreign exchange availability to the project are generally accepted by the host government as being necessary for the project to run smoothly. However, some of the specific provisions aimed at this objective are felt by host governments not to be sufficiently balanced. An example is the usual provision in the agreement relating to the foreign exchange retention account which obliges the government to permit the sponsors to convert into hard currency local currency held by them in sufficient amounts to make up for shortfalls in the account as a result of a significant fall in product price. Lenders, on the other hand, often resist the inclusion of a balancing provision which would result in transferring back to the host country foreign exchange in excess of what is strictly required to meet all foreign exchange obligations (which situation can occur in a period of high product prices). The Government of Ghana has now become quite firm on this matter and requires that the percentage of gold proceeds which is allowed into the retention account be subject to frequent reviews to see whether it may be reduced.

III. DEBT FUNDING SOURCES

Most international commercial lenders continue to be cautious about lending to projects located in African countries. As a consequence, sources of project financing for mining ventures in Africa are quite limited. A survey of available lending sources is presented below.

Commercial Banks

International commercial banks continue to have a wait-and-see attitude regarding new long-term credit exposure in Africa. The few international commercial banks still active in the region are very selective in their project financing decisions and generally prefer to co-finance under syndication agreements with established lenders such as the International Finance Corporation (IFC).

Gold Loans

Gold loans are a recent development in mining finance. Gold loans are loans denominated in gold which are repaid, principal and interest, in gold bullion. They provide several distinct advantages over traditional currency denominated loans. With a gold loan, the mine realizes the benefits of: (i) hedging; (ii) low nominal interest rates (3-4% p.a.); and (iii) avoids the consequences of adverse currency fluctuations associated with loans denominated in currencies other than US dollars. Gold loan financing remains limited in Africa for the same reasons commercial bank lending is scarce. No major gold bullion lenders have up to now been active in Africa but it is believed that the situation will improve when the general business and investment climate improves.

Governmental and Intergovernmental Credit Agencies

Bilateral and multilateral institutions are a major source of project financing for mining ventures in Africa. Some of the more active institutions in the region include IFC (Washington), European Investment Bank (Brussels), Caisse Centrale (France), DEG (Germany), CDC (U.K.), OPIC (USA) and FMO (The Netherlands). These institutions specialize in project finance and are normally prepared to lend on a non-recourse basis. Generally, they offer long-term credit (5-10 years) and competitive fixed and variable interest rate terms.

Export Credit Agencies

Export credit agencies continue to be active providers of medium and long-term finance for mining ventures in Africa. Their credits tend to be of shorter duration than those of the bilateral and multilateral institutions, averaging 3-5 years maturity. Financing is tied to equipment originating from the host country of the credit agency. Some of the more active export credit agencies in the region include ECGD (U.K.), COFACE (France), KfW (Germany), Exim Bank (USA) and JEXIM (Japan).

Suppliers' Credits

Suppliers' credits are an additional source of debt financing for mining projects in Africa. However, most suppliers' credits require a bank guarantee or a guarantee from the project sponsor. This requirement limits the attractiveness of this source of financing.

Debt Conversion

An investor can minimize the cost of his equity participation in a mining venture by utilizing debt conversion mechanisms, where available. Although not yet popular, several African countries have accepted the principle of debt conversion. In all forms of debt conversion, a hard currency loan is purchased on a secondary market for hard currency but at a substantial discount to face value. Discounts of up to 80% of the face value of the loan are not uncommon. In exchange for cancellation of the loan, the central bank of the debtor country agrees to make payment in local currency. The payment is usually close to the hard currency equivalent face value of the loan, say 85%. Proceeds of the local currency provided by the Central Bank in exchange for cancellation of the loan can be used by the investor as part of their equity contribution to the mining venture. The discounts involved can substantially reduce the investor's reaal cost of equity in a mining venture.

IV. RECENT GOLD FINANCING EXPERIENCE IN GHANA

Mining, primarily gold mining, has a very long history in Ghana. Since 1900, gold production has gone through various boom and bust cycles with gold output peaking in the early 1960s at approximately one million ounces per annum. Since then gold production steadily declined mostly as a result of difficult economic conditions in the country. In the mid-1980s, Ghana embarked on an Economic Recovery Programme and, at the same time, introduced a new Minerals and Mining Law designed to provide incentives for new foreign investment. These changes have attracted substantial investments in gold mining projects in the country and a significant increase in gold production. Three new mines, expected to produce over 200,000 ounces per annum, have recently (end 1990) come into production, and the country's largest gold mine, Ashanti Goldfields Corporation, is undergoing a substantial expansion programme. Financing arrangements for several of these projects are described below.

Ashanti Goldfields Corporation

Ashanti Goldfields Corporation (AGC) operates a large underground mine which has been in production continuously since 1897. In 1989, it produced nearly 340,000 ounces of gold. The mine is owned by the Government of Ghana (55%) and Lonrho Plc of the U.K. (45%). In late 1989, AGC embarked on a US$93 million expansion programme designed to increase the mine's output by 180,000 ounces per annum, or 45%. Approximately 65% of the cost of this expansion project was financed via non-recourse, long-term loans provided by IFC (US$30 million) and a syndicate of commercial banks (US$30 million). The balance of financing, i.e. US$33 million, was provided from cash flow from AGC's current

operations. AGC is operating an active hedging programme to provide protection against downward movements in gold prices.

This expansion programme was preceded by an earlier one which was started in 1985. That programme which cost US$160 million was also funded under similar terms with loans totalling about US$80 million from the same sources but including ECGD of U.K.

Canadian Bogosu Resources Ltd

Canadian Bogosu Resources Ltd (CBRL) is a joint venture company owned by Billiton B.V. (62.5%), Sikaman (14%), IFC (13.5%), and the Government of Ghana (10%). Construction of a US$86 million open-pit mining facility began in late 1989 and was ready for production near the end of 1990. Production will average approximately 100,000 ounces per annum of gold. The financial plan of the project was based on a debt/equity plus quasi-equity ratio of 60:40. The equity portion of the financial plan included a share capital of US$10 million and shareholder loans of US$25 million. The balance of financing, US$51 million, was provided in the form of long-term, non-recourse debt. Lenders included IFC (US$14 million), a commercial bank syndicate (US$24 million), and DEG of Germany (US$13 million). The maturity of the loans was 8 years. A significant portion of the early gold production was sold forward to ensure that all operating costs and debt service requirements would be met even in the event of depressed gold prices.

Goldenrae Mining Company Ltd

Goldenrae Mining Company Ltd (Goldenrae) is a Ghanaian registered company jointly owned by ITM International s.a (51%), Sikaman of Canada (34%), the Government of Ghana (10%), and a local Ghanaian company (5%). In early 1990, the company began the construction of a US$5.9 million alluvial mining facility. Construction was completed in September of 1990. Production would average 20,000 ounces per annum of gold. The financial plan was based on a debt/equity plus quasi-equity ratio of 60:40. Shareholders provided US$1 million in share capital and subordinated loans of US$2.6 million. The balance of the financing was provided in the form of long-term, non-recourse debt by two

bilateral institutions, DEG of
Germany (US$2.7 million) and FMO of
the Netherlands (US$2.7 million).
The maturity of the loans was 6
years. Interest on the loans was
fixed at competitive rates. No
hedging programme was felt to be
necessary because of the low
operating cost of the mine and the
relatively small amount of annual
gold production.

Small-scale mining to commercial operations—requirements for successful development

J. G. Park
M. P. Martineau
SAMAX, Ltd, London, England

SUMMARY

SAMAX Limited is a private company formed by substantial and experienced investors in Africa to participate in the renaissance of mining activities throughout the continent. Through its investors SAMAX has existing bases from which to operate in a wide range of countries in both English and French speaking Africa. This paper illustrates through the use of a hypothetical gold mine the criteria by which a Mining Company may make a selection between areas in which to invest and details those features which a mining company finds most essential in the fiscal and operating regimes. Most sub-saharan African countries have major resource potential so to a large extent countries showing a greater awareness of foreign investor concerns about the risks and rewards of long term mining investment and introducing pragmatic mining and investment codes tailored to bringing together both national and investor interests will be preferred.

These countries are likely to offer investment security and rates of return on investment comparable to those available in mineral - rich developed countries with which African countries are competing for investment and will continue to attract a disproportionate share of the limited funds available from private and public sector sources.

INTRODUCTION

Much has been said and written about the lack of foreign mining investment in Africa and many papers have been published, from a theoretical stand point, about the development of special fiscal regimes designed to both encourage investment and at the same time to extract the maximum resource rent by the taxation of unusual project gains.

Little has been written about the actual costs or risks facing Mining Companies in Africa and hence the comparative effects of these taxation regimes on decisions made by Mining Company managements are not recorded in the public domain.

This paper seeks, firstly, to look at some of the misconceptions which surround Investors' perception of Africa and secondly to examine, from the Mining Companies standpoint, the effect on project economics and investment decisions, of the variations between different fiscal regimes, investment codes and mining policies.

The results of a multivariable computer analysis are presented in a series of graphs from which the relative effect of changes in each of the major fiscal and project variables on the returns to the Investor and the Government and on the target needed to justify exploration can be compared directly. This type of analysis is an essential element of the communication between the Investor and both the Government negotiators and the International Agencies who are frequently brought in to advise and who, although highly skilled, are sometimes unfamiliar with the development requirements of mineral deposits and the nature of exploration to which the fiscal terms are tailored.

THE MISCONCEPTIONS

Firstly sub Saharan Africa (excluding South Africa) is not a dark continent in terms of mining activity, although it is frequently portrayed as such. The region produces for example 60% of the world's cobalt, 40% of the diamonds, 21% of the phosphate, 20% of the uranium, 14% of the copper, 12% of the manganese, 6% of the chromite, 4% of the gold, 4% of the lead, 3% of the zinc and 3% of the iron ore amongst others.

Secondly Africa is not particularly poorly explored. The published geology and mining reports of Zimbabwe are justifiably respected but they are not unique and comparable reports exist for a major part of Africa. The French, British, Belgian and Portuguese colonial authorities carried out extensive surveys which have since been augmented with surveys carried out by, amongst others, the United Nations and East European Agencies. There has been extensive prehistoric and artisanal mining of gold evident on aerial photography, remote sensing or by word of mouth. There is, of course, much exploration to be done and many new mines to be discovered but not notably more so than in Northern Canada, Australia, or South America where major discoveries are still being made in comparable terraines. Often data exists in great detail; Tanzania, for instance, has total coverage of low level magnetics and detailed electromagnetic surveys of the principal diamond and gold fields. Similar data exists in several other African countries; yet comparable surveys are not available for much of the USA, and only recently for Western Australia. The problem is as much one of underdevelopment as one of underexploration.

Thirdly, and to the surprise of some potential investors, most African Countries have long had workable mining codes, inherited in many cases from pre-independence time, but nonetheless more modern and directly relevant than the 1897 Acts governing mineral exploration in the USA or the even more archaic and difficult to work acts of much of Western Europe.

Finally Sub-Saharan Africa does host major high grade orebodies. The base information however, is often recorded in French or Portuguese and hence is inaccessible to the average anglophone geologist this is hardly unreasonable but it does compound the problem of its location in obscure filing cabinets in remote capitals or musty European museums. Anglophone geologists tend not to have heard of the Kilo-Moto gold deposits, the Nimba Iron ore deposits or the bauxite deposits of Guinea despite them being world class resources. They were unfamiliar until recently with the gold exports of the Malian empire, the importance of the Proterozoic sedimentary hosted gold deposits of West Africa, or the abundance of West African heavy minerals deposits. Promotion of potential is as much a need as is further exploration or improvement in mining codes or fiscal regimes.

WHY THE LACK OF INVESTOR INTEREST

If the basis for successful exploration and development exist in much of Africa why then do foreign companies not devote more resources to the area? The following reasons are considred to be the most significant.

a) **Lack of Availability of Mineralized Ground**
which has frequently, even normally, been granted in large blocks for long periods to non developing companies, Government and International agencies and inefficient or cash constrained parastatal organisations.
In the developed world mining concessions are typically small, of limited duration and carry stringent work commitments. Failure to perform results in forfeiture and consequently good ground frequently becomes available to investors with different ideas or technology. Often these are smaller, hungrier companies with fewer resources perhaps, but with less time to wait or sit on portfolio acreage. In contrast in parts of Africa vast tracts, often of 1000-10000 km² of the most favourable ground have been conceded for periods of 10 years or more with no requirements for forfeiture. As one of those smaller hungrier companies SAMAX is well aware of numerous viable deposits held but not developed within such concessions in both East and West African countries.

b) **Expectations of Unrealistically Large Projects.**

The majority of mines, particulary gold mines, in the developed world are quite small. Of over 500 developed gold projects in SAMAX's database fully 75% produce less than 2 tonnes per year and 75% mine and process annually less than 1 million tonnes. 75% of recent projects cost less than US$20 million in 1988 terms. It is illogical then to suppose that the early development of mining in African countries should be on the basis of larger projects than this. The perception, however, of mining as a large scale business persists and access to many development projects has been restricted on the basis of "all or nothing" rather than on the consideration of a staged development with controllable risks and manageable expenses. Such large projects frequently exceed the capacity of the road, rail and power infrastructure and would dominate the national economy to such a scale as to be impracticable. Regrettably this view has often been encouraged by the major international aid and financing agencies. A parallel perception also exists that the majority of deposits are developed by the larger mining companies who must be attracted in at any cost. In reality, in both the developed and less developed world, and particularly in gold, the majority of developments have been made by small or medium sized companies many of whom have grown larger in the process.

c) **Bureaucratic Delays.**
The recent establishment in several African countries of the "one stop shop" approach to handling new investment has gone along way to alleviating the delays and frustration of dealing with several Ministries, and has permitted the assembling of small groups of people who understand the needs of the mining industry and who are able to assist. Where these have not been established the lack of investment in new technology in the departments themselves and the ignorance of a new industry frequently leads to an unwillingness to make any decisions at all and to the departure of would be investors.

d) Separation of the Right to Explore From the Right to Develop.

In Francophone Africa mining companies often need to return to the Government after exploration in order to define the fiscal regime governing the final development of a project. This subjects Mining Companies to the double jeopardy of uneconomic development terms following expensive exploration success. The existence in some countries of the "developed world model" granting, br right, subject to performance, progressively longer and more secure tenure for each of the exploration and development stages of a project eliminates this. In such a system the fiscal terms are negotiated prior to the granting of any tenure and holder of the first (usually exploration) Licence has the exclusive right to apply for subsequent mining authority on known and fixed terms. Over the past few years the right of appeal to International Arbitration has also lessened investor concern over exploration and arbitrary change in conditions.

e) Fear of Exploitation

by the host Government is a natural and long standing concern arising from a combination of past experience, real, anecdotal, and apocryphal, misconception of the industry through lack of recent experience, and ignorance of technical changes affecting mineral markets. As a consequence many countries place unrealistic restrictions on the retention of foreign currency earnings for the purchase of spares, servicing of loans and repayment of dividends, thereby forcing the move to inefficient development, highgrading and transfer pricing. This situation can only be improved through mutual trust and the mining companies have much to do in promoting their case through better communication and by example.

f) Lack of infrastructure

Including the high cost of transport and port/railway facilities limits development potential of low value, high tonnage commodity projects inland. This is often coupled with a deterioration in those facilities which did exist and with a lack of ancillary industries providing equipment, spares and support to the industry. The lack of or decline in a country's skills and technology requires higher capital and operating investment costs through the need for additional equipment, spares and the temporary secondment of expatriate personnel while locals are trained. The development or recovery of a country's mining skills base will generally be more efficiently made by the concurrent development of several smaller high value projects than by the establishment of a single large project especially one of the low unit value and high tonnage.

g) Ignorance of Africa in the Investment Communities of the Developed World of the Investment Climate

This is as often due to the lack of promotion as to lack of development "The "Africa - It's a big country" view is real and pervasive as is the not wholly incorrect perception of country risk. In many cases investors and financiers, ignorant of the differences between countries place Africa in the too difficult basket on the strength of limited or sensational Western media coverage.

ONE COMPANIES EXAMPLE

In order to give a specific example of what one Investor seeks in evaluating alternatives, SAMAX's perspective in selecting countries for investment is based on looking for:

a) Clear title to the ground

b) Unambiguous and fair fiscal terms applicable to mining from the granting of the first exploration licence

c) International arbitration in the event of irreconcilable disputes.

d) The right to utilize foreign currency earnings unrestricted for mine expansion, purchase of spare parts and replacement.

e) The right to manage a mine efficiently without bureaucratic interference.

f) The right to repatriate foreign currency earnings with no restrictions on loan repayment and to pay a fair return on our own investment.

g) Access to end markets for our products and in particular the ability to use price mechanisms such as forward sales, product linked loans, or sales contract equity investment.

and in return SAMAX looks to offer

a) Training of local staff at all levels up to the highest based on ability and selection

b) Participation in projects at a minority level whether by Governments or local partners provided it fairly reflects the value added through investment and skills free of bureaucratic or managerial interference.

c) The payment of taxes and imposts provided that they are not more penalising than those applicable to other export earning industries.

d) Reinvestment of a share of earnings in expansions and new mine developments.

e) Rapid exploration and efficient development of mines.

f) Access through Related Companies to mineral markets in Europe, Asia and North America.

SAMAX does not seek

a) Special concessions, outside of those which reflect miners' special risks, unavailable or inapplicable to other industries.

b) Abnormally large areas or concessions in excess of 400km^2.

c) Abnormally long tenure in excess of two years prior to selection of core areas.

In short SAMAX seek a legislative and fiscal environment in which it can work competitively relative to the developed mining world with further incentives only where justified by the extra difficulty of working in areas of poor infrastructure and adverse living conditions and which reflect the mining industry's high front end risk and development lead times.

THE FISCAL REGIME

In practice, the differing perceptions of host governments and investors have to be reconciled before a project can go ahead and perhaps the most contentious issue is that of the fiscal regime within which the potential investor is required to operate.

It is axiomatic to an economist that the mining industry as a whole cannot be more profitable in the long term than other industries. If it were new entrants would come in and surplus production would restore the balance through metal prices. It is equally axiomatic that mining in one part of the world will only be more profitable than another to the extent that is needed to compensate for the extra price of capital, development cost infrastructure and risk. A comparison of the profit record of companies operating in Africa versus those operating elsewhere would confirm this. In the modern world it is evident that that the fiscal regime can have a critical factor on the investor's willingness to invest in a particular country.

In order to demonstrate the impact of various fiscal regime components for a project from the investors viewpoint, SAMAX have chosen to model a gold resource for simplified project analysis. The orebody in question is open pit-able in Archean or Proterozoic terrain and is treatable by conventional carbon in pulp extraction following fine grinding. Capital and particulary operating costs reflect a remote location with comparatively poor infrastructure, and exploration costs assume access to a known historic mining site or one located and currently being worked by artisanal miners. Table 1 lists the Base Case assumptions for operating data against which the effect on investment of varying fiscal and economic regimes have been examined. Figures 1A & B illustrate the distribution of project revenues on both discounted and undiscounted bases.

The fiscal model is based on an hybrid regime involving a fixed royalty, import duties on capital equipment and operating supplies, 100% investment allowance and the carry forward of losses (not required in this model), a corporate tax regime which allows a five year grace period at a lower tax rate and the imposition of an additional profits tax (APT) on realised profits above a given threshold. The model allows for a relatively

early payback of the initial investment thus minimizing the risk of absolute loss and is believed to be more than equitable in providing the Government with 55% of the value of the resource at the 20% discount applied. Monte Carlo sensitivity analysis and the effects of inflation, exchange rate variations and metal prices cyclicality have been ignored in order to generate simple relationships which indicate the impact of changes on the projects viability and apportionment of value.

TABLE 1.

OPERATING DATA

RESERVE SIZE	4 000 000	TONNES
GRADE	5	GMS/T
STRIP RATIO	4	WASTE/ORE
RECOVERY	85	%
THROUGHPUT	1 300	TPD
MINE LIFE	9	YRS
ANNUAL THROUGHPUT	450 775	TPA
ANNUAL PRODUCTION	61 594	OZS PER YR
DISCOVERY COSTS	3 000 000	US $
CAPITAL	17 500 000	US $
OPERATING COSTS	25.5	US $/ORE
EXPLN + PRE-PRODN	3	YEARS

FISCAL AND ECONOMIC DATA

GOLD PRICE	380	US $/OZ
ROYALTY	2.5	% FIXED
IMPORT DUTIES	10	%
GOVT EQUITY	0	%
INVESTMENT ALLOWANCE	100	%
TAX RATE 1	20	% FOR 5 YRS
TAX RATE 2	45	% THEREAFTER
APT RATE	25	%
APT TRIGGER POINT	30	%
W'HOLDING TAX ON DIVIDENDS	0	%
DISCOUNT RATE	20	%

SUMMARY ECONOMICS

PROJECT NPV PRE FISCAL	14 115 000	$
PROJECT IRR	45.0	%
INVESTORS NPV	6 334 000	$
IRR	31.7	%
SHARE	44.9	%
GOVERNMENT NPV	7 781 000	$
GOVERNMENT SHARE	55.1	%

The model's cash flows have been discounted at 20%; however this rate does not imply that the Investor accepts a 20% Rate of Return as a criteria for entering the country. Rather it implies that, once the technical risks have been constrained by the completion of a successful feasibility study and the negotiation of project financing, the Investor will proceed with a project yielding a 20% ROR in preference to abandoning it and losing his investment.

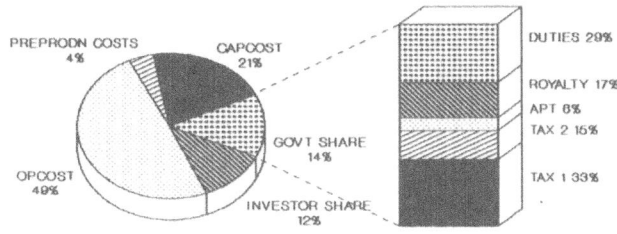

FIG 1 A :DISTRIBUTION OF PROJECT REVENUE WITH COMPONENTS OF GOVERNMENT SHARE

DISCOUNT RATE 20 %

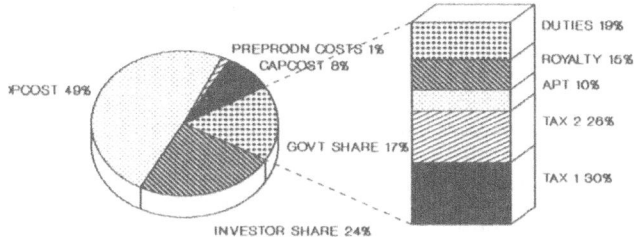

FIG 1 B :DISTRIBUTION OF PROJECT REVENUE WITH COMPONENTS OF GOVERNMENT SHARE

UNDISCOUNTED

passing on of subsidized loan funds or through the access to Discounted Debt is substantial but would need to be evaluated on a case by case basis.

The model is representative of many gold and base metal deposits in being most sensitive to those variables which most influence revenue (notably grade, recovery, gold prices - see Fig 2, and royalty, Fig 3) and to those variables which impact on the distribution of revenue (operating costs and the level of Governmental free carried equity). One feature which is specific to the high grade and poor infrastructure assumed in the model is the greater sensitivity to operating costs and revenue variables and lesser sensitivity to capital cost than would typically be encountered in a well developed infrastructure such as that of Zimbabwe. This immediately implies that a

Bearing in mind the high failure rate of exploration projects and feasibility studies the Investor in Africa will normally require a target yielding a Rate of Return of about 35% thereby covering the risk and expense of the unsuccessful projects. This model yields a Rate of Return on the Base Case of 32% and would be at the low end of those considered acceptable by investors to initiate an exploration programme.

In evaluating the Government's return a strong case can be made for discounting the Government's cash flows by less than the Investor's or not discounting them at all because the Government bears no primary risk and the Government's risk (on its non equity portion) is confined to receiving lower and later revenues. The model considers an all equity case only. The ability of Government to assist a project through

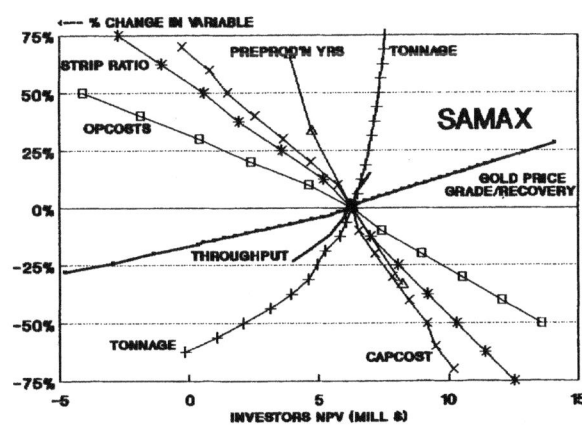

FIG 2 : PROJECT SENSITIVITY TO CHANGE IN OPERATING VARIABLES FROM BASE CASE

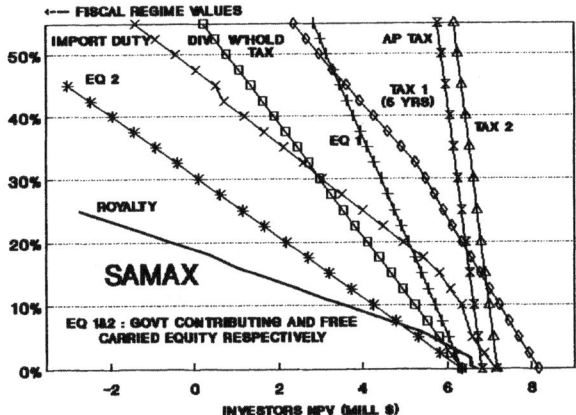

FIG 3 : PROJECT SENSITIVITY TO CHANGES IN FISCAL VARIABLES FROM BASE CASE

Country seeking to develop its Mining Industry should seek first to reduce the operating cost and revenue related variables and specifically sales taxes, import duties, transport rates and ad valorem Royalties.

It is erroneous to believe that by increasing royalties or taxes the Government's total share of revenue from the same resource increases. The Investor seeks a consistent threshold Rate of Return; hence for an increase in Government take the Investor will seek compensation through raising the grade or size of deposit he requires to justify development and continued operation. In general this will so reduce the size and number of deposits meeting the Investors Rate of Return criteria that the total Government take will actually be reduced.

Whilst high margins and bonanza profits can and do occur in gold and other mining operations in the case of outstanding orebodies and special situations, it is often assumed that this is the normal state of affairs. Examination of the impact of operating, particulary grade, problems (Fig.2) shows how this might occur (and equally how well profitability can be completely eroded) but if taken together with a study of both mining company failures and the profitability levels of mining companies relative to other industries it can be demonstrated that this is the rare exception not the rule.

That projects are sensitive to size and grade is of course obvious but of interest here is the impact on the grade and size of the resource which it is necessary to find and develop successfully in order that both the Investor's and Government's expected returns requirements are met. With the chances of exploration success never high it is

relevant to consider the odds against finding orebodies which meet investment criteria as fiscal impositions become more stringent.

Fig 4. illustrates the distribution by size and grade of a number of developed Archean and Proterozoic gold deposits similar to the model chosen. Figure 5 shows, more specifically, the grade-size relationships for both Archean and other types of gold deposits. In general Archean deposits whilst smaller are of higher grade. In terms of grade it can be seen that if orebody target criteria are, or have to be, raised (see below) then the chances of meeting these criteria are diminished. Fig. 4 demonstrates that only 25% of developed Archean deposits contain greater than 7g/t while 35% contain less than 4.2g/t, which is the breakeven grade required for the model considered. Only 20% of developed Archean deposits meet both the criteria of the tonnage of ore and the contained gold required by the model to justify development. If, as is generally the case, the explorer cannot assess the size or grade of a deposit prior to drilling, then it can be deduced that if 20 tonnes of gold is the required target (within operating grade and tonnage constraints) then 80% of the targets tested will fail to meet the development criteria. In the absence of new technological changes or significant increase in metal price exploration funds spent on these targets will have been lost irrecoverably. Figure 5 also illustrates that as grade requirement increases the incidence of deposits with reserves of the same size (tonnage) falls away rapidly and vice versa. On the basis of these figures the impact of a change in weighting attributed to any

single component of the
terms imposed to collect
"resource rent" can be
expressed and understood in
terms of the likelihood of
an exploration programme
achieving success.

FIG 4 : GOLD DEPOSIT FREQUENCY DIST'N
ARCHEAN TYPE DEPOSITS

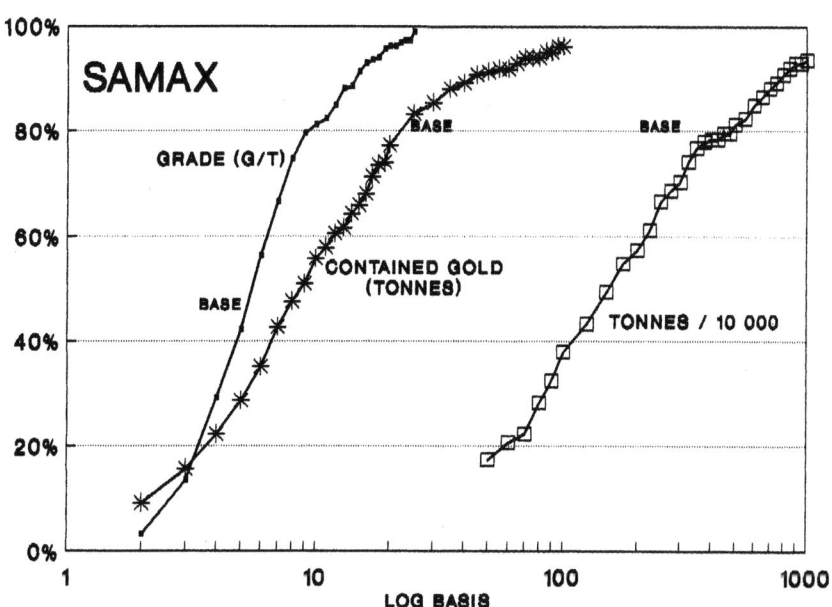

FIG 5 : SIZE - GRADE RELATIONSHIP : GOLD

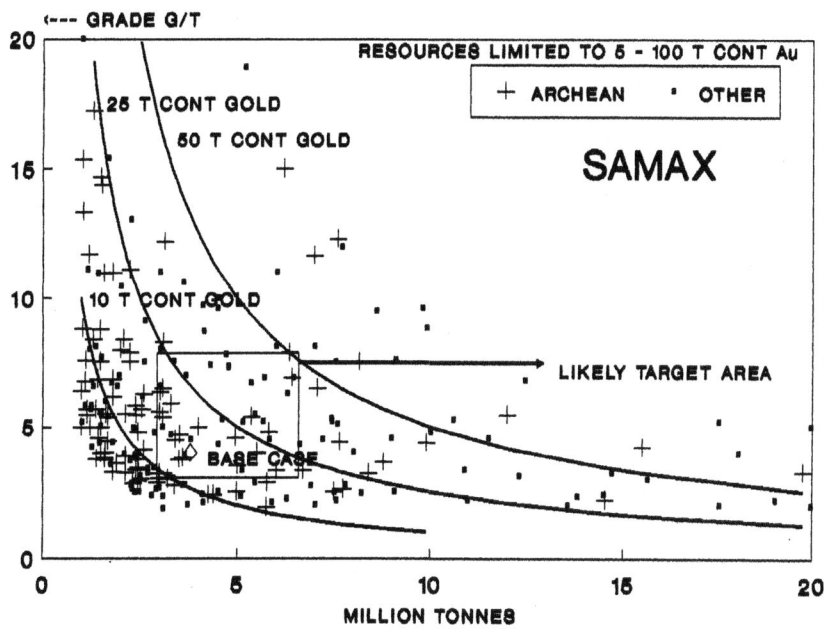

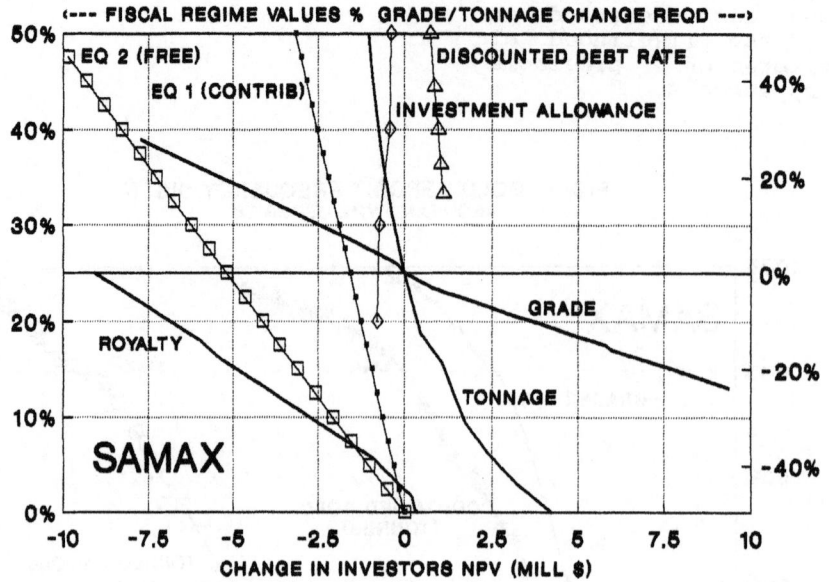

FIG 6A : EQUIV CHANGES IN RESOURCE TO
COMPENSATE FOR CHANGES IN FISCAL REGIME

Figure 6A and B show changes in the Investors NPV versus changes in the principal taxation or revenue earning components of the fiscal regime and relates these decreases to the extra grade or tonnage, required to compensate the investor in NPV (if not IRR, payback or equitable share of the project) terms. The conversions are, of necessity, simplified.

Figure 6C illustrates how these diagrams may be used. For instance if, in order to encourage investment Government chooses to reduce the concessional tax rate for first five years from 20% to 10% (point A) this would permit development of a deposit having 18% less tonnage (points B and D) or a 4% lower grade (points C and D). Were the Government to declare a five year tax free holiday (as in Tanzania) the tonnage requirement would be reduced by 35% or the grade requirement by 6%. From Figure 4 it can be deduced that a 35% decrease in tonnage needed will increase the number of deposits which can be developed by 50%

From the investors point of view an increase in Royalty beyond say a 3% rate, a free

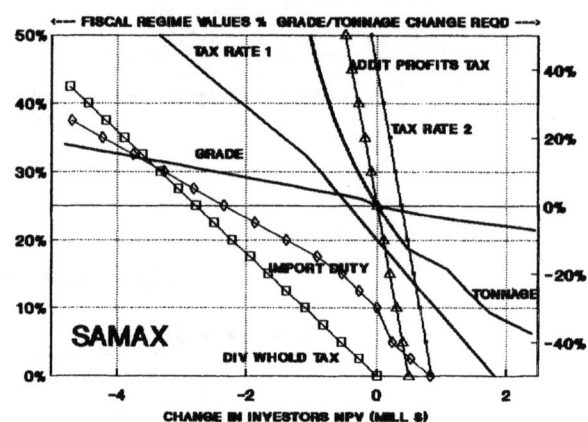

FIG 6B : EQUIV CHANGES IN RESOURCE TO
COMPENSATE FOR CHANGES IN FISCAL REGIME

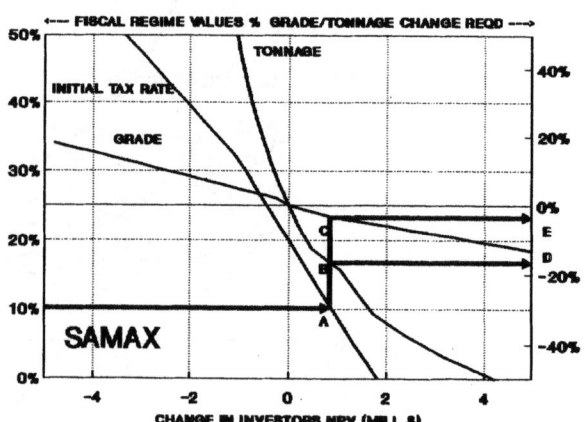

FIG 6C : EQUIV CHANGES IN RESOURCE TO
COMPENSATE FOR CHANGES IN FISCAL REGIME

carried Government equity, a dividend withholding tax and the payment of import duties and sales tax on capital and operating items have the most serious effect in deterring investment and raising the threshold for deposit consideration. However as figure 1B shows in undiscounted cash flow terms the influence of these taxes on the total funds received by the Government is relatively minor with the major cash contributors being income taxes and APT. In this particular model APT applies for only two years. However Were the removal of duties or a five year tax holiday to apply then the concomitant reduction in grade treatable coupled with the increase in tonnage mineable at the lower cutoff grade could permit the extension of the mine life by a further one to two years. Under such circumstances the proportion of Resource Rent extracted through the APT and final rate tax would increase dramatically and the overall receipts of the Government would be far higher.

Figure 8 illustrates the distribution of Resource rent by fiscal component received by the Government over time. It clearly illustrates the relative insignificance of royalties and duties in the overall receipts of the Government and the overwhelming importance of the final tax rate and APT. If as indicated above the project life is increased by the elimination of those taxes most deleterious to the Investor's Control of risk then the total revenues received by the Government will more than pay for the change.

As a further example from Figure 3 or 6, an increase in Base Case Royalty from 2.5% to 10% decreases the investors NPV by $2.7 million over the life of the mine. Under these circumstances the Investor will not initiate exploration for the Base Case target because his modelled Rate of Return has fallen below his critical level. For the Investor to restore his model Rate of Return and so justify

FIG 8 : GOVERNMENT RESOURCE RENT
UNDISCOUNTED

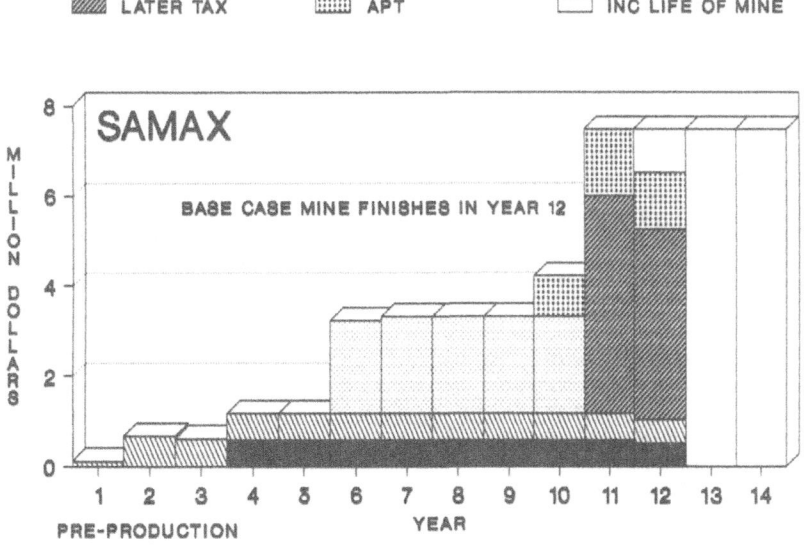

| ▓ ROYALTY | ▨ DUTIES | ☐ INITIAL TAX |
| ▨ LATER TAX | ▦ APT | ☐ INC LIFE OF MINE |

exploration an increase in recovered grade of 11% will be necessary requiring an increase in the target reserve grade from 5 to 5.6g/t. From Figure 4 the chances of exploration achieving this decrease from 50% to 42% on a grade basis and from 20 to 15% on a contained gold basis. The corollary to this of course implies that if the Royalty were still 2.5% and successful exploration had found an orebody grading 5.6 gm /t (or if as is equally probable simple metallurgy had permitted a higher recovery of gold in the mill which would have a similar effect on project economics), then maintaining the same NPV would allow the Government's share to be increased. The Government's interests might be better served, however, were this increased share to be obtained through the taxation of realised profits rather than through royalty. If the Government took all the gain then the Government's share of the project's resource rent would increase from 55% to 75%, an inequable exchange in the eyes of the investor who, not sharing in the good fortune, would have no incentive to improve the project further.

Of further significance in these figures especially 6A and 6B is the understanding gained of the equivalence to the investor of different single components of the fiscal regime. Thus an equal reduction in the investors return of $2.5 million would be brought about if
- Royalty increased from 2.5 to 9%
- Import Duties increased from 10 to 22.5%
- Dividend withholding tax were to be introduced at a level of 26%
- The Government was to be free carried for 12.5% of project equity or were to fully contribute to a level of 40%.
- The initial concessional tax rate was increased from 20 to 42.5%
- OR were grade or recovery to reduce by as little as 10%

The balance of the risk to the Investor of these alternatives, however, is not equal and he will evaluate their imposition differently. Obviously later stage imposts assist the investor in achieving an earlier payback thus reducing exposure by increasing front end revenue share. They do however discount to a smaller value as the time value of money asserts itself.

Since one may expect a Government to be primarily concerned with the long term development of a mining industry and tax base there is a strong argument for discounting low risk Government cash flows at a lower rate when evaluating a Government's requirements of the Project. In the Base Case model Additional Profits Tax in undiscounted terms amounts to $4 million (10% of Government income compared to 14% from Royalties) over the life of the project but because APT does not apply until year 8 of the period the value of the APT is discounted heavily in the evaluations.

The impact of the return on investment at which APT is triggered can be seen in Figures 7A and B. SAMAX have arbitrarily set this at 30% to reflect the type of return the Investor will require. It has, however been suggested by a number of advisers whose experience lies in the oil industry that APT should apply at a lower rate. Whilst perhaps appropriate in the oil industry, from where APT's derive, and where high capital, low operating costs are the norm, the nature of the high operating costs in mining would, if the APT trigger were lowered, push this and other projects to an unacceptable rate of return and stifle development.

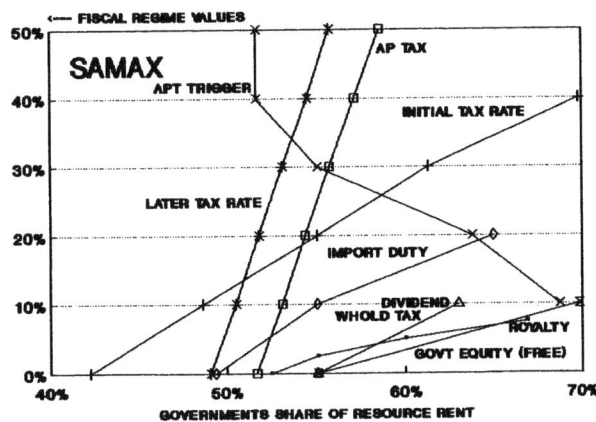

FIG 7A : GOVERNMENT SHARE
DISCOUNT RATE 20 %

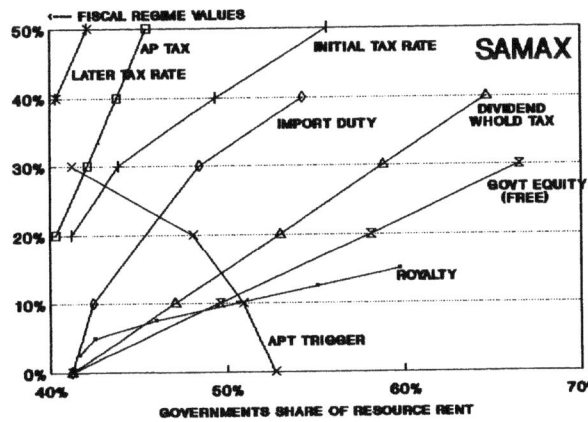

FIG 7B : GOVERNMENT SHARE
UNDISCOUNTED

CONCLUSIONS

The Investors perception of a project and a Government's differ. The Investor is primarily concerned with controlling his medium term risk, ensuring that he recoups his initial investment and hopefully receiving a reasonable rate of return upon it. The Government is assumed to be primarily concerned with developing employment in remote areas and with receiving a long term resource rent capable of funding a gradual long term increase in GDP. The two objectives are best achieved through short term concessions in the form of relief from operating and revenue cost variables such as import duties, sales taxes, royalties and dividend withholding taxes coupled with long term imposts in the form of an Additional Profits tax (levied at a high threshold) and a contributing Government equity share.

Role of small-scale mining in Africa: building on the informal sector

J. Hollaway
Harare, Zimbabwe

SYNOPSIS

'Formal' mining in sub-Saharan Africa has declined over the past decade as a consequence of reduced exploration, poor commodity prices and investment disincentives. However in contrast to this trend there has been a marked growth in the informal sector, mining gold and gemstones on a small-scale or artisanal basis due to the black market value of such goods.

It is shown that, even if legitimised, the national contribution of the this sector in Africa is normally fairly small, both in terms of its proportion of the GDP and in numbers of persons employed. It is suggested that its main long-term value to the state, and a principal reason for supporting it, is that a healthy small-scale mining sector encourages prospecting by local people, who historically have tended to be the main discoverers of certain minerals and metals, leading to the establishment of major mines.

To achieve this objective, it is proposed that lessons drawn from the successes and failures of small-scale mining elsewhere in Africa be incorporated in guidelines promoting this sector.

It is suggested that the current trend to make special provision for small-scale mining in the administration is unnecessary, and that experience has shown that systems for marketing, law and technical support can conveniently be the same for every scale of mine. In the case of marketing and technical support, it can be left to the mines themselves to decide whether they wish to avail themselves of these Government services.

INTRODUCTION

Over the last decade small-scale mining has become a fashionable area of concern to agencies concerned with the development of sub-Saharan-Africa (SSA - i.e. the non-Arabic speaking countries, excluding South Africa). Arguments over a universally acceptable definition of such mines are unlikely to achieve a firm conclusion; in this paper a definition is avoided as it is believed unnecessary for reasons which will become apparent.

This paper seeks to clarify a number of perceptions about this sector and to establish its real importance in the context of African mining. Following on from these considerations it also sets out proposals for guidelines that might be used to take advantage of the achievements of the small-scale miners to create the major mines that are the real generators of wealth in this domain.

The need for increased exploration and mine investment in Africa is not doubted. A World Bank survey[1] has suggested that the current exploration expenditure in SSA is of the order of US$ 100 million annually. By contrast Australia was spending about US$ 280 million and Canada US$ 900 million annually in this sector during the late eighties. In constant dollar terms SSA non-fuel mineral production increased by only 1.5% annum between 1960 and 1987[2] compared with about 3% in Asia and Latin America. At the same time the SSA population growth rate was about 3.1% p.a.

However to create the major mines needed to start a recovery in this sector requires time as well as money. The estimate of the World

35

Bank is that it would not be before the late 1990's that a useful growth rate (in this case about 5% a year) could be reached even if the exploration expenditure was raised to about half a billion dollars annually between now and then. Realistically this is not going to happen, particularly as the world appears to be slipping into recession at the beginning of the nineties.

On this assumption, therefore, there is an urgent need for African Governments to reconsider their policy regarding exploration and mining investment. For the last 25 years the answer to decreasing mining investment has been the creation of state mining companies. Their record has been poor; to quote the World Bank again[1] "State mines have generally failed to mobilise the investment funds needed for a steady growth of production. Constant political intervention in mine operation and management has jeopardised efficiency".

It is the contention of this paper that one alternative remains. This is that the way to the rapid creation of many large mines is the vigorous encouragement of very small ones.

CURRENT MINING POLICIES

Legal Distinctions Between Large-and Small-Scale Mines

In general, mining law in most SSA states divides miners into two classes, large-scale and small-scale. This derives from three assumptions:

a) that Government should decide who prospects (and when and where)

b) that large-scale miners have exploration techniques available to them making them capable of evaluating extensive areas of land

c) that large-scale miners have the money - and in particular the time - to undertake the sequence of reconnaissance, target selection, detailed exploration, drilling out, development and production.

By contrast, small-scale miners are typically allowed, on sufferance as it were, to claim for and work at a deposit they have discovered provided that they do not get too big. However their title is one which is under the supervision of government rather than a legal property right in itself, and they are obliged to surrender it to a larger scale mining organisation if the Government thinks the latter will do a better job of development.

The long-term view necessarily taken by large-scale miners and provided for in this legislation arises from the lengthy gestation and troubled childhood of mining ventures. In undeveloped countries these hazards are often increased through infrastructural failings and a irresolute bureaucracy. For example, in the relatively well-ordered environment of Zimbabwe it has been shown that the average duration between discovery and production for eleven medium sized base metal mines (milling rates of between 300 - 3000 tons/day) was about 10 years[3].

A review of the start-up achievements of mines and mills in both developed and undeveloped countries concluded that ore output from new mines cannot be expected to meet design capacity in less than two years and for new treatment plants the equivalent figure is three years.[4]

These figures do not take into account the time spent on exploration. Here it is difficult even to give an order-of average. However, to take the example of Botswana's diamonds, which are the only truly major new mineral development in SSA for the twenty years, prospecting began in the mid-fifties but production did not begin until fifteen years later in July 1971, at Orapa. Letlhakane was found shortly after Orapa and began production in 1977. Jaweng was discovered in 1973 and production commenced in January 1983.

These 20 - 30 year time spans for mining projects are pertinent to any consideration of the development of mining in SSA, because currently the only view taken by investors in the continent is an extremely short-term one. Very few major mining houses are engaged in grass roots exploration in Africa on their own account; since the early 1970's they have transferred their attention to Australia, Canada, South America and the Pacific. Such development as has occurred has arisen mainly from small or medium size firms using risk money raised on stock exchanges. These firms are under a compulsion to make quick profits and have tended to re-open old mines than find new ones.

Frequently their objectives are to return their investors money to them in 2 - 3 years, which is completely unrealistic in Africa. Negotiations to get to the starting post - i.e. to obtain a concession area for exploration - have frequently taken longer than this.

Ad-Hoc Financial Arrangements

A major contributing factor to the delays in getting a concession comes about because investors want to have the conditions of their profit remittance and taxation clarified before spending money. As non-official (i.e. non-aid) investment is so unusual in many SSA countries (about US$3 billion compared to US$81 billion in South America in 1988 according to the OECD[5]), ad hoc agreements must often be laboriously negotiated.

A lack of understanding by Governments of the risk/reward structure of mining contributes greatly to delays during such bargaining. It seems axiomatic to many finance officials (and perhaps to all politicians) that ownership of, for instance, a potential gold mine was a sure way to wealth. As a result the entry and rent (royalty) prices were often initially set at extremely high levels compared with say, investment in the manufacturing sector.

In fact the average real returns on gold as an investment were better than average share returns (5.96% against 4.06%) over the period 1968-88[6] only because of the huge leap in the price in the early 1970's when the Bretton Woods Agreement was finally abandoned. An investor in gold mining shares in 1988 would have seen his holdings lose value on average by about 22%; in that year such shares were the worst performers of all the categories traded on the world's stock markets.[7]

Bias Towards Downstream Beneficiation Options

A further complicating factor for mining investors has been a general desire to ensure the maximum upgrading or treatment of mined products in the host country. This direction was reinforced by a natural tendency of aid organisations to look for areas where the manufacturing skills and equipment of their own countries could be used. In fact the record of attempts to upgrade mineral products has been poor and the emphasis

placed on this wrong, a fact recognised by a United Nations study[8].

This study was commissioned by the UN Secretary-General and undertaken at a high level by a team of experts chaired by the former Australian Prime Minister, Mr. Malcolm Fraser. It concluded that 'the failure in the commodity sector has been central to the economic crisis facing Africa.' The report calculated that by 1988 Africa's market shares for cocoa, coffee, cotton and copper had fallen by between 20 and 40% of the 1970 market share. Fig.1 shows how, over the last thirty years, Africa has lost ground overall in the major metal minerals mined there. If the continent is unable to produce the basic commodities, there is little merit in investing in sophisticated downstream industries based on them.

The idea dies hard, however, and policy proposals for encouraging fabricating facilities for iron, steel, aluminium and copper were made by the Economic Commission for Africa in 1990.[9]

The Stagnation of Investment in Mining

The forgoing has shown that there is a mis-match of investor requirements and investment conditions which has led to mining companies rating Africa low in their assessment of countries worthy of mineral exploration. This has been quantified by two recent studies. In the first[10] 32 mining companies were asked to rank nations for their attractiveness, and the replies matched against a 1969 study. While the results are not strictly comparable due to political changes (Mocambique and Angola were amongst the top-ranked countries in 1969), in the earlier study a number of African countries appeared well up on the list; in the 1990 study only Botswana was considered to have a good investment climate and only Zaire was rated as having sufficient geological potential to possibly overcome investment drawbacks.

The second survey was undertaken by the World Bank Mining Unit[2]. The views of forty-five international mining companies currently operating in developing countries were canvassed. In Africa only Botswana, Ghana and Zimbabwe rated as favourable (together with, for example, Chile, Papua New Guinea, Indonesia and Mexico). Analysis of the data suggested that good geological potential in a number of African countries was outranked by such factors as lack of mineral resource information, poor infrastructure and unfavourable economic policies.

Coupled with this gloomy evaluation by the larger mining companies, SSA is running out of time to recover. This is principally because of the population explosion. A graphic example is Zimbabwe, where job creation has resulted in perhaps 10,000 openings a year for over 150,000 school leavers.

There is therefore an urgent need in Africa not just to get new mining investment but to compress the time span taken for the creation of new mines.

THE ROLE OF SMALL-SCALE MINES

Financial Status

The importance of the small-scale mining sector is not because of the potential value of its output. Zimbabwe is the foremost small-scale mining country in Africa. Thus, for instance, it has several hundred small gold operations producing less than 500 ounces (15.5 kg) a year. However production from the half-a-dozen big gold mines in the country which produce above 35 000 ounces (1125 kg) a year each far exceeds the small mines output, by a ratio of about 4:1.

In Zambia the revival of the small gold mining industry there is only just starting, and small-scale mining is concentrated on the pegmatite minerals. Because of smuggling (of emeralds, aquamarine and amethysts) it is difficult to ascribe a reliable figure to the value of the output. Guesses as high as US$ 200 million annually have been quoted. However, one informed estimate of the current value[11] is that it is of the order of US$20 million. By contrast the gross revenue of Zambia Consolidated Copper Mines (ZCCM) is about a billion US$ annually.

The value of production from pegmatite minerals in Zimbabwe was about US$13 million in 1989, out of a total mineral production value of US$562 million that year. The multitudinous small and small-scale gold operations accounted for perhaps 20% of the US$ 195 million total revenue from this metal in the same year. Compared with Zambia, these are fairly 'hard' figures, as the amount of smuggling is a great deal less in Zimbabwe.

Employment Generation

Nor is the small-scale sector particulary important because of the employment opportunities it offers. In Zimbabwe small mines employ perhaps 10,000 of the 56,000 workers in the mining industry, excluding seasonal panners of alluvial gold in river beds. The numbers of the latter are a matter of contention; this author estimates that their number is well under 10,000.

An interesting point is that while the fairly sophisticated small-scale gold mining sector in Zimbabwe is more labour-intensive than the larger mines, they are not greatly so; the nature of most reef gold deposits makes mechanisation difficult and there are few economies of scale.

Such gold operations are small mines rather than small-scale mines. Elsewhere the sight of a milling crowd of people working alluvial gold deposits or scattered over precious stone claims might suggest a labour force of tens of thousands. However actual counts tend to give more modest figures and it is estimated by this author that in Zambia about 4,000 people are working regularly on the small-scale gemstone mines. In Zimbabwe employment on comparable pegmatite operations is about 2,000, out of a total of 50,000 for the mining industry as a whole.[12]

Small-Scale Mines and Exploration

It is argued here that the real importance of having a thriving small-scale mining industry is that big mines develop from small mines. In Zimbabwe, of the 4,000 or so gold mines recorded, almost all were found and worked as small mines, often centuries ago. The local people guided European prospectors to these abandoned operations. Some of these 'ancient workings' became amongst the world's richest mines - such as the Globe and Phoenix.

These features do not only apply to gold mines. The lead and zinc deposits of Kabwe and many of the copper deposits of the Copperbelt, Lumumbashi and Mhangura were found by local people prior to the colonial period. This process is continuing. In Zimbabwe in 1988 the biggest emerald producer was Chigumba's Mine, which began as a small-scale mine a few years before. The Kafubu (Ndola) emerald field in Zambia was found by local people in 1974 after years of unsuccessful prospecting by formal exploration operations in the area. In Tanzania, the Bulyanhulu gold deposit (also known as Kahama) was located by local small-scale miners (wasokomoto) in 1982, and has been established as having reserves of over 2 million tons grading at 12 grams a ton.

This is not to denigrate the value of formal geological exploration in Africa. Base metal deposits in particular have been successful located by geochemical and geophysical techniques, while the search for large, low grade epithermal or weathered gold deposits suitable for heap leaching has as its objective grades that are normally below the interest of the small-scale miner.

However, sophisticated exploration methods are of little use in locating most resources of reef gold, of pegmatite minerals and of precious and semi- precious stones. Such deposits are found normally only by direct prospecting on foot. It is no coincidence that it is in these minerals that small-scale miners are usually most active. They are also those most favoured by mining investors in Africa as well.

The Resurgence of Small-Scale Mining

The small-scale mining sector has seen a marked growth in a number of SSA countries in the 1980's. Tanzania, Moçambique, Zambia, Kenya and Zimbabwe all now have much larger numbers of such workers in the bush than before. The economic incentive for the resurgence of small-scale mining in Africa has been the deteriorating economic situation in the continent. Currencies have failed to fall in line with their real value and have become overvalued, and smugglers have appeared, prepared to pay black market prices on the spot for the small-scale miner's output. As a consequence there has developed a tremendous incentive for local people to search actively for valuable minerals and a corresponding boom in the informal mining sector.

Until the mid 1980's Anglophone countries in Africa were the principal areas of this activity; the linkage of the CFA franc with the French franc in most Francophone countries meant that miners there were effectively dealing in a hard currency. However the Anglophone countries were all in financial crisis by that time and were forced (under IMF urging) to devalue

massively in order to recover their competitiveness. In the meantime the French franc has been appreciating, dragging the CFA franc up with it, until by 1990 it was seriously overvalued. For example, a hand of bananas bought at the roadside in Gabon, costs the equivalent of between two and three US dollars, or rather more than they would cost in the United States.

As a result gold mining and smuggling has now become a lucrative activity in the Francophone states. One estimate has it that there are more than 50,000 people involved in alluvial gold winning in Burkina Faso. Recorded output from small-scale mines there went up from 83 kg (2670 ounces) in 1985 to 831 ounces (26,720 ounces) in 1988.[13]

This uncontrolled development is conventionally frowned upon by Governments. Nonetheless it is suggested here that it represents an opportunity rather than a threat, and the discoveries of small-scale miners can, as shown, result in economically important formal mining operations. By encouraging this sector decades can be taken off the slow process of exploration and development by major companies.

THE REQUIREMENTS FOR A SUCCESSFUL SMALL SCALE-MINING SECTOR

Markets and the Small-Scale Mining Sector

Prospecting and mining are high risk enterprises even in the formal sector. In the informal sector the failure rate must be much higher. In the case of the Bulyanhulu deposit mentioned earlier, for example, the strike is surrounded by deep pits which failed to intersect the weathered reef.

Such laborious and unrewarding work can only be sustained if there is the knowledge that an assured economic benefit will result if a find is made. This presupposes a ready market; in the case of Bulyanhulu it was provided by the presence of illegal private gold buyers (easily identifiable by the umbrellas they carry), who had on them large sums in cash.

Smuggling is endemic: it encompasses gold and cassiterite in Tanzania, emeralds and amethysts in Zambia and tantalite and other pegmatite minerals in Moçambique. Zimbabwe is not immune from this problem; much of the alluvial gold panned in the dry season is thought to be smuggled out of the country rather than passing through the banks to the Reserve Bank.

The provision of formal markets for the informal mining sector thus must either wait on currency reform or adopt one or more of a number of ways of overcoming this major obstacle. The options tried here include:

a) Payment in kind of items otherwise not locally available (Tanzania).

b) Payment in hard currency from the Central Bank (Uganda, very briefly, and Moçambique in 1990).

c) Encouraging more formal mining by local mining groups on gemstone deposits found by the small-scale miners. (Zambia - ZCCM)

d) Payment in local currency in competition with the illegal buyers. (Tanzania again)

None of these attempts have been very successful. Bearing in mind that the value of the output of small scale miners is, as shown above, nothing like as great as often presumed, it is argued that there are only two practical choices for Governments wishing to encourage this sector with a view to making it a short cut to the creation of a large mines:

a) marketing arrangements for the products of small-scale miners must be established that are able to out-bid the black market purchasers

b) a policy adopted that the state is non-interventionist in this matter, but reserves the right to revoke the mining title of small-scale miners who sell their products illegally and to reassign it to formal mining companies.

Mining Title and the Small-Scale Mining Sector

The local prospector typically precedes the small-scale miner(s). Prospecting is often an activity undertaken by subsistence farmers during a lull in agricultural work; towards the end of the dry season, for instance. He or she almost certainly has no formal training, no mining skills in the conventional sense and no access to capital.

As a result in many cases the person finding the deposit is not seeking to become a miner, although if that is the only route for making money out of the deposit, then he or she will undertake this. If the person has a long-term commitment at all to mining, it is as a prospector, in the hope that he or she will be able to find a deposit that others, with more need and/or skill and/or capital will work on for a fee.

This important distinction between the ambitions of the typical seeker of a deposit and those who work it applies to mines of every size. For this reason the need for an effective legal property right over a discovery by such a person is therefore very important, and not just for 'formal' sector exploration. This right should enable the title to the deposit to be sold, tributed or optioned in a way that is routinely administered, and requires little discretionary action on the part of the authorities.

This assurance of a legal property right is the driving force behind legal prospecting by local people for economic deposits, and such a right must be sustainable against the interests of large mining companies wishing to exploit the discovery. A primary requirement for a successful small-scale mining sector is therefore a simple, uniform mining code that applies to every scale of potential discovery. Since, as has been shown, it is local people that find the majority of deposits of many economic minerals, it might be said that this requirement is a sine qua non for a successful mining sector.

As mentioned earlier, mining codes in Africa are often split into two sections, one for the large-scale formal sector and the other directed to controlling the small-scale or 'artisanal' sector. It is possible that the only country in the region that does not have this dichotomy between large- and small-scale mining in its legal system is Zimbabwe. As it is also the country with arguably the most successful, and certainly the most diverse, mining industry in the region, the system in force there (based on 200m by 500m claim blocks) is worth serious consideration.

A system of large scale concessions does exist in Zimbabwe, however. This is the Exclusive Prospecting Order (EPO) which gives exclusive rights to prospect for specified minerals or metals in an area. This system was introduced in 1950 and after its introduction there was, perhaps coincidentally, much exploration activity, particularly geochemical work, directed at locating base metal deposits. Several were found, for example Avondale (copper), Perseverance (nickel) and Epoch (nickel). However the majority of the base metal mines that were developed in the period between 1950 and 1970 were on deposits which had been discovered and pegged on the claims system some years before.

The EPO areas are not necessarily large, but they are numerous. Between their introduction in 1950 and 1983 some 600 EPO's had been awarded, with an average life of two years. At a guess this activity resulted in the creation of about 20 new formal mines of every size.

It can be concluded that a uniform, simple, mining code giving full title to the discoverers of mineral deposits, whatever their circumstances, will do more to encourage small-scale mines than attempts to make special provision for this sector in the law.

Technical Support for Small-Scale Mines

While 'bare-foot prospectors' are amongst the worlds most successful discoverers of certain high value deposits, the small-scale miners who follow them are, inevitably, inefficient in terms of getting the best return for the energy expended. This is principally because of their lack of equipment that will effectively undertake the necessary rock breaking and crushing to liberate the minerals sought.

This situation leads to high grading by the small-scale miners; the great amount of effort needed to recover any minerals at all tends to make mining activity directed at seeking a bonanza rather than at systematic development of a resource. Other factors contribute to this. For instance the mining method of the small-scale miner is by development rather than by stoping. Larger areas which may be lower grade are disregarded in favour of narrow high grade zones that can be burrowed through. Another constraint that leads to high grading is that typically beneficiation is by hand-sorting. In the case of gold this means that only rock with visible metal is selected. Such material can be assumed to have grades of the order of

about one ounce (31 grams) a ton, or about five times the grade of a large- or medium- scale gold mine.

In broad terms therefore technical support for small-scale miners should be directed at enabling them to mine profitably a greater tonnage at a lower grade - in other words make better use of the resource available. To be effective such support will therefore attempt provide the following:

a) Geological services (drilling, assaying, interpretation of intersections, simple mapping) that will enable the small-scale miner to mine less blindly than before, and allow him to appreciate the potential of lower grade zones.

b) Mining services that will enable him to undertake safe simple stoping, and thus achieve some bulk extraction from a deposit instead of 'picking the eyes out'.

c) Metallurgical services that enables lower grade material to be extracted or beneficiated than previously.

Other Support for Small-Scale Mines

Financial assistance should also be extended to the small-scale sector. In Zimbabwe a series of seven different loans are available for specific purposes at low rates of interest. Such loans are made on the recommendation of the Chief Government Mining Engineer and the Directors of Geology and Metallurgy.

The training aspect should also be considered. Graduate mining and metallurgical training is not as important in a nation's small- to medium-scale mining industry as the training of competent technicians who are capable of initiating and managing such mines. Graduates normally prefer to work on large mines or in the larger centres; the skills needed to run smaller mines in the bush are practical rather than academic and need a different sort of training to that offered to University entrants.

Who Utilises the Support Services?

Here again there is no need to make a distinction in law between those eligible for support and those who are not. Certainly the principal users will be small mines, but there may be occasions when formal mining operations wish to make use of the facilities. If the demands on the services offered become too great, then in principal it will be the wealthier formal mining operations that will have to go to the back of the queue.

CONCLUSIONS

In this paper it is pointed out that, historically, it is local people who have made the majority of discoveries of certain minerals and metals in Africa. As a consequence it is suggested that a principal reason for promoting small-scale mining is that it serves to encourage prospecting by local people, for the success of such prospecting can dramatically shorten the lead time for the creation of large mines.

It is proposed that this promotion be undertaken through a three-point plan:

  a) the government either provides purchasing arrangements for the products of small-scale miners that are capable of out-bidding the black market purchasers, or a policy is adopted that the state is non-interventionist in this matter, but reserves the right to reassign the mining title of small-scale miners who sell their products illegally

  b) creation of a legal system that gives full title to any finders of deposits of minerals, so that they have security of tenure (provided they work the discovery), and can sell, option or tribute it if necessary

  c) government technical support for miners which is provided at no or nominal cost to holders of mining title.

In none of the above sections should an attempt made to differentiate between large-scale and small-scale mines. It will be up to the miners themselves to decide to what extent they make use of the marketing and technical services provided by the state, while the mining code would apply equally to all prospecting and mining activities, regardless of the size of the operations concerned.

References

1. Sub-Saharan Africa: from crisis to sustainable growth. The World Bank, Washington D.C. 1989. p. 124 - 125.

2. Investment in African Mining Development. E. Bolte. Workshop on African Mining Development, organised by the Economic Commission for Africa, Harare. 1990.

3. Viewing K.A., Phimister G. and Jourdan P.P. A review - past, present and future - of Zimbabwe's mining industry. African Mining 1987 Harare, IMM. p. 401 - 424.

4. Charles River Associates. Start-up of new mine, mill-concentrator and processing plants for copper, lead, zinc and nickel. Prepared for the World Bank, November 1979.

5. The Economist, December 8th 1990, p. 75

6. Batchelor. R. Gold as an Investment: Ten (Stylised) Facts. Supplement to the Mining Journal. London. November 23rd, 1990.

7. The Economist. November 12th 1988, p. 124.

8. Africa's commodity problems: towards a solution. Task Force on UN-PAAERD, UNCTAD Secretariat, United Nations, Geneva, 1990. Johnson C.H.

9. Enhancement of the contributions of African aluminium, copper, iron and steel towards the regions' economic advancement. Workshop on African Mining Development, organised by the Economic Commission for Africa, Harare. 1990.

10. Ranking Countries for Minerals Exploration. Natural Resources Forum, UN, New York, August 1990.

11. J. Williams (RTZ). Pers. Comm.

12. Chamber of Mines, of Zimbabwe, 1990.

13. Enhancement of the contributions of precious and semi-precious minerals produced by small-scale mining organisations to the economic progress of Africa. Workshop on African Mining Development, organised by the Economic Commission for Africa, Harare. 1990.

Mining 1

Geotechnical aspects of vertical crater mining method in a deep mine

S. C. Goel
Zambia Consolidated Copper Mines Ltd, Zambia

SYNOPSIS

Vertical Crater Retreat is compared to other open stoping methods. Various geotechnical factors have been considered including the effects of stress concentration, jointing, dynamic stresses superimposed by blasting vibrations, amount of development required and shrinkage draw. All these factors favour a VCR method in a deep mining environment. The joint orientations, however, need to be studied prior to accepting suitability of the method. Mindola Mine of Nkana Division of Zambia Consolidated Copper Mines Limited, Zambia is the deepest mine on the Zambian Copperbelt. The method has been tried against the background of the currently used variations of open stoping. The VCR method has significantly improved upon various problems experienced with other open stoping methods.

INTRODUCTION

The Nkana Division of Zambia Consolidated Copper Mines Limited, Zambia produces close to 4 million tonnes of Copper-Cobalt ore per year. Mindola Mine is the largest underground mine of the Division and is producing at the deepest levels on the Zambian Copperbelt. The orebody is generally tabular and steep dipping and is a sedimentary deposit. At deeper levels of Mindola Mine the orebody dips between 55° to 90° and the thickness varies from 8 m to 14 m. Stoping, at present, is active down to 4180 L (1274 m below surface) and the shaft is being deepened to produce down to 5220 L (1591 m). A vertical longitudinal projection of the mine is shown in Fig. 1.

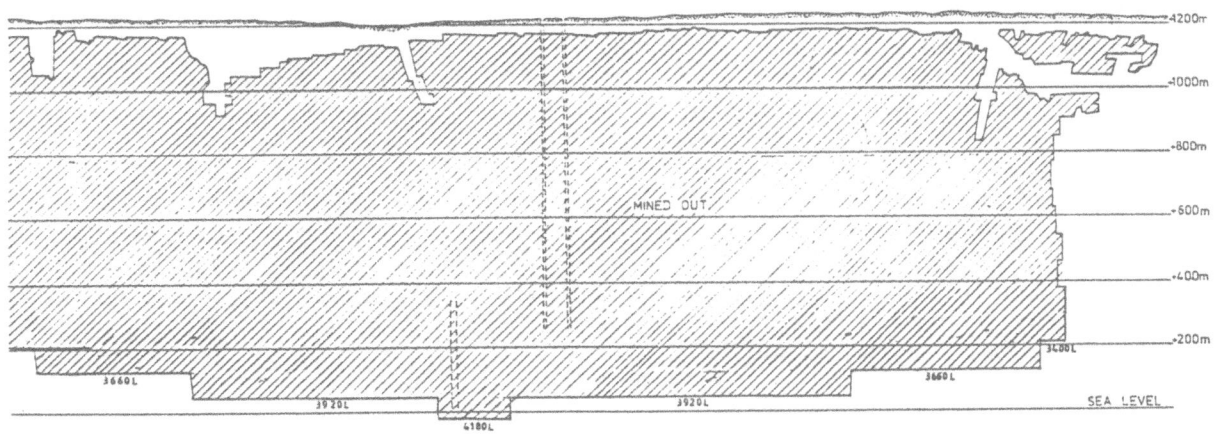

Fig. 1   Vertical Longitudinal Projection of Mindola Mine

GEOTECHNICAL ENVIRONMENT

A typical section through the orebody is shown in Fig. 2. The strata is generally classified between fair to good. Bedding is generally prominent and is the main plane of weakness. There can be up to 3 other joint sets, out of which one set is prominent, i.e. set 'B' shown in Fig. 3. The uniaxial compressive strengths of various strata in and around the orebody are shown in Fig. 4.

There are some important features which influence the geotechnical environment in the area. Firstly, there is a thin stratum known as schistose ore which has low uniaxial compressive strength and can yield to reduce and distribute stresses in high stress concentration areas. Secondly, jointing plays an important role in the failure of the hanging wall and the exposed faces of the orebody. They result in poor fragmentation, dilution and, at times, even loss of access for blasting a stope.

MINING METHODS

Sublevel Open Stoping with total extraction, using 57 mm diameter long hole fans, was the method extensively used at Nkana Division until mid 1970's, Fig. 5. The method was successful at shallow depths down to 2880 L (878 m). The method posed a number of mining problems below this depth at Mindola mine. The major problems are given below:

instability of strike drifts

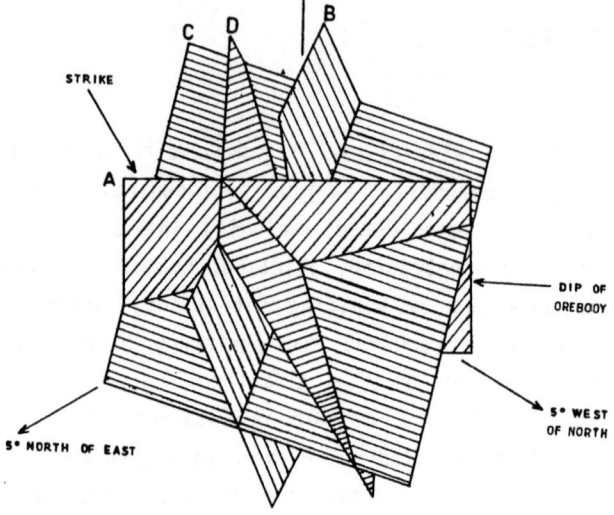

Fig. 3   Joint Sets

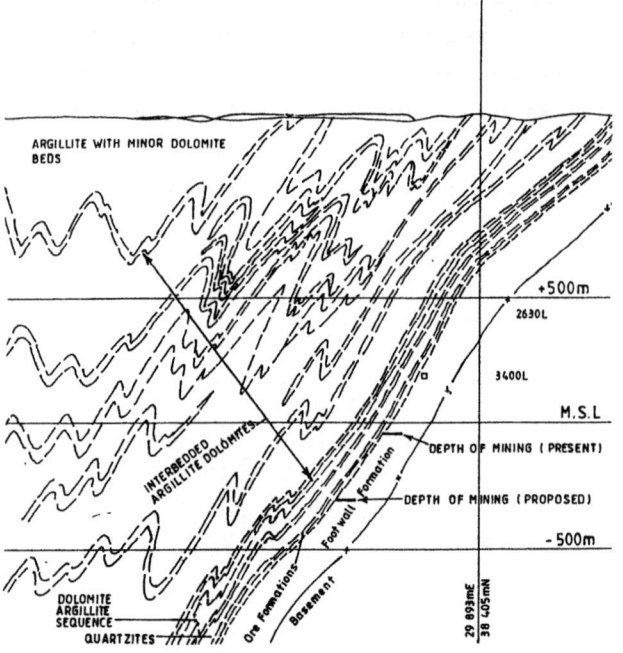

Fig. 2   A Typical Geological Section of the Orebody at Mindola Mine

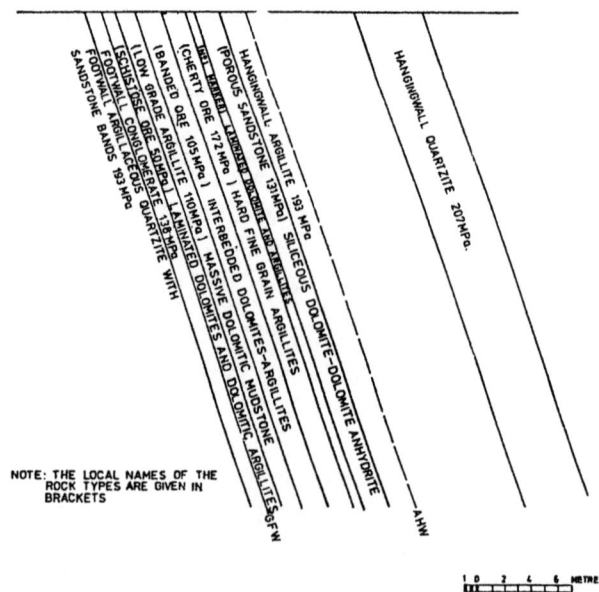

Fig. 4   Rock Quality in and Around the Orebody

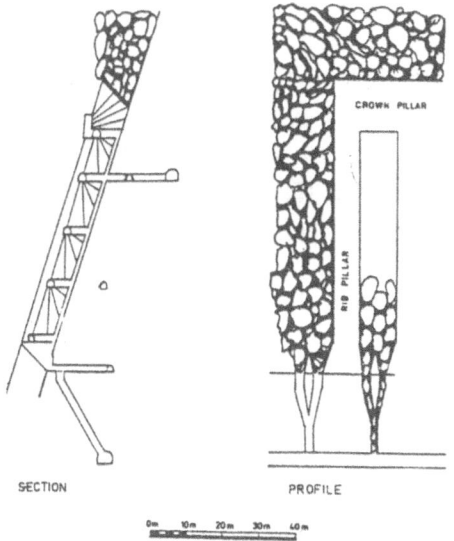

SECTION   PROFILE

Fig. 5   Conventional Sublevel Open Stoping

. premature closure of blast holes due to movement on weakness planes and due to stress concentration effects

. need to precharge the blast holes which sometimes detonated prematurely

. high rock temperature with humidity was deleterious to the precharged explosive

leading to misfire of the blast holes

. reduced recovery of metal and increased waste dilution of ore

. poor fragmentation due to premature falls of stopes and large waste slabs from the hanging wall resulting in reduced rate of production and excessive secondary blasting

. interruptions in production due to poor ground conditions

To reduce these problems, especially the closure and misfire of holes, bench blasting from sublevels[1] was introduced in the late 1970's, Fig. 6. The method used 127 mm diameter parallel holes along the dip of the orebody which improved stability as the drilling was done from crosscuts which are more stable than strike drifts. Since bedding forms the prominent weakness planes, strike drifts pose more instability problems. In cross cuts the exposure of bedding planes remains limited to the width of the crosscut only. The premature closure of holes was also minimised as they were drilled along stratification and were of larger diameter. The 57 mm diameter fan holes,

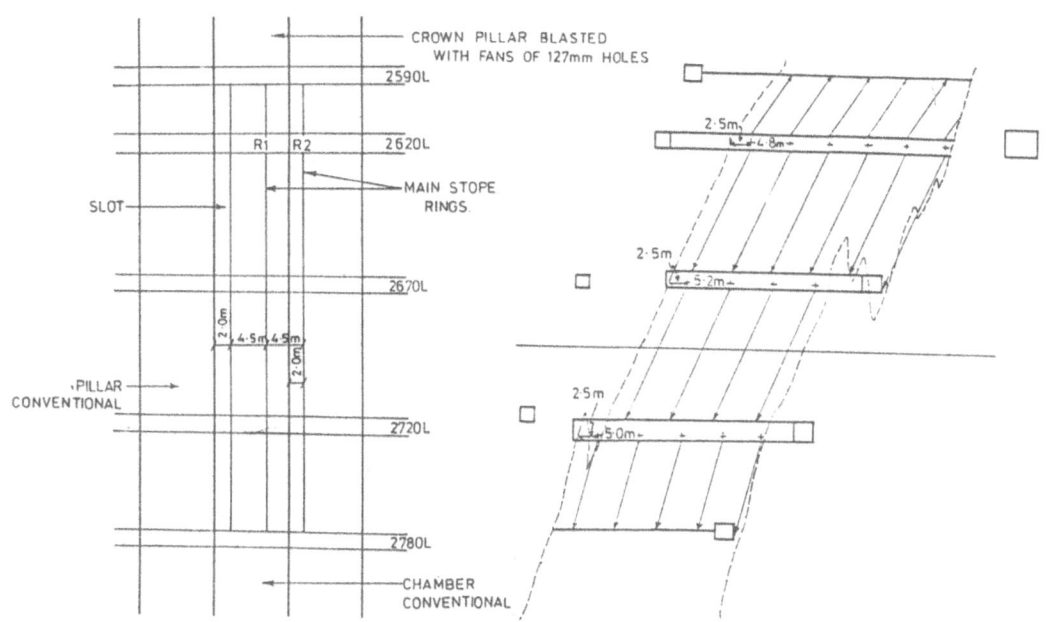

Fig. 6   Large Diameter Bench Blasting Sublevel Open Stoping

used earlier, cut across the stratification and movement on any bedding plane closed the blast holes. The method made it possible to successfully mine down to 4180 L (1274 m) with much reduced mining problems.

The large diameter reamed holes were drilled from sublevels at 12 to 15 m intervals. The blasting sequence from sublevels made it possible to keep the stopes nearly full up to the last blasted sublevel. The method of shrinkage draw is practised for the reduction of waste dilution from the hanging wall. The success of the method proved the potential of large diameter blast holes. The method, however, suffered from the following disadvantages:

. high development requirement

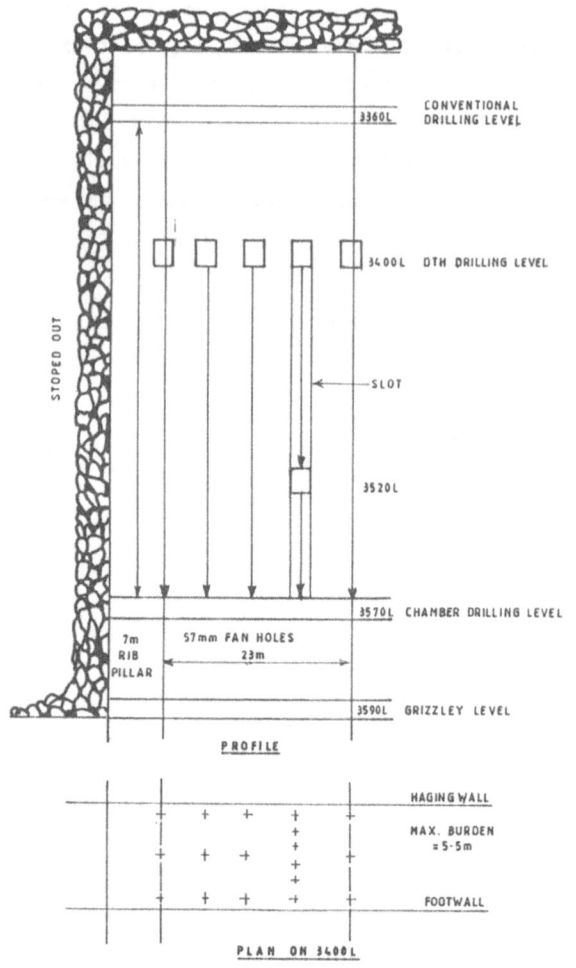

Fig. 7   Blast Hole Open Stoping

. slow blast hole drilling as the holes were drilled in two stages, pilot 57 mm diameter holes drilled with a bar and arm mounted drifter and then reaming them to 102 or 127 mm diameter.

. larger blasts of normally 180 kg explosive per delay

. premature fall of ore after every blast resulting in poor fragmentation.

It was therefore realised that the next logical stage of development should be the use of down the hole machines for drilling longer holes. Longer holes will reduce the need for multiple sublevels[2]. The modified layout for the blast hole open stoping method is shown in Fig. 7. All levels indicate feet below surface. The method reduces the sublevel development requirement to nearly half. The method also increases the charge density from 0.5 kg/m^3 to 0.8 Kg/m^3 due to the increase in the hole diameter and should therefore improve ore fragmentation.

Initially to reduce charge per blast and length of holes in a slot, an intermediate sub level was established, 3520 L as shown. It allowed blasting below 3520 L to be divorced from that between the 3520 L and the 3400 L. The undercut surface of the stope, of 13 m x 23 m area, stood quite flat without any overbreak after mining the chamber/trough on 3570 L. When the slot was blasted from the 3520 L down, the ore started to unravel prematurely and the slot widened up to 12 m. After blasting the complete stope below 3520 L, an undercut was again created of the same size and remained stable. The instability after blasting the slot below 3400 L was even more severe. The outcome was initially put to high charge density in the blast holes and the narrowness of the slot. Inefficient utilisation of explosive energy into a narrow slot opening can cause wide spread damage and thus causing unravelling of the stope ore. The ground conditions deteriorated even in the access drifts.

Detailed investigations showed that the stability of the flat undercut and instability of vertical faces created by slot mining were mainly due to joint orientations in the area and the unravelling of ore was joint controlled. All the above open stoping methods, however, require the creation of at least one open vertical face by mining a slot before the bulk of the stope is blasted and therefore it was considered essential to find a more suitable mining method. As the orebody is steep, of uniform thickness and remains stable with flat undercuts, the Vertical Crater Retreat mining method was chosen as a potential alternative. Visits were made to a number of North American mines in an effort to determine various parameters for implementing the method.

VCR Method

This mining method uses spherical charges as against column charges[3,4,5,6] used in most other methods of stoping. To create spherical charges, the method uses short lengths of high strength explosives in large diameter blast holes. An available down the hole machine was used to drill 165 mm diameter holes. The spherical charges are blasted into an undercut and the vertical slot is not required. Only the swell is drawn after each blast to provide support to the walls of the stope and to the undercut of the stope. After completion of blasting, the stope is fully drawn.

Before changing to full scale implementation, it was planned to try the principle in an area with some existing development. A schematic stope layout for the adaptation of the method is given in Fig. 8. The layout shows that the adaptation was only to test the blasting method and only a part of the stope was blasted with this method, the remainder of the stope was blasted conventionally. The results[7] of the first stope blast were excellent and some of the important observations are given below:

. The blasting was easy to control. The height of break in each blast was approximately 3 m as expected.

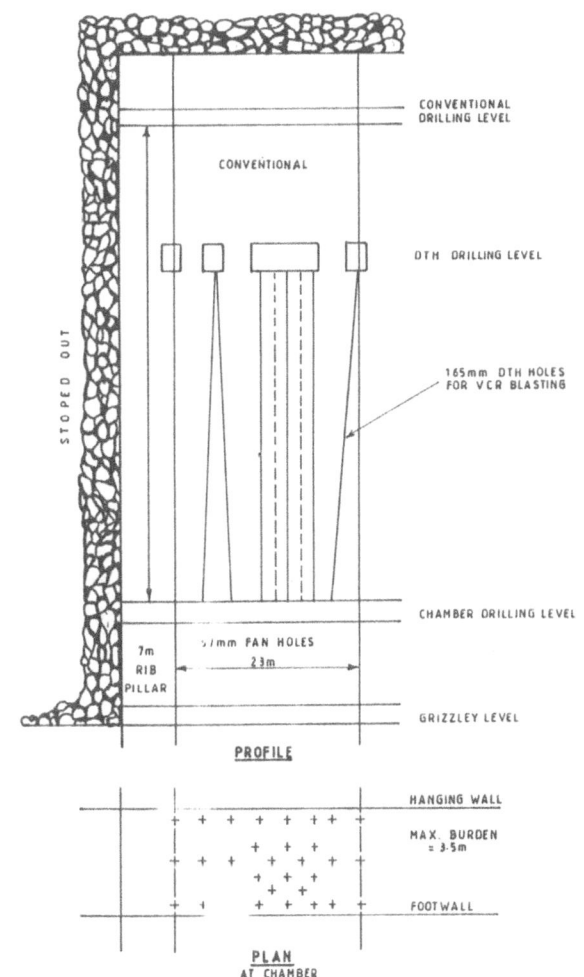

Fig. 8   VCR Layout

. The problem of premature ore falls were not observed except when the face reached very close to the DTH drilling level.

. Initial worry of not being able to blast the last cut efficiently proved unfounded

. The blast holes remained open until the final blast and no precharging was necessary

. The accesses to the stope stood extremely well, far better than in any other method used previously and they remained accessible even after the stope was fully blasted.

. The general ground conditions around the stope were excellent

. The back of the stope after blasting below the DTH drilling level or the bottom of the crown/sill pillar was clearly seen not to have overbroken at all. All the support in the back on the DTH drilling level could be seen not to have overbroken at all even after the stope below was fully blasted.

. Rib pillar remained intact and could be drilled and blasted successfully as against sub level open stopes where it gets badly damaged and due to poor conditions of holes the efficiency of blasting the rib pillar remains suspect.

. The conditions in the crown pillar again deteriorated when it was blasted conventionally

. the fragmentation of the ore was far superior and the secondary blasting of boulders at the grizzleys was more than halved.

. A number of stopes have since been blasted successfully repeating a similar performance giving confidence into the method. The method is therefore being expanded to other areas.

The remainder of the paper compares the geotechnical aspects of a sub level open stoping method with a VCR method as used at Mindola Mine.

GEOTECHNICAL CONSIDERATIONS

The ground stability improved with the application of the VCR method of mining as compared to the other variations of Open stoping used earlier at the same depth. This observation showed possible application of the method at deeper levels[7]. A combination of factors contribute to the improvement in stability with the VCR method. These factors are analysed and discussed below.

Effect of Stresses

Stress analyses were carried out using a three dimensional boundary element computer model for a thin seam, called BESOL3D. The model is used to carry out elastic analyses for a VCR stope and a conventional sublevel open stope at equivalent stages of blasting for comparison purposes. Failure Index is then calculated using Mohr Coulombs strength criterion to delineate potential failure zones for the two cases. The Failure

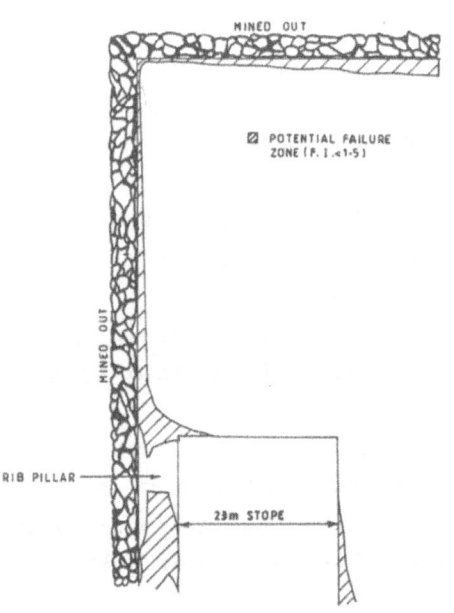

Fig. 9   Potential Failure Zone for a VCR Stope

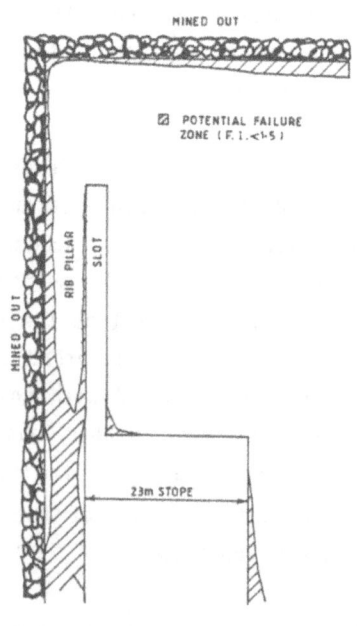

Fig. 10   Potential Failure Zone for a Conventionally Blasted Sublevel Open Stope

Index is defined as the ratio of strength and stress and is similar to the factor of safety. The potential failure zones for the two cases are plotted in Fig. 9 and 10. The potential failure zone is considerably reduced in a VCR stope. It is mainly due to a reduction in exposed faces. Each exposed face reduces lateral restraint or stresses in one direction and thus reduces triaxial strength. Failure zones at the convex corners formed at the slot opening in sublevel stoping can start a process of unravelling of blocks formed by the pre existing joints. This phenomenon was extensively observed at Nkana Division Mines.

Often the effect becomes so pronounced that parts of the stope falls prematurely and fully opening the slot becomes difficult due to the continued unravelling. At times the accesses to stope also deteriorate. The rib pillar reduces in its effective size and thus can breach prematurely and cause excessive waste dilution. The stress effects are more favourable for VCR stopes and therefore the above mentioned problems are minimized.

Dynamic Stresses

Dynamic stresses in a stope are generated by the blasting vibrations. The intensity of these stresses at a point mainly depends upon its distance from the blast and the amount of explosive charge blasted at one time. In conventional sublevel open stoping, at least one full column charged blast hole is blasted on one detonator delay. At Mindola mine, the explosive per delay varied from 180 kg to 800 kg depending on the hole length and its diameter. The charge being full column i.e. to the collar of the blast hole, the distance of the access development is equal to its distance from the collar of the blast holes which is only a few metres. While in the case of a VCR stope, the blast holes are charged in short lengths with limited explosive per delay, 37.5 kg at Mindola Mine. Also the charge is close to the bottom of the hole which can be at a large distance from the access development near the collar of the hole. The dynamic stresses in the

access development are, therefore, likely to be much smaller than those in a conventional open stope. The values of the Peak particle velocity for various combinations are calculated using the relationship[8]

$$V = a \left(\frac{D}{\sqrt{E}} \right)^b$$

where V = peak particle velocity in mm/sec
  D = Distance of the blast in metres
  E = weight of the explosive in kilograms

a and b are constants depending on rock characteristics and are assumed to be:

  a = 900
  b = -1.6

The calculated values of 'V' are tabulated in Table I. For comparison, the advised maximum peak particle velocity for heavily reinforced concrete structures is 120 mm/s. The value for an adjacent blast hole reduces to nearly half for a VCR stope as compared to a conventional sublevel stope and to less than a sixth for a blast hole open stope. Therefore, the adjacent blast holes are more likely to remain open for subsequent blasts and as observed at Mindola Mine most holes remained open and did not need precharging as in holes for all previous methods. 'V' also reduced dramatically for both the access development and the next rib pillar in VCR stopes. This helps in maintaining stability of development drifts and the next rib

Table I: Peak Particle Velocity for Various Mining Methods at Various Points

Description	Distance	V = Peak Particle Velocity
Conventional sublevel Stoping:		
Next blast hole	5 m	4366 mm/s
Access drift or next pillar	10 m	1440 mm/s
Blast hole open stoping		
Next blast hole	5 m	14400 mm/s
Access drift or next pillar	10 m	4750 mm/s
VCR method		
Next blast hole	3.5 m	2203 mm/s
Next pillar	10 m	400 mm/s
Access drift	30 m	71 mm/s

pillar. Similar effects will be observed for the hangingwall and the footwall. Reduced 'V' in these areas will help reduce the waste dilution from the stope walls and that due to premature failure of the rib pillar. This general reduction in dynamic stresses in the stope and its surroundings generally increases stability and thus the control over stope blasting operations.

Density of Drifts

The VCR method required development mainly on one level as against many drilling levels in sublevel open stoping. The reduction in number of drifts increases effective bearing area for transferring stresses from hangingwall to footwall. This reduces the stress concentrations in the VCR method and therefore improves general stability. As shown in the sublevel open stoping layout, Fig. 5, there are 2m x 2m drifts at nearly 12 m interval. Assuming that a 0.5 m annulus around a drift is damaged by blasting during development, the bearing area is reduced by atleast 25%. This will increase stress concentration in sublevel open stoping at least proportionally. This effect is much reduced in VCR stopes due to less development drifts.

In addition, most of the development in a VCR stope is generally cross cutting as against strike drifts in conventional open stoping. Cross cuts remain more stable due to the fissile nature of the orebody.

Effects of Shrinkage Draw

In sublevel open stopes, the stope needs to be opened a number of times during the blasting cycle for creating a free face for the next blast. It, therefore, allows the dismantling of the blocks already formed by weakness planes, blasting cracks etc, in the hangingwall and in the other exposed surfaces of a stope. This factor can add to waste dilution and general instability in the area. In case of a VCR stope, it is easily possible to keep the stope full with ore to within 2-3 m from the undercut face. This broken ore can provide some lateral restraint to such blocks

delineated by weakness planes and thus restraining further loosening of such blocks.

Interaction of Stope Geometry and Jointing

There are upto 4 joint sets as shown in Fig. 3 but the most prominent joint set after the bedding planes is joint 'B'. 'A' is bedding which is closely spaced and plays the most significant role in stability of all openings. Joint set 'B' is quite oblique to the bedding and dips steeply. It therefore, does not form free falling wedge with bedding on the undercut face of a VCR stope. In a conventional open stope joint set 'B' does provide a free sliding surface day lighting on the vertical exposed face of a slot. Apparently this joint set has the most significant effect on relative stability of the horizontal undercut and instability of the vertical slot cut.

Joint set 'C' can form a wedge with bedding in the undercut surface but this joint set is not so prominent and continuous. In addition the wedge formed is thin and high. It has, therefore, not played such an important role in causing instability of the undercut. Joint set 'D' is even less prominent and continuous, and is much less frequent. Therefore it is also not of such a significance.

This example shows the need to study the joint pattern in the orebody in establishing the stability of an undercut face in a VCR mining method. Since this method continues to create a new undercut face after each blast, any instability of the undercut can have serious repercussions on the fragmentation of the ore.

CONCLUSIONS

VCR method has distinct geotechnical advantages over the conventional open stoping method in steep dipping orebodies. It reduces stress concentration on the mining face and thus improves stability and reduces extent of failure zones in the stope. Therefore, the method is likely to be more suitable than the open stoping method at deeper levels.

The limited charge per delay reduces damage to the walls of the stope and other adjacent areas. The premature closure of adjacent blast holes can be reduced or avoided and therefore the need to precharge some of the blast holes. The dynamic stresses in access development are also significantly reduced.

Shrinkage draw can reduce dilution by providing some support to the stope walls and the rib pillar.

The geometry of the stope and the joint pattern can play a very significant role in deciding the suitability of a method. The VCR method requires stability of the undercut horizontal face while the stability of a vertical face is crucial for an open stoping method.

Due to the reasons given above, VCR method is considered suitable for continuing mining at even deeper levels at Mindola and the method is planned to be expanded to other mines of the Division. Other variations of the method such as updip mining with fill are also being considered for optimising the benefits.

References:

1. Page C H and Goel S C. An adaptation of big-hole blasting underground, Trans.Instn.Min.Metall. (Sect A: Min. Industry), 91, April 1982, p A63-A70

2. Goel S C, Implementation of blast hole open stoping using a down-the-hole machine, Internal report no. RM56,04/86, 1986

3. Mineral Resources Development Limited, Nkana Division Vertical Crater Retreat stoping, Internal Report, Dec 1988

4. Eric Kossatz, Inco boosts productivity using bulk mining methods, Canadian Mining Journal, December 1983, p 23-25

5. Jansseurs J J and Percival J J, A practical approach to vertical crater retreat mining in soft ground at Whitehorse Copper, CIM Bulletin, June 1981, p 72-77

6. Nangle J B, The mining of open stopes at the Carolusberg Mine O'okiep, Proceedings, 12th CMMI Congress, (M W Glen editor), Johannesburg, S Afr Inst Mine Metall. 1982, p 435-465

7. Goel S C. Vertical Crater Retreat (VCR) 5050 S 3660 L trial stope at Mindola, Report No. 65/09/89, Sept 1989

8. AECI, The efforts, measurement and control of ground vibrations, Explosives Today, 2, No. 27, March 1982

Stability predictions based on back analysis of collapsed crown pillar, Epoch mine, Zimbabwe

T. R. Stacey
Steffen, Robertson and Kirsten Inc., Johannesburg, South Africa
J. A. C. Diering
Gemcom Services (Pty) Ltd, Vancouver, Canada
N. Rigby
Steffen, Robertson and Kirsten (UK) Ltd, United Kingdom

SYNOPSIS

Talc-carbonate ore is mined at Epoch Nickel Mine near Filabusi in Zimbabwe using open stoping and sub-level stoping methods. In 1978 the crown pillar above one of the ore bodies collapsed suddenly, resulting in an air blast which caused considerable damage to the shaft system. These shafts are located close to the orebody, and there was concern that the cave crater which was formed might develop further and endanger the stability of the shafts. A series of analyses was carried out in 1979 with the following aim, to:

. determine the nature and cause of the crown pillar failure
. evaluate the changes in stress, as a result of future mining, in the shaft region and on a major geological contact
. assess the likelihood of instability of the footwall
. obtain an indication of shaft deformations, and potential misalignment and failure, due to future mining
. assess the stability of the crown pillar above a second orebody
. assess the potential effect of future mining on surface mine structures.

Mining has progressed substantially in the past 10 years and it is now possible to compare the observed behaviour with that predicted in 1979.

1 INTRODUCTION

Epoch Nickel Mine is situated near Filabusi in Zimbabwe. Mining commenced in 1976 at a rate of 25000 tonnes per month, and is currently about 42000 tonnes per month, from three main orebodies and various minor, hangingwall orebodies. Early production was from open stoping operations which resulted in the formation of large open excavations in close proximity to the surface. In 1978 an 18m thick crown pillar above the largest orebody collapsed suddenly, resulting in an air blast which caused considerable damage to the shaft system. A cave crater was formed at the surface. Following this event there was concern that, with mining progressing to deeper levels, the stability of the shaft system might be at risk, and that collapse of the crown pillar above one of the other orebodies might also occur. Consequently an investigation was carried out in 1979 with the following aims, to:

- determine the nature and cause of the crown pillar failure
- assess the stability of the crown pillar above a second (north) orebody
- evaluate the changes in stress in the footwall rock adjacent to the shaft region and on a major geological contact as a result of future mining, and hence to assess the stability of the footwall
- estimate possible shaft deformations and potential misalignment due to future mining.

2 GEOLOGY, GEOLOGICAL STRUCTURE AND ROCK CHARACTERISTICS

The geology of the Epoch nickel deposit has been described in detail by Baglow[1]. The Epoch ultramafic intrusion into the metabasalt country rock (greenstones) is some 600m by 300m. The orebody host rock is talc-carbonate schist, which is massive though strongly foliated, the foliation dipping steeply to the south. There are three major orebodies which are in close proximity to the sheared contact between the talc-carbonate rocks and the footwall greenstone, Fig. 1.

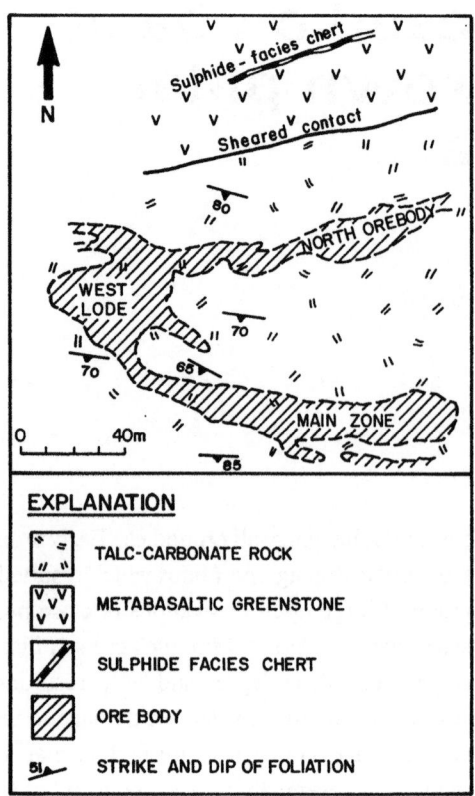

Fig. 1 Plan of Epoch Nickel Deposit on 7 level (approximately 250m below surface)(after Baglow, 1986).

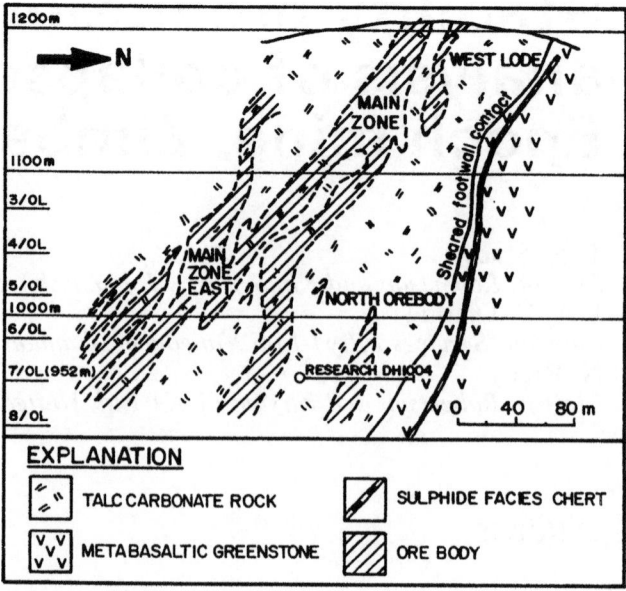

Fig. 2 Section through Epoch Nickel Deposit (after Baglow, 1986).

The dip of this contact is in the range 60° to 80° to the south, over a depth of more than 200m, Fig. 2. Compared with the talc-carbonate rock the greenstone exhibits a well-developed pattern of jointing. Joint mapping in the greenstone was carried out by Oldroyd[4], and the results indicate the existence of three sets, Fig. 3. The average strike of the talc-carbonate/greenstone contact is N75°E, and there is therefore little likelihood of major planar sliding in the footwall towards the stopes. However, as can be seen from Fig. 3, there is a wide scatter in the joint orientations, and therefore it was decided to deal with potentially unstable orientations.

To obtain an indication of the strength and deformation properties of the two rock types (uniaxial compressive strength, UCS, modulus of elasticity, E and Poisson's ratio, v), and, more importantly, their relative properties, three samples of each were tested in uniaxial compression. Further, to gain an indication of the shear strength (or cohesion C) of the intact talc-carbonate parallel to its foliation, simple punch type shear tests[5] were carried out. The results are summarised in Table I.

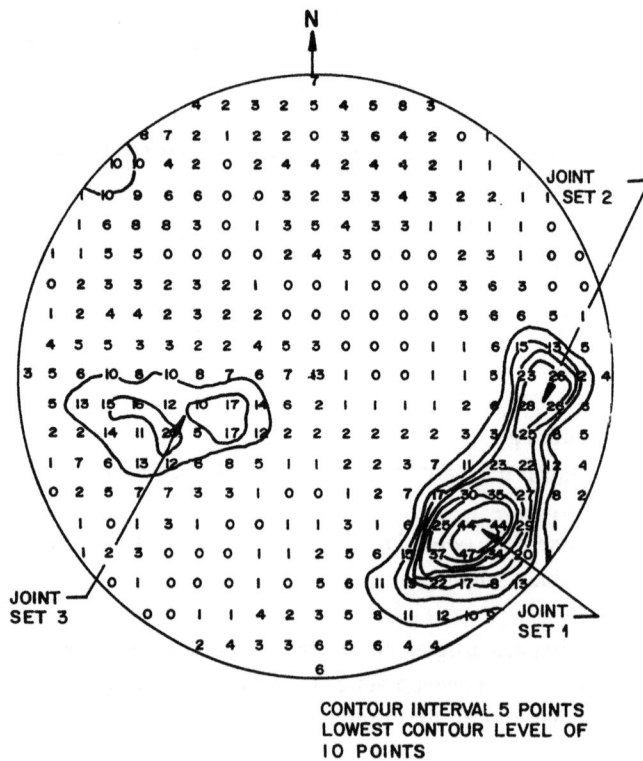

Fig. 3 Stereoplot of poles of joints in greenstone (lower hemisphere projection) (after Oldroyd, 1976).

Table I Results of laboratory tests on samples of greenstone and talc-carbonate

Rock Type	UCS (MPa)	E (GPa)	υ	c (MPa)
Greenstone	251	108	0.27	-
	157	112	0.24	-
	249	107	0.28	-
Mean	219	109	0.26	-
Talc-carbonate	62	102	0.38	2.7
	48	100	0.39	2.7
	41	101	0.43	2.4
Mean	50	101	0.40	2.6

From the above the talc-carbonate rock mass may be described as competent, relatively weak, and strongly foliated. The shear strength of the intact material is low in the direction of the foliation. The greenstone rock mass is well jointed, consisting of high strength rock material.

3 PROBLEM DEFINITION AND ANALYSIS

The situation at Epoch Mine represents a complex problem since it is three dimensional in geometry and rock properties, contains the major greenstone/talc-carbonate contact, and the rock mass structural characteristics are significantly different in the two rock types.

The most important items of data available for the evaluation of the problems were the geometry of the collapsed area and the geometry of the mining at the time of the crown pillar collapse. The approach adopted was to carry out several stress analyses so that the collapse could be evaluated. The philosophy of the practical application of theoretical stress analyses to real mining situations has been described by Diering and Stacey[3], and it was this philosophy that was followed in the investigation. In 1979 there were not the sophisticated methods of analysis available that there are today. The approaches adopted were therefore simple by today's standards. Four methods of analysis were used in an attempt to provide cross checks. These were three-dimensional boundary element stress analysis, finite element axisymmetric stress analysis, and three-dimensional and two-dimensional displacement discontinuity element analysis. The main analyses were carried out with the first two approaches, and these will be dealt with in this paper.

The geometry of the stopes at Epoch was simplified for the purposes of analysis, but it was considered that this would not detract significantly from the realism of the simulations, Fig. 4. The program used for the analysis was developed by Diering[4].

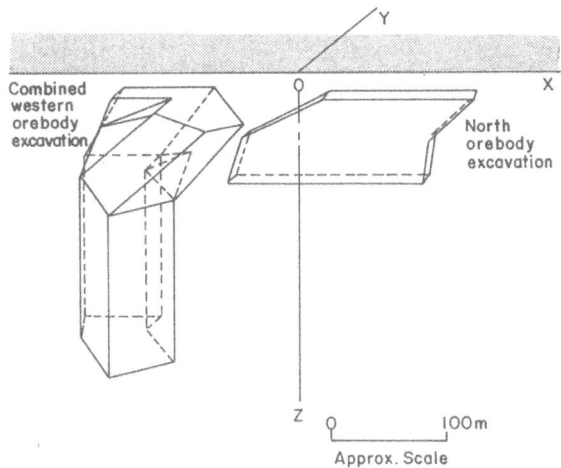

Fig. 4 Three dimensional sketch of model used in the boundary element analysis.

4 RESULTS OF THE ANALYSES

The three-dimensional and axisymmetric simulations both indicated the existence of tensile and shear stresses in the weathered talc-carbonate crown pillar to the western orebodies. The extent of the tensile stresses is shown in Figs. 5 and 6. The vertical displacements at the surface calculated from the three-dimensional model indicated that uplift would occur as a result of the excavation of the stopes. Nevertheless, the results of the analyses also indicated that the surface above the western orebodies would subside, and it is possible to compare the extent of this prediction with the observed area of the surface crater. The initial crater formed when the depth of mining was about 125m. Subsequently, the shape and extent of the caved area developed with the progression of mining in depth. Thus the actual outline for mining at a depth of 280m maybe compared with the extent of the crater predicted from the three-dimensional stress analysis for mining at a depth of 255m. The talc-carbonate/greenstone contact was not modelled in the three-dimensional analysis, but the axisymmetric analysis indicated that the tensile zone occurred up to the greenstone contact, but was not continuous across it. The predicted extent of the crater was therefore extended towards the location of the contact. It can be seen that, notwithstanding the coarseness of the analyses, the agreement between the predicted and observed crater outlines is remarkably good.

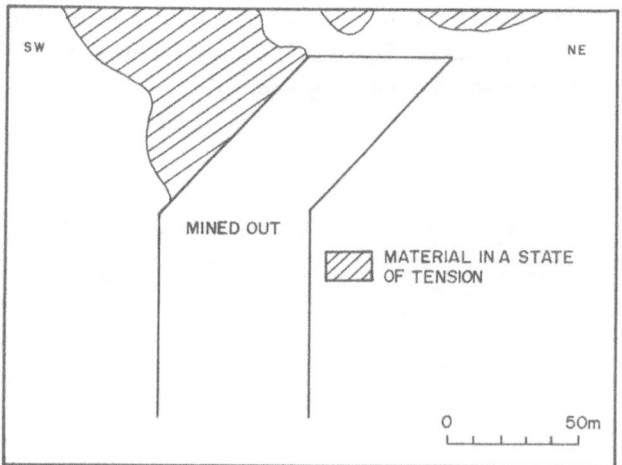

Fig. 5 Section through three-dimensional boundary element model showing zones of tension above the stopes.

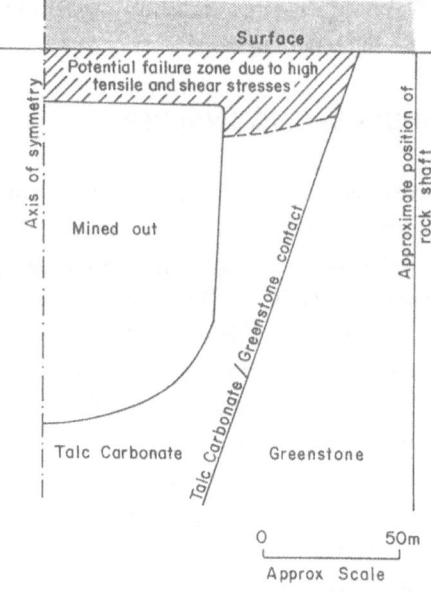

Fig. 6 Simplified section through western orebodies used for axisymmetric finite element analysis showing the extent of the potential failure zone.

Resistance to shear within the rock mass in the tensile zone would be provided solely by cohesion of the rock mass. Frictional resistance could not be generated since in the tensile zone there were no compressive stresses normal to the potential failure planes. Maximum shear stresses predicted from the analyses in this region were of the order of 2 MPa. The average "indicative" cohesive strength measured on the samples of unweathered talc-carbonate taken from a depth of about 200 m below surface was 2.6 MPa. The cohesive strength of the weathered rock would be less than this value, and it is therefore expected that the shear strengths and calculated shear stresses would be of very similar magnitude.

It was concluded from the results of the analytical work that the crown pillar failed as the stresses in the pillar reached the strength of the pillar rock mass. Specifically, the failure was due to the:

- presence of a large tensile zone above the western orebodies
- the low shear strength of the talc-carbonate rock mass parallel to the direction of foliation; and
- the shear stresses induced above the stopes, which attained the level of the shear strength of the talc-carbonate.

It was concluded that the methods of analysis could be utilised to analyse potential problems in the future mining programme.

North orebody crown pillar stability

The three-dimensional stress analysis indicated that a tensile zone occurs in the crown pillar of the north orebody. However, the magnitudes of the tensile and shear stresses in this zone would be considerable. No significant "subsidence" of the surface was indicated by the analysis. It was therefore concluded that the north orebody crown pillar would be stable.

Effect of future mining on stability of the footwall

The close proximity of surface installations and shafts are situated close to the greenstone contact, Fig. 7, and the stability of the footwall was considered to be of paramount importance.

It is often a reduction, rather than an increase, in stress that causes the problems in mining operations. The relaxation of the rock mass arising from a reduction in confinement loosens rock blocks and so causes unstable conditions.

Changes in normal and shear stress acting on potential failure planes in the footwall of the Epoch Mine were determined from the analytical work. Potential failure planes with inclinations between 35° and 71° were considered. The results showed that there was a large margin of safety against sliding on any of these potential shear planes in the footwall. The angle of friction required for stability was predicted to be less than 10°. The analyses also indicated that there would

59

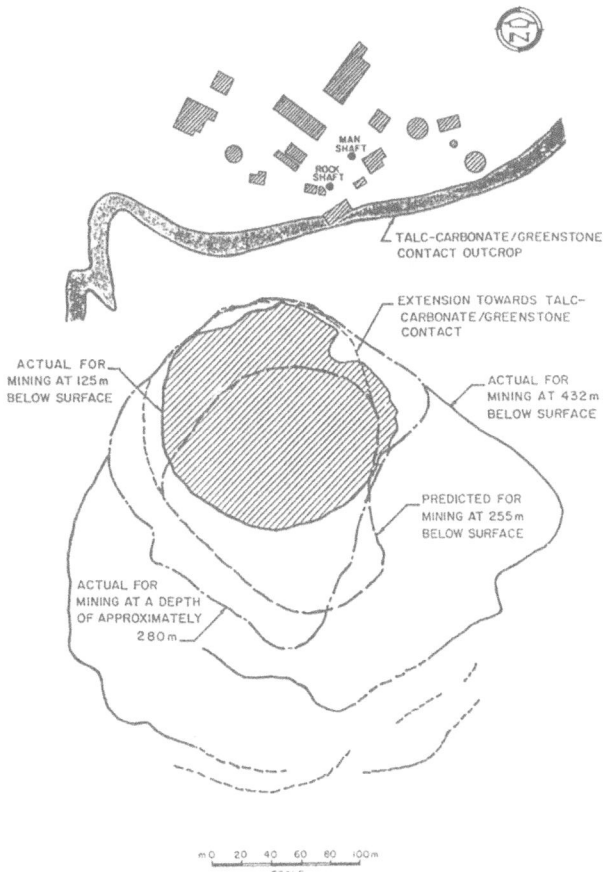

Fig. 7 Outlines of cave areas.

be little likelihood of toppling failure in the footwall. It was therefore concluded that the footwall would be stable, and that the safety of the shafts and surface structures was not in jeopardy.

Potential shaft deformations

The analyses conducted enabled the magnitudes and directions of the potential footwall displacements to be predicted at several locations relative to the stopes and shafts. The results indicated that very small displacements of the footwall towards the stopes would occur which would reduce with distance away from the stopes. These data were presented for comparison in the future with the results from three extensometer installations. The expected horizontal displacements of the shaft for mining to depths of 125m and 255m are shown in Fig. 8. It can be seen that predicted displacements are small, and no significant misalignment would be predicted. These results were based on the rock deformation properties obtained from tests on small scale samples. However, even if the modulus of deformation of the rock mass was only one

tenth of the intact rock modulus, relative footwall displacements and shaft misalignment would still be small, of the order of 25mm. It was concluded therefore, that shaft movements would not represent a problem with future mining.

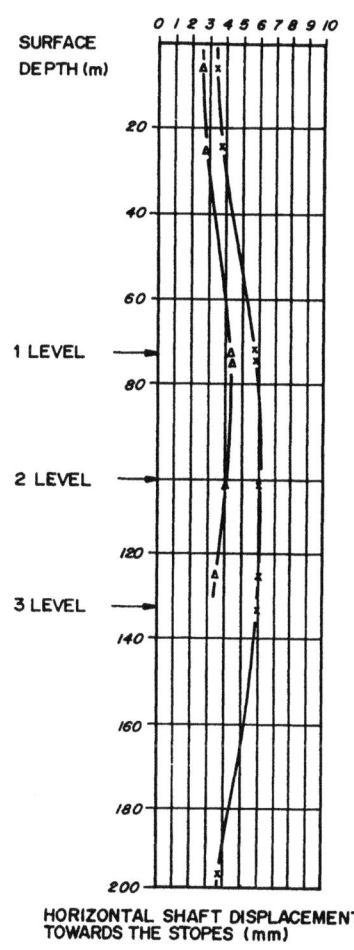

HORIZONTAL SHAFT DISPLACEMENT
TOWARDS THE STOPES (mm)

x x x   HORIZONTAL SHAFT DISPLACEMENTS WHEN
        STOPES MINED OUT TO A DEPTH OF 255 mm
Δ Δ   HORIZONTAL SHAFT DISPLACEMENTS WHEN
        STOPES MINED OUT TO A DEPTH OF 125mm

Fig. 8 Predicted horizontal shaft displacements.

5 CONCLUSIONS

The case study has demonstrated that the use of stress analysis models can provide adequate detail for design purposes. The agreement between the observed crown pillar failure and the failure predicted from the analyses was very good. The same models could therefore be used with confidence for the evaluation of the effects of future mining. From a practical point of view, stability of the footwall rocks, the surface installations and the shafts was predicted, which has been proved over the past ten years of mining operations.

ACKNOWLEDGMENTS

The permission of the Managements of Epoch Mine and Bindiva Nickel Corporation to publish this paper is gratefully acknowledged.

REFERENCES

1. Baglow, N. (1986) The Epoch nickel deposit, Zimbabwe, in Mineral Deposits of Southern Africa, ed Anhaeusser, C.R. and Maske, S., Geological Society of South Africa, Johannesburg,
pp255-262.
2. Diering, J.A.C. (1982) Further developments of the boundary element method with applications in mining, M Sc Thesis, University of the Witwatersrand.
3. Diering, J.A.C and Stacey, T.R. (1987) Three-dimensional stress analysis: a practical tool for mining problems, APCOM 87: Proc. 20th International Symposium on Application of Computers and Mathematics in the Mineral Industries, v 1:Mining, Johannesburg, South African Institute of Mining and Metallurgy, pp 33-42.
4. Oldroyd, D.C. (1976) Special Projects Report No 14/76, August 1976, Anglo American Corporation Technical Development Services.
5. Stacey, T.R. (1980) A simple device for the direct shear strength testing of intact rock, Journal of the South Africa Institute of Mining and Metallurgy, v 80, no 3, pp 129-130.

Metallurgy 1

Two gold heap leach operations in Zimbabwe

W. P. Channon
Anglo American Corporation, Harare, Zimbabwe
(currently BSR Ltd, Bindura Nickel Corporation, Bindura, Zimbabwe

SYNOPSIS

Anglo American Corporation has commissioned two gold projects in Zimbabwe since Independence, both utilising heap leach technology. This technique has enabled these relatively low grade, low tonnage properties to be economically developed; something which may not have otherwise been possible by more capital intensive routes.

Two variations on heap leach technology are employed:

i)   Isabella is a run-of-mine leach situated in the dry western part of Zimbabwe.

ii) Hopefield Gold relies on the agglomerated heap leaching of old mine tailings. In this case, the tailings are on an environmentally sensitive site, therefore careful consideration had to be given to process selection.

INTRODUCTION

With the closure of the Champion Mine in late 1976, Anglo American Corporation lost its active association with gold production in Zimbabwe. Although some geological work continued, it was not until comparatively recently that gold production recommenced using heap leach technology.

The Isabella Mine was developed as a run-of-mine dump leach with prime objectives being to establish a financially viable property while gaining technical expertise in both design and operating practice. The first gold was poured in late December 1989 with anticipated output being achieved by April 1990.

Hopefield Gold was developed as an agglomerated tailings heap leach. These tailings were on an environmentally sensitive but compact site, completely surrounded by arable and pastoral land. Process selection was of paramount importance and the agglomerated heap leach route was considered to be the most appropriate form of treating this 80%-75 μm tailings product. Agglomeration of tailings, followed by heap leaching, is relatively new to Zimbabwe and while it perhaps presents more technical challenges it is regarded as a treatment option which should be considered seriously in many applications. Hopefield Gold was in the project development stage at the time of writing with commissioning scheduled for December 1990.

ISABELLA

General Background

Isabella Mine is situated in the Bubi District approximately 80 km due north of Bulawayo. The mine location is shown in Figure 1. Reserves had been delineated in two separate orebodies and a further two potential orebodies had been discovered before geological work was terminated in January 1987. The escalation of dissident activity in the immediate area impeded further progress on the ground. However, the property was evaluated on the metallurgical results obtained from the defined orebodies.

Figure 1 : LOCATIONS OF HEAP LEACH PROPERTIES

63

Geology and Reserves

Geologically, the claims cover a mineralised shear zone in felsitic schists of volcanic origin. The east-west trending suite of silicified shears runs parallel to the contact between mafic volcanic greenstones to the north and phyllitic sediments to the south. These silicified shears are irregularly impregnated with pyrite and arsenopyrite and host the gold deposit. Ore oxidation extends to 50 m depth but the oxide/sulphide interface is generally at 27 m.

The shears hosting the Castile and Maria orebodies dip southwards at approximately 45° and ore zones average 10-12 m true width over strike lengths of 450 m and 270 m respectively. Ore reserves, to the planned mining depth of 27 m are 650 000 tonnes at an average in situ grade of 2,3 g/t.

Geological investigations have recommenced in the immediate area and additional reserves appear likely.
Mining

Both pits are small and narrow in relation to equipment size. For each pit a simple end-on road follows the working benches. It was initially thought that the ore and waste would be relatively soft 'free-digging' material. However early operations indicated that blasting would be required in both pits as the material was just too hard to dig. Initially paddock blasting was tried but the mine has now settled on blasting to a free face. Holes are drilled on a 2,5 x 2,5 m grid with the light blast working across the shear for the 2,5 m benches. This procedure has maintained good fragmentation without the generation of major amounts of fines.

Prior to mining, sample rip lines are established across strike and at 10 m intervals. Samples are taken at 1 m intervals along each line and these are used to delineate the ore body using bottle roll analyses. Blast drill holes can also be sampled.

Production was planned at 10 000 tonnes/month ore at a 2,5:1 waste:ore stripping ratio. Allowing for dilution, expected heap leach grade was 1,9 g/t. Earthmoving equipment comprises a Cat 225 excavator, two Cat D25D articulated dump trucks and a Cat D6D bulldozer, which is used for both mining and heap construction. Both the drill and blast as well as the haulage are undertaken by contractors.

During the early production months, it soon became apparent that the limits of the ore body near surface were considerably wider than those extrapolated from the exploratory drilling but the grades were lower than forecast. This was explained as a 'halo' effect probably caused by weathering and erosion of overlying material. This phenomenon significantly lowered the grade of ore to the pad to an average of 1,0 g/t for the first six months of operation. While this 'halo' effect was present in both pits, the grade was much higher in the Maria orebody. Mining of the Maria orebody began in May 1990, and by blending ores, average grade improved to the pad.

Metallurgical Testwork

Bulk samples were obtained from two old shafts on the Castile orebody, crushed to -25 mm and column tested at two independent metallurgical laboratories. Recoveries of 70% were indicated from 1,5 g/t ore. Although these column tests were encouraging, there remained some doubt that heap conditions were not simulated and there was also some discussion on the merits of crushing the ore or not.

Consequently a field heap leach was planned using run-of-mine ore. However, increased dissident activity in the proposed mine area led to the cessation of all work.

By April 1988, the political climate had improved considerably and the field test was carried out

1 000 tonnes of run-of-mine material were leached for 53 days. Percolation was excellent and a recovery of 70% was achieved from a 1,73 g/t ore.

Parallel column leaches were also run on 2 tonne lots of crushed and uncrushed ore from the same material. In the laboratory tests, the run-of-mine ore performed better than the -25 mm product where some degree of solution channelling was suspected.

A summary of metallurgical testwork is given in Table 1. From this work, a run-of-mine leach, on high density polyethylene (HDPE), was considered to be the appropriate treatment route.

Table 1 : Heap Leach testwork results

	Sizing	Leach Period (Days)	Grade (g/t)	Recovery (%)
Column Test No. 1	R.O.M.	53	1,46	74
Column Test No. 2	-25mm	32	1,58	67
Field Test	R.O.M.	53	1,73	75
Column Test No. 3	R.O.M.	49*	2,11	71
Column Test No. 4	-25mm	49*	2,04	66

*Available time for test columns; leaching incomplete.

General Process Description

Leach Pad

An initial pad area of 130 x 270 m has been prepared and this has been divided into four separate cells, by means of half metre high berms, to provide for metallurgical accounting and control. The pad liner is 0,5 mm HDPE laid on prepared ground of generally 1% slope. The compacted sub-base includes a thin layer of old mill tailings immediately below the liner. Further liner protection is provided by a 200 mm thick layer of old mill tailings on which 100 mm diameter drainage collection pipes have been placed at 10 m centres.

This perforated earthenware piping has proved to be both successful and cost effective. Provision has been made for future pad extensions.

Heap Construction

Ore is hauled directly to the pad area and dumped at the end of a narrow haul strip on the heap using two Cat D25D haul trucks. Spreading and ripping is carried out using a Cat D6 bulldozer. Compaction due to equipment operating on the heap is not excessive and percolation rates generally average 8 litres/m²/hr. Dicalcium silicate slag is used at 3 kg/tonne ore for pH control. Current practice is to stack in 5 m lifts with 15 m being regarded as the ultimate heap height.

Solution Circuit

Barren solution of 0,04% NaCN equivalent strength is pumped through an in-line 150 µm filter to the heap at an average rate of 35m³/hr. Drip emitters, at 0,8 m grid spacing are used to distribute the solution. These drip emitters have helped to reduce water evaporation loss in an area where water supply is of concern. In addition drip irrigation helps reduce the surface area of solution exposed to the atmosphere and reduce reagent consumption.
Calcium cyanide consumption is 0,35 kg/t ore.

The primary leach cycle is generally 60 - 70 days, however leach cycles have been extended and, in the latter stages of leaching, restricted solution flows have been maintained to older sections of the heap. This procedure helps maintain pregnant solution tenor while reducing pumping and antiscalant costs.

Leachate, from the various sections of the heap, passes through V-notch weirs and monitoring of both solution flow and gold tenor on a twice per shift basis gives metallurgical performance data for each section. A double drainage channel arrangement allows heap underflow to be directed to either the pregnant or barren ponds depending on solution tenor.

These ponds are each of 900m³ capacity and lined with 1 mm HDPE. An emergency pond of 5400m³ and lined with 0,5mm HDPE, is available for stormwater collection.
Gold recovery from Isabella ore averaged only 40% for the first six months of operation. This was attributed in part to the low grade ore fed to the pad and to the slow leach characteristics exhibited by the commercial heap. However, gold extraction from ore over the latter six months of 1990 has reached 70% and the upward trend continues.

Gold Recovery

Pregnant solution is pumped from the solution pond to five expanded bed adsorption columns arranged in series; solution grade and volume is generally 0,8 g/t and 30m³/hr respectively. Each column contains 0,2 tonnes of activated carbon. The carbon typically loads to 5000 g/t. Recovery through the adsorption circuit is virtually 100%. Adequate capacity is available within the existing circuit to increase adsorption flow rate by 50% or more if so required. Calcium cyanide additions are made to the barren pond. Antiscalant is added prior to the carbon columns and also at the barren pond to achieve a dose rate of 15ppm in solution pumped.

Prior to elution, the loaded carbon is subjected to a cold hydrochloric acid wash. Carbon desorption and electrowinning are carried out simultaneously in a proprietary vessel over a 24 hour period. The strip solution contains 5% sodium cyanide and 3% sodium hydroxide at 100 kPa and approximately 120°C and gold is electrowon onto a mild steel wool cathode. Overall elution efficiency is generally 99%.

Thermal regeneration is not undertaken at the mine. Carbon activity is monitored on a regular basis and carbon is sent for custom regeneration when required. During the first year of operation however there has been little need for thermal regeneration. Carbon consumption has been 6 g/tonne of ore treated.

The stripped cathode is acid washed, fluxed and melted in a small diesel fired furnace. Bullion is generally 82% gold and 11% silver.

HOPEFIELD GOLD

General Background

Hopefield Gold tailings, situated 20 km north west of Harare, had been the subject of considerable debate since 1986. In 1988 it was decided that these tailings dumps should be processed in the interests of the national economy. The old mine was first worked in 1913, with the main activity taking place between 1940-57; after this period, all milling operations ceased. Agricultural activities since this time have led to the

Figure 2 : HOPEFIELD GOLD : GENERAL PLANT LAYOUT

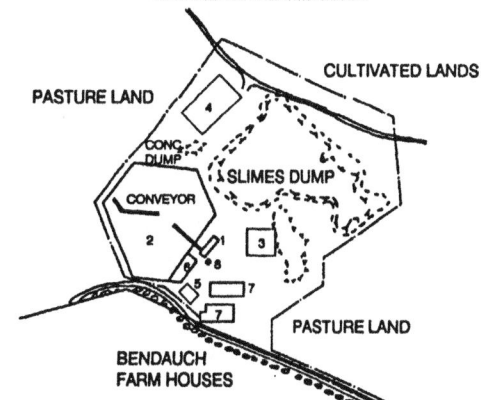

1 Agglomeration Plant with associated stacking conveyors
2 Main Pad Area   7 500m²
3 Pregnant Pond   20 x 20 x 3m
4 Emergency Pond   40 x 30 x 4,5m
5 Existing Building to house Carbon Absorption Tanks, Smelting and basic laboratory facilities
6 Barren Pond
7 Farm Buildings
8 Abandoned inclined shaft

tailings being totally surrounded by pastoral and cultivated lands and residential properties belonging to the farm are also at the project site. The project location is shown in Figure 1 with some detail of site constraints given in Figure 2.

Metallurgical testwork was pursued with these abovementioned constraints in mind. After a considerable amount of investigation, heap leaching of agglomerated tailings was chosen as the option believed to have advantages in achieving an acceptable gold recovery for a modest capital and operating expenditure while satisfying the environmental constraints imposed.

At the time of drafting this paper, project construction was at an advanced stage, therefore this written presentation will cover only the design and general philosophy adopted. It is expected that comparisons will be made between anticipated and actual project performance at African Mining '91.

Reserves

The Hopefield tailings were auger sampled between October 1986 and April 1987. Tailings had been deposited in three separate dumps and results of an auger sampling programme on the dumps indicated reserves of 60 500 tonnes of 1,7 g/t and a further 2 500 tonnes of weathered concentrate material at 11.3 g/t.

The tailings were found to be extremely variable in size distribution and this factor also had some influence on process selection.

In addition to the reserves outlined by the geologists, recent construction work on site has revealed that tailings are more widespread than originally expected. With the passage of time, a considerable area had been covered with top soil and growth of both trees and grass has been prolific, giving no indication of the earlier mining activity. Farm buildings have also been erected on areas underlain by tailings. The proposed operation will consider recovering these more widespread tailings.

Metallurgical Testwork

Initial metallurgical testwork explored the possibility of subjecting these low tonnage dumps to vat or static leaching. Conclusions drawn from this work were that the tailings varied considerably in grade and that poor percolation characteristics were exhibited. Gold recovery from the 1,7 g/t material was 55% while the weathered concentrate, which varied in grade from 8,5 - 13,2 g/t, had corresponding recovery variations of 65 to 84%.

Further metallurgical investigations on the 32 auger samples confirmed that while the dumps were particularly fine, being typically 80% -75 μm, there were portions of the main tailings dump, which indicated very wide particle size distribution with the -75 μm fraction varying between 35 and 95%. Such variations in ore sizing would have been problematic in any agitation plant and these findings signalled some cautionary warning to this higher capital investment approach which would also be environmentally sensitive.

Earlier work on the concept of agglomerated tailings heap leaching indicated favourable results during April 1987. Although agglomeration conditions were not optimised, strong durable and porous agglomerates were produced using cement additions of 20 kg/tonne of tailings and 21% moisture. A gold recovery of 66% was achieved from a 1,6 g/t sample in 14 days.

Bulk samples were excavated from the tailings dump and subjected to more extensive heap leach testwork in December 1988. Results from small diameter (200 mm) columns indicated consistent gold recoveries of 82 - 84% in 20 days with cement additions of 6 - 10 kg/tonne tailings. Larger column tests (500 mm x 4 m high) highlighted that the agglomeration procedure was critical. Agglomerates formed with cement additions of only 5 kg/tonne tailings did not possess the required qualities, solution percolation rates were low and consequently gold recovery was low at 60% from a 1,3 g/t head value over a 40 day leach period. At the 10 kg/tonne cement addition level, stable agglomerates were formed and percolation rates were good. The gold recovery achieved was 82% from a 1,2 g/t sample. Earlier work, using agitated leaching, had not significantly improved on this recovery rate. Experience gained from other operations has shown that the leach cycle may be many times longer than that indicated by testwork and this should be taken into consideration at the design stage.[1,2]

Several advantages were seen in applying this technology to the Hopefield tailings project:

i) Low capital and operating costs.
ii) Gold recovery comparable with other processes.
iii) Low labour requirement.
iv) Environmentally sound.

The agglomeration and heap leaching of a total mill tailings product is a new concept to Zimbabwe. However several operations, including Gooseberry Mine in Nevada and Mount Gaines in California, have applied this technology successfully. [2,3]

General Process Description

Mining and Agglomeration

From an early stage in the project development, it was understood that tailings moisture content would be a significant factor in materials handling. While the dump was essentially dry near surface, at depth moisture content was 20% and above. Consideration had also to be given to the effect of seasonal rainfall on mining operations. Provided the moisture content is kept below the 10 - 12% level, tailings are actually friable and crumble easily. However at higher moisture contents a sticky mass can develop. These observations suggested that thin slices of tailings be skimmed progressively from the dump in order to maintain as low a moisture content as possible. With this in mind, tractor-drawn dam scoops were selected as the appropriate equipment for this material handling problem. It is envisaged that daily air drying of a large surface area of tailings dump will generally maintain tailings moisture content at an acceptable level. A combination of discing and scraping can also be used.

A 3,6 m diameter rehabilitated pelletising disc has been selected for agglomeration. Tailings are fed to this pelletising disc via a bin at a rate of approximately 30 tonnes/ hour on a single shift basis. A one year project life has been assumed. This feed rate satisfies both the disc and the mining capacities. Two 1,5m³ capacity dam scoops adequately cope with feed requirements and have proved to be cost effective when compared to other more conventional material handling approaches.

The pelletising disc is generally angled at 50° to the horizontal and rotates at 5 rpm. Portland Cement (PC 15) is discharged from a vibrating feeder onto the tailings feed conveyor to maintain a 12 kg cement/tonne tailings mix. This discharges to the pelletizer where a coarse spray of circuit cyanide solution is used to form pellet nuclei. Moisture is generally 18% in the pelletised tailings.

Regular checks are made on tailings and cement additions. Moisture addition can vary depending on the condition of the feed, and this is generally left to the experience of the operator. However, it is known that pellet moisture content can be the key to producing a strong agglomerate which exhibits good percolation characteristics.

Agglomerate strength and stability tests are performed on the final product after a 72 hour curing period. Agglomerate samples are subjected daily to a shear stress test in water; -1 mm fines generation gives a measure of agglomerate quality.

Final agglomerate product is conveyed directly to the pad where it is stacked to 5 m using a mobile stacking system. Agglomerates are allowed to cure for at least 72 hours prior to solution application. Uncured agglomerates can be covered with shade cloth as a means of protection against driving rain. It is anticipated that maximum heap height will be 10m.

An irregularly shaped initial pad area of 7 500m² has been established on the extremely constrained site. This small area did not warrant subdivision as is the case at Isabella Mine but the complete pad area is surrounded by a 1 m high berm. The pad liner is 0,5 mm HDPE laid on fine pit sand. The liner is covered by a 150 mm layer of tailings which serves as a protective covering. Again perforated earthenware piping is used for heap drainage the piping being 100 mm internal diameter and placed at 10 m centres.

Solution Circuit

Drip emitters at 0,8m centres, are to be used for solution distribution. Barren solution will be pumped through a 150 µm in-line filter and to the heap at a rate of 15m³/hr. This should provide for a 60 day continuous leach if so required. It is anticipated that the leach may be extended intermittently for longer periods; this will be within the capability of the circuit.

Cyanide make-up and antiscalant additions will be made to the barren pond. It is anticipated that little correction will be needed to maintain solution pH.

Heap underflow will be directed to the 500m³ pregnant pond via a V-notch weir. The pad and pregnant pond elevations have been so arranged that in the likely event of a major rainstorm, excess solution will back-up in the pad area and overflow to an emergency pond of 3 400m³ capacity. This solution would be used for process needs when required.

It is likely that the storage capacity will be adequate for stormwater collection even over a sustained period. However provision will be made for the destruction of cyanide in solution if so required. Hydrogen peroxide will be the preferred chemical although some calcium hypochlorite can be used. Cyanide destruction could be carried out under control in the small 300m³ capacity barren pond. The barren and pregnant ponds are lined with 1,0 mm HDPE while the emergency pond has a 0,5 mm HDPE liner.

Gold Recovery

Pregnant solution will be pumped to four adsorption columns, each containing 80 kg of activated carbon, with carbon loading likely to be 4 000 - 5 000 g/t. A simple valve arrangement allows for the interchange and by-passing of columns.

It is envisaged that a hot acid wash will be used before loaded carbon is eluted in a 160 kg capacity pressure vessel similar to the Isabella circuit. Electrowinning, onto mild steel wool, is followed by acid washing and smelting.

CONCLUSIONS

A wide spectrum of gold bearing ores can be successfully treated using variations on the heap leach theme. At Isabella, where the ore is porous, but liable to create a major fines problem on crushing, dump leaching was selected as being appropriate, while at Hopefield Gold the percolation of mill tailings was enhanced by agglomeration.

Although metallurgical testwork is used to assess ore amenability to heap leach, commercial heaps generally result in significant delays in metal recoveries. For this reason it is advantageous to make use of a permanent leach pad where the ore leach cycle can be extended as required.

Projects of low tonnage and short duration employing low capital and operating costs make heap leaching an attractive option in Zimbabwe where many gold ore reserves are relatively small. The simplicity of the concept can allow for great flexibility of operational approach and through careful design considerations a heap leach facility can accommodate the ever changing production requirements.

ACKNOWLEDGEMENT

The author wishes to record his thanks to the Manager (Technical Services) of Anglo American Corporation Services Limited for permission to publish this paper.

REFERENCES

1.   VAN ZYL D. The Design of Heap Leach Facilities. Heap and Dump Leaching, Vol 5, No. 1, Jan - Mar 1989, p3

2.   BUTWELL J.W. Heap leaching of fine agglomerated tailings at Gooseberry Mine, Nevada. Advances in Gold & Silver Processing. Proceedings of the symposium at Gold Tech 4, Reno, Nevada, Sept 1990, p. 3 - 13

3.   McCLELLAND G.E. Agglomerated and un-agglomerated heap leaching behaviour is compared in production heaps. Mining Engineering, July 1986, p. 503.

Heap leaching of clayey ore—Three Cheers project, Selous, Zimbabwe

P. T. Simpson
Peacocke, Simpson and Associates (Pvt) Ltd, Harare, Zimbabwe

1. INTRODUCTION

The Three Cheers Gold Prospect, which lies approximately 70 km south-west of Harare, is currently under investigation by the Zimbabwe division of Falconbridge Gold Ltd. The ore body, dipping approximately 70 degrees to the south east, lies within sheared Bulawayan felsites, highly altered to chlorite and muscovite.

The 13 m x 700 m strike contains approximately 1 200 000 tonnes of oxide ore to depths of 25 – 30 metres, with an average grade of 1,37 g/t gold. A further 700000 tonnes of sulphides grading on average 3,42 g/t gold have been identified to date, at depths of up to 200 m.

2. TESTWORK

Testwork on trench and percussion drill samples from the potential ore body began in late 1986 and incorporated numerous grind/cyanidation tests and column heap leach simulations. Preliminary tests indicated minimal extraction differences from upper level ore via these two processes, and further investigation of the heap leach route was initiated. Bulk testing in 600 kg columns established such parameters as optimum crush size, leach period, percolation rate, reagent consumption, etc, and a small pad was built on site to further investigate process parameters.

Concurrent with this, grind/cyanidation versus heap leach tests were carried out on 51 deeper level samples derived from an exploratory shaft and crosscut exercise to the 29 m level.

SYNOPSIS

Heap leaching of gold ores with a high clay content presents a number of problems, the most obvious of which is that of heap permeability. More obscure, but equally important, are problems relating to heap construction methods, irrigation rate and method, and excessive silting.

Certain of these problems and possible solutions are discussed here in the light of experience gained from the Three Cheers Prospect, some 70km south-west of Harare. This ore body contains up to 40% silt and clay in the upper levels, while the low gold content dictates the need for low cost treatment methods such as heap leaching.

Extensive heap leach column tests showed the desirability for agglomeration with strong cyanide solution. Application of this process on a number of 200-tonne trial heaps further indicated the need for drip irrigation rather than sprays, while also demonstrating the suitability of conveyor/stacker heap loading methods.

The two-year program has culminated in the installation of a 10 000 tonne pilot heap-leach plant which is currently under assessment.

As could be expected, the more competent rock derived from deeper than 16 m yielded better extraction via grinding/cyanidation than by heap leaching (82,8 % on average versus 63,1 % extraction via heap leach).

Although grind/CIP was demonstrated to be more economically viable than heap leaching for the ore body as a whole, a corporate decision was nevertheless taken to still further investigate heap leaching.

3. BULK TESTING

Bulk testing was carried out on site using a 200 tonne pad and parallel zinc and carbon recovery systems. Although it was appreciated that wall effects in a heap of this size would affect ultimate recoveries to some extent, it was nevertheless felt that valuable comparative information could be gained. Four bulk tests were carried out, according to the parameters listed in Table 1.

The tests demonstrated:

3.1. Greatly enhanced extraction rate and recovery from surface ore after crushing and agglomeration with high strength cyanide solution and hydrated lime.

3.2. Somewhat decreased recovery from deeper, more competent ores.

3.3. No apparent need for incorporation of a coarse particle filter base.

3.4. An anomalous drop in permeability after agglomeration. Enhanced permeability with deeper level ore.

3.5. Suitability of activated carbon despite high silt levels.

The tests also revealed less obvious important information:

3.6. Formation of what was apparently an hydrated iron oxide suspension which persisted through siltation and sand clarification circuits. This problem was alleviated to some extent by the reduction of alkalinity levels (provided by lime addition) to a bare minimum.

3.7. Formation of a largely impervious slime film on the heap surface, resulting apparently from solution drop impact. This was evident even when using drip irrigation, but was overcome by placement of drip lines some 30 mm below the heap surface.

3.8. Releaching of the fourth test heap after a "rest" period of 20 days resulted in a further 3 % recovery in 20 days, and a second releach after 200 days rest realised a further 13 % recovery in 60 days. Total extraction from heap 4 was thus 73 % over a one-year period (incorporating 160 days irrigation).

TABLE 1: BULK TESTWORK PARAMETERS AND RESULTS

TEST	1	2	3	4
Ore source	SURFACE	SURFACE	PIT	U/G
Grade (g/t gold)	0,53	0,90	1,91	2,06
Nominal crush size (mm)	R.O.M.	19	19	19
Agglomeration *	No	Yes	Yes	Yes
Filter base	No	No	Yes	No
Irrigation	Spray/Drip	Drip	Drip	Drip
Extraction 20 days	24,9	42,6	37,6	47,3
(%) 40 days	38,6	55,9	49,2	53,3
60 days	44,3	70,3	52,1	57,7
80 days	45,6	72,7		60,0
Final extraction (g/t gold)	0,242	0,657	0,996	1,237
(%)	45,6	73,0	52,1	60,0
NaCN consumed (kg/t)	1,12	1,58	1,31	2,19
Lime addition (kg/t)	5,5	3,75	2,0	2,0
Permeability (l/m2/day)	75	25	38	93

* Using 1 % sodium cyanide solution to 14 % moisture.

4. PILOT HEAP LEACH

Parameters had at this stage been largely established by laboratory and bulk testwork, which had taken considerable time and development owing to the clayey, non-permeable nature of the ore.

Although heap leaching appeared progressively less promising vis a vis "conventional processing", establishment of a 10 000 tonne test heap was decided upon for final evaluation of the process. The unusual nature of the ore dictated several modifications to conventional heap leach, as depicted in fig 1 and discussed below:

4.1. AGGLOMERATION

Work to date had shown greatly enhanced extraction rate and recovery subsequent to high-strength cyanide agglomeration, despite lower permeability and total solution flow.

The latter was apparently due to a number of factors which were investigated in the laboratory subsequent to anomalous behavior on the 200 tonne test pad:

4.1.1. Crushed as opposed to ROM ore obviously contained considerably more fines, with consequently reduced permeability.

4.1.2. Application of excess moisture either during or subsequent to agglomeration resulted in fairly rapid collapse of agglomerates and slumping/sealing of the heap.

4.1.3. Agglomeration method played an important role. Drum agglomeration produced a very much more competent product than pug mill homogenising and agglomeration.

The 10000 tonne pilot heap was thus subjected to drum agglomeration at nominal - 19 mm. Agglomerate moisture, provided by 1 % KCN equivalent solution (using calcium cyanide), was controlled to +/-14%, this having been determined as optimum for agglomerate stability. At the time of writing, some 6 months after completion of heap loading, competent agglomerates are still evident.

4.2. PAD LOADING

Soil mechanics tests on the crusher product prior to agglomeration indicated that the material comprised only about 20 % "rock", as incompetent schist, the remaining 80 % being accounted for by "silt, clay and sand". Further tests showed immeasurably low permeability on the compacted material.

Obviously, pad loading had to be achieved by methods which would reduce compaction of any sort, particularly that produced by dump trucks or other vehicles on the heap surface. It was at the same time necessary to construct a full - height (6 metre) heap in order to determine various physical parameters, including heap stability.

Conveyor stacking utilising a retreating extending-nose radial

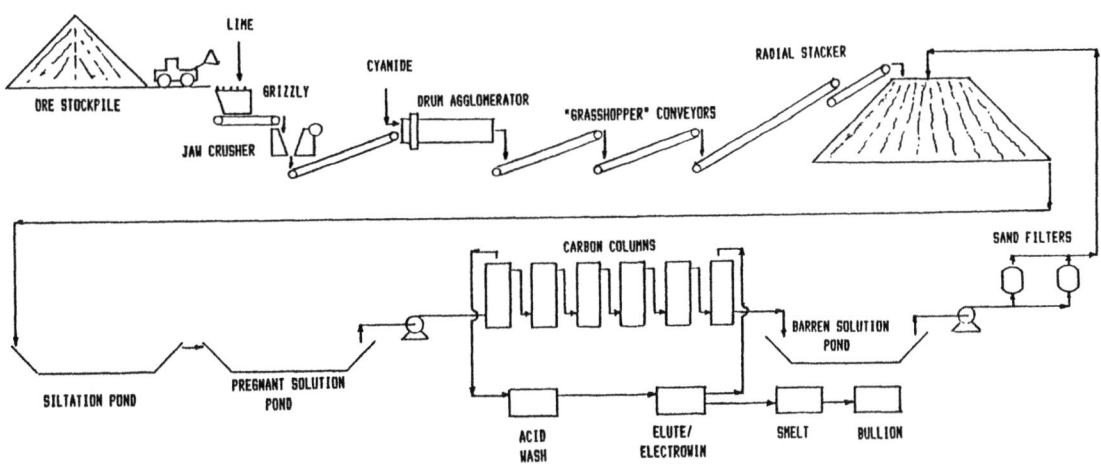

FIG 1: PROCESS FLOWSHEET - PILOT HEAP LEACH

stacker was selected as the obvious loading method. In order to minimise costs, the stacker was fabricated with a fixed discharge height of 8 m. Reduction of drop height and agglomerate impact/compaction was achieved by means of a flexible chute fitted to the stacker nose, down which the ore rolled (also assisting agglomeration).

Once a full-height retreating edge was attained, the flexible chute was removed and ore discharged directly onto the upper heap slope. Rolling of ore down the slope again assisted agglomeration and caused minimal segregation.

4.3. IRRIGATION

The first 200 tonne test heap had demonstrated a detrimental side effect of spray irrigation, ie the impact of droplets which apparently caused the formation of a largely impervious silt film on the heap surface.

Drip irrigation, as applied on subsequent bulk tests, was thus selected for the pilot heap. Drip emitters, in addition to being laid directly on the heap surface, have the advantage of being locally manufactured in Zimbabwe.

Unfortunately, silt and suspension persisted throughout the solution circuit, despite the inclusion at design stage of a pressurised sand filter. Consequent scaling in drip emitters was removable only by vigorous acid washing, and the drips have recently been replaced by sprays.

Minimal film formation and ponding has been observed, probably owing to optimised agglomerate competence and removal of the bulk of surface silt by prior drip irrigation.

4.4. PERMEABILITY

Permeability of the pilot heap has to date averaged approximately 60 l/m2/day. This is considerably below the accepted design figure of 240 l/m2/day (10 l/m2/hr, 0,004 USGPM/ft2), but is nevertheless reasonable for clayey material of this nature.

Augers driven into the heap show good wetting throughout, with little evidence of chanelling.

4.5. EXTRACTION

Extraction from the heap has to date (4 months) has averaged only 0,004 g gold/t/day, at which rate some 250 days will be required to achieve target recovery. No appreciable enrichening of solutions was observed during early leach stages.

This behaviour contradicts that of both column and bulk pad tests, wherein average extractions of 0,008 - 0,016g gold/t/day were experienced, with extractions of 0,016 - 0,034 g gold/t/day during the first 30 days of leaching.

The reason for the slow extraction is unclear, as cyanide agglomeration and water leaching tests have indicated that dissolution is practically complete after curing, and removal of dissolved gold species, essentially by osmosis, is all that is required. Oxygen starvation within the heap can thus be effectively ruled out.

Presumably, therefore, low extraction rate is due largely to slow removal rate of dissolved species. The significantly higher clay content of the pilot heap (as opposed to test samples) would allow for higher moisture retention by agglomerates, and reduced osmotic effect and extraction rate.

This moisture retention effect could presumably also allow permanent entrapment of pre-dissolved gold, and consequent reduced extraction. A rest period to allow the heap to dry out, followed by further solution application with enhanced concentration/moisture gradients should thus realise further recovery, as was the case with bulk test No 4.

4.6. SOLUTION TREATMENT

Treatment of solutions from the pilot heap has proved to be the major problem area. High silt levels have been found to persist throughout the solution circuit, even beyond a pressurised sand filter on the solution return line to the heap.

The silt has given rise to abnormally high levels of blinding on the activated carbon recovery medium, necessitating regular and thorough acid washing and resulting in elution problems.

The silt also caused operational

problems arising from blocking of drip irrigators prior to replacement of the latter by sprays.

Flocculants have little or no effect in decreasing silt levels, and should the project proceed to full scale heap leaching, it will become necessary to incorporate a more rigorous solution filtration circuit in addition to the generally accepted settlement ponds.

5. DISCUSSION

A good deal of important - and in some cases relatively obscure - information has been revealed in the ongoing heap leach testing of the Three Cheers ore body:

5.1. When evaluating a rubble type deposit for heap leach performance the sample is all important, more from the physical character than the chemical dissolution of gold. In the case of 3 Cheers the initial samples originated from the upper horizons which were enriched in pebbles and therefore gave a better percolation. The underground mining trial produced an entire cross section including some deep level ore and was therefore more realistic physically, but totally unrepresentative. The large scale 10 000 tonne trial used excavated ore in which the pebble horizon was insignificant. As it was impractical to mine to the competent ore horizon the resulting feed to the heap leach contained an anomalously high clay content and was possibly therefore a very negative basis for this trial.

5.2. Injection of excess moisture during agglomeration can lead to subsequent rapid agglomerate deformation and slumping. Although agglomerates may appear physically competent at production stage, deterioration on application of leach solutions takes place more rapidly than when less moisture is injected at agglomeration stage.

5.3. The use of strong cyanide solution during agglomeration results in the dissolution of the majority of extractable gold during the curing period prior to the application of leach solutions. Thereafter the removal of dissolved or soluble gold cyanide complexes is essentially all that is required, to the extent

that this ore can be largely leached to completion by water alone.

5.4. The driving force for migration of gold into the leach solution is apparently controlled largely by moisture and (to a very much lesser extent) cyanide concentration gradients. The Three Cheers ore is highly thixotropic, with the ability to absorb a great deal of solution and thereby reduce moisture gradients. Drying out periods between successive leach cycles restore these gradients and realise accelerated extraction during initial releach stages. Further investigation is however indicated to determine whether this cyclic process realises a more rapid overall extraction rate than that of a slow but continuous single leach.

5.5. "Silt" in solutions comprises in this case both fine suspended particles originating from the ore body, and a gelatinous suspension of what is apparently some form of hydrated iron oxide. Occurrence of the latter can be limited to some extent by reduction of alkalinity levels; however that portion which persists is extremely fine grained, to the extent that it does not settle in silt ponds and passes through sand filters.

REFERENCES

1. Bourhill, PE, Personal Communication November 1990.
2. DeMull, TJ, Heap Leaching - Keep it Simple. Randol Gold Forum, Arizona 1988.
3. Graham, NJG, et al. Development of Heap Leaching at Royal Family Mine, Zimbabwe. Proceedings African Mining, Institution of Mining & Metallurgy, Harare, 1987 pp 153 - 159.
4. Mashozhera, G, Personal Communication November 1990.
5. Muthadi, O, Personal Communication January 1988.
6. Thorndycraft, RB, Personal Communication February 1988.

Improving the recovery of gold and copper in a CIP operation: flotation of a sulphide copper-bearing ore

U. G. Mugoro
W. P. Furusa
Lonrho Zimbabwe Ltd, Harare, Zimbabwe

ABSTRACT

The paper presented outlines the scientific approach taken to improve the metallurgical efficiency of a modern plant at Athens Mine. The plant is based on the CIP method of gold recovery with copper concentrate as a major by-product.

In the early stages of the operation the overall gold recovery achieved averaged 61% and copper recovery was 74%. The plant was experiencing problems in the leach, adsorption, elution and electrowinning sections mainly as a result of the presence of copper in the ore. The changes made to the operating standards, the response of the circuit to these changes and the final circuit layout and operating conditioning resulting in a 90% gold recovery are discussed.

It was concluded that the distribution of oxygen and cyanide along the leach stream determined the extent of gold dissolution. A substantial amount of cyanide (200 - 250 mg/l NaCN equivalent) was necessary to avoid preferential adsorption of dissolved copper onto carbon in the CIP section. Rinsing carbon in a cold caustic cyanide solution resulted in good elution of copper prior to the gold elution stage. The copper-free loaded carbon was easily treated for gold with very high elution and electrowinning efficiencies of 96 and 98% respectively. A bullion of high fineness (in the region of 98% gold and silver) was obtained and the activity of carbon after elution was reasonably high at 80% of virgin carbon.

One detrimental effect of the changes made to improve the gold recovery was a high residual cyanide content in the CIP tails. This was overcome by the use of the Inco process at a cost of 20 cents per tonne ore treated (current cost is 35 cents/tonne) (Zimbabwe currency).

Introduction

The fouling of mill solutions by copper present in gold ores has been experienced in many gold treatment plants. The copper has been found to interfere with the recovery of gold by CIP cyanidation.

Following investigations on a 4 500 tpm CIP plant with a copper flotation circuit for the cyanidation tailings, operating conditions resulting in 90% gold dissolution, 98% adsorption, high elution and electrowinning efficiencies at 96 and 98% respectively and a 70% copper recovery at a concentrate grade of 20% copper were established at Athens Mine. A similar plant capable of treating 14000 tpm of the underground gold and copper ore was designed and erected.

On commissioning the plant, recoveries failed to reach the expected levels. Gold dissolution in the leach section was as low as 52%. Carbon in the adsorption tanks lost its activity rapidly, such that within 48 hours the carbon was almost ineffective. Solution losses were as high as 0,38 g/t whilst up to 8% copper was loaded on the carbon. Elution and electrowinning efficiencies averaged 62 and 63% respectively. However, between 70 and 80% of the copper reporting to flotation was recovered at a concentrate grade of 26% copper.

The flowsheet of the plant designed to treat 14 000 tpm is presented as Figure 1. The different sections of the plant are described below.

1.1 Crusher Sections

Run of mine ore at -300 mm was drawn from a rough ore bin, past an Outokumpu magnetic detector into a 129,5-23 cm primary crusher. The opening of the crusher was set at 35 mm The crusher product was conveyed onto two 3700 x 1520 mm mild steel screens with an arpeture size of 25 mm. Screen overflow was conveyed into a 145-15 cm secondary gyratory crusher with a discharge setting of 25 mm. The crushed ore joined the primary crusher product to the screens. A 25,4 x 53,3 cm jaw crusher fed from a seperate rough ore bin discharged its product onto a conveyor feeding into the secondary gyratory crusher. The screen underflow was stored in 900 t silos, of which 500 t was live load.

1.2 Milling

Two 2,15 x 3,38 m ball mills were each driven by 250 kw motor through a gear and pinion reduction at a speed of 23 rpm, being of 77% of the critical speed. Each mill had a capacity of 10 tph. Grinding media consisted of 100, 80, and 70 mm steel balls charged every eight hour shift at a rate of 4 kg per tonne milled in the ration of 3;1;1; respectively. Mill discharge at 75 to 80 % solids by mass was diluted to 50% solids and pumped into a 15 inch cyclone. A circulating load of 400% was required. The cyclone overflow had 20% solids and grind of 80% minus 75 microns. The overflow passes over a 840 micron Delkor linear screen which removes trash before the pulp was pumped into an 8 metre (diameter) high rate supaflo thickener. Thickener underflow was drawn at 60 % solids and aerated in Outokumpu (OK 8) flotation cells. The pulp was then diluted to 45% solids, 1,4 SG prior to the addition of lime (for protective alkalinity), lead nitrate and aerobrand cyanide and pumped to the leach section.

1.3 Leach Circuit

The leach section consited of two sets of four down draught agitators in series. Each tank had a capacity of 150m^3. Supplementary blower air could be introduced in to the tank through a sparge.

1.4   Adsorption Section

Six 80 m^3 down draft agitators, each with a 20 mesh interstage wedge-wire cylindrical screen were used in a typical carousel set up. The CIP tailings were thickened to 60% solids in a second 8 metre highrate supaflo thickener. The overflow of which diluted the pulp to leach.

1.5   Flotation Circuit

Thickener underflow was diluted to 45% solids and pumped to bank of OK 8 flotation cells to remove talc. Flotation tailings gravitated into three 20 m^3 Wallace agitators for oxidation of residual free cyanide prior to copper flotation. Talc concentrate gravitated into a slime dam pump sump.

The copper flotation circuit consisted of four OK 3 rougher and scavenger cells. Rougher concentrates went through two cleaning stages in OK 1.5 flotation cells. Flotation tailings were thickened to 60% solids on a third supaflo thickener before disposal. Thickener overflow diluted feed to talc flotation cells.

1.6   Carbon Transfer

Pulp was pumped using a Vasa vertical pump from the leading adsorption tank onto a 840 micron sieve bend. The sieve bend underflow gravitated back into the leading tank. The overflow went through a washing trommel screen into a mild steel storage tank. The amount of carbon (800 kg) required for elution was drawn from the storage tank into a stainless steel blowcase. The carbon was acid washed using a 2% m/v hydrochloric acid solution circulated through the blowcase. The blowcase was pressurised to 300 kpa using compressor air to blow the carbon into an elution vessel. A fresh charge of carbon was put into the leading adsorption tank which was then made the last tank on line.

1.7   Elution and Electrowinning

A Zadra elution system was employed. The elution vessels were designed to accommodate 800 kg of carbon per charge. The eluant was heated to 90°C in heat exchangers using steam

from a Del Monego boiler. Hot solution was introduced from the top of the vessel. The pregnant solution gravitated through three electrowinning cells in series. Each cell had a stainless steel anode and a wire wool cathode with a specific surface of 120 cm^2/g. Forty eight hour elution tails were water washed and regenerated in two Del Monego rotary kilns. Each kiln had a capacity of 6 - 9 kg/hr.

Metallurgical Operations Prior to Changes

The plant was treating an underground sulphide ore at 3,36 g/t gold and 0,65% copper. The major copper mineral was chalcopyrite.

Fine ore was drawn into the ball mills at 12,5 tph. A grind of 73% minus 75 microns was achieved. Oxygen levels in the pre-aeration cells were increased from 0,8 mg/l to 5,1 mg/l as measured by a Yellow Spring Instrument´s (YSI) Oxygen meter mode 54A. The pH of pulp before cyanide addition was adjusted using dicalcium silicates and measured using a Corning 220 pH meter. The pH was found to fluctuate between 8,5 and 11,5. Calcium (aerobrand) cyanide was added continuously by means of a vibrating feeder to give a strength between 0,10 and 0,15% free NaCN equivalent in the first leach tank. The cyanide strength was determined every hour using conventional silver nitrate titration with potassium iodide as indicator. A retention time of 27 hours was achieved. A maximum of 1,5 mg/l oxygen was recorded in the leach tanks.

Leach pulp flowed though all six carbon tanks. The carousel system was used. The pulp had a retention time of 11 hours in the adsorption tanks, allowing for a further 2% dissolution of gold. The profile of gold in solution in the adsorption tanks did not follow the expected decreasing trend and only 83% of the dissolved gold was adsorbed.

Following talc flotation the pulp was conditioned with copper sulphate. 70 to 80% of the copper in the CIP tailings was recovered as a concentrate assaying $\pm$ 26% copper. Flotation tailings contained about 10 mg/l NaCN.

Loaded carbon was acid washed before elution. This removed 75% of the calcium compounds. However, copper assays remained as high as 8,0%. Most of the copper was eluted together with gold in a 48 hours Zadra elution cycle. The eluant was circulated at 20 - 25 litre per minute, being 43 to 54 bed volume per charge. The elution reagents were maintained at 1% NaOH and 2,0% NaCN and the eluant at a temperature of 90°C.

The advance electrolyte averaged 22,68 g/t gold and 1 500 g/t copper. After passing through the electrowinning cells with a voltage of 2,5 volts across each cell, 8,48 g/t gold was detected in the spent electrolyte. The corresponding copper assay was 1 200 g/t. The bullion produced after acid treating, calcining and smelting the wire wool assayed $\pm$ 62% gold, $\pm$ 15% silver and the balance was base metals.

Only one of the two regeneration furnaces was operating. This managed to reactivate only 9% of the carbon to be charged to the CIP. The activity of the regenerated carbon was only 83% of virgin carbon.

Investigations and Plant Changes Leading to Solution of Problems

Factors affecting recovery of precious metals in the different sections of the plant were varied until the optimum operating conditions were established. The activity chart of the events discussed below is presented.

Calcium cyanide was added to the third leach tank to maintain a strength of 0,03% NaCN equivalent. This was followed by stage addition of lead nitrate at 200 g/t into the second leach tank.

For a period of seven days, leach retention time was maintained at 14 hours so that the effect of retention time or cyanide consumption could be ascertained. Following the poor recoveries experienced, the retention time was stepped up to 27 hours again. In the seven days that followed free cyanide in the first leach tank was stepped down to an average of 0,087% NaCN equivalent, whilst cyanide in the third tank was increased to 0,042% NaCN equivalent.

Oxygen in the pre-aeration cells was stepped up to 8,3 ppm by introducing lime to maintain a pH between 9,5 and 10 in the pre-aeration cells. Air lances placed outside the draught tube were replaced by longer lances inside the draft tube such that the air flow was concurrent to pulpflow. Oxygen levels between 5 and 6 ppm were detected in the leach tanks. Leach results indicated 67% gold dissolution were obtained.

A second attempt was made to economise on cyanide consumption. In this case cyanide strengths were reduced to (seven day average of) 0,059 and 0,012% NaCN equivalent in the first and third leach tanks respectively. This did not affect gold dissolution.

Sodium cyanide and caustic soda were used for leaching instead of calcium cyanide and dicalcium silicates respectively. Despite causing a further 2% gold dissolution, the resulting increase in operational cost was too high. However, Rholime was found to be a better substitute for the dicalcium silicates as it was easier to handle and use to maintain the pH in the required range of 9,5 to 10,0 for an extra 7 cents per kilogramme. The last change made to the leach section was to step up the cyanide in the third leach tank to 0,03% NaCN equivalent.

From the assays of gold in solution in the adsorption tanks, it was apparent that carbon was desorbing gold in the last two carbon tanks. These tanks were then isolated from the adsorption circuit.

Tests conducted to establish the cause of low carbon activity and gold adsorption included :

Establishing (i)   the effect of humic acids and flotation reagents.

    (ii)   the effect of acid washing unregenerated carbon

    (iii)   the effect of reducing carbon retention in the adsorption tank to 12 days.

The effect of humic acids and flotation reagents on adsorption was investigated by conditioning virgin carbon in fresh water and flotation tailings water respectively for 14 hours. The activity of the carbon was then tested using a MINTEK carbon activity analyser. From the results obtained (Figures 3 and 4) it was evident that the latter had some effect on carbon activity. Use of re-circulated water was therefore minimised.

Eluted and unregenerated carbon was acid washed before charging back into the carbon tanks. The retention time of carbon was reduced by maintaining a carbon concentration of 30 g/l. All this did not help the activity of carbon in first adsorption tank which remained at 40% of virgin carbon. The operation of the CIP section was changed from carousel system to the conventional counter current transfer of carbon at a reduced carbon concentration of 20 g/t. Gold adsorption remained as low as 76,4%.

Carbon was drawn from the first adsorption tank and activity tests conducted on weekly composite sample. A relationship between carbon activity and copper loaded on the carbon was established. This is presented as Figure 4. An exercise aimed at preferentially eluting loaded carbon was initiated. An eluant with 2% free NaCN equivalent and 2% NaOH was air agitated at ambient temperature in a stainless steel pressure vessel for two hours. Fresh water was then pumped from the bottom of the tank. The water was allowed to overflow for 30 minutes, thereby washing the eluted copper. The operation achieved 90% copper elution and about 1% gold loss. Carbon was drawn from the adsorption tanks and treated in this manner. Carbon activity improved significantly such that adsorption efficiency increased to 97%. The last two carbon tanks were then brought back on line.

To avoid further adsorption of copper, aerobrand cyanide was added to the carbon tanks. Analysis of plant assays (Figure 5) indicated that 200 - 250 mg/l NaCN equivalent was sufficient to inhibit further adsorption of copper.

In the Zadra elution circuit the following operating options were tried -

  i) increasing elution time to 72 hours (82 bed volumes) for a period of two weeks,

 ii) eluting carbon prior to acid washing,

iii) reducing eluant cyanide strength from 2% to 1% NaCN.

The surface area avaluable for gold deposition in the electrowinning cells was varied by changing the mass of wire wool cathode from 0,5 kg to 1 kg per cell. The mass was increased further to 1,5 kg and finally 2 kg per cell. Though there was an improvement in electrowinning efficiency from 88% to 85%, the expected efficiency of 96% was obtained only after the elution of copper prior to gold elution. It was also at this stage that elution efficiencies as high as 98% were achieved.

Following the introduction of high cyanide levels in the CIP plant, recovery of copper by flotation dropped to 60%. The concentrate grade assayed $\pm$ 18% copper. The poor flotation performance was attributed to insufficient neutralisation of cyanide in the conditioners. An average of 90 mg/l NaCN equivalent at a pH of 9,5 was reporting to the flotation section. Elemental sulphur was therefore burnt in a cylindrical mild steel pressure vessel to produce sulphur dioxide which was forced into the first conditioner to help oxidise cyanide and reduce the pH as in a typical INCO-SO_2-Air process. A second sulphur burner was installed in the second conditioner. This helped the cyanide oxidation process greatly as shown in Table 5.

After restricting the use of recirculated water, there was a build-up at the slimes dam water collection pond. A bench test was carried out using a Wemco flotation cell to establish if it was possible to oxidise the free and complexed cyanide in the water using the INCO process with sodium sulphite as a source of sulphur dioxide. Results of the test are presented as Figures 6 and 7. Following the test, a plant consisting of four self aerating Wemco flotation cells was installed. Each cell had a capacity of 0,75m^3, making the plant capable of treating 90 tpd of effluent. The sodium sulphite was mixed in a 0,5m^3 mixing tank and dosed using a micrometering pump. The composition of the effluent produced is presented in Table 6.

Leach Section

	12.01.89			01.02.89			01.03.89		01.04.89
Ret. time hrs	+ 27	+ 27	+ 14	+ 27	+ 27	+ 27	+ 27	+ 27	+ 27
No 1 % NaCN	0,125	0,114	0,114	0,097	0,087	0,074	0,059	0,045	0,061
No 3 % NaCN	0,020	0,033	0,03	0,051	0,042	0,019	0,012	0,020	0,03
ppm O2	1,5	1,5	1,5	1,5	1,5	6,5	6,5	6,5	6,5
g/t Pb(NO3)2	300	500	500	500	500	500	500	500	500
Gold Dist. %	52	56	59,7	34,5	50,7	66,7	74,6	80,2	83,1

Adsorption Section

	12.01.89		01.02.89	01.03.89	01.04.89	26.04.89
Tanks on line	4	4	4	6	6	6
Carbon g/l	30-35	30-35	25-30	20	20	20
NaCN mg/l	200-40	200-40	200-40	200-40	250-200	250-200	250-200
System	Carousel	Carousel	Carousel	Counter Current	Counter Current	Counter Current	Counter Current

Tot Au dissol %	55	59,2	71,8	57,3	62,3	72,8	79,9	76,4	82,4	86,0
Au Adsorption %	81,9	85,0	66,3				97,3	98,0

Elution and Electrwinning Operating Conditions

Eluant: % NaCN 2   % NaOH 1   % NaCN =1   Cathode Mass

Temp 00°C Flowrate l/min 22   Retention changed to 72 hours   Elution Prior to Acid Treating   1kg Caustic Cyanide Wash Before Elution   1st Cell Cathode 1,5 kg   1st Cell 2kg Bal. 1,5kg   All Cath 2kg each

Wirewool kg 0,50   Retention, hrs 48

	12.01.89	01.02.89	01.03.89		01.04.89	26.04.89
Elution Efficiency %	61,7	65,5	34,2	83,1	92,5	92,9
Electrowinning %	62,6	94,7	71,5	86,1	88,6	87,5

Cu-Flotation

	12.01.89	01.03.89	10.08.89
pH	9,5 - 10	9,5 - 10	8 - 9,0
NaCN mg/l	± 10	90	18

Table 1   Performance of the Adsorption Section

Activity	Carbon Addition g/l	Regenerated %	New %	Eluted %	Leach Tails g/t	Solution & (Carbon) Au g/t Profile — Carbon Tanks 1	2	3	4	5	6	Adsorption %
Before Changes	30-35	32,7	9,6	57,7	2,26	0,87 (2420)	0,55 (1360)	0,47 (1190)	0,39 (960)	0,49 (910)	0,41 (700)	81,9
Reducing Flotation Reagents	30-35	7,6	16,0	76,4	1,27	0,77 1610	0,47	0,29	0,19 400			85,04
Acid Treat Carbon After Elution	25-30	0,0	9,8	90,2*	1,75	1,10 (1390)	0,64	0,61	0,59 (890)			66,29
Counter Current Carbon Transfer and Caustic CN Wash	20	19,4	3,2	77,4	1,78	0,91 (1470)	0,75	0,69	0,63	0,44	0,42 (860)	76,40
Increased Cyanide in Carbon Tanks 250-200 mg/l	20	19,4	3,2	77,4	1,85	0,89 (1940)	0,65	0,43	0,23	0,10	0,05 (360)	97,30
Present Operation	20	19,4	3,2	77,4	2,00	0,63 (2120)	0,26 (1540)	0,13 (1230)	0,06 (900)	0,04 (440)	0,03 (270)	98,6

Table 2   Performance of the Elution and Electrowinning Sections

Activity	Carbon Composition					Wire-Wool per Cell	Temp	Flow Rate	Eluant Composition							Efficiency	
	Fe	Cu	Ca	Au Heads	Au Tails				pH	NaCN	NaOH	Cu Head/Tail		Au Head/Tail		Elut	Electro Winning
	%	%	%	g/t	g/t	kg	°C	l/min		%	%	%	%	g/t	g/t	%	%
Before the changes	0,18	0,23	1,51	1490	570	0,5	90	22	12,7	2,00	0,99	-	-	22,68	8,48	61,74	62,62
72 hours elution time	0,21	0,28	0,43	1480	510	0,5	92	21	12,5	1,98	0,94	-	-	44,25	2,34	65,54	94,71
Reduce free NaCN to 1% and Elute before Acid Wash	0,20	7,46	0,50	1200	790	0,5	95	20	12,9	0,82	1,07	1,99	1,72	12,13	3,46	34,17	71,48
Introduce Caustic																	
Cyanide Wash	0,23	0,10	0,10	1950	330	1	89	19,5	13,1	0,77	1,21	0,90	0,84	22,63	3,15	83,08	86,0
Increased first cell to 1,5 kg	0,18	0,08	0,35	1610	120	-	90	22	13,1	0,86	1,22	0,17	0,17	33,81	3,87	92,5	88,55
First cell Cath. 2kg. The rest 1,5kg	0,53	0,15	0,31	2110	150	-	93	21	12,8	0,98	0,98	0,12	0,12	44,15	5,53	92,9	87,47
Current operation with 2kg wirewool in each cell				2560	110	-	95	22	13,0	0,93	0,97			65,2	1,30	96,0	98,0

Table 3   Characteristic Assays of Carbon and Eluant during carbon
          preparation and elution stages

	Carbon			Solution		
	g/t Au	% Cu	% Ca	g/t	% Cu	% Ca
Before Acid Washing	2270	1,15	1,39	–	–	–
After Acid Washing	–	–	0,35	–	–	0,85
After Water Washing	–	–	–	–	–	0,15
After Caus. Cyanide Washing	2260	0,10	0,35	–	0,13	–
After Water Wash	–	–	–	–	0,02	–
Elution Heads	2260	0,10	–	–	–	–
Elution Tails (48 hours)	260	0,05	0,01	–	–	–
Electrowinning Heads	–	–	–	39,09	0,12	–
Electrowinning Tails	–	–	–	2,32	0,12	–

Table 4   Operating Parameters of the Copper Flotation Circuit

	Before Plant Changes	Before the INCO Process	After the Inco Process
pH	9,5	9,5	8 – 90
NaCN (mg/l)	10	90	18
SIBX Collector (g/t)	50	70	50
Acrol Talc Depressant (g/t)	–	30	50
TEB Frother (g/t)	50	50	50
CuSo$_4$ Activator (g/t)	–	100	–
Collector Addition Stages	1	2	2
Depresant Addition Stages	1	1	1
Frother Addition Stages	1	1	1
Activator Addition Stages	–	1	–
% Cu in Concs	26	18	18
% Cu Recovery	80	60	80

Table 5   Performance of Sulphur burners (INCO Process)

	mg/l NaCN	pH
One Burner on Line		
Conditioner 1 Feed	71	9,25
Conditioner 3 Discharge	43	9,10
Tails Thickener u/f	25	
Two Burners on Line		
Conditioner 1 Feed	87	9,48
Conditioner 3 Discharge	36	8,97
Tails Thickener u/f	18	
Slimes Dam Discharge	18	

Table 6   Performance of the INCO Process using 400g/m^3 Sodium Sulphite

Run	Laboratory Results				Plant Results			
	Feed		Discharge		Feed		Discharge	
	mg/l CN_T	pH	mg/l CN_T	pH	mg/l CN_T	pH	mg/l CN_T	pH
1	49	9,84	Tr	8,72	22	8,64	Trace	8,52
2						9,42	Trace	8,64

Discussion

It was evident that the poor recovery initially experienced in the leach section was due to insufficient free cyanide. The introduction of stage addition of aerobrand cyanide produced an increase in gold dissolution from 56 to 59%. Further improvement was achieved by dosing lead nitrate indicating that there was a deficiency of lead nitrate for

(i) precipitating sulphide-ions which were otherwise reacting with oxygen and free cyanide to form thiosulfates (3)
(ii) the formation of a lead-gold "alloy" which is more susceptible to cyanide attack than elemental gold (4).

Reducing the retention time to 14 hours did not achieve the aim of reducing cyanide consumption at 5,5 kg/t. Instead, dissolution dropped to 72%.

The drop in gold dissolution to 62% after readjusting the cyanide distribution to 0.087 and 0.042% NaCN equivalent in the first and third leach tanks respectively seemed to suggest that the presence of free cyanide had taken over the role of a rate determining factor. This was supported by the sharp increase in recovery to 80% experienced when the oxygen levels were stepped up from 1.8 to 8.2 ppm. A further decrease in free cyanide to 0.059 and 0.012% NaCN equivalent, with the same oxygen levels could be afforded with no effect on gold dissolution.

The ore is known to consist of pyrrhotite, pyrite, chalcopyrite and pentlandite. The downdraught agitators though efficient aerators did not supply the oxygen demanded by the unstable sulphides - present especially pyrrhotite. It was necessary to introduce an air sparge into the centre of each draught tube below the impeller in order to cope with this air demand and to saturate the pulp. This was in addition to preaeration before leach using efficient Outokumpu flotation cells.

Using pure reagents in the form of sodium cyanide and caustic soda improved gold dissolution by 2%. However, the high operating costs made this option a disadvantage. Similarly, the purity of Ro-lime gave it an advantage over dicalcium silicates, making it easier to control pH between 9.5 and 10. Increasing free cyanide in the third leach tank to 0.03% NaCN equivalent improved gold dissolution to 88%. It is this concentration with 0.06% NaCN in the first leach tank that are currently recommended as the best operating parameters for the leach section.

Carbon activity results obtained after testing virgin carbon conditioned for 14 hours in plant fresh water and flotation tailings water (Figures 2 and 3) suggest that the latter affected carbon activity. This was expected to have an even bigger impact in the plant because of the high circulation of unregenerated carbon experienced as one of the regeneration furnaces was down. Mixing of slimes dam return water and flotation tailings solution with leach solutions was discouraged and in response, gold adsorption increased from 82% to 85%.

Washing out calcium on carbon in an acidic solution prior to recharging to the carbon tanks was expected to provide more adsorption sites for gold. The presence of H^+ ions was expected to help reduce the pH of the pulp and enhance adsorption of gold by replacing Ca^{2+}, Mg^{2+} and K^+ ions in the aurocyanide ion (M^{n+}) (AuC^{CN2}) (5, 6, 7). Despite all these speculations, gold adsorption decreased to 66%. The carbon ended up loading up to 8% copper.

Reducing carbon retention time in CIP did not improve the recovery. According to Figure 4, carbon activity decayed exponentially as copper loaded on the carbon increased. Rinsing carbon in a cold caustic cyanide solution resulted in a very good preferential elution of copper. Nicol (8) attributed this good elution to excess cyanide allowing the formation of higher order complex of copper, $Cu^{(CN)}_4{}^{2-}$. This complex has poor adsorption properties than lower complexes $Cu^{(CN)-}_2$ and $Cu^{(CN)}_3{}^{2-}$. Similarly a stage addition of aerobrand cyanide to the CIP section provided excess free cyanide for the formation of the higher complex of copper, thereby inhibiting adsorption of copper and maintain high carbon activity. The optimum concentration of 200 - 250 mg/l was established as shown in Figure 5.

The poor recoveries in the Zadra system could not be attributed to deficiency in operating time or free cyanide. In fact 0,13g NaCN carbon 1kg was saved by reducing cyanide concentration. It was established that eluting carbon before acid washing did not have significant impact on process efficiency, except occasional interruption of eluant flow by blockages in screens and pipes caused by precipitation and scaling of calcium compounds.

It was after copper had been removed before gold elution that good results were obtained. At that time, the advance electrolyte had very little copper (0,08%), allowing gold to be deposited before copper (9). The spent electrolyte, being very low in gold was 98% efficient in stripping loaded carbon in the elution vessels. As a result the wire wool cathodes were easily digested in a hydrochloric acid solution on clean up operations before smelting. The bullion produced assayed at most 4% base metals, an appreciable improvement from the 23% originally experienced.

The critical pH curves for copper minerals indicate that flotation of chalcopyrite is possible at a pH of 9.5 when up to 10 mg/l NaCN is present. Therefore, when the level of free cyanide reporting to flotation increased to 90 mg/l NaCN, poor flotation results were obtained. To overcome this problem, the INCO/SO_2 process was installed. Free and complexed cyanide then reacted as follows

$$SO_2 + CN^- + H_2O + O_2 = CNO + H_2SO_4 \dots\dots\dots\dots\dots (i)$$
$$Cu\,(CN)_4{}^{2-} + 2H_2SO_4 = 4\,HCN + Cu^{2+}\,SO_4{}^{2-} \dots\dots\dots (ii)$$

The primary aim of achieving low cyanide levels and a reduced pH are achieved in the first reaction. It was found necessary to use two burners for good results to be achieved. Reaction (ii) help generate copper ions from up to 720 mg/l complexed copper for catalysing the first reaction. The results obtained are presented in Table 5.

The bench tests carried out indicated that it is possible to generate sulphur dioxide from sodium sulphite. At a dosage of 400 g/l a retention time of 50 minutes was required for complete oxidation of cyanide in the slimes dam water. Table 6 shows a comparison of the expected and the actual results obtained which are in fair agreement.

Recommendations

It was recommended that the pulp pH be maintained between 9.5 and 10 using Ro-lime, preferably during the preaerating stage. The cyanide concentration in the first leach tank should be maintained between 0.05 and 0.06% NaCN equivalent whilst between 0.02 and 0.03% NaCN be kept in the third leach tank. Oxygen levels between 5 and 6 ppm would be sufficient.

From the daily solution assays if the carbon was found not adsorbing, normally indicated by irregularities in the solution profile, then the following course of action should be followed.

(i)   On transferring, rinse the carbon in a cold caustic cyanide solution to remove copper or

(ii) If all above is done and problem still persists, the other reasons and remedies are -

    (a) Too much flotation reagents in the circuit, hence reduce these.
    (b) Returned carbon not efficiently eluted hence still highly loaded to charge to the last tanks (eluted carbon assay to be compared to last carbon tank assay).

Normally it is advisable to check if copper is adsorbing on carbon by doing a solution profile for percent copper for leach tails, carbon tanks 1, 2, 3, 4, 5 and 6.

Copper should be removed from the loaded carbon prior to gold elution by an agitated caustic cyanide rinse followed by water wash. If the eluant has a high copper content, then it is necessary to bleed it off to lower the content. Ideal elution pH is 13 and cyanide strength is 1.0% NaCN equivalent. If poor elution persists with the above conditions at a temperature higher than 90 C, then the elution time can be increased to 72 hours. Carbon should be acid washed before charging to elution vessel and enough gold plating surface should be available all the time.

Flotation of chalcopyrite is very poor at a pH above 9 in the presence of cyanide. Both the cyanide and the pH can be reduced by burning more sulphur in the burners. A pH of 7 is best for good results.

6.0 <u>REFERENCES</u>

1. Habashi F. Principles of Extractive Metallurgy: Hydrometallurgy, Volume 2, Gordon and Breach (1990) 24-33

2. Pryor E.J. Mineral Processing, Third Edition (1978), Applied Science Publishers Limited, 439-441

3. Nagy I, Mrkusic P., McCulloch H.W. Chemical Treatment of Refractory Gold Ores. Literature Survey, National Institute of Metallurgy. Report 38 (1966)

4. Sheveleva L.D. and Kakovskii I.A. Lead Compounds and the Dissolution of Gold in Cyanide Solutions, Tsvenye Metally, July 1979, 100-102

5. Adams M.D. and Fleming C.A. The Mechanism of Adsorption of Aerocyanide onto Activated Carbon. Metallurgical Transaction B, Volume 20B, June 1989, 315 - 325

6. Adams M.D., McDougal G.J. and Hancock R.D. Models for the Adsorption of Aerocyanide onto activated carbon. Part II Extraction of aerocyanide ion pair by polymerc Adsorbends, Hydrometallurgy, 18 (1987) 139 - 154

7. Adams M.D, McDougall G.J. and Hancock R.D. Part III. Comparison between the Extraction of Aerocyanide by Activated Carbon, polymerc adsorbents and 1 - pentanol. Hydrometallurgy, 19 (1987) 95 - 111

8. Nicol M.J. Elution Theory, Council of Mineral Technology, (August 1985)

9. Paul R.L. Gold from Eluants (Electrowinning), Council of Mineral Technology.

10. Conrad B.R. and Devuyst E.A. Application of Inco's SO_2/Air Cyanide Removal Process to Industrial Effluents, J.R. Gordon Research Laboratory, Inco Ltd, Canada.

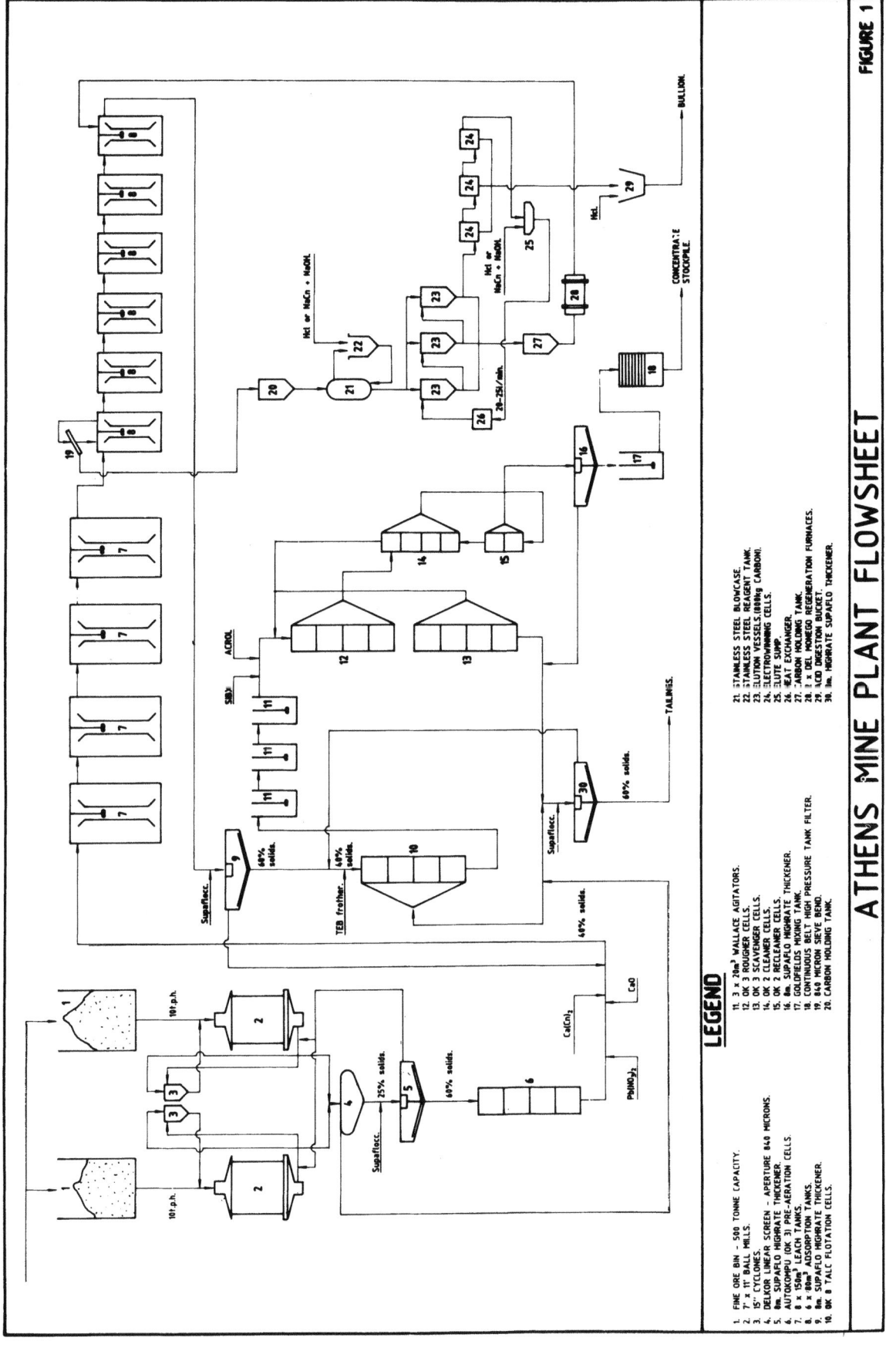

LEGEND

1. FINE ORE BIN - 500 TONNE CAPACITY.
2. 7' x 11' BALL MILLS.
3. 15" CYCLONES.
4. DELKOR LINEAR SCREEN - APERTURE 840 MICRONS.
5. 6m. SUPAFLO HIGHRATE THICKENER.
6. AUTOKOMPU (OK 3) PRE-AERATION CELLS.
7. 8 x 150m³ LEACH TANKS.
8. 6 x 60m³ ADSORPTION TANKS.
9. 6m. SUPAFLO HIGHRATE THICKENER.
10. OK 8 TALC FLOTATION CELLS.

11. 3 x 20m³ WALLACE AGITATORS.
12. OK 3 ROUGHER CELLS.
13. OK 3 SCAVENGER CELLS.
14. OK 2 CLEANER CELLS.
15. OK 2 RECLEANER CELLS.
16. 6m. SUPAFLO HIGHRATE THICKENER.
17. GOLDFIELDS MIXING TANK.
18. CONTINUOUS BELT HIGH PRESSURE TANK FILTER.
19. 840 MICRON SIEVE BEND.
20. CARBON HOLDING TANK.

21. STAINLESS STEEL BLOWCASE.
22. STAINLESS STEEL REAGENT TANK.
23. OK 3 ELUTION VESSELS. (1000kg CARBON).
24. 3 ELECTROWINNING CELLS.
25. ELUTE SUMP.
26. HEAT EXCHANGER.
27. CARBON HOLDING TANK.
28. 2 x DEL MONEGO REGENERATION FURNACES.
29. ACID DIGESTION BUCKET.
30. 6m. HIGHRATE SUPAFLO THICKENER.

ATHENS MINE PLANT FLOWSHEET

FIGURE 1

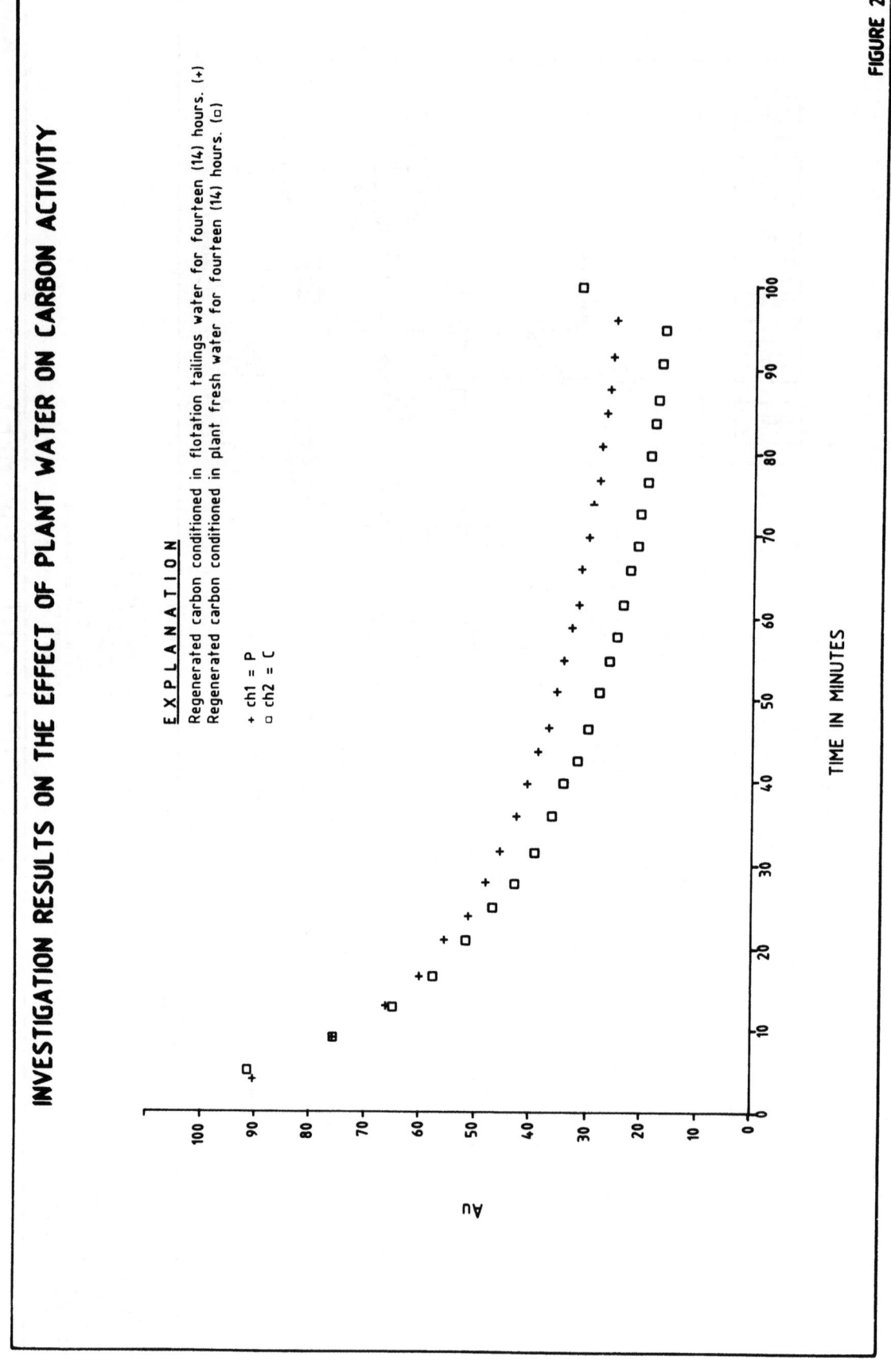

INVESTIGATION RESULTS ON THE EFFECT OF PLANT WATER ON CARBON ACTIVITY

EXPLANATION

Regenerated carbon conditioned in flotation tailings water for fourteen (14) hours. (+)
Regenerated carbon conditioned in plant fresh water for fourteen (14) hours. (□)

+ ch1 = P
□ ch2 = C

Au

TIME IN MINUTES

FIGURE 2

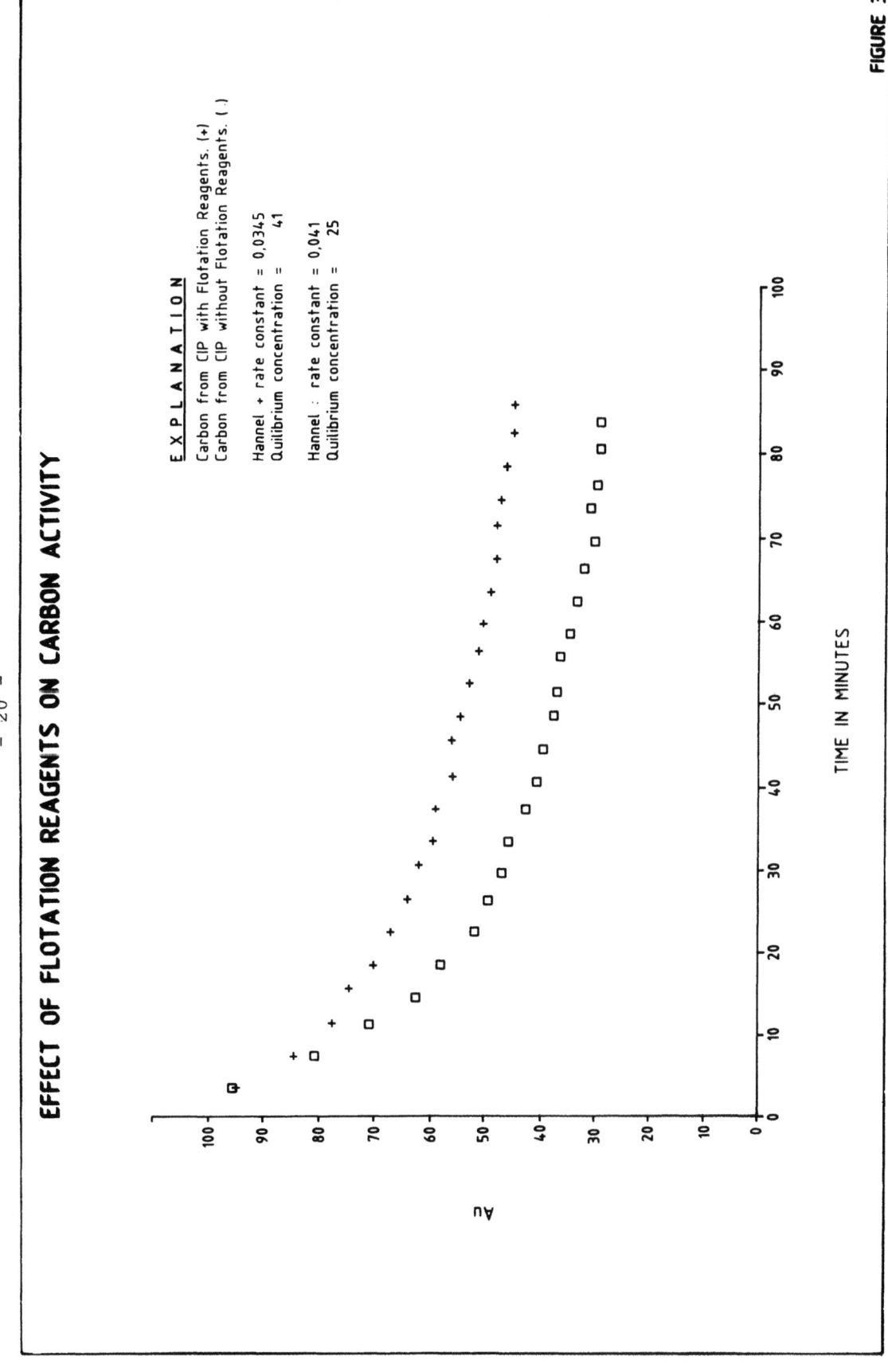

EFFECT OF FLOTATION REAGENTS ON CARBON ACTIVITY

EXPLANATION

Carbon from CIP with Flotation Reagents. (+)
Carbon from CIP without Flotation Reagents. ( )

Hannel + rate constant = 0,0345
Quilibrium concentration =   41

Hannel : rate constant = 0,041
Quilibrium concentration =   25

Au

TIME IN MINUTES

FIGURE 3

– 20 –

94

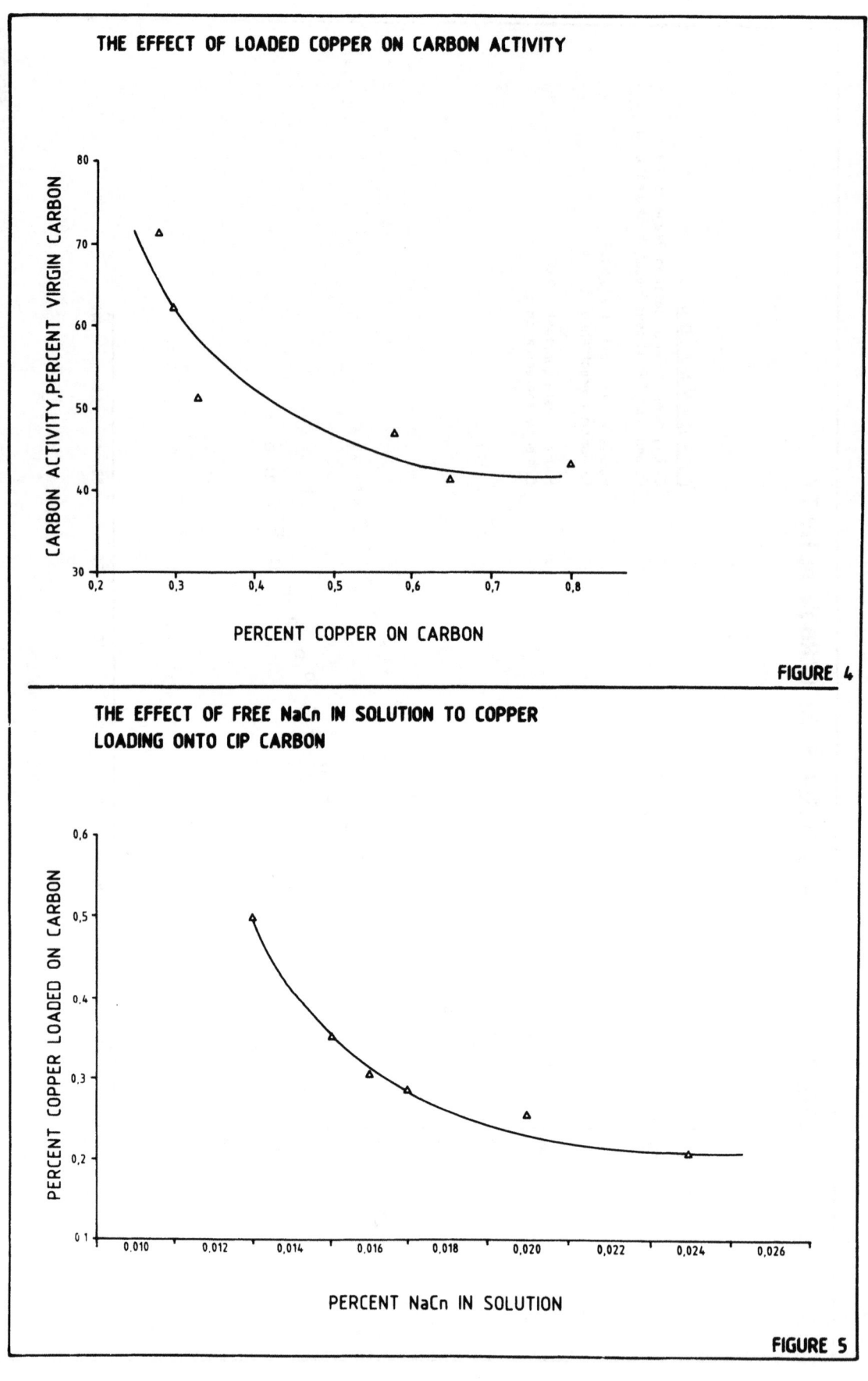

THE EFFECT OF LOADED COPPER ON CARBON ACTIVITY

PERCENT COPPER ON CARBON

FIGURE 4

THE EFFECT OF FREE NaCn IN SOLUTION TO COPPER
LOADING ONTO CIP CARBON

PERCENT NaCn IN SOLUTION

FIGURE 5

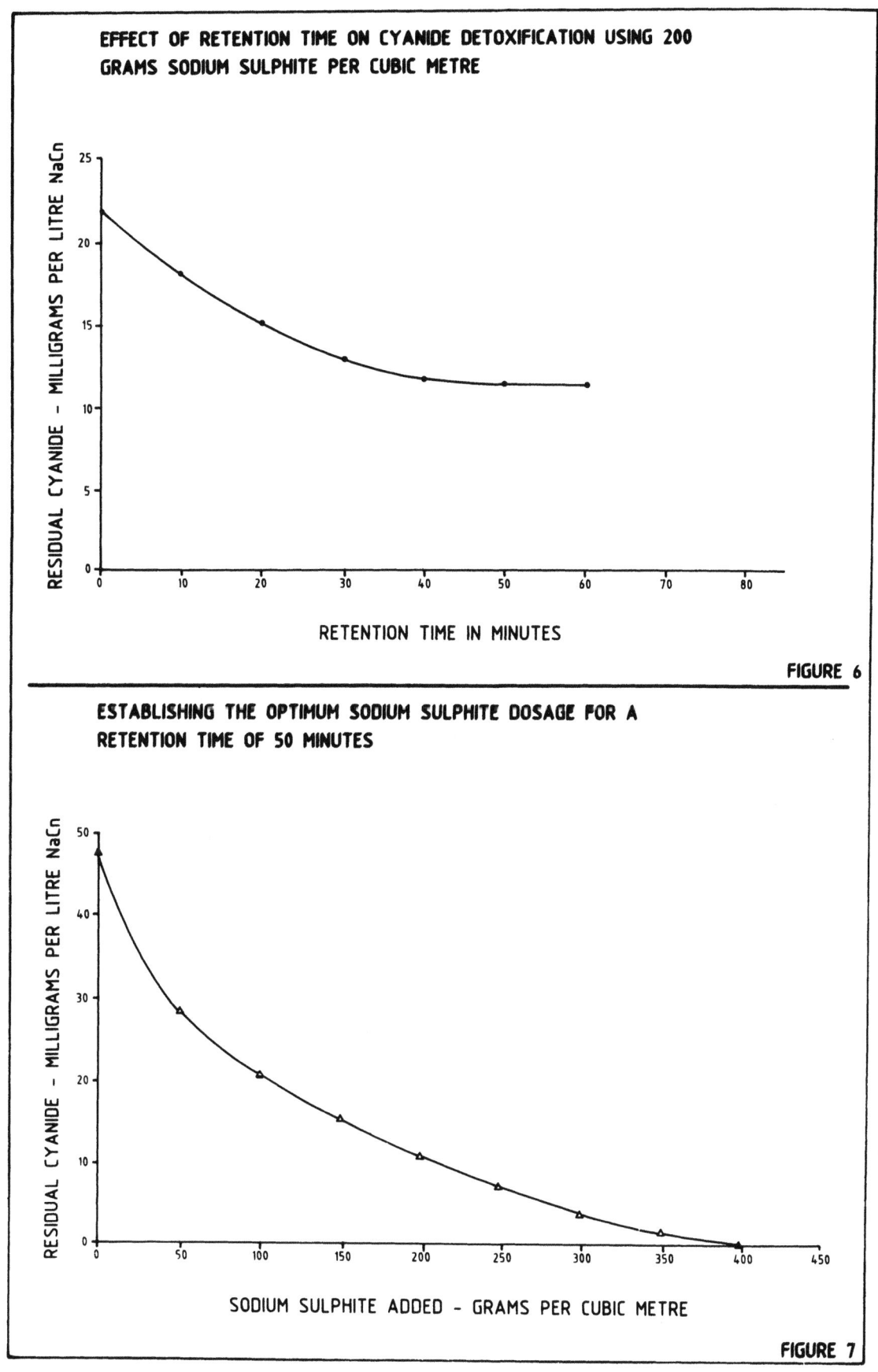

EFFECT OF RETENTION TIME ON CYANIDE DETOXIFICATION USING 200 GRAMS SODIUM SULPHITE PER CUBIC METRE

RETENTION TIME IN MINUTES

FIGURE 6

ESTABLISHING THE OPTIMUM SODIUM SULPHITE DOSAGE FOR A RETENTION TIME OF 50 MINUTES

SODIUM SULPHITE ADDED - GRAMS PER CUBIC METRE

FIGURE 7

Geology—Exploration

Exploration for gold by use of cyanide leach analytical techniques on soil samples in the Greenstone Belts of Zimbabwe

G. D. Collis
Masasa Mines (Pvt) Ltd, Harare, Zimbabwe
C. M. E. Moles
Reunion Mining (Pvt) Ltd, Harare, Zimbabwe
I. Mazaiwana
Anacal Laboratory, Harare, Zimbabwe

ABSTRACT

The detection of gold at very low levels of concentration using cyanide leach analytical techniques has been commercially available in Zimbabwe only since 1988. Since then the method has been used effectively on soil samples to locate and assess gold mineralisation within Archaean greenstone terrains.

Standard sampling techniques are used to collect bulk, coarse-fraction soil samples. These are agitated in a cyanide solution for 24 hours. The pregnant solution is filtered and the gold concentrated in an insolvent organic phase for measurement by atomic absorption.

The results of orientation surveys are presented in which the cyanide leach (bottle-roll) method is compared with standard fire assay analysis of different soil size fractions. This work demonstrates the effectiveness of the method in defining gold-soil anomalies. Peak values for the two methods broadly coincide but detection limits are much lower for the cyanide leach method.

A number of case histories are presented and the advantages and disadvantages of cyanide leach are demonstrated and discussed.

1. INTRODUCTION

Zimbabwe has reasonably well developed residual soils over most of its metallogenic regions. This soil provides an excellent sampling medium in the search for new mineral deposits and soil sampling techniques generally present the explorationist with a cost effective tool for locating these deposits.

In Zimbabwe soil geochemistry has been used successfully to discover base metal deposits [1] but little success has been reported in the search for Archaean gold deposits. Precious metal deposits are less amenable to discovery by soil geochemistry as they are generally smaller, contain less actual metal and the mobility of gold within the soil is considerably lower. These factors combine to ensure that any analytical technique used in the search for gold must be capable of very low levels of detection.

The exploration for gold deposits using soils goes back to pre-historic times. The panning of soil samples, known as "loaming" was almost certainly common practice in Zimbabwe more than a thousand years ago [2]. The technique was considered to be the only reliable method of determining gold in soils until quite recently. However, the geochemical analysis of "pathfinder" elements to identify target areas had been used since the late 1950s.

Improvements in the direct analysis of gold in soils using fire assay led to a decline in the popularity of "loaming" in the early 1980s.

An alternative to the fire assay method is the hot extraction of gold by acid digestion, this has the advantage of dissolving a number of other elements, the values of which can be determined by atomic absorption spectroscopy.

A further development in the analysis of gold in soils has been the use of cyanide solvent extraction (bottle-roll method). This method was recently introduced into Zimbabwe by Delta Gold N.L. in conjunction with the Institute of Mining Research at the University of Zimbabwe. A commercial method was developed in 1988 which has proved to be a very successful reconnaissance tool for defining target areas as well as for guiding the evaluation of existing prospects.

This paper is presented to illustrate both the effectiveness and limitations of the bottle-roll method without attempting a rigourous appraisal.

All the work discussed here is confined to the Archaean greenstone terrain in Zimbabwe and the reader is referred to the following authors [3,4,5] for details on the general geological and pedological environments found in the country.

2. DESCRIPTION OF THE METHOD

2.1 Sample Collection

Under dry conditions about 1 kg of soil is collected and sieved in the field to minus 1 mm (30 mesh). If the soil is wet approximately 2 kg of material is collected and taken to the laboratory where it is dried and sieved to minus 30 mesh before analysis. The soil horizon at which the sample is collected is usually specific to a particular area and is generally determined by a geologist during an orientation survey.[6]

2.2 Analytical Technique

Soil samples are screened, homogenised and split to obtain 500 g of product passing through a 1 mm sieve. This is followed by leaching in 1000 ml of 0.2% KCN solution containing 4 g of slated lime. The agitation is done on a rotating rig at 45 rpm for 24 hours. Two variations of the rotating rig have been designed. The first involves rotation of the bottle on a roller system. Extra agitation is provided by the addition of compressed air which is blown into the mixture. This procedure adds oxygen to the mixture which facilitates dissolution of the gold. With the second design the bottle is secured to the side of a rotating spindle which enables a greater agitation to be achieved and also promotes the absorption of atmospheric oxygen.

Following leaching of the soluble gold by cyanidation, a 200 ml aliquot of the solution is filtered before being agitated with 10 ml of Di-isobutyl Ketone (D.I.B.K.) solution containing about 2% of trioctyl methyl ammonium chloride (Aliquat 336). The filtration process is lengthy and only 200 ml of the pregnant solution is filtered to economise on time. Eighteen grammes of sodium chloride are added to the solution to enhance the separation of the aqueous and organic phases. After separation of the phases a sample of the organic layer at the top of the container is aspirated into an atomic absorption spectrophotometer (AAS) for gold determination.

In cases where the inductively coupled plasma (ICP) instrument is used the D.I.B.K. concentration procedure is omitted and a direct determination of the gold in solution is made. Approximate detection limits for the AAS and ICP methods are 2 ppb Au and 0,5 ppb Au respectively.

3. COMPARATIVE WORK

3.1 Test Area One

A traverse line, 1 km long, was established across a zone of known gold mineralisation. Soil samples of different weight and size fraction were collected every 25 m along this line. Table 1 summarises the samples taken and the methods of analysis employed.

Table 1: THE SAMPLES COLLECTED

SIZE FRACTION	WEIGHT	ANALYTICAL METHOD
(MESH)	(g)	(FA = Fire Assay) (BR = Bottle-Roll)
- 30	50	FA
- 80	50	FA
- 120	50	FA
- 200	50	FA
- 30	500	BR
- 80	500	BR

The bottle-roll method employed was that described in Section 2.2. The fire assay involved dissolution of the prill in aqua-regia with final estimation by atomic absorption. The authors' experience is that on a production basis, in Zimbabwe, this fire assay method is only capable of a detection limit of approximately 20 ppb Au.

The study area is underlain predominantly by massive and pillowed basaltic rocks containing layers of komatiitic-basalt and numerous thin beds of sulphide facies banded- iron formation (BIF). The gold occurs in one of the BIF horizons and is associated with disseminated and stringer-type sulphide mineralisation. No discrete quartz veining is evident in the mineralised zone. The soil depth varies between a half and two metres and in most places exhibits a well developed 'B' horizon. All of the samples were taken below a rubble zone at the top of this 'B' horizon, some 25 to 30 cm below surface.

Figure 1 illustrates the results obtained along the traverse line and is split into two parts for ease of interpretation. Observations from this data are summarised below:

(a) Three main anomalies are defined by the results.

(b) The fire assay results of the coarser size fractions (minus 30 and minus 80 mesh) produce very erratic profiles probably due to the effects of coarse gold.

(c) To obtain reliable results from the fire assay method a minimum sieve size of minus 120 mesh is required.

(d) The two separate bottle-roll data sets compare favourably with no significant difference between the minus 30 mesh and the minus 80 mesh size fractions. However the minus 80 mesh results do show a higher peak value over the main anomaly located at 325 metres south of the base line.

(e) Background values for the fire assay method are about 20 ppb gold which is equivalent to the detection limit of the method employed. The bottle- roll method provides a background of 8 ppb gold which is more than twice the detection limit of the method used.

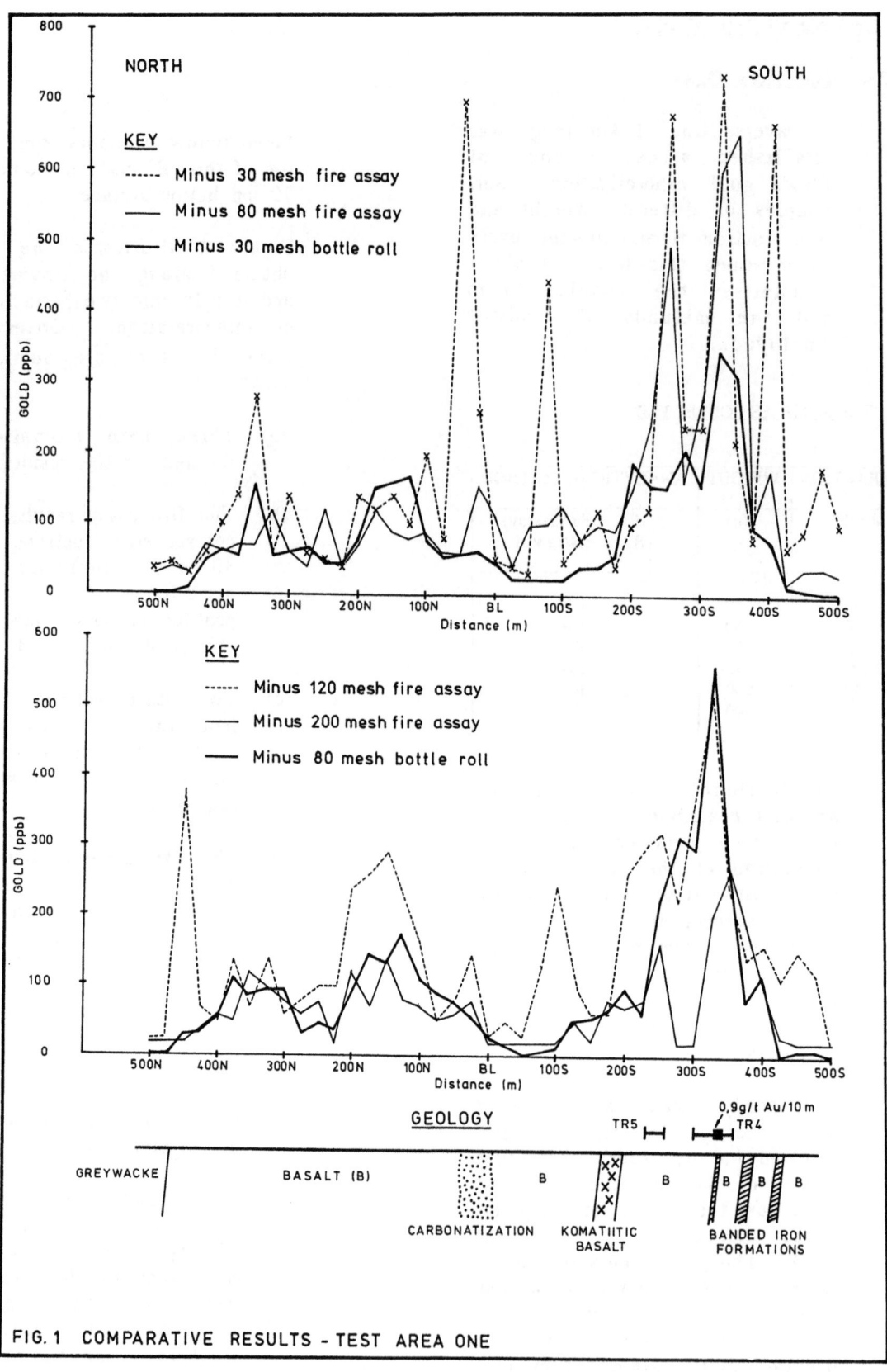

FIG. 1 COMPARATIVE RESULTS - TEST AREA ONE

(f)   A threshold value of about 50 ppb gold is indicated. This is approximately twice the detection limit of the fire assay method but over 10 times the detection limit for the bottle-roll method.

(g)   Table 2 summarises the various peak-to-background ratios.

Table 2:   PEAK TO BACKGROUND RATIOS

MESH SIZE	ANALYTICAL METHOD	PEAK/ BACKGROUND
- 30	FA	14,8
- 80	FA	13,2
- 120	FA	17,3
- 200	FA	13,5
- 30	BR	43,8
- 80	BR	70,0

The peak-to-background ratios obtained from the bottle-roll results are more than 3 times larger than those obtained from the fire assay method.

(h)   The bottle-roll data produces a much smoother profile and the anomalies generally show better definition.

3.2   Test Area Two

A number of traverse lines, 100 metres apart, were run across a carbonate filled shear zone which was known to contain sulphide hosted gold mineralisation. These were extended beyond the shear zone to test for parallel mineralised bodies.

Two separate samples were collected at 25 metre intervals so that two methods of analysis could be compared. One sample was tested using fire assay of a minus 200 mesh size fraction, the other was sieved to minus 30 mesh and then bottle-rolled using the cyanide leach method. Blind duplicates of every tenth sample were also submitted to test the precision of each method.

After an interval of nine months a third sample was collected from certain traverse lines. These were sieved to minus 80 mesh and either fire assayed or bottle-rolled.

The fire assays were carried out on a 50 g subsample, and gold values were determined using atomic absorption spectroscopy. Bottle-roll tests were conducted on a 500 g subsample using the method described in section 2.2.

In all cases the sample was collected, using a pick and shovel, from the lower part of the 'A' horizon generally located 20 to 30 cm below the surface. Soil cover in the test area varied between 1 and 3 m thick.

The style of mineralisation suggested that most of the gold in the known mineralised shear would be in the fine fractions. However a secondary event appeared to have remobilised some of the sulphide-hosted gold into quartz- hosted 'free' gold which was expected to be concentrated in the coarser fractions.

Results obtained from the various size fractions and assay methods are plotted in fig. 2.

The sample results from **line 1**, taken across the western extension of the main shear zone show low gold values along strike from the known mineralisation, 100-175 m south of base line. Higher values are evident 200-300 m further south which suggest a hitherto unrecognised zone of mineralisation.

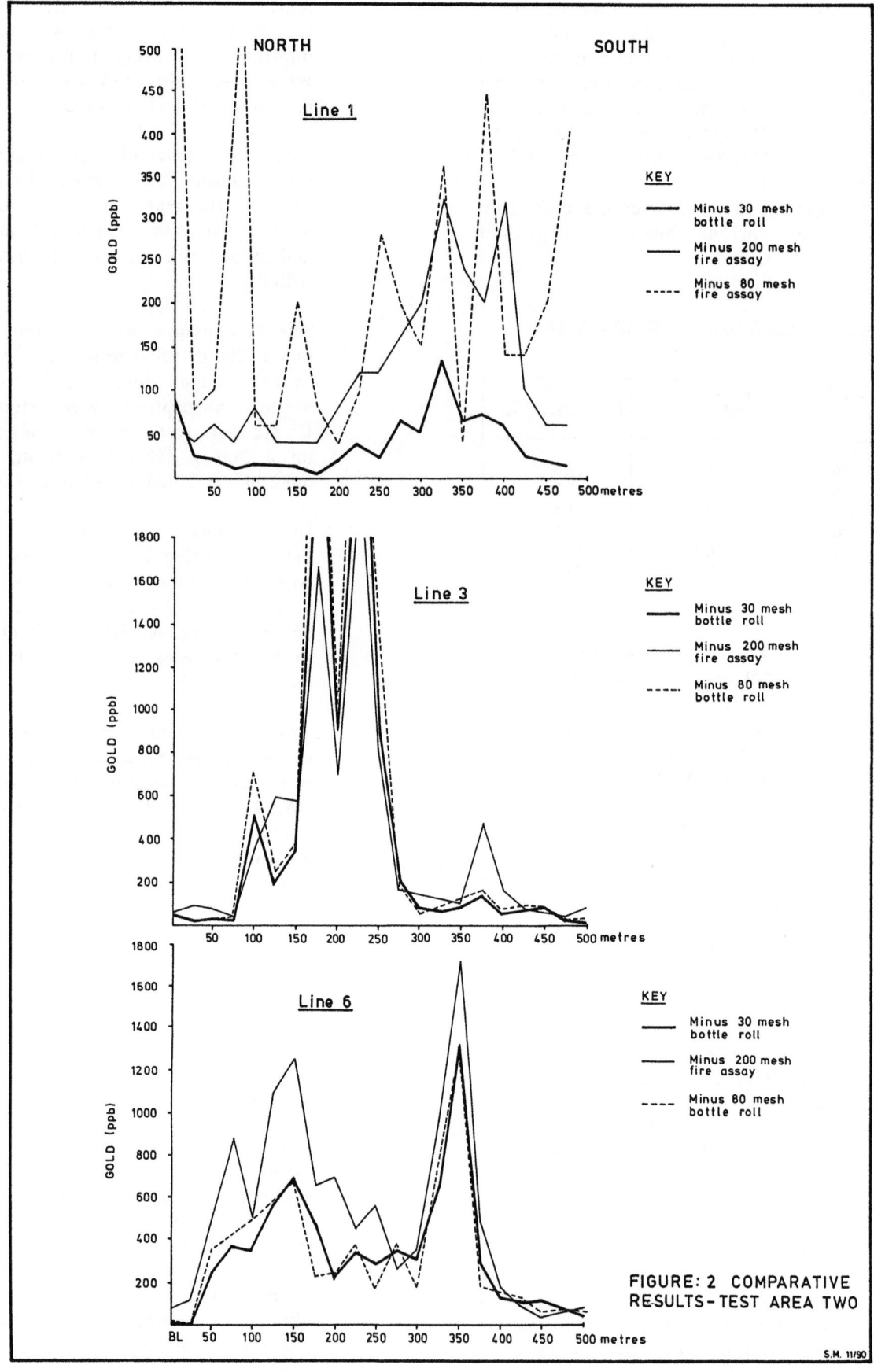

FIGURE: 2 COMPARATIVE RESULTS - TEST AREA TWO

S.M. 11/90

The minus 200 mesh fire assay results show much higher absolute values, suggesting that the gold is concentrated in the finer fraction. The fire assay values show a peak-to-background ratio of only 8 to 1 while the minus 30 mesh bottle-roll results exhibit a peak to background ratio of 44 to 1.

The minus 80 mesh fire assay values tend to be erratic and uncontourable. This is probably due to the presence of coarse gold causing a bias in the small sub-sample.

On line 3 both the fire assay and the bottle-roll methods define the main zone of mineralisation. A second peak between 225 and 250 m from the base line is also recognised and may represent a parallel body. At this point the bottle-roll values are higher than the minus 200 mesh fire assay values. This could be due to the inclusion of coarse gold in the bottle-roll samples.

The anomaly at 375 m is barely defined by the bottle-roll values but is well marked by the fine fraction fire assay.

On line 6 the two zones of mineralisation are well defined by both methods, although absolute values are somewhat higher for the fire assay. The difference between the values of the minus 30 mesh and the minus 80 mesh bottle-roll is very small and suggests that in this instance there is little to be gained by sieving down to minus 80 mesh.

Check sampling revealed poor precision for both the bottle-roll and the fire assay. Of 29 checks carried out on bottle-roll determinations only 13 were within 50% of the original value. Precision of fire assays were better with 20 out of 32 of the checks falling within 50% of the original value.

4. CASE HISTORIES

4.1 Prospect A

Cyanide leach analytical techniques were used on soil samples taken from Prospect A. The target consists of an isolated banded ferruginous chert horizon with an east-west strike of just over three kilometres. It is on average 60 m thick and dips 85º to the north. The chert is sandwiched between basaltic lava in the footwall and andesitic lava and volcanoclastics in the hanging wall.

Gold had previously been mined from an open pit which exploited the northern contact of the chert horizon. A total of 35,33 kg of gold was won from 17687 tonnes of ore giving an average recovered grade of 2 g/t gold. The soil sampling exercise was conducted to try and define further mineralisation along strike from the old working.

The soil thickness varies between 30 and 100 cm and consists of a very immature red loam mixed with scree from the chert formation. The chert horizon forms a prominent ridge and gradients are steep on both sides. Topo-corrected traverse lines were established every 100 m along the strike and soil samples were collected every 25 m along each line.

A significant gold-in-soil geochemical anomaly was defined along the northern contact of the chert horizon (Fig. 3). Anomalous values often exceed 1000 ppb Au and a background of 30 ppb Au can be established. A threshold value of 100 ppb Au gives an anomaly width of about 250 m. Trenches were dug and sampled over the geochemical anomaly and these results are also shown in Fig. 3. The trenches define a zone of gold mineralisation grading approximately 1 g/t, over

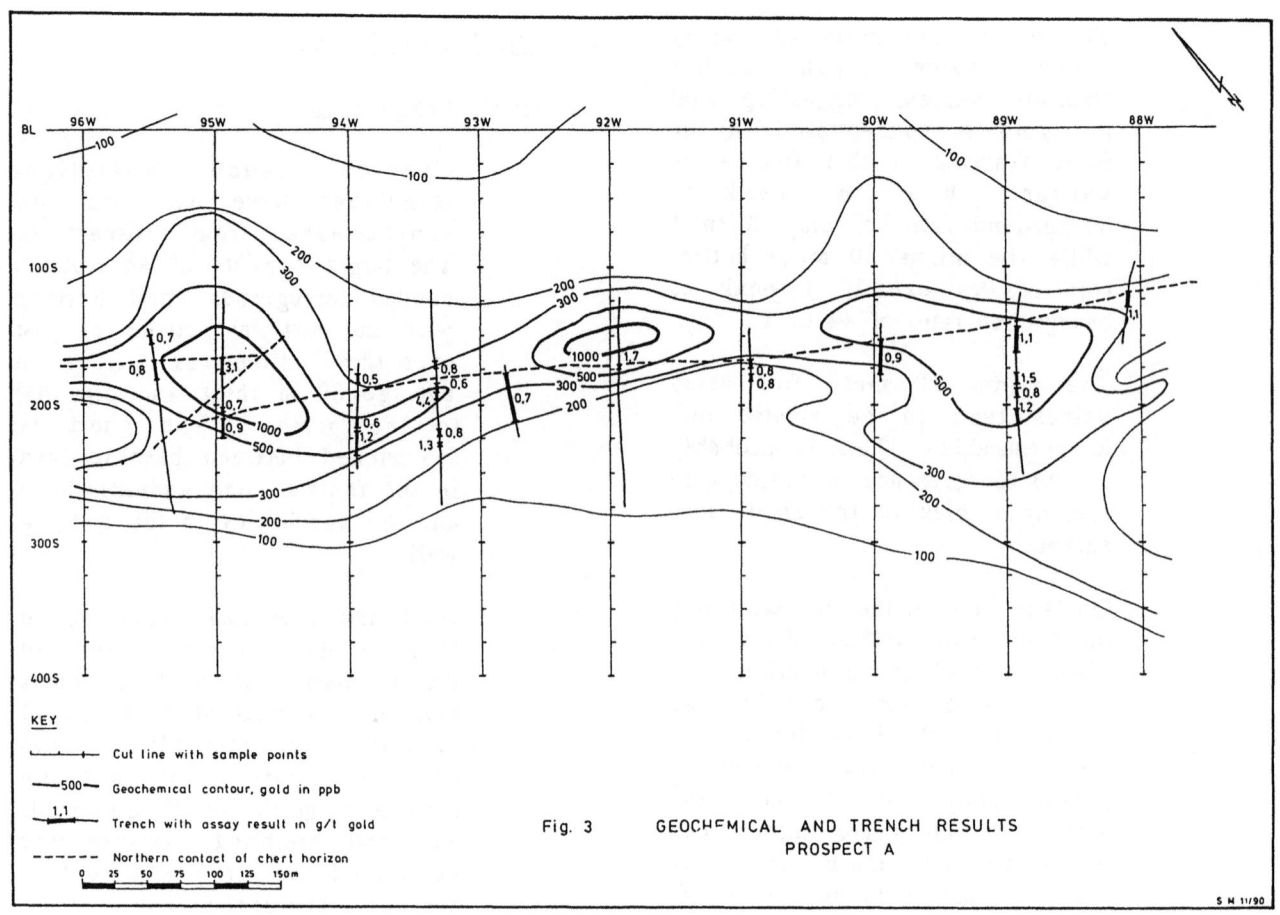

KEY

⊢———┼——— Cut line with sample points

———500—— Geochemical contour, gold in ppb

⊥1,1 Trench with assay result in g/t gold

– – – – – Northern contact of chert horizon

0   25   50   75   100 125 150m

Fig. 3   GEOCHEMICAL AND TRENCH RESULTS
PROSPECT A

S H 11/90

a true width of 20 m, running along the northern contact of the chert horizon. This result confirms the existance of bedrock hosted gold mineralisation beneath the soil geochemical anomaly.

4.2   Prospect B

This prospect is centred on a prominent ridge within Shamvaian sediments in a greenstone belt in the north-east of Zimbabwe. On both sides of the ridge gradients are steep and soils poorly developed, often comprising little more than scree or talus. The crest of the ridge is composed of highly silicified volcanoclastics which contain disseminated sulphide mineralisation. A line of old workings, adits, shafts and trenches run along the top of the ridge within the silicified zone (Fig. 4).

On either side of the silicified zone there is a zone of extensive bleaching, sericitisation and carbonatisation. This alteration, within grits and volcanoclastics, extends for 10m to the north of the silicification and nearly 200 m to the south. Pods of silicification and disseminated sulphide exist within this zone.

A soil sampling grid was constructed to test for gold using the cyanide leach method. Samples were collected at 50 x 100 m intervals. Soil was taken from just below surface and sieved to minus 1 mm in the field. Duplicates were taken at every twentieth sample point and submitted in separate batches.

Before the results of the soil samples were obtained, a limited programme of channel sampling old trenches and adits was undertaken along the silicified ridge which was thought to be the focus of mineralisation.

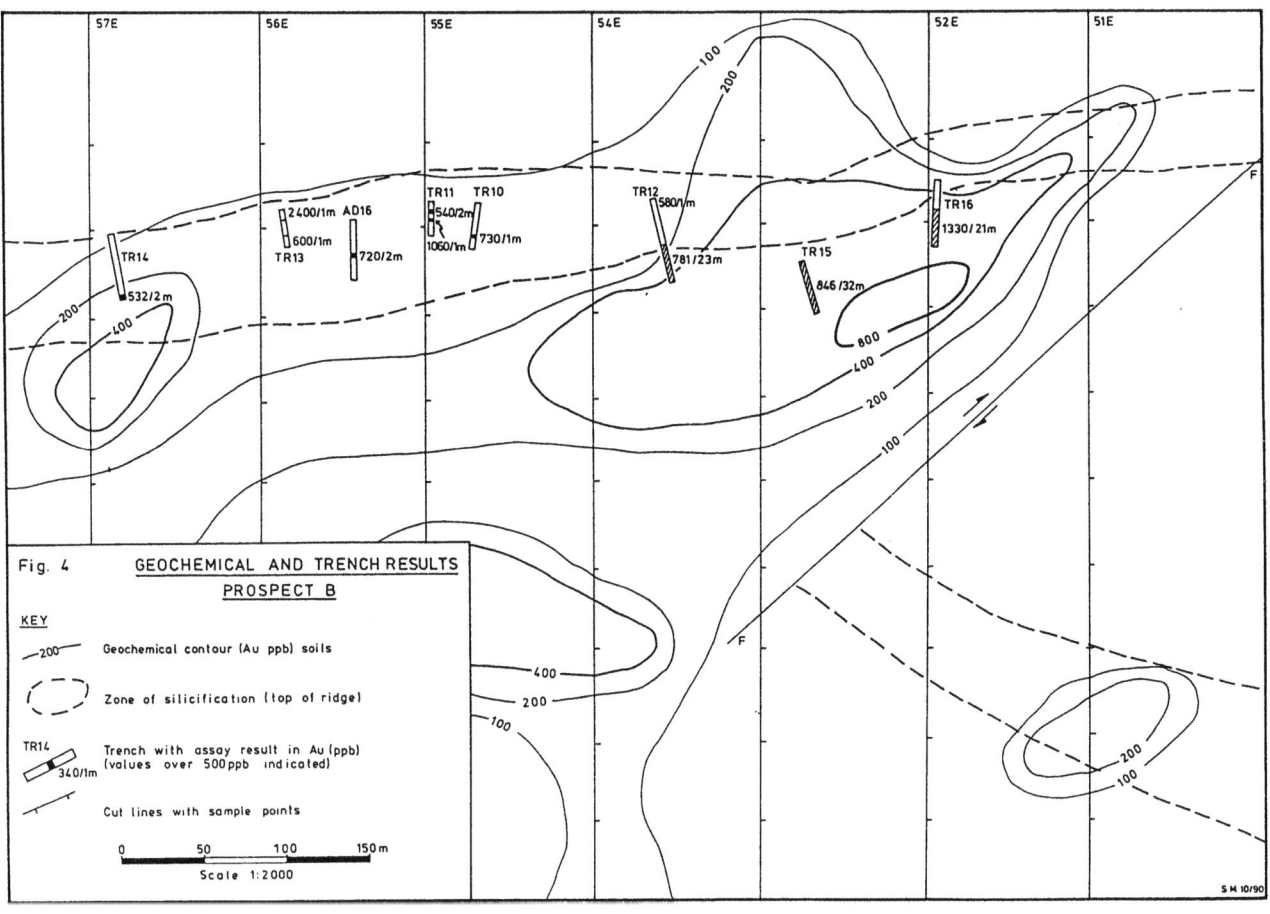

Fig. 4   GEOCHEMICAL AND TRENCH RESULTS
PROSPECT B

KEY

—200—   Geochemical contour (Au ppb) soils

Zone of silicification (top of ridge)

TR14   Trench with assay result in Au (ppb)
340/1m   (values over 500 ppb indicated)

Cut lines with sample points

0   50   100   150 m
Scale 1:2000

S M 10/90

Results of the trench sampling along the crest of the ridge were generally disappointing. However, considerably better values were found where the trenches passed into the zone of bleaching and sericitisation to the south of the main silicified zone.

When the soil geochemical results were plotted it became apparent that the better trench results were associated with a geochemical soil anomaly which was running parallel to and south of the main zone of silicification.

Trenching before the soil sampling results are known is clearly unwise and in this case demonstrates the effectiveness of soil sampling, even in areas of poorly developed soils on steep gradients.

5.   DISCUSSION

The cyanide leach (bottle-roll) method has a number of advantages over other methods of analysing the gold content of soils, but there are also a number of potential hazards. These advantages and disadvantages are discussed below.

(a)   Perhaps the most important advantage is the ease with which a bottle-roll facility can be set up and operated at relatively low cost. This is particularly appropriate in underdeveloped or developing countries where sophisticated analytical facilities may not exist and where skilled technicians are scarce. A bottle-roll facility can be set up using some fairly basic mechanical engineering, a simple AA machine and a few chemicals. To demonstrate this point it has been noted that there are now a number of small dump retreatment projects in Zimbabwe which have set up their own bottle-roll units as a cheap, fast and reliable method of assessing gold extraction. Those that do not have their own AA machines simply take their filtered solutions to the nearest laboratory for analysis.

(b) The comparative work outlined in Section 3 of this paper suggests that a minus 30 mesh soil fraction for a bottle-roll determination is quite adequate for recognizing gold anomalies. Little appears to be gained by reducing the size to minus 80 mesh. At least one major exploration company in Zimbabwe does not sieve the soil samples at all. Sieving down to minus 30 mesh in the field is quick and easy and many samples can be collected in a day. On the other hand we have demonstrated that the fire assay method requires the soil to be screened down to at least minus 120 mesh before reliable results can be achieved. The collection of even 100 grammes of this fraction can take three to four times as long as a 1 kg sample at minus 30 mesh. Expensive and more delicate screens are required and there is a greater temptation for field crews to cheat. Screening the soil in the laboratory is a possibility, but this seems to be done with some reluctance where thousands of samples are concerned and again 'cheating' by laboratory staff can be a concern.

(c) Even where good fire assay facilities exist, many exploration companies favour the bottle-roll method because it is far less vulnerable to contamination both in the field and in the laboratory. In areas such as the greenstone belts of Zimbabwe, there are very few gold exploration targets which do not have extensive old workings in the area. If for example a 500 g bottle-roll charge was contaminated by a gramme of old tailings with a gold content of 2 g/t the effect on the final result would be insignificant. While a similar contamination of a 50 gramme fire assay charge could result in a serious error especially where the underlying soil anomaly is subtle.

(d) When conducting broad scale reconnaissance soil geochemistry for gold, anomalies are often indicated by values which are not much above background levels. In this case peak-to-background ratios are most important. We have attempted to show that the bottle-roll method produces far greater peak-to-background ratios primarily because the detection limits are much lower.

(e) The bottle-roll method utilises a relatively large sample of a coarse soil fraction, it is therefore far more representative than the small sample of a fine soil fraction required for fire assay. Using a fine soil fraction is advantageous where the gold particle size is known to be small because the sample will be biased and anomalies will be exaggerated. However in reconnaissance sampling where the particle size of the gold is unknown, exclusion of the coarse gold particles from the sample could lead to mineralisation being overlooked.

(f) Finally, the relative costs of the two methods need to be compared. Fire assays of soil samples cost in the region of Z$8-15 per sample while bottle-roll determinations cost Z$5-7 in commercial laboratories in Zimbabwe.

It is necessary to note that there are a number of drawbacks and unsolved problems with the bottle-roll method.

(a) One minor drawback is that large samples are heavy and may have to be carried long distances.

(b) A more serious problem is that of assay precision. Exhaustive comparative tests have not yet been carried out, but our limited experience suggests that repeatability is poor amongst the laboratories in Zimbabwe which are carrying out bottle-roll determinations. The repeatability of the fire assays that have been checked is not good but is considerably better than that of the bottle-roll. We are unable at present to identify the cause of this and hope that the laboratories will attempt to investigate.

(c) Another problem for the bottle-roll method is that gold is often concentrated in the very fine soil fraction. Collection of this size fraction will enhance any anomalies while "dilution" by the coarse fractions in a minus 30 mesh sample will surpress them. It is possible therefore for an anomaly to be missed or to be considered unimportant when the results are plotted. This problem suggests that an orientation survey should be carried out before an area is soil sampled. If it can be established that the bulk of the gold is in the fine fractions it may be prudent to fire assay a minus 120 or even a minus 200 mesh fraction. The orientation survey should also try to establish the optimum depth at which the soil samples should be taken as this also helps increase the quantity of gold in the sample. Results from areas with different soil profiles, underlying geology or topography should be treated as separate populations and be assessed independently of each other.

6. CONCLUSIONS

The test data and case histories presented in this paper serve to demonstrate the effectiveness of the bottle-roll method in analysing for gold in soils. It has numerous advantages over other analytical techniques commercially available in Zimbabwe. These include a lower detection limit, greater representivity, lower susceptibility to contamination and a lower cost. Unfortunately the samples are bulky and the method seems to exhibit poor precision. The bottle-roll method is highly recommended for exploration in developing countries as the laboratory is easily constructed and the method relatively simple to perform. However one should be aware of its limitations and be prepared to conduct orientation surveys to establish the effectiveness of the method in specific areas.

ACKNOWLEDGEMENTS

We would like to thank the management of Masasa Mines (Pvt) Ltd and its parent, Delta Gold N.L., for permission to publish this data which was collected during routine exploration work. Mr and Mrs S.A. Mawson provided invaluable assistance in helping to prepare the paper. Thanks also to Professor G.J.S. Govett who encouraged us to put pen to paper and to Dr. D.C. Gellatly for sharing with us his expertise and experience during the early days of experimentation with the method in Zimbabwe.

REFERENCES

1. RUGMAN, G.M., 1982. Perseverance Mine - a prospecting case history. Mining Magazine London., Vol. 146 : 381-391.

2. SUMMERS, R., 1969. Ancient mining in Rhodesia. Mem. natn. Mus. Sth. Rhod., 3.

3. STIDOLPH, P.A., 1977. The Geology of the country around Shamva. Zimbabwe Geological Survey Bulletin 78.

4. STOCKLMAYER, V.R., 1980. The Geology of the Inyanga-North-Makaha Area. Zimbabwe Geological Survey Bulletin 89.

5. ANHAEUSSER, C.R., 1976. The nature and distribution of Archaean gold mineralisation in Southern Africa. Mineral Science and Engineering, 3 : 46-84.

6. VIEWING, K.A., 1987. Geochemical orientation studies for gold in Zimbabwe. African Mining, 1 : 385-399.

7. ANSARI, H., WITKIN, B., KATSANDE, D., MAZAIWANA, I., VIEWING, K.A., 1989. The analysis of gold in geochemical prospecting samples by cyanidation, solvent extraction and atomic absorption spectrometry. Institute of Mining Research 20th Annual Report, 77 : 26-41.

8. VIEWING, K.A., 1989. Geochemical exploration for gold. Institute of Mining Research 20th Annual Report, 77 : 56-64.

Design and development of a lithogeochemical database for regional exploration using dBASE IV

K. P. Fox
Rössing Exploration, Windhoek, Namibia

ABSTRACT

The paper describes the design and development of a PC based lithogeochemical database system, using dBASE IV, for use in regional geochemical exploration. The system is arranged in an hierarchical structure of menu-driven modules. Although primarily developed as an aid in data capture and management, selection modules to interrogate the database can be designed so as to reflect geological models appropriate to the type of mineralisation being sought. Data is stored in the database in the form of a lithological description for each sample, its unique number, its location and relevant geochemical analyses. Data entry is facilitated by the use of a menu-driven capture program and the user is offered a range of options to enter, update and delete information as well as to selectively abstract and print data. dBASE IV was chosen for its ease of use on a PC and its relational capacity in holding, retrieving and matching both descriptive and numerical data.

The system was developed "in house" in Rossing's Exploration department and is used to store and access lithogeochemical data from a wide range of geological environments. The system is flexible enough to allow a variety of elemental suites to be incorporated in the database so it can be applied to a variety of exploration programmes.

INTRODUCTION

Rossing Exploration is involved in a wide variety of mineral exploration programmes in Namibia. See Figure 1. Targets include industrial minerals and base and precious metals. A broad spectrum of exploration techniques is applied in mineral search but geochemistry is one of the most widely used because of its appropriateness in the arid environment of Namibia in which rocks have undergone long periods of weathering. Soil, rock and drainage geochemistry are all applied at various stages of exploration programmes but are mainly used to highlight areas for detailed investigation. Use of the term "regional" in the title therefore, is interpreted as implying areas of > 1km or several square kms. (1,2) By and large, area selection is on the basis of geological criteria, with geochemistry being applied to test or confirm favourable environments for mineralisation in the area as a whole. Individual grant areas up to 100 sq.kms in size can be held under the Namibian prospecting licence system.

Once selected, a regional reconnaisance programme employing lithogeochemistry is often undertaken on an entire grant area in order to highlight particular locations for initial detailed work and to obtain a preliminary geological and geochemical picture of the area. Elements selected for analysis are predominantly the target and associated trace elements, and pathfinder elements are employed where warranted.

One such programme in which the author was involved was a search for epithermal gold mineralisation in central Namibia. Having withdrawn a large block of ground considered suitable for this type of mineralisation, an initial geological and geochemical reconnaisance was begun. This involved geological mapping and simultaneous rock sampling of numerous lithologies throughout the area. The approach called for the collection of large numbers of rock samples and their subsequent analysis for up to ten elements. It soon became obvious that this approach, though useful, could be more valuable if a method of integrating the large amounts of geological and geochemical data thus generated could be developed. The idea of a database to incorporate the two complimentary data sets was therefore put forward. This paper outlines the approach taken in the

111

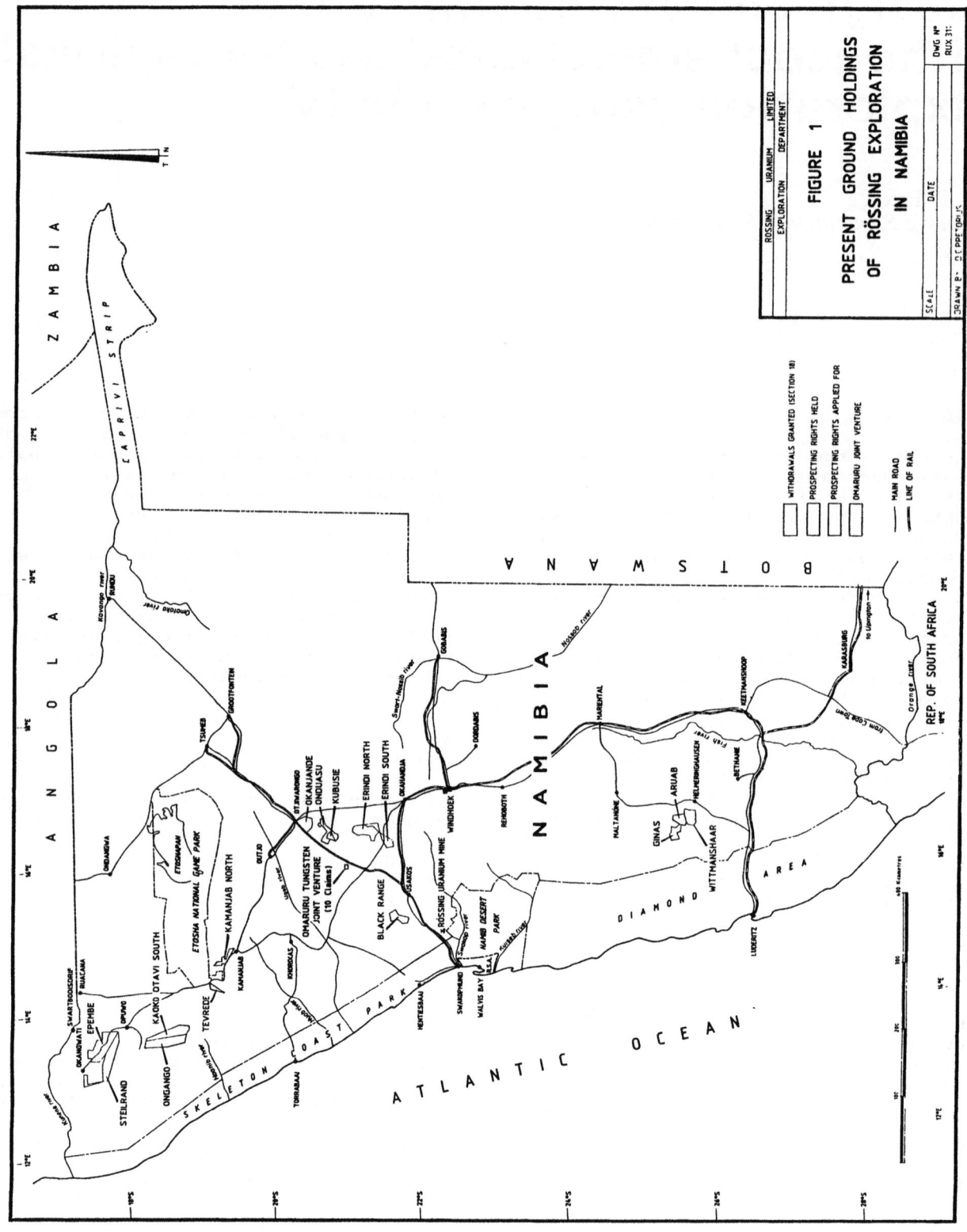

FIGURE 1

PRESENT GROUND HOLDINGS
OF RÖSSING EXPLORATION
IN NAMIBIA

design and development of the database using a commercially available "off the shelf" database package and describes the ongoing work involved in taking the concept beyond its initial role as a data management system.

DESIGN OF THE DATABASE

Design of the database had to take account of the following factors:

1. The need to incorporate geological and geochemical data from several areas, so that its usefulness would not be limited by being restricted to a particular mineral search in one area, e.g. epithermal gold in central Namibia. This became necessary when it was decided to expand the original idea to incorporate relevant data from all the areas in which Rossing is carrying out prospecting.

2. Deriving from this, the need to incorporate data from each area in separate database files, each appropriate to that particular area in terms of number and nature of elements analysed for, means of storing and retrieving data and several other criteria which render data from different areas incompatible with that from others. This was considered necessary because:

   a) each area is unique in terms of its geological and geochemical character- istics and its potential for various types of mineralisation.

   b) the approach taken in exploration reflects this uniqueness and therefore data sets should be stored and accessed separately while maintaining a relational capacity between the sets.
This requirement, in effect, dictated the system architecture shown in Figure 2, which illustrates the design, in which each database file is separate and all of the databases (there can be as many as required, one for each project area) are separate from the hierarchical set of modules used to access and operate on the database files.

3. Data to be stored for each sample would need to consist of:
   a) a unique sample number
   b) a sample location
   c) a farm name describing where the sample came from (optional since Rossing also operates in parts of Namibia where there are no designated farm boundaries).
   d) a lithological description for the rock sample which would accurately summarise features considered relevant in the search for mineralisation.
   e) analyses for up to ten elements for each rock sample. The composition of this element suite could vary from area to area.

4. Data for any and all the samples would need to be accessed using any of the above criteria or any of them in combination.

5. The system should be able to satisfy these requirements and retain a user friendly approach in order to encourage its use and to avoid complex coding systems for e.g. rock names which tend to discourage usage by imposing rigid criteria for data input etc.

6. The system needed to be PC based in order to be compatible with other departmental hardware and be appropriate to the needs of the department's staff.

DEVELOPMENT OF THE SYSTEM

With these requirements in mind it was felt that a freely available commercial database software package would be appropriate and offer advantages in terms of quick set up time and ease of use. dBASE IV from Ashton Tate was chosen mainly for its power and ease of use on a PC. It was appreciated at the time however that use of such a non dedicated database package could impose limitations in such a specific end use and this indeed has proven to be the case (3, 4). Nevertheless, it is felt that such limitations are out-weighed by ease of use in both programming and user applications.

In order to satisfy the requirements described above and the many others that became apparent with time, it was felt that a modular system design would offer the greatest flexibility in development, testing and operation of the system. The need to store data from each operational area separately was easy to implement by simply using uniquely titled and designed database files for each area. Storage and access facilities for data in these files was however more difficult to develop since the various programme modules were to be common in order to avoid unnecessary duplication and therefore the system needed to be flexible enough to recognise and tolerate unique features of each area's database files.

Database files.

As described above, these needed to be designed to reflect the storage and retrieval requirements of each particular area and to allow capture of data and its storage using pre-defined titles or fields. This was achieved by simply titling each component field of the file with the required parameters, e.g. (see Figure 3)
      Field 1 = Sample No.
      Field 2 = Sample location

114

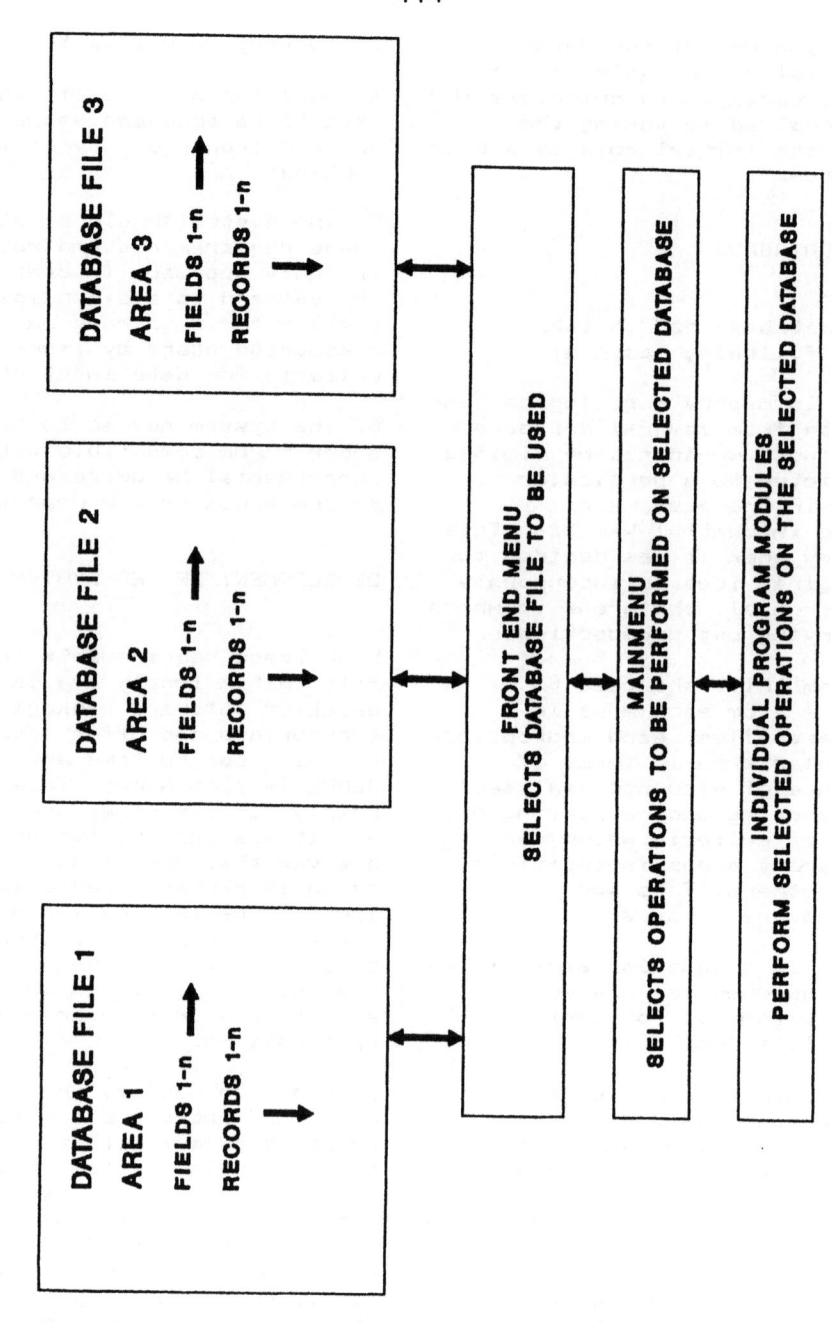

FIGURE 2. LITHOGEOCHEMICAL DATABASE
SYSTEM ARCHITECTURE

DATABASE FILE 1
AREA 1
FIELDS 1-n
RECORDS 1-n

DATABASE FILE 2
AREA 2
FIELDS 1-n
RECORDS 1-n

DATABASE FILE 3
AREA 3
FIELDS 1-n
RECORDS 1-n

FRONT END MENU
SELECTS DATABASE FILE TO BE USED

MAINMENU
SELECTS OPERATIONS TO BE PERFORMED ON SELECTED DATABASE

INDIVIDUAL PROGRAM MODULES
PERFORM SELECTED OPERATIONS ON THE SELECTED DATABASE

FIGURE 3. LITHOGEOCHEMICAL DATABASE SYSTEM
STRUCTURE OF DATABASE FILES

DATABASE FILE 1

AREA 1

FIELD 1	FIELD 2	FIELD 3	FIELD 4	FIELDS 5-n
SAMPLE NUMBER	SAMPLE LOCATION	FARM	LITHOLOGICAL DESCRIPTION	ANALYSES
CHARACTER	CHARACTER	CHARACTER	CHARACTER	NUMERIC
EACH SAMPLE HAS A UNIQUE NUMBER	OPTIONAL	OPTIONAL	SUMMARISES THE RELEVANT FEATURES OF EACH SAMPLE	ANALYSES RELEVANT TO EACH AREA

Field 3 = Farm
Field 4 = Lithological description
Field 5 - n = Analyses

Each area for which data needs to be stored and accessed has such a field structure, or a variant of it, depending on the number of element fields required. Each record in the database files has entries in the appropriate fields. A feature of dBASE IV which makes it particularly appropriate for this application is the facility to mix both numeric fields e.g. analyses and character fields e.g. lithological descriptions. This feature becomes particularly useful when interrogating the database since the ability to match numeric and character data for each sample allows the combination of geological and geochemical data to be extracted. This matching of data types therefore satisfies one of the main requirements of the explorationist's use of such databases i.e. the need to associate and combine data from different sources (5). An example of the usefulness of this would be the discovery, on interrogating the database that consistently high arsenic values are recorded where the lithological description includes the word "alteration".

Program modules

The system specifications were interpreted to mean that the database files containing the geological and geochemical data were to be accessed using program modules to allow:

a) addition of data
b) editing of data
c) deletion of unwanted or incorrect data
d) display of data on screen or on hard copy, using a variety of selection criteria, either uniquely or in combination.

A modular system design was therefore adopted to satisfy these requirements and a "top-down" approach taken with the writing of the program modules. The flow chart of the system (Figure 4) displays the layered hierarchical structure of the program modules which resulted. The advantages of this structure have become very apparent during the development of the system and in solving the problems (forseen and unforseen) which have appeared. Perhaps the most obvious benefit is the ease of writing and testing each module when that particular program module is dedicated to a particular task. This not only simplifies writing and debugging but also allows addition and subtraction of modules as required. A further advantage of this design is that it allows the evolution of the system along with ideas on its use and further development. A

useful illustration of this evolutionary approach was provided when it was realised that the concept of interrogating the database could be taken beyond its main role as an information management system and used as a testing ground for geological and geochemical ideas. This further development of the system is ongoing and is described in more detail below.

The program modules are common for the entire system and therefore had to be developed to allow for variations in the structure of the database files according to the specifications applicable to each area. In practice this did not cause major difficulties since the major source of variations between areas was the suite of elements assayed for, reflecting the minerals sought in that particular area.

Usage of the system

In use, limitations and weaknesses of the system have become apparent. However, in the main, the system works well and satisfies the main design requirement i.e. to allow capture, storage and retrieval of large quantities of geological and lithogeochemical data in a user friendly way. In practice the system functions by having lithological descriptions and lithogeochemical data captured using the capture and validation modules and stored in the appropriate database files. The user of the system then uses the menu driven modules to access, display, edit, add to or print out the required data. These operations are largely transparent to the user and data corruption and loss is guarded against by the validation modules and transferral of data to back-up storage systems after a capture session.

The system functions therefore, primarily as an information management system. This applies particularly to the upper level modules, levels 1 to 4, which are, in effect, file maintenance facilities. Data can be stored and accessed using each of the field names, in level 5, which in itself provides a useful means of keeping track of the categorised data. It is from this level down however, that the system began to evolve in a more useful way when it was recognised that the selection criteria modules in level 5 could be employed to reflect current thinking on geological or lithogeochemical ideas. A simple illustration of this was outlined above where selection using the Analysis module (No.12) with As values set above a certain threshold, combined with selection employing the lithology module (No. 11) searching for the description "alteration" leads to the abstraction of a particular data set which matches the criteria imposed on the selection. The framing of this selection can be seen as the incorporation of an idea of

117

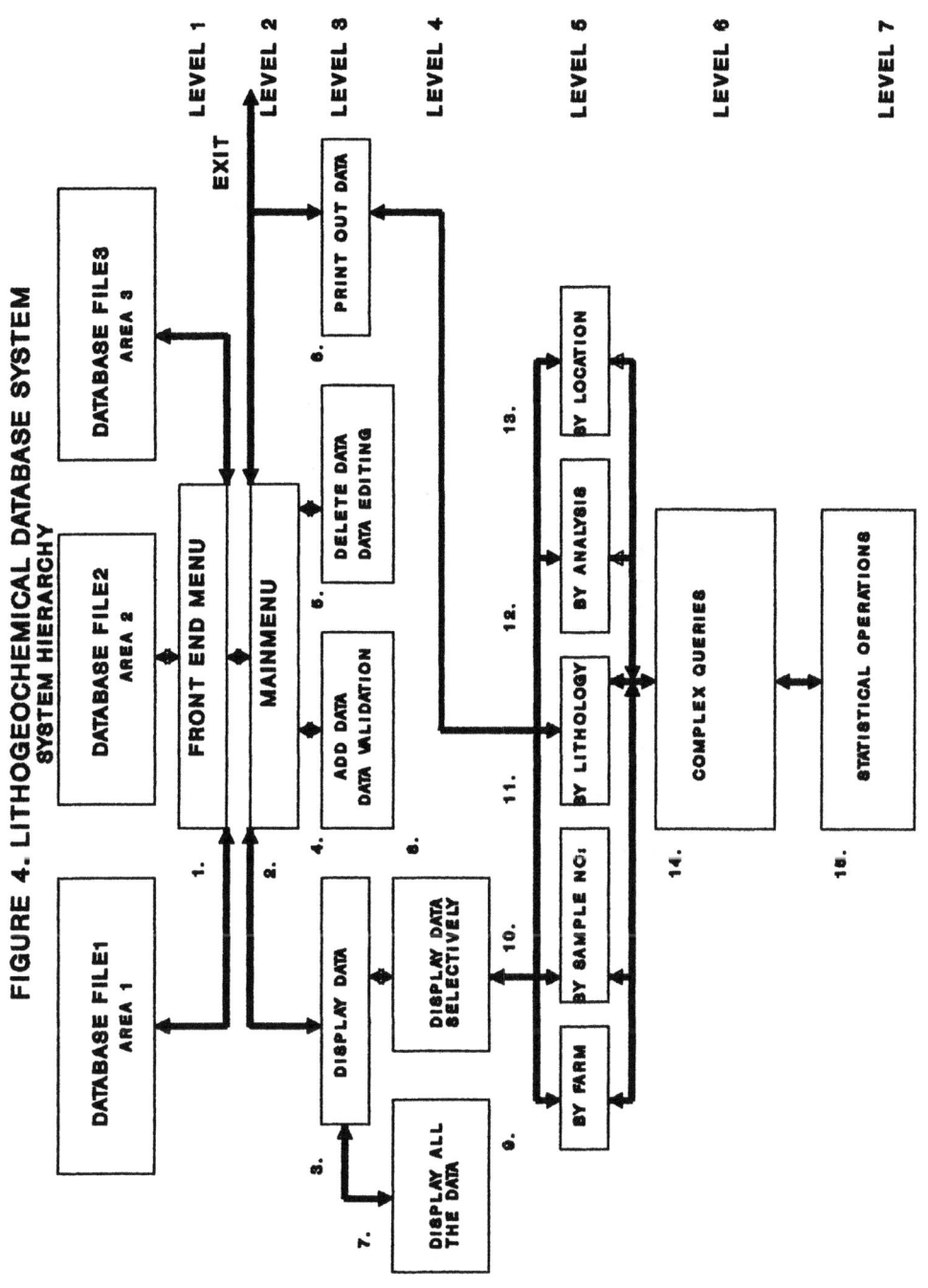

FIGURE 4. LITHOGEOCHEMICAL DATABASE SYSTEM
SYSTEM HIERARCHY

association of features, in this case, of alteration with high As values. Even such a simple combination of selection criteria can be a very useful filter to select data that support an idea as encapsulated in the selection qualifiers. Without the use of such a database and associated means of data abstraction, such simple queries would be very time consuming.

More complex queries

As noted, it was realised that using a commercially available database package such as dBASE IV in such a specific end use could impose limitations on the application. This became particularly apparent when more complex query modules came to be developed. A module to frame and select using a query such as "Display all rocks with Cu > 200ppm, As > 100ppm and Au > 300ppb that are described as marbles" requires considerable coding. This type of query is possible within the system and the modules in level 6 allow any combination of the relevant fields to act as filters of the data. In addition, the choice of any analysis field allows subsequent designation of upper and lower limits for each of the chosen analyses so that various thresholds can be applied to the selections. See Figures 5 and 6 for examples of this type of query. This use of several filtering criteria provides a useful means of scanning the data and thereby assessing the extent to which the ideas embodied in the query are supported by the data.

Obviously, care must be applied in any interpretation of such an exercise since the size of the sample population, the way in which it was selected (which includes the bias of the samplers) and the degree to which it is representative of the rock mass from which it comes will all influence the outcome of the test. Also, there is considerable debate in the literature over the extent to which rock sampling can be expected to pick up haloes or indicators of mineralisation and, indeed, the areal extent over which hidden bodies of mineralisation have a recognisable and detectable signature (1, 6). The raw data is therefore viewed as being qualitative and "anomalous" assay values are always regarded in conjunction with the associated lithological descriptions. Difficulties in interpretation of what constitutes anomalous values derive from the possible presence of several geochemical populations in the data. Multivariate statistical analysis has been employed elsewhere in an effort to discriminate this type of data and success has been claimed for this approach (7, 8, 9). Use has been made of lithogeochemical sampling to identify geochemical regions in the carbonates of the Otavi Mountain Land in Namibia and success in delineating both regions and the style

of Vanadium mineralisation has been described (10).

Nevertheless, being aware of the limitations provides a counter-weight to over-reliance on such a system and appropriate use of such complex queries is a useful aid in data examination and reduction. It would be impossible to claim that such a system can "find" mineralisation, however, the grouping of the data that results from the application of relevant selection criteria can lead to the identification of associations which would not otherwise have been apparent. Interpretation of these remains the task of the geologist concerned who must then decide on their significance and whether they represent a target to be followed up. It is important to remember too, at the interpretation stage, the objectives in mind when the actual sampling was done, e.g. the aims behind a detailed programme of sampling a layered intrusive would be very different to those influencing the sampling of an iron formation along many kilometres of strike length. This requires that use and development of such a database must be flexible since, inevitably, ideas on mineralisation models change and interpretations vary according to who is using the system.

Experience to date suggests that the more obvious large scale trends become apparent quite readily and several associations between elements and rock types, for example, can easily be detected.

PROPOSED DEVELOPMENT OF THE SYSTEM

At present the system does not incorporate modules to allow statistical examination of the entire data set or selected subsets of it. Modules to provide, e.g., a) simple ratios between selected elements from designated samples and b) discriminant analysis would obviously provide a useful and more objective means of examining subsets of the data (3, 11). A further limitation of the system at present is the lack of a digitising and plotting facility to locate and plot the data to a variety of useful scales, according to the selection criteria, if required. These enhancements to the system are being developed and it is hoped that they will further encourage its use.

FIGURE 5

```
              ENTER SELECT INFORMATION BELOW

  ENTER THE SELECTION CRITERIA YOU WISH TO USE MIN 2:MAX 5

  ENTER FARM OR LITHODES FIRST THEN ANALYSES
  SELECTION USING WHICH FARM?            JOUMBIRA

  SELECTION USING WHICH LITHODES?        MARBLE

  WHICH ANALYSIS?                        ZN

  ENTER LOWEST VALUE YOU WANT DISPLAYED?   100

  ENTER HIGHEST VALUE YOU WANT DISPLAYED? 10000

  SECOND ANALYSIS?        PB        THIRD ANALYSIS?   CU

  LOWEST VALUE?   200                   LOWEST VALUE?   100

  HIGHEST VALUE? 10000                  HIGHEST VALUE?   500
```

FIGURE 6

```
              ENTER SELECT INFORMATION BELOW

  ENTER THE SELECTION CRITERIA YOU WISH TO USE MIN 2:MAX 5

  ENTER FARM OR LITHODES FIRST THEN ANALYSES
  SELECTION USING WHICH FARM?            ERINDI

  SELECTION USING WHICH LITHODES?        MARBLE ALTERED

  WHICH ANALYSIS?                        CU

  ENTER LOWEST VALUE YOU WANT DISPLAYED?   200

  ENTER HIGHEST VALUE YOU WANT DISPLAYED?   500

  SECOND ANALYSIS?        AS        THIRD ANALYSIS?   AU

  LOWEST VALUE?   100                   LOWEST VALUE? **0.30

  HIGHEST VALUE?   500                  HIGHEST VALUE? **0.50
```

120

Acknowledgements

This paper is presented with the permission of the management of Rossing Uranium Ltd. The author would like to record his gratitude to Ken Hart, Exploration Manager of Rossing and to his other colleagues in the Exploration Department, in particular Ed Cunion who provided considerable practical assistance.

References

1. Govett G.J.S Bedrock geochemistry in mineral exploration. In: Exploration 87: third decennial international conference on geophysical and geochemical exploration for minerals and groundwater, Toronto, Canada. pp 273 - 299.

2. Govett G.J.S. Rock geochemistry in mineral exploration: Handbook of exploration geochemistry, volume 3. Elsevier, Amsterdam 1983 p. 461.

3. Garret R.G. The role of computers in exploration geochemistry. In: Reference 1 pp 586 - 608.

4. Holroyd M.T. The relevance of data base technology to resource exploration data. In: Reference 1 pp 811 - 821.

5. Pride D.E. Exploration Geochemistry An Integrative science. In: Explore. The Association of Exploration Geochemists newsletter, Number 69. September 1990.

6. Plant J.A., Hale M., Ridgway J. Developments in regional geochemistry for mineral exploration. Transactions of the IMM Section B. Vol. 97 B116 - B140.

7. Selinus O. Lithogeochemical exploration data in sulphide prospecting in Northern Sweden. Journal of Geochemical Exploration 15 (1981) pp 181 - 201.

8. Selinus O. Factor and discriminant analysis to lithogeochemical prospecting in an area of central Sweden. Journal of Geochemical Exploration 19 (1983) pp 619 - 642.

9. Amor D., Nichol I. Identification of diagnostic geochemical alteration in the wallrocks of Archean Volcanic-Exhalative Massive Sulphide Deposits. Journal of Geochemical Exploration 19 (1983) pp 543 - 562.

10. Van Der Westhuizen W.A., Tordiffe H., De Bruiyn H., Beukes G.J. The composition of descloizite-mottramite in relation to the trace-element distribution of Pb, Zn, Cu and V in the Otavi Mountain Land, South West Africa / Namibia. Journal of Geochemical Exploration 34 (1989) pp 21 - 29.

11. Barnes S.J. The use of metal ratios in prospecting for platinum group element deposits in mafic and ultramafic intrusions. Journal of Geochemical Exploration, Volume 37, 1990, pp 91 - 99.

Ruby and garnet gemstone deposits in southeast Kenya: their genesis and recommendations for exploration

R. M. Key
British Geological Survey, Edinburgh, Scotland
J. O. Ochieng
Mines and Geological Survey Department, Nairobi, Kenya

SYNOPSIS

Commercial ruby and green grossular garnet deposits in East Africa are mined directly from bedrock. Consequently exploration programmes for new deposits have to be based on a sound knowledge of the geological controls on the growth of these two minerals.

Rubies in, or immediately adjacent to, chromiferous ultramafic bodies in the Mozambique Orogenic Belt formed in areas where the regional metamorphism attained granulite facies conditions. Green garnets are also confined to these areas of high metamorphic grade as disseminations in vanadiferous graphitic schist and gneiss associated with marble. The superb body colours which makes the rubies and garnets so valuable are due to high contents of chromium and vanadium respectively, derived from their host-rocks during metamorphism.

The host-rocks are themselves ideal targets for direct and indirect prospecting methods. Exploration programmes should be based on the use of satellite imagery, geological mapping, soil and stream geochemistry, and airborne or land geophysical methods.

Current exploitation of gemstones, especially the coloured corundums (sapphires and ruby) and the various varieties of coloured garnet, has established East Africa as a major new gemstone province. In this arid area there is only local secondary dispersion of the mineralization: gemstones are mined directly from bedrock after initial eluvial extraction. Therefore the scientific search for new deposits has to be be based on a sound knowledge of the geological controls on the growth of each type of gemstone.

As such the nature of the host rock has to be known as well as the geological- and especially metamorphic-conditions that were necessary for gemstone formation. However, it is relevant to prospecting programmes to note that the first discovery of gem-quality sapphires in SE Kenya, announced in the Annual Report for 1936 of the Geological Survey of Kenya, was from soils and gravels[1]. These were locally derived from the corundum-bearing wall rock of an ultramafic intrusion.

Gem-quality ruby and garnet in East Africa formed during regional metamorphism within the Mozambique Orogenic Belt. An understanding of the controls on mineral growth is needed in order to effectively plan exploration programmes for new gemstone deposits. Four controls are apparent:[11,13,22].

1. Host rock lithology: gemstone varieties are confined to specific lithologies or to specific lithological associations such as the altered contacts of intrusive bodies.

2. Stratigraphy: host lithologies often occur within lithostratigraphic units which are mappable.

3. Metamorphism: each gemstone variety formed within a narrow range of physical conditions (pressure, temperature, activity of CO_2 and H_2O fluid phases) during regional metamorphism.

4. Chemistry: the body colours of ruby and the garnets are due to the presence of traces of certain transition group elements, especially chromium and vanadium, in the mineral matrix. Of equal importance is the absence of other transition group elements which can spoil the body colour, especially iron, traces of which can produce ugly brown tints in ruby considerably depreciating their value.

REGIONAL GEOLOGY

The late Precambrian (Neoproterozoic[34]) meridional Mozambique Orogenic Belt underlies much of East Africa between Ethiopia and Mozambique. It developed as a result of Neoproterozoic collision between a western continental plate (Tanzanian Craton) and an eastern "Kibaran" plate. Shelf sediments and oceanic volcanics were laid down during an initial extensional phase and subsequently disrupted by major folding and associated ductile shearing and thrusting. This collision-related deformation resulted in tectonic interfingering of different lithologies and lithostratigraphic units. Regional metamorphism at amphibolite to granulite facies, as well as crustal melt igneous intrusion, accompanied deformation.

At the end of the Precambrian, meridional transpressive shearing and folding was again accompanied by regional metamorphism (greenschist to amphibolite facies) with crustal melt igneous intrusion.

Regional uplift throughout Cambrian times accompanied orogenic cooling. Recent reviews of the Mozambique Orogenic Belt are given by Cahen and others[2], El Gaby and Greiling[3], Shackleton[4], and Key and others.[5].

Metamorphic and igneous rocks of the Mozambique Orogenic Belt underlie most of Kenya although exposure is confined to a central north-south belt. To the east, these rocks are covered by various Phanerozoic sedimentary strata. The extensive volcano-sedimentary sequences of the East Africa Rift System mantles much of west and north Kenya (Fig. 1).

CONTROLS ON RUBY CORUNDUM FORMATION IN SE KENYA

Rubies were first located in SE Kenya, in the Taita Hills area during 1973 by two American geologists, John Saul and Elliot Miller. These deposits at Mangari are now known to be amongst the World's richest so that East Africa may become the future centre of World ruby mining[6].

Lithological and stratigraphic control

According to Pohl and Horkel[7] the bedrock corundum deposits of SE Kenya are of four types:

1. Desilicated plumasitic pegmatites in ultrafamic bodies.

2. Desilication zones at the contacts of the ultramafics and metasedimentary country rocks.

3. In aluminous metasediments (not economically important).

4. In marbles, associated with red spinel (not economically important).

The lithological control on the economic ruby deposits is clearly the ultramafic intrusives, of which recent mapping [5,7] has shown that there are two main suites. Early sheets tectonically interleaved with metasediments (Fig. 2) are regarded as slices of ophiolite complex which pre-date two regional metamorphisms. Later discordant ultramafics emplaced into already metamorphosed rocks, pre-date the second regional metamorphism. At Mangari, the chemistry of the ultramafic hosts to the ruby mineralization indicates that they were derived from continental crust (J. Saul, pers. comm.). However, these ultramafic intrusions pre-dated the main regional metamorphism and probably belong to the early ultramafic suite. In Northern Kenya (at Baragoi), corundum (sapphire) deposits occur close to ultramafics which are definitely parts of disrupted ophiolite complexes[8].

Similar corundum deposits in central Kenya (west of the Samburu Game Reserve) have, however, formed by desilication processes associated with an ultramafic stock belonging to the younger ultramafic suite[9], indicating that corundum formation is independent of the age of the ultramafic intrusions.

In SE Kenya ultramafics are poorly exposed except in hilly country (Fig. 2). The Mangari ultramafics are not naturally exposed and are consequently not shown on the original geological map produced by the Kenyan Geological Survey[10]. The high calcium and magnesium contents of the ultramafics means that they commonly have a cover of calcrete (kunkar) as well as the ubiquitous sandy soil.

There is no stratigraphic control on ruby growth. In SE Kenya the rubies, marginal to the ultramafics, form in metasediments assigned to the Kurase Group [11, 12]. In central Kenya, the marginal corundum deposits occur in a desilicated diorite[9], and in northern Kenya they occur in various lithologies of the Siambu Complex which is a volcano-sedimentary unit.

Metamorphic control

Regional metamorphism in SE Kenya within the Mozambique Orogenic Belt reached amphibolite or granulite facies [7, 13, 14, 15, 16]. The peak temperature of the progressive metamorphism exceeded 550°C and possibly 750°C. Arneth and others[14] deduced from carbon isotope analysis of graphitic metasediments that the metamorphism took place between 550°C and 650°C. Studies of the grossular garnet (tsavorite) deposits at Mgama Ridge (Fig. 2 and next section) have established that the peak metamorphic temperature was greater than 650°C[13,15].

In a study specific to the ruby-bearing mineral assemblages at Mangari, Key and Ochieng[16] found that the metamorphism reached temperatures in excess of about 630°C. This was based on two-feldspar geothermometry. The presence of kyanite as the Al_2SiO_5 polymorph in these mineral assemblages indicates that pressures must have exceeded 7 Kbars in this temperature range[17]. Similar PT conditions have been suggested for the development of the corundum deposits further south in Tanzania[18].

At Mangari, rubies occur in assemblages of the following minerals: tourmaline, plagioclase (albite to labradorite), K feldspar, muscovite, phlogopite, margarite- paragonite and kyanite (some secondary sillimanite) with accessory amounts of graphite, xenotime, zircon, rutile, pyrite, spinel, amethyst, and arsenopyrite. Ruby exhibits marginal margarite alteration[6] and there is some sericite after feldspar. The absence of a fluid phase of the rocks possibly prevented retrogression during cooling after the granulite facies metamorphism. Local replacement of kyanite by sillimanite[7] indicates that the cooling was not isobaric but took place as pressure dropped.

Chemical control

Corundum is allochromatically coloured (ie. coloured by impurities) by traces of transition group elements. The red colour of rubies is due to the presence of chromium[19]. Rubies from SE Kenya have splendid red body colours which rival those of the best stones from SE Asia. Rubies from Mangari contain 0.39 to 0.96% by weight of Cr_2O_3[16], compared with a maximum of 0.32% $Cr_2 O_3$ in Nepal rubies[20]. Significantly the FeO values for the Mangari rubies is very low at less than 0.017% by weight[16]. The presence of TiO_2 in the ruby crystal matrix can also

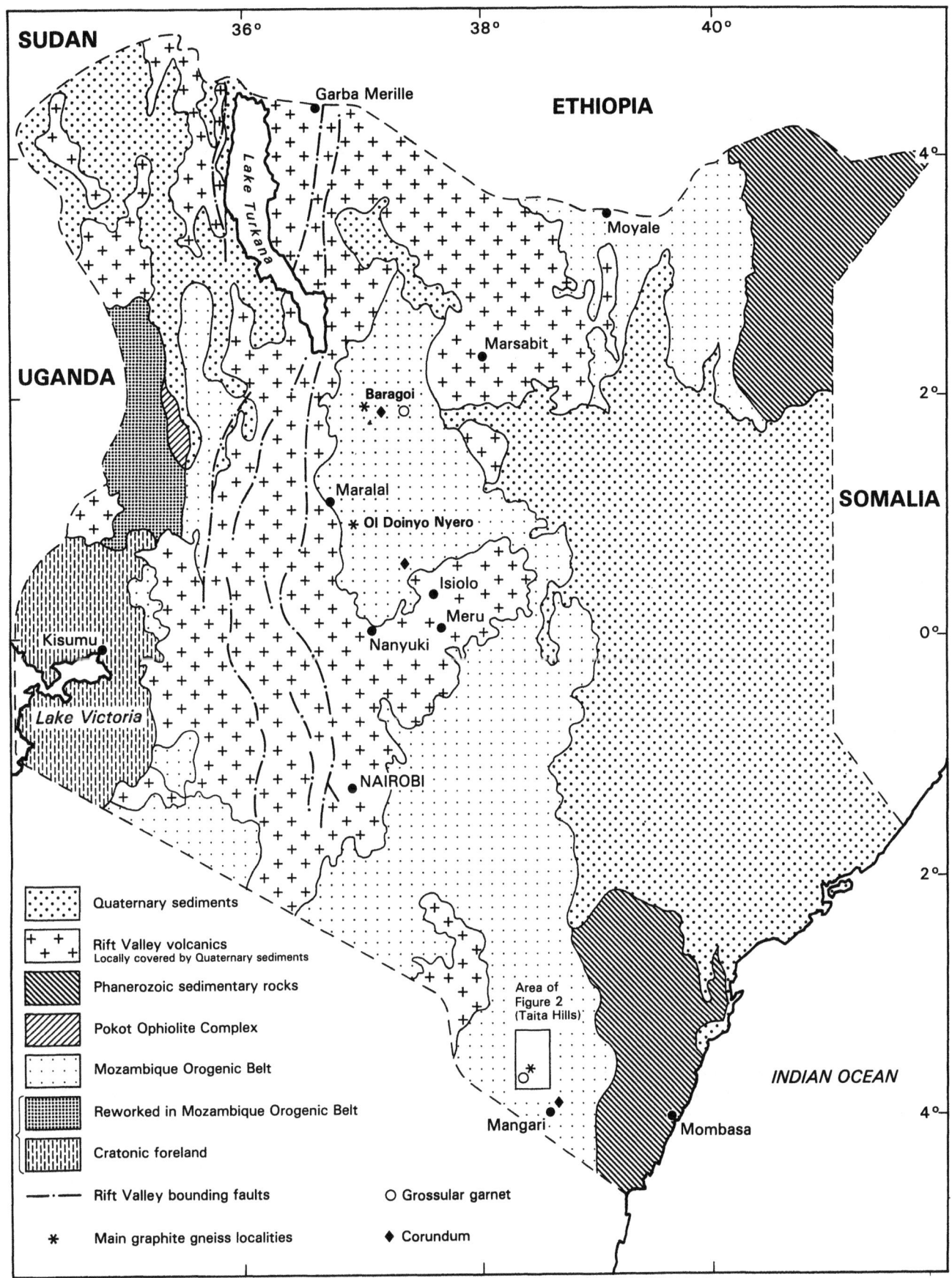

Fig. 1   Geology of Kenya with the locations of the main ruby and tsavorite deposits

dull the body colour. Nepal rubies contain up to 0.30% TiO_2, whereas, in the crystal matrix of the Mangari stones there is a maximum TiO_2 content of 0.03% by weight[16]. Most of the titanium is removed from the matrix as ubiquitous orientated, exsolved rutile needles which formed during cooling associated with lowering pressures.

The source of the chromium in the rubies is thought to be the ultramafic intrusives; similar rocks elsewhere in Kenya can host chromite pods[8]. Hughes[6] tabulates chemical analysis of rubies from the main mining areas; Kenyan, Tanzanian and Burmese rubies contain the highest Cr_2O_3 contents (all shown to contain up to 1.89% by weight).

Economic ruby deposits in Kenya clearly occur next to, or in, chromiferous ultramafics. Ruby growth by desilication reactions occurred during regional metamorphism under granulite facies conditions. The nature of the host rocks adjacent to the ultramafic pods is not important. However, the absence of fluids after the peak of the regional metamorphism was probably important as this prevented retrogressive mineral reactions during cooling.

CONTROLS OF GREEN GROSSULAR GARNET (TSAVORITE) FORMATION

Green grossular garnets, described as tsavorites by Tiffany & Co[33], were first discovered in Tanzania in the 1960's, and gem quality stones mined in the south-east of the country at Lelatema Hills[21]. In 1971, better quality garnets were also found on the border of Tanzania and Kenya. Mining on the Mgama Ridge, where the main Kenyan deposits are found, started in 1973. Four controls on the growth of the green garnets are now established[11, 13, 22].

Lithological control

The garnets are confined to graphitic schists and graphitic gneisses immediately adjacent to marble beds, or containing thin marble seams or pods, a control first noted by Bridges[22]. In the Mozambique Orogenic Belt of Kenya the host graphitic rocks form units up to 1000m thick. Individual beds wedge out laterally. In SE Kenya there is a ubiquitous association of marbles with the graphitic gneisses[11] which host the best tsavorite deposits. In central Kenya there are no marbles immediately next to the main graphitic gneisses which lack garnets[8] (see below). However, in northern Kenya, around Barogoi, although the garnetiferous graphitic rocks lack internal marble seams, there are immediately adjacent marble beds[8, 23]. The graphitic schists and gneisses are enriched in vanadium and considered to represent altered bituminous black shales [11, 13,21].

Stratigraphic control

In SE Kenya tsavorite is confined to the Lualenyi Member of a thick succession of graphitic gneisses marbles and various biotite gneisses referred to as the Kurase Group[11, 12]. In central and north Kenya a similar metasedimentary sequence is mappable over several hundred kilometres along strike as the Ol Doinyo Ngiro Gneisses[9,24]. The SE and central to north Kenya parts of the Mozambique Orogenic Belt are separated by a cover of rift valley volcanics (Fig. 1). In the Mozambique Orogenic Belt, it is impossible to correlate disconnected lithological

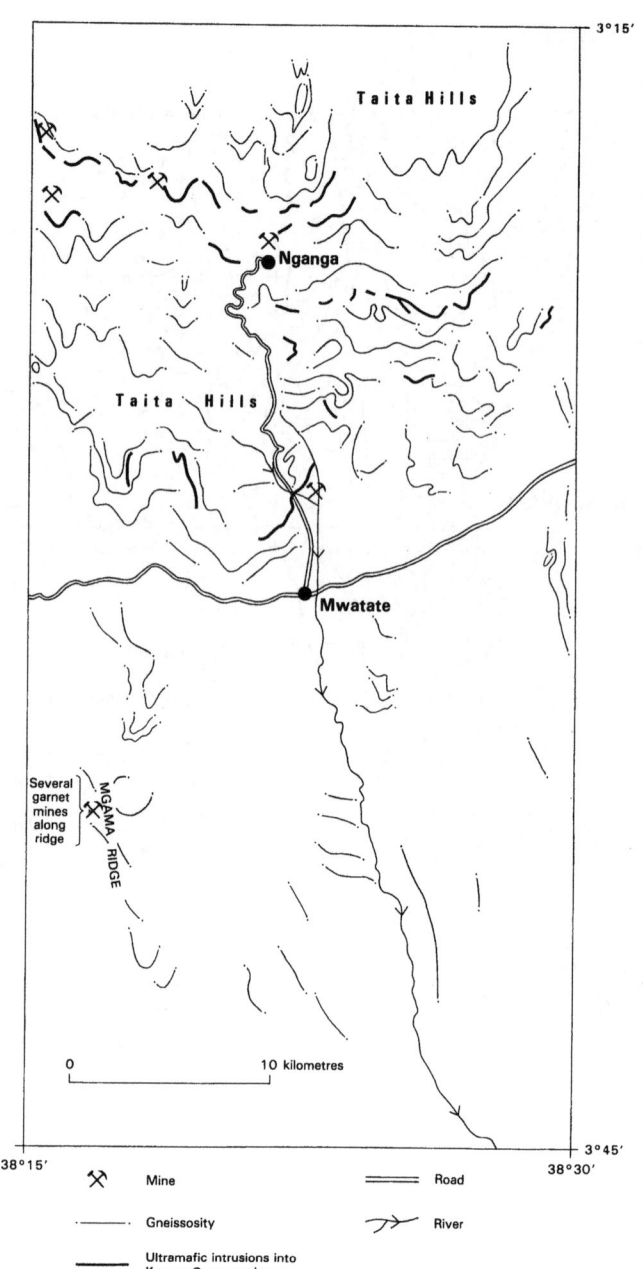

Fig. 2   Details of the geology of Taita Hills area to show the concordant ultramafic sheets in the metasediments

sequences exactly (member and formation status) due to its polyphase tectonothermal history. The Kurase Group and Ol Doinyo Ngiro gneisses are grossly correlatable as successions characterised by thick marbles and graphitic gneisses, within the metasedimentary pile of the Mozambique Orogenic Belt.

Metamorphic control

In SE Kenya tsavorite occurs in randomly distributed calcareous pods or as disseminated idioblastic crystals in graphitic gneisses. On Mgama Ridge (Fig. 2) the garnet pods are enclosed by kelyphitic rims of epidote, scapolite, quartz, clinopyroxene and spinel. Within these rims there are symplectite intergrowths of vanadiferous grossular garnet and meionitc scalpolite[13] which formed by the reaction,

Vanadium-grossular garnet + (tsavorite) + quartz + CO_2 + Na^+ = Meionite + V-enriched grossular garnet.

Experimental work suggests the reaction took place at temperatures in excess of about 750°C and 5 Kbars pressure[25, 26, 27, 28]. The garnets therefore formed under granulite facies conditions and the rims, which protected the valuable garnet pods from breakdown, formed due to an increase in Xco_2 possibly accompanied by sodium metasomatism.

The regional metamorphism also reached granulite facies in northern Kenya around Baragoi[8,23]. Two pyroxene-bearing gneisses are associated with graphitic gneisses which host small, disseminated tsavorites. No tsavorites are found in the graphitic gneisses near Ol Doinyo Nyero in central Kenya where the regional metamorphism reached only amphibolite facies[8].

Chemical control

The deep-green colour of the tsavorites is due to the presence of vanadium derived from their rock matrix[21]. The host graphitic gneisses throughout East Africa are enriched in vanadium (up to 1697 ppm V[13]). By comparison up to 5000 ppm V occurs in unaltered bituminuous shales and limestones[29]. Mean values of only 56 and 59 ppm V are recorded for USA surficial deposits[30] and rocks of the Canadian Shield[31] respectively. Vanadium is immobile during high grade metamorphism and will remain in its original host to enter the metamorphic mineral phases. The tsavorites of SE Kenya contain up to 3.30% V_2O_5, with the symplectite garnets in the kelyphitic rims containing up to 11.45%. The vanadium replaces aluminium in the general formula, $Ca_3Al_2 (SiO_4)_3$.

MINING AND FUTURE PROSPECTING

Of the two ruby mines at Mangari, developed in separate, altered ultramafics just over 3 km apart[6], an eastern open pit, known as the Penny Lane Mine, removes rubies from pockets within ultramafic slivers following mica-rich veins[32], whereas the western John Saul Mine is at the contact of a rhomb-shaped ultramafic about 200m across[6]. Plumasitic pegmatites in the contact zone are strongly enriched in ruby in association with chlorite[13]. Considerable eluvial reserves of ruby are thought to exist around the two mines[6].

The main tsavorite mines, mostly on Mgama Ridge[11] consist of small open cuts into mineralised graphitic schists and gneisses; some underground mining takes place with short inclines following the local dip. The best garnets form in irregularly spaced pods commonly associated with weathered, sulphide-bearing lenses in the host rock. There are local, near surface secondary concentrations of garnet in clayey pods which may be extremely rich e.g. at Lualenyi Mine Number 3[11].

Direct prospecting for gemstones in SE Kenya is handicapped by the small size of the deposits and the generally poor level of rock exposure. The local pasturalists nevertheless are fully aware of the value of the gemstones and adept at spotting unusual minerals both in bedrock or superficial deposits so that new occurrences are unlikely be found in exposed rocks. However, the area of poor exposure is so vast (thousands of square kilometres) that undiscovered, exploitable gemstone deposits probably still exist. Pohl and Niedermayr[11] note that the optimum method of examining the unconsolidated overburden for gemstones (and any pathfinder minerals - tourmaline, fuchsite and various calcsilicates) is to separate the 0.2 to 2.0mm fraction by wet-sieving. In the presence of water the red and green body colours of the ruby and garnet are more obvious. Soils can be examined by means of grids sited over areas targetted by other exploration methods or on regional grids.

Pilot studies on streams draining known gemstone deposits could define which fractions to analyse in the layered deposits of sand and gravel infilling the dry river beds in new prospecting areas. Maps on scales of 1:50 000 to 1:125 000 are published for much of SE Kenya revealing regional structural trends as well as the main areas of exposure, focusing attention on areas worthy of detailed, large-scale mapping for ruby and garnet exploration.

The anomalously high vanadium and chromium contents of the graphitic gneisses hosting the tsavorites are targets for soil/stream sediment geochemical exploration programmes. Similarly the ultramafics which control ruby mineralization have anomalous chromium contents as well as high nickel, copper and cobalt values. Ridgway, (in Hackman and others[9]), describes a successful geochemical exploration programme within the Mozambique Orogenic Belt of Kenya using sieved and panned stream sediments. The graphitic gneisses are also obvious targets for ground and airborne electromagnetic surveys as well as for radiometric surveys, due to their relatively high uranium contents[7]. The larger ultramafic bodies should also produce gravity anomalies.

Enhancement of high resolution satellite images is a means of locating concealed graphitic gneiss/marble and also concealed ultramafic bodies. This technique, using the computerized I^2S system has been applied to the Mozambique Orogenic Belt in northern Kenya[8]. The marbles have distinct spectral signatures which can be classified and highlighted to show their full extent - exposed and concealed beneath thin soil cover. Unfortunately, recent surface limestone (calcrete or kunkar) has a similar signature to the marbles so some ground control is essential. The ultramafics also have distinctive spectral signatures and are readily recognized on the satellite images.

Exploration for new ruby and tsavorite deposits should use as many of the exploration methods outlined here as possible. Geological maps and satellite images could be used to identify target areas worthy of detailed ground investigations. Detailed mapping involving trenching and pitting should always accompany geochemical and geophysical prospecting.

ACKNOWLEDGEMENTS

This paper is published with the permission of the Director of BGS (NERC) and the Commissioner of the Mines and Geological Survey Department, Kenya. The authors are grateful to Dr Mike Gallagher for his valuable comments on an early draft of this paper.

The British Council and the Overseas Development Administration funded much of this research.

References

1. Parkinson J. Outlines of the geology of the Mtito Andei - Tsavo area, Kenya colony. **Report Geological Survey Kenya**, 13, 1947.

2. The geochronology and evolution of Africa. Lucien Cahen and others. Oxford: Clarendon Press, 1984. 512pp.

3. The Pan-African belt of northeast Africa and adjacent areas: tectonic evolution and economic aspects of a late Proterozoic orogen. S. El Gaby and R. Greiling (editors). Braunschweig/Wiesbaden:F. Vieweg and Sohn, 1988. 369pp.

4. Precambrian collision tectonics in Africa. R.M. Shackleton. In: M.P. Coward and A.C. Ries (editors), **Collision tectonics.** Oxfords: Blackwell, 1986. p. 329-349.

5. Key, R.M. and others. Superimposed Upper Proterozoic collision controlled orogenies in the Mozambique Orogenic Belt of Kenya. **Precambrian Research,** 44, 1989, p. 197-225.

6. Corundum. R.W. Hughes. London: Butterworth-Heinemann, 1990. 314pp.

7. Pohl W. and Horkel A. Notes on the geology and Mineral Resources of the Mtito-Andei-Taita area (Southern Kenya). **Mitteilungen der Osterreichischen geologischen Gesellschaft, 73,** December 1980, p. 135-152.

8. Key, R.M. Geology of the Maralal area. **Report Mines and Geological Survey Department, Kenya,** 105, 1987. 93pp.

9. Hackman B.D. and others. Geology of the Isiolo area. **Report Mines and Geological Survey Department, Kenya,** 104, 1988. 88pp.

10. Walsh J. Geology of the area south of the Taita Hills. **Report Geological Survey of Kenya,** 49, 1960.

11. Pohl W. and Neidermayr G. Geology of the Mwatate quadrange (Sheet 195/2) and the vanadium-grossularite deposits of the area. **Report Kenya-Austria Mineral Exploration Project,** 11, 1978. 89pp.

12. Saggerson E.P. Geology of the Kasigau-Kurase area. **Report Geological Survey of Kenya,** 51, 1962.

13. Key R.M. and Hill P.G. Further evidence for the controls on the growth of vanadium grossular garnets in Kenya. **Journal of Gemmology,** 31, 1989. p. 412-422.

14. Arneth J.D. and others. Graphite content and isotopic fractionation between calcite-graphite pairs in metasediments from the Mgama Hills, southern Kenya. **Geochimica and Cosmochimica Acta,** 49, 1985. p. 1553-1560.

15. Sarbas B. and others. Zur genese ostafrikanischer Grossular vorkommen. **Zeitschrift der Deutschen Gemmologischen Gessellschaft,** 33, 1984. p. 48-62.

16. Key R.M. and Ochieng J.O. The growth of rubies in south-east Kenya. **Journal of Gemmology,** 1991 (in press).

17. Holdaway M.J. Stability of andalusite and the aluminosilicate phase diagram. **American Journal of Science,** 271, 1971. p. 97-131.

18. Altherr R. and others. Corundum and kyanite-bearing anatexites from the Precambrian of Tanzania. **Lithos,** 15, 1982. p. 191-197.

19. Gems. Their sources, descriptions and identification. R. Webster. 3rd edition. London: Butterworth. 938pp.

20. Harding R.R. and Scarratt K. A description of ruby from Nepal. **Journal of Gemmology,** 20, 1986. p. 3-10.

21. Gubelin E.J. and Weibel M. Green vanadium grossular garnet from Lualenyi, near Voi, Kenya. **Lapidary Journal,** 29, 1975. p. 402-414 and 424-426.

22. Bridges C.R. Green grossular garnets (tsavorites) in east Africa. **Gems and Gemmology,** 1974. p. 290-296.

23. Baker B.H. Geology of the Baragoi area. **Report Geological Survey of Kenya,** 53, 1963. 74pp.

24. Shackleton R.M. Geology of the country between Nanyuki and Maralal. **Report Geological Survey of Kenya,** 11, 1946. 54pp.

25. Baker J. and others. Corona textures between kyanite, garnet and gedrite in gneisses from Errabiddy, Western Australia. **Journal of Metamorphic Geology,** 5, p.357-370.

26. Warren R.G. and others. Wollastonite and scapolite in Precambrian calc-silicate granulites from Australia and Antartica. **Journal of Metamorphic Geology,** 5, 1987. p. 213-223.

27. Oterdoom W.H. and Gunter W.D. Activity models for plagioclase and CO_3-scapolites - an analysis of field and laboratory data. **American Journal of Science,** 283-A, 1983, p. 255-282.

28. Goldsmith J.R. and Newton R.C. Scapolite-plagioclase stability relations at high pressures and temperatures in the system

$NaAlSi_3O_8$ - $CaCO_3$ - $CaSO_4$. **American Mineralogist**, 62, 1977. P. 1063-1081.

29. Wedepohl K.H. Untersuchungen am Kupferschiefer in Nordwest Deutschland. Ein Beitrag zur Dentung bituminoser sedimente. **Geochimica and Cosmochimica Acta**, 28, 1964.

30. Shacklette H.T. and others. Elemental composition of surficial materials in the conterminous United States. **U.S. Geological Survey Professional Papers**, 574-D, 1971.

31. Eade K.E. and Fahrig W.F. Regional, lithological and temporal variations in the abundance of some trace elements in the Canadian Shield. **Geological Survey of Canada Paper**, 72-46, 1973.

32. Bridges C.R. Gemstones of East Africa. **Proceedings International Gemmological Symposium**, 1982. p. 266-275.

33. Gem Testing. B.W. Anderson. London: Butterworths, 1980, 434pp.

34. Key, R.M. Classification of the Earth's oldest rocks. **NERC News**, 1989, 10, p8.

Geology and exploration of the Bokitiso Concession, Central Region, Ghana

D. D. Chikohora
Cluff Mining Ghana, Kumasi, Ghana

SYNOPSIS

The Bokitiso Concession is located ten miles west of Dunkwa-On-Offin in the Central Region of Ghana. The country rocks are mainly phyllites, greywackes and tuffaceous beds believed to be of the lower Birimian. These rocks strike northeast and dip steeply to the southeast. The phyllites have been intruded by granitoid porphyries which occur as lenticular bodies. The major structures in the concession have a northeasterly trend and are defined by a series of enechelon quartz veins infilling fissures.

The first registered title holder on the property was Bokitiso Goldfields Limited who commenced operations in 1900. However, by 1905 the company had gone bankrupt. The Atta Gold Company Limited then operated the property from the early 1930's through to 1947. Little production detail is available about these operations.

Cluff Taywood Mining Limited a joint-venture partnership between Cluff Resources PLC and Taylor Woodrow International, both of the United Kingdom became involved through an agreement with Canon Farms and Mining Industry, a Ghanaian Company who are the holders of the prospecting licence. Detailed exploration commenced at the beginning of October 1988. At the end of 1988 Cluff Resources PLC purchased Taylor Woodrow International's interest in Cluff Taywood Mining Limited and the company changed its name to Cluff Mining Ghana.

Exploration focused on two aspects:-

1. Geological mapping, surface trenching and sampling of known gold occurrences in the Concession with a view of elevating the viable prospects to drilling stage. In less than six months a decision was reached to drill two targets: the old Bokitiso prospects and the Esuaja old workings.

2. Regional studies on the concession were undertaken in order to locate previously unknown gold mineralisation. This was achieved by regional soil sampling, aerial photography, use of local knowledge aand old records and general reconnaissance. To date a total of ten prospects have been identified. Three of these are now at feasibility drilling stage, one is at prefeasibility drilling stage, another three at trenching and sampling stage and the remainder proved not viable.

Details of the exploration techniques are described.

The company has established a Field Camp near Ayanfuri where Expatriate and Ghanaian staff are accommodated. Despite the previous 'Galamsey' history in Ayanfuri village, a sound working relationship has been established between the company and the inhabitants of Ayanfuri as a result of the company's willingness to assist in community development projects.

INTRODUCTION

Location and Physiography

The Bokitsi licence area of 40km² is located ten miles west of Dunkwa-On-Offin (Fig.1) in the Central Region of Ghana. The area is one of high rainfall. Over the principal portion of the prospect the bush is short and thin having been thoroughly cleared in the past. The area is generally of low relief with numerous streams and creeks that cut across the general trend of the country rocks.

Previous Work and Results

The first registered title holder on the property was Bokitiso Goldfields Limited who commenced operations in 1900. However, by 1905 the company had gone bankrupt. The Atta Gold Company Limited then operated the property from the early 1930's through to 1947.

No production has been recorded during the time. Extensive underground development was carried out during the same time of exploration by the Atta Gold Company. One 300ft inclined and two vertical (total 440ft) shafts were sunk on the Main reef. From these shafts a total of 1800m of drives were developed on 6 levels.

The records indicate that this development outlined 93,000 tons of reserves at a grade of 9.5 g Au/t between levels 1 - 3. Additional prospective reserves of between 60,000 - 70,000 tons of relatively lower grade were blocked out between 3 and 5 levels.[1]

GEOLOGY

The country rocks are typical phyllites, greywackes and tuffaceous beds believed to be of the lower Birimian succession of Proterozoic age.[2] These rocks strike northeast and dip steeply to the southeast and have undergone low grade metamorphism. The phyllites have been intruded by granitoid porphyries which occur as lenticular bodies (Fig.2). The intrusion of the granitoid bodies appear to have been controlled by the northeast-southwest deep seated strike faults.[3] In the northwestern edge of the map, the phyllites form erosion resistant ridges possibly due to higher degree of metamorphism.

The Granitoid Bodies

These are coarse grained with euhedral plagioclase and irregular quartz surrounded by a fine grained ground mass, in places forming a well defined foliation. The plagioclase is of albite to oligoclase composition with inclusions of white mica and carbonate. In thin section, the ground mass surrounding the quartz and feldspar appears as finely grained randomly orientated, white mica, feldspar and quartz. No biotite has been observed in the limited thin sections made. However, in some hand specimen an evenly distributed pale mineral believed to be biotite at this stage has been observed.

The intrusive nature of these bodies is readily demonstrated by a well defined contact aureole usually rich in andalusite. Work done so far suggests a protolith of sodic plagioclase, potash feldspar and quartz composition. The potash feldspars have been preferentially altered to white mica with time. This mineralogy suggest a protolish of adamellite to granodiorite composition. The mica and the carbonate could have been a result of weathering or hydrothermal activity.

The mineralisation is generally related to quartz veins, stringers, stockworks and disseminations within the granitoid. The alteration which is easily observed in trenches has been predominantly sericitisation and carbonatisation.

WORK DONE

Introduction

The work programme by CLUFF MINING GHANA was divided into two phases.

Phase 1

The initial work in phase 1 concentrated on two aspects; assessment of old workings and regional studies to ascertain overall trends within the Bokitiso Licence Area.

131

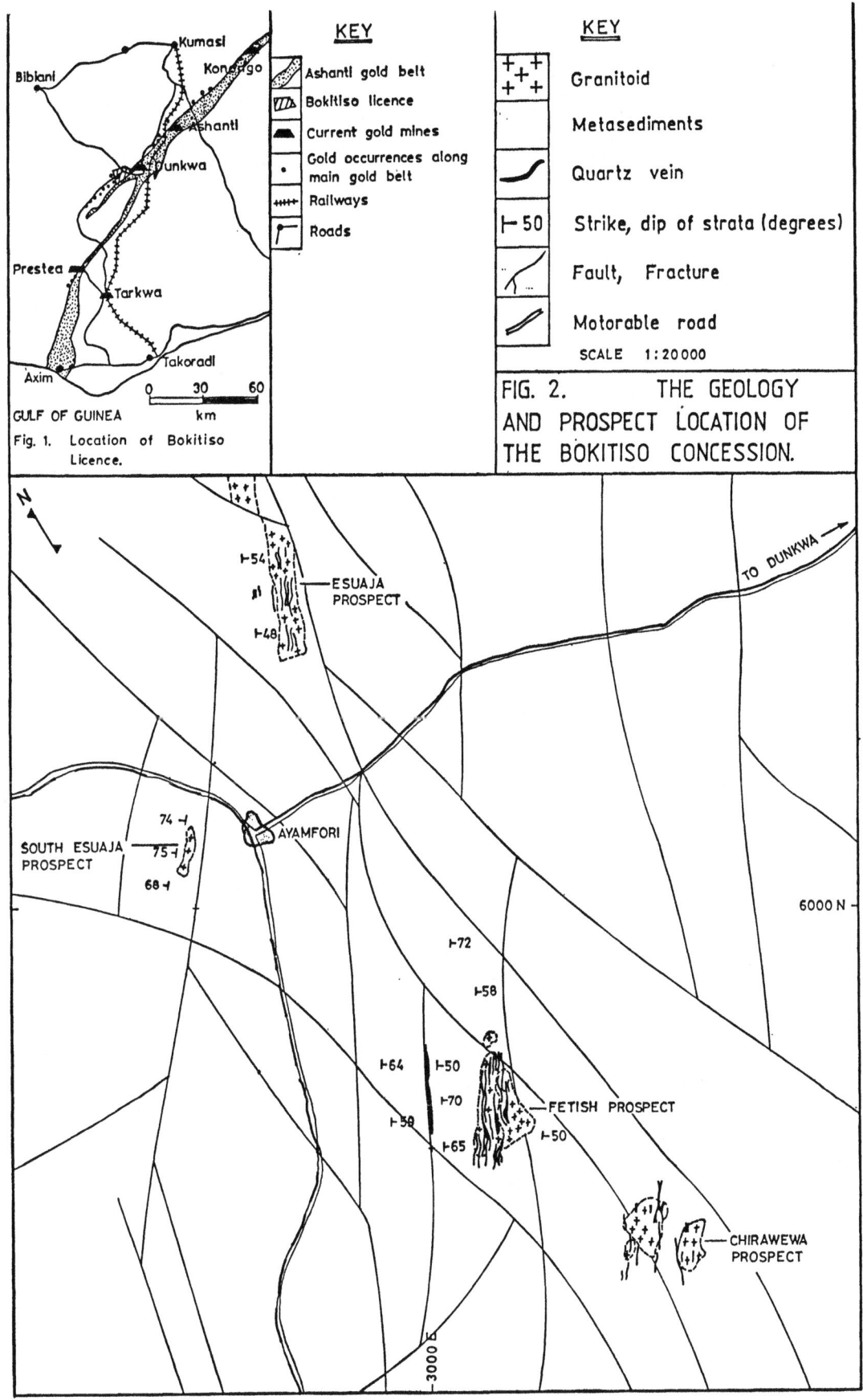

KEY

- Ashanti gold belt
- Bokitiso licence
- Current gold mines
- Gold occurrences along main gold belt
- Railways
- Roads

Fig. 1. Location of Bokitiso Licence.

KEY

- Granitoid
- Metasediments
- Quartz vein
- ⊢50 Strike, dip of strata (degrees)
- Fault, Fracture
- Motorable road

SCALE 1:20 000

FIG. 2.   THE GEOLOGY AND PROSPECT LOCATION OF THE BOKITISO CONCESSION.

Assessment of old workings

This included sketch geological mapping, selective grab sampling of dumps, outcrops and tailings initially on the Bokitiso gold prospects and subsequently on the Esuaja and Mitchell's Adit gold prospects. Old pits in the vicinity of these prospects were cleaned, logged and sampled.

An extensive trenching and sampling programme was then initiated on the basis of the results of the initial work. This programme concentrated on the Bokitiso and Esuaja gold prospects.

Line Cutting

Initially a 1000m x 500m base line was established over the old Bokitiso gold mine and the Esuaja prospect. This was later extended to cover most of the Licence Area for regional studies. Using compass and tape a base line was cut (045° magnetic) parallel to the strike of the ore zones at the old Bokitiso prospect. Section lines were staked every forty metres at right angle to the established baseline, 3000E. The section lines were numbered at 80 metre intervals increasing northwards from the 5000N section line.

Beyond the 1000 x 500m baseline over Bokitiso and Esuaja, section lines were staked every 240m. A team of three men could cut up to 500 metres a day through the secondary bush at Bokitiso and the line cutting was on a contract basis. Every square metre of bush cleared attracted a payment of 3 cedis, This proved very popular and a total of 56.9 kilometres of line cutting were completed during the six months. An additional two kilometres of old roads clearing and rehabilitation were completed for vehicular access to the greater part of the Licence Area during the regional prospecting phase.

The established grid formed the frame-work of the detailed prospect assessment and regional prospecting.

Geological Mapping

The rock outcrop in the Bokitiso Licence Area is poor and the initial geological mapping was concentrated on the old workings in the area and was plotted on 1:1000 field map. The data were later combined with trench mapping information to produce a provisional geological map over the old Bokitiso gold mine.

Main Zone

The main reef which was extensively developed underground outcrops on top of an ill-defined ridge. The weathered outcrop has been extensively mined by illegal gold miners. During the early phase of the programme, examination of such outcrops was, at times obstructed by the activities of these workers and in order to combat this, a team of security guards was engaged to patrol the main "Galamsey" activity area. The outcrop which measures up to 10 metres in width can be traced for a distance of about 320 metres. This is composed of quartz vein up to 5 metres thick occupying a fissure running parallel to the strike of the phyllites with an average dip of 60° to the southeast.

The footwall and hanging wall of the quartz vein is composed of phyllite liberally veined with quartz which forms stockworks. The weathered outcrops of the phyllites have a pock-marked appearance, due to weathered pyrite and arsenopyrite. Gold occurs as coarse and finely disseminated grains in the quartz veins and in phyllite. Assay results from the channel samples gave an average grade of 5.68 g/t over 10.8m over 320 strike length using a cut off grade of 1 g/t.

Fetish Zone

This is a parallel zone dipping 60° southeast and about 240 metres in the hanging wall of the Main Zone. It is composed of enechelon quartz veins up to 5 metres wide with footwall and hanging wall quartz vein stockworks up to 10cm thick. Pyrite and arsenopyrite occur ubiquitously in the zone. Associated with this is a granitic body which in places hosts low grade disseminated mineralisation. The zone is extensively pitted and where accessible, the pits were grab sampled before trenching commenced in areas of potential.

Esuaja Porphyry

This is exposed as a ridge 500 metres northeast of the Ayanfuri village. The ridge with a 035° strike (magnetic) extends for about 400 metres and is composed of a lenticular granitoid porphyry with quartz veins whose strike range from 035° to 050° (magnetic). The quartz veins range from being flat lying to gently dipping to the southeast. Zones of quartz veining appear on top of the ridge and are associated with high gold values.

The porphyry has an intrusive contact with the phyllites and appear to have been intruded along the phyllite bedding planes. Trench assay values across the ridge yielded an average grade of 1.3 g/t over 70 metres along 300 metres strike length.

Phyllites

This term refers to greywackes, and ashy beds in the area. These range from grey to black rocks with little bedding to well bedded ashy horizons with a brown purplish colour. This forms lateritic cappings on ridge slopes in the Licence Area where the phyllite is resistant to erosion. The phyllites exhibit a strong crenulation cleavage which forms a lineation with a 105° trend (magnetic).

Throughout most of the Licence Area, the phyllites are spotted and show recrystallisation near the granitic porphyries. Arsenopyrite and pyrite are ubiquitous in zones within metres of the porphyry-phyllite contacts.

Metamorphisms

The granitic intrusions resulted in low grade contact metamorphism in the area. The contacts range from gradational wide zones of fine grained contact hornfels to sharp thin zones of coarse grained contact hornfels with large acicular crystals of andalusite.

Alteration

The quartz veining is associated with pervasive sericite and carbonate alteration.

Grab and Selective Sampling

This was undertaken over most of the old workings in the Licence Area. The assay results were used as an indication of the potential of the various zones. The trenching programme which followed was to establish the widths and strike of the mineralized zones. Where possible old workings (adits and pits) were rehabilitated, mapped and grab or channel sampled. In places grab samples were collected in duplicate and one was sent for assay while the other was panned for coarse gold count. The panning results provided a quick, though limited, indication of the potential of an area.

Main and Fetish Zone

Grab and channel selective samples were collected from the exposed main zone. In the old "Galamsey" pit, 2 metre 3-5kg channel samples were collected. Where possible short trenches were dug to expose both the footwall and hanging wall of the exposed quartz veins before sampling. Duplicate samples were collected after every twenty samples.

On the Fetish Zone where the reef was not exposed on the surface, grab samples were only collected from old pits.

Mitchell's Adit

An adit more than fifty metres long was rehabilitated before 3-5kg grab samples of the representative lithologies were collected for assay.

Esuaja Prospect

Accessible old pits were cleaned, mapped and channel or grab sampled. In addition soil samples up to 5kg were scooped from near surface.

Results of Selective Sampling

Encouraging results were obtained from the Fetish and Main Zones on the old Bokitiso gold mine, and the Esuaja old workings. Results from Mitchell's Adit did not warrant immediate further work.

Trenching

On the basis of the results of the selective sampling, a trenching programme was drawn up to cover the Bokitiso gold mine and the Esuaja prospect. Initial trenches were sited adjacent to sample locations with encouraging results and at right angles to the strike of the ore zones.

Later during the programme, trenches were located every 80m on a grid centred on the Bokitiso gold mine and the Esuaja prospect. This phase of trenching covered an area of 500m x 500m and 500m x 250m at Bokitiso and Esuaja respectively.

Infill trenches were sited every 40 metres depending on results. Trenching was done by contract manual labour using pick-axes and shovels.

Trenches were dug to bedrock or to a maximum of three metres where overburden was thick. Generally in the area, bedrock was reached at about two metres depth. To date a total of 6334 metres have been excavated.

Trenching Logging and Sampling

Trenches were mapped and information plotted on 1:1000 plans. Most of the trench logging was done immediately after the trenching was completed before the heavy rains in the area flooded the trenches.

Trench side walls were cleaned and levelled using a pole digger or broad chisel. A horizontal channel, one inch by two inches was then cut using a sharp chisel and samples collected in rectangular plastic containers before being transferred to polythene bags and labelled. Samples were collected over two metre intervals and channels were tailored such that sample weights averaged 3-5kg for the interval. Where lithological changes dictated, shorter sample intervals were used but the channel size increased so that the sample size was constant. Duplicate samples were collected routinely to test for variation in results due to inherent variation in the rocks and also to monitor the laboratory's precision. Samples were stored at the Bokitiso Field Camp before shipment overseas for analysis.

Sample Preparation and Analysis

The 3-5kg samples were airfreighted to Omac Laboratories, Ireland for analysis. This was undertaken at a cost of seventeen pounds sterling (£17.00) per sample inclusive of airfreight, sample preparation and analysis. In the laboratory, the sample were treated as follows:

DRIED
↓
PULVERISED TO MINUS 1MM
↓
RIFFLE SPLIT TO 1KG
↓
PULVERISED (TO -200 MESH)
↓
RIFFLE SPLIT TO 50 g
↓
FIRE/AAS AU ANALYSIS

REGIONAL STUDIES

The base line was extended to cover the greater part of the Licence Area. Section lines were staked every 240m and were used in the soil sampling programme.

Soil Sampling

An orientation soil sampling programme was designed to determine:

i. The minimum sample size,

ii. Soil anomaly expression.

Four samples were taken at each site, ten metres apart over hundred metres adjacent to known trench anomalies. The hundred metres traversed both mineralised and barren ground. Two traverses, ten metres away from trenches (to avoid contamination) were undertaken. The following samples were collected at each site from a depth of thirty centimetres:

3 kilograms, one kilogram, half a kilogram and a small sample packet.

At the laboratory the whole samples were dried, pulverised, split and fire assayed for gold. The results of the survey are being used in the regional soil sampling at Bokitiso.

Aerial Photography

Air Charters Limited was engaged to undertake Aerial photography of part of the Bokitiso Licence Area at a cost of US$159.00 per square kilometre. The work produced aerial photographs at a scale of 1:20 000. These were used to establish regional trends and structures in the area.

RESULTS

Trench Sampling Results

The sample location and assay results were used to calculate trench intersections using a cut off grade of 1 g/t. The assay values were weighted by the sample widths to get the average grade along the trench intersecions.

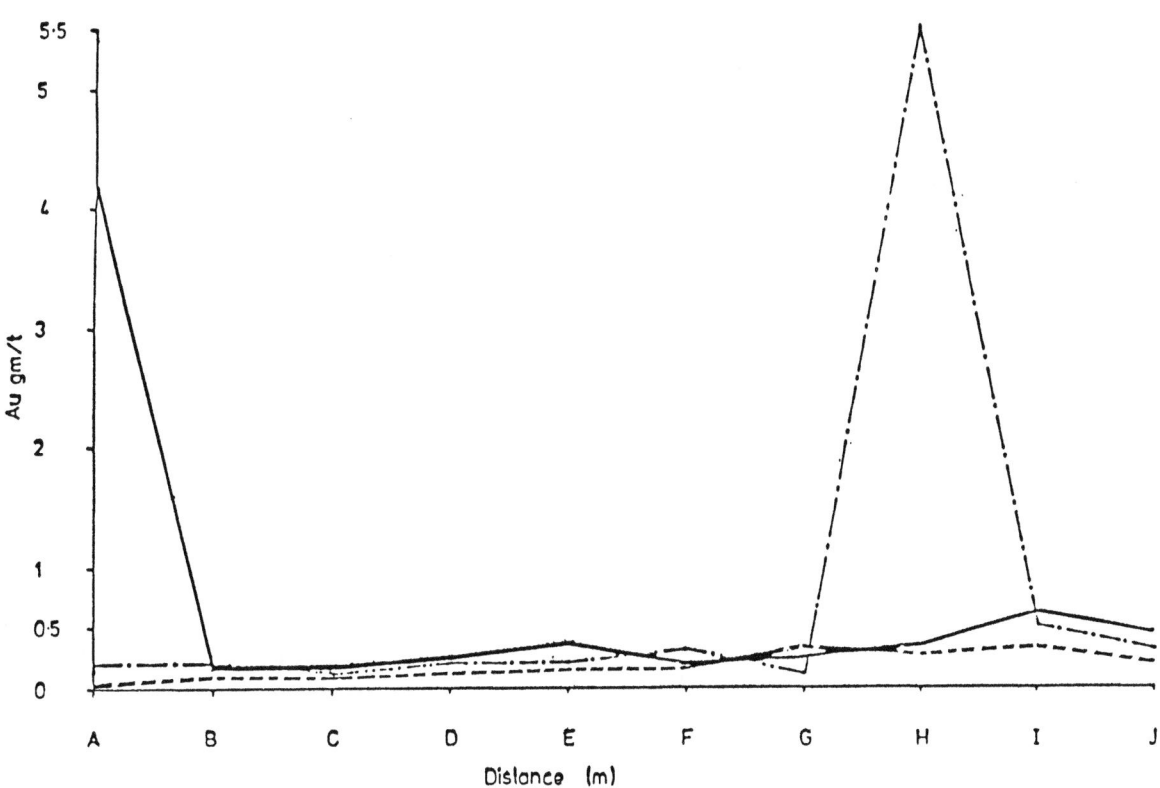

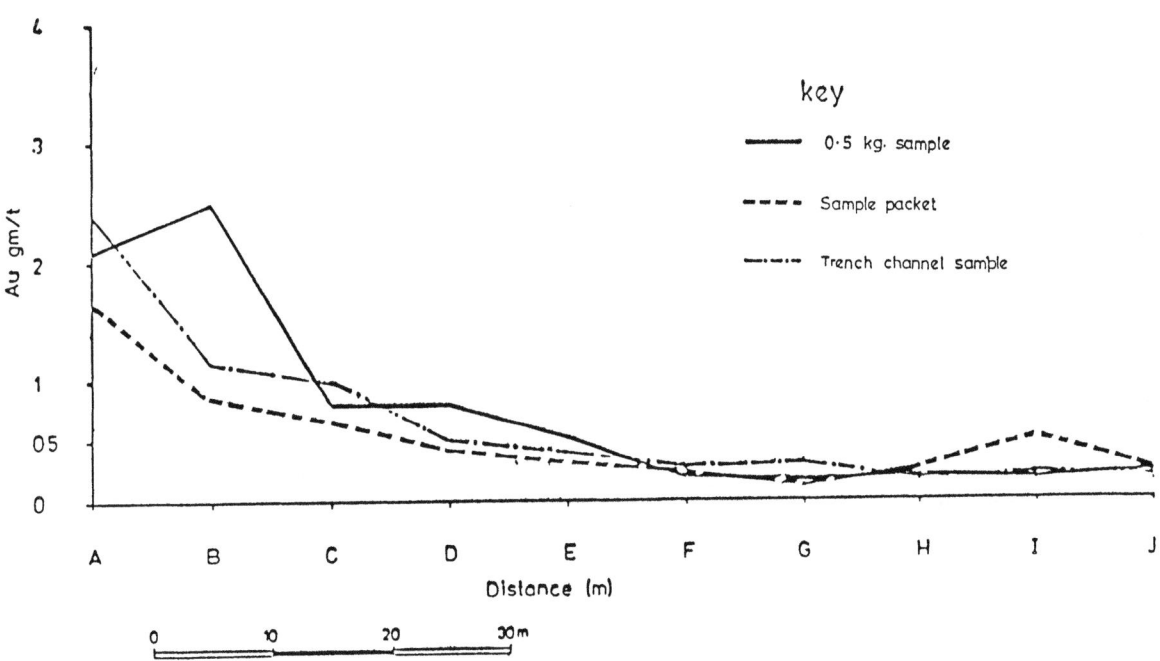

Fig. 3a . Soil sampling orientation survey results .

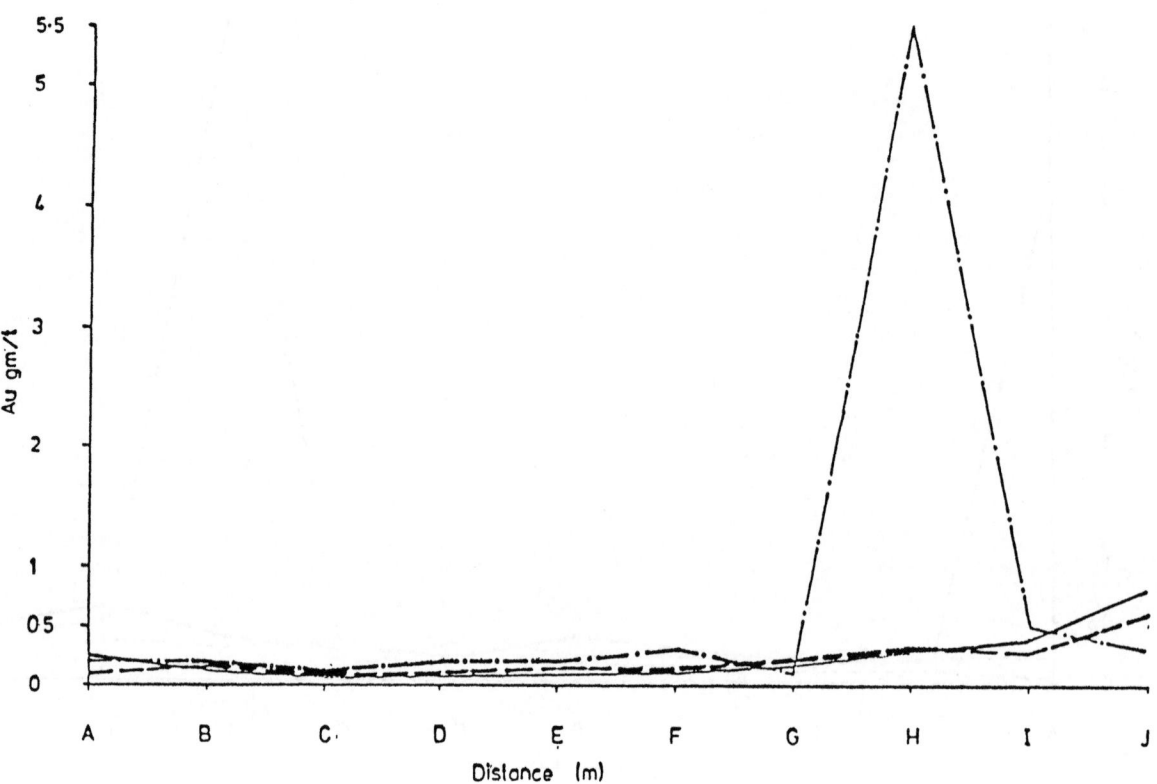

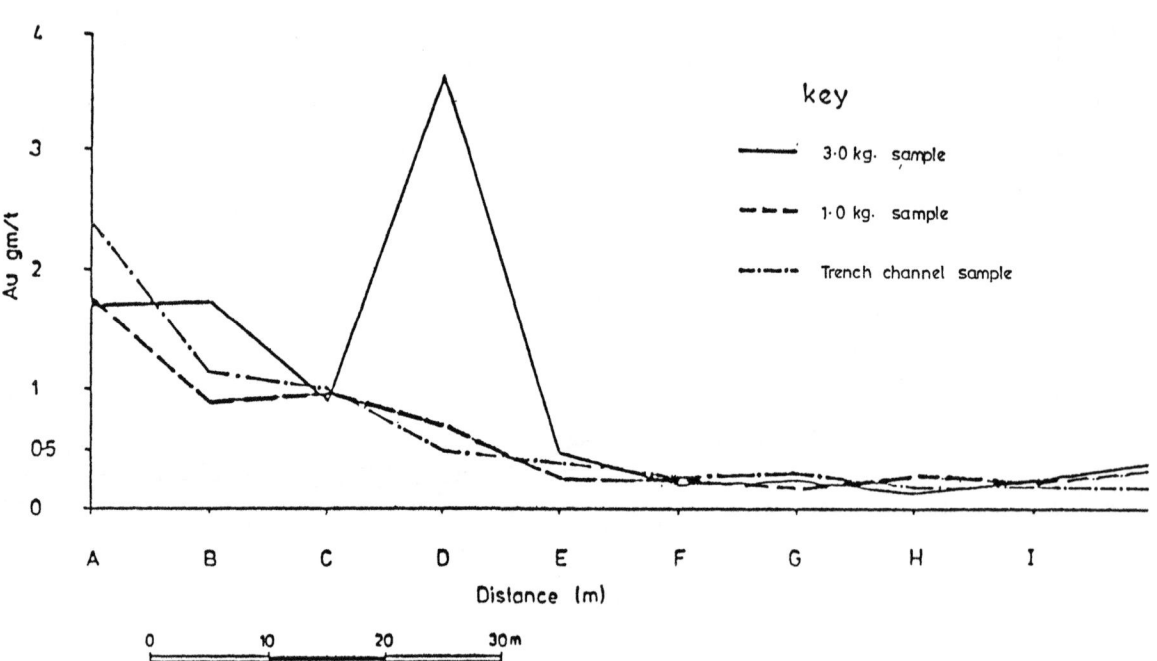

Fig. 3b. Soil sampling orientation survey results.

Assay results were contoured to produce a contour map at an interval of 0.5 g/t.

DISCUSSIONS

BOKITISO PROSPECT

The assay contour map indicated two main zones: the Main zone and the Fetish zone.

Main Zone

This has an indicated strike length of 320m grading 5.68 g/t over an average width of 10.8m. The inferred strike length is 500m.

Fetish Zone

The mineralisation is granitoid hosted and at its widest intersection it has an average grade of 1.89 g/t over 84.7m. This has been extensively pitted in the past and some of the pit grab samples yielded the following:-

PIT	DEPTH OF SAMPLE (M)	ASSAY (g/t)
1	4	12.24
1	7.5	24.11
2	30	30.88
3	7.5	6.60

These few grab samples results, though inconclusive, suggest a better grade with depth. This will hopefully be resolved by the proposed drilling in phase 2 of the work programme.

Esuaja Prospect

The Esuaja trenching has indicated a mineralised porphyry with an average surface grade of 1.3 g/t over 70 metres and has a strike length of 300 metres with both ends still open.

Orientation Soil Sampling

The results indicated that at a depth of thirty centimetres 0.5 kg sample size provides good contrast between mineralised and country rock. These results are being used in regional soil sampling programme at Bokitiso, Fig.3 a and b.

CONCLUSION

The initial phase of detailed prospect assessment and regional studies yielded four drilling targets and another three targets for detailed sampling on the Bokitiso Concession.

These results were achieved through dedicated team work among the field based staff and total commitment of the company to Ayanfuri community development programmes.

ACKNOWLEDGEMENTS

Thanks are due to all Cluff Mining Ghana employees whose hard work has made it possible to put together this paper.

REFERENCES

1. The Atta Gold Mines Reports 1928. (unpublished).

2. SERVICE, H. (1937) Geological Reports on the Bokitiso Concession in the Dunkwa District, Gold Coast Colony. (unpublished report).

3. Boshier, PR. (1989). Provisional Photogeological interpretation of the Ayamfuri Area, Ghana (unpublished report).

Metallurgy 2

Pre-designed modular plants for small-scale mining operations in Third World countries

R. Abate
Van Eck & Lurie (Pty) Ltd, Johannesburg, South Africa

SYNOPSIS

In this paper we have attempted to describe the development of the Modular Plant concept which has come about mainly from the development of Modular Heavy Media Separation (HMS) units for the Diamond Mining Sector which have been developed by Van Eck & Lurie (VEL) over the past 25 years, and which have been proven to be very successful in Africa and other areas of the world.

WHAT IS A MODULAR PLANT?

A Modular Plant is a self contained-unit or a number of self-contained units or modules which when assembled in the correct sequence make up a small scale treatment facility or plant.

The size of the module is normally determined by transport limitations and normally governed by road transport regulations and dimensions.

In some cases these modules have been designed to be totally dismantled enabling them to be packed into standard shipping containers. The concept behind the modular plant design is to transport the largest possible module with as much equipment pre-installed in it as practically possible, resulting in minimal re-erection time on the plant site.

DEVELOPMENT TO DATE

In 1967 VEL developed its first Modular Plant. The unit was a 10 tph HMS mobile plant mounted on a trailer ordered by Brazil Diamante in Cape Town and was to be used for prospecting in diamondiferous gravel in the Hondeklipbaai area in north-western Namaqualand. The fee associated with this order was R4 000 (US$4 800). (Refer Fig 1).

From this humble beginning it was realised by VEL that there was a need in the marketplace for transportable or modular HMS plants or units. Since then VEL has supplied some 90 units, either self-contained or associated with complete installations. (Refer Fig 6). Seventy per cent of the plants designed and supplied have been into Africa.

The modular framed concept developed into a full range of suitably sized units. As mentioned previously, the modular concept idea was based on being able to fit as much equipment as possible into a transportable framed structure. It was along these lines that a full range of HMS modular units were developed. A modular X-Ray Diamond Recovery Plant was developed to complement the range of modular HMS units. This modular X-Ray Recovery Plant utilises a 10w Wet Sortex x-ray machine, eliminating the requirement of the HMS plant gravel concentrate to be dried.

A simple modular jig prospecting plant was designed and developed complete with variable speed scrubber, water supply pump, Pleitz jigs and power generation. In other words, a fully-contained plant for prospecting of diamonds.

At a later stage a gold concentrator was added to the above unit to recover gold from the -3mm fraction. This plant is particularly suited for the small scale mining operator for the exploration and recovery of diamonds and/or gold and, in particular, alluvial deposits in Africa. The unit is simple to operate and is ideally suited for a small unsophisticated operation with low operating costs.

The full range of modular plants which have been developed to date are:

Modular HMS Units 5, 10, 15/20, 30, 40 and 50 tph units.

Modular Diamond Recovery X-Ray Plant (including 1 x 10w Wet Sortex machine).

Modular Diamond Recovery X-Ray Plant (including 2 x 10w Wet Sortex machines).

Modular Prospecting Jig Plant for Diamonds/Gold.

Modular Gold Elution/Electrowinning and Regeneration Plant

Modular Drum Coal Washing Plant.

APPLICATIONS

The modular concept has been utilised by VEL mainly for Mineral Processing Plants such as Diamonds, Iron Ore, Gold/Diamonds, Coal and Andalusite.

On some projects the total plant has been designed and constructed on the modular concept incorporating a Feed Preparation Section, HMS and Final X-Ray Recovery and Sorting facility.

The modular concept does not suit plant installations with large tankage incorporated into the design. Multi-storied structures are also not suited for the modular concept due to the doubling up of all structural members in the box design concept of the module, thus making the structural cost rather expensive.

VEL did supply two 65 tph HMS Coal Plants when the cost implications and doubling up of the structural steelwork was accepted by the client against the benefits of the modular concept which will be discussed later. (Refer Fig 2).

CURRENT DEVELOPMENTS

For the diamond industry the smallest HMS unit yet built by VEL, a 5 tph HMS Unit, has been developed, designed and supplied to MIBA in Zaire for the geological testing of Kimberlitic ore. This unit comes complete with its own spillage sump and utilises only one vibrating screen divided into three sections, ie feed preparation, tailings and concentrates.

The bigger HMS units with specific reference to the 40 to 50 tph modules are designed in a manner that the feed preparation module can be located in various positions to the main module making this design extremely flexible for fitting into existing installations.

Special modular HMS units have been developed for ocean diamond mining operations. The standard units have been modified and adapted to operate on a sea-going vessel. (Refer Fig 3).

Modular pump-fed HMS modules have been re-designed to accommodate gravity-fed cyclones. Two 40 tph HMS modules were exported to Red China for the processing of diamond gravels. (Refer Fig 4).

In Zaire at MIBA, a proposal has been made to replace their existing washing plant by a 4 x 40 tph gravity-fed module making their outdated washing plant redundant. The existing plant plan dimensions are 78m x 46m and are replaced by a housing containing the four modules with a plan dimension of 20m x 12m.

Units supplied with their own catch sumps eliminate complicated civil works on site as well as expensive civil construction.

A continuous elution pilot plant (non-modular design) for the stripping of gold from carbon has been operating successfully since trials started in February 1990. (Refer Fig 7). The above design concept used for the pilot plant has been extended to a commercial modular design unit treating 1,5 tons of carbon per day which incorporates continuous acid wash and continuous elution.

The elution design has considerably lower operating costs than conventional designs.

A modular one ton/batch carbon treatment plant incorporating acid wash, elution, carbon regeneration and electrowinning sections for the recovery of gold has been designed for Zimbabwe.

The equipment used in the modular plants has been specifically selected for reliability and low maintenance, ie:

Vibrating screens with out-of-balance drive motors have been used, together with polyurethane screen decks and side liners. The out-of-balance motors eliminate the combination gearbox v-belt drives which required far more attention;

Gravel and medium pumps: tried and proven Warman International Pumps;

Piping-rubber hoses on the cylone feed lines and mild steel piping replaced by polyurethane piping on sea water applications;

Painting preparation of plant and equipment to

special painting specifications;

Rubber lining of all platework on inside surfaces to prevent corrosion from sea water on sea water operating plants;

The use of proven high-wear plate liners, 3CR12 steel;

Electrical switchgear housed in Polycarbonate enclosures, mounted on module or remote; and

Nuclear automatic density controllers and recorders.

ADVANTAGES OF MODULAR PLANT DESIGN

The concept of the modular processing plant is generally enhanced by the following factors:

Units are totally assembled in the works, cutting erection time on site down to the minimum and ensuring that all components have been supplied and pre-fitted.

The units are completely tested mechanically, electrically and put through a water test in the works. (Refer Figs 5 and 8).

The units are fully cabled up electrically.

Units are very versatile and able to fit into existing installations, in particular the 40 and 50 tph HMS modules.

The costs of the units are extremely attractive due to the pre-design.

The units are tried and tested in the field and are updated continually, based on feedback from operating experience which is closely monitored.

Reliability of the equipment with low maintenance costs.

Delivery of a pre-designed unit can be supplied in as little as 14 or 16 weeks.

CONCLUSION

The modular design concept of constructing modular units or total modular plants has been proven to be very successful in Africa and in other areas of the world providing the scale of the plant is kept in proportion to the tonnage being treated through the plant. (Refer Fig 6).

First modularheavy media plant constructed by VEL in 1967.
Fig. 1

Plant Designed in container sized Modules with equipment installed in each unit

Overall Dimensions (1 unit): L 14,8m, W 2,8m, H 15,8m. Weight 95 tonnes

2x65 t/h HMS Modular Unit Combinations. Fig. 2

A ship-mounted modular 10 t/h HMS unit complete with feed system. Fig. 3

View of two 40 t/h Gravity Feed Modules. Fig. 4

Modular X-Ray Recovery Plant - undergoing test in works. Fig. 5

View of complete Modular Diamond
Plant during Construction on Site.
Fig. 6

Continuous Elution Pilot Plant
undergoing test in field. Fig. 7

2x40 tph Modular HMS Units under test in works.   Fig. 8

Critical evaluation of small-scale gold plants in Africa, with special reference to Geita gold plant, Tanzania

C. P. Kinabo
Department of Geology, University of Dar-Es-Salaam, Tanzania
(currently Institute for Mineral Processing and Chemical Technology,
Technical University of Clausthal, Germany)

SYNOPSIS

The Buck Reef gold mine at Geita in the northwestern part of Tanzania has been producing gold since the early 1930's. The method applied is direct cyanadation of the milled ore. Gold recovery is at present less than 50 % from the ore of head grade 8 g/t Au. Critical mineralogical studies associated with metallurgical process evaluation have been conducted. The purpose of the studies was to identify the nature of gold and gold-carrier minerals, determine their size distribution and rank them on the basis of amneability to process options. Ore Microscopy, X-Ray Flourecence, X-Ray Diffraction and Mössbauer Spectroscopy methods were used. The ore samples contain sulphides (mainly pyrite), oxides, carbonates and silicates. Almost all native gold particles (< 15 µm) were locked in pyrite. Applying semi-batch flotation, overall gold recovery was above 96 %. The experimental data collected under various flotation operating conditions suggest a flowsheet that incorporates bulk sulphide flotation followed by oxidation-leaching of the flotation concentrates.

Recently, dynamic research and studies, discussions and new innovations have been devoted to the subject of treatment of refractory gold ores. The reason for the interest is clear:- high prices of gold in the world precious metals market since 1970. As a result, many research papers have been published on the subject of gold mineralogy in relation to recovery processes[1-4].

An outstanding feature of these studies reflects the fact that factors such as gold-carrier minerals, grain size, host minerals and the association of host minerals with gold bearing minerals determine the metallurgical process to be applied. Application of such techniques can lead to economic processing of previously untreatable deposits containing traces of invisible and refractory gold.

This paper outlines the mineralogical aspects of the Geita gold deposit in Tanzania and the possible metallurgical methods which can be applied for the effective recovery of gold.

Geology

Buck Reef gold deposit is situated about 100 km southwest of Mwanza town, on the southern shore of Lake Victoria, in the north-west of Tanzania. The Archean greenstone terrane in Geita area is a part of the Nyanzian system of the Tanzanian Craton and constitutes one of the oldest known fragments in the world. The gold mineralization, although low in quantity, nevertheless is widely distributed in the terrane, very similar to that of Archean gold mineralization of Zimbabwe but on a much smaller scale[5,6].

MINERALOGICAL STUDIES

Although some mineralogical studies have been conducted by some geological surveys in Tanzania, it was decided that no further metallurgical testwork should be carried out before systematic mineralogical examination of the deposit. The following mineralogical parameters were evaluated for development of metallurgical testwork:

♦ *determination of minerals present;*
♦ *mode of occurence of the potential ecomomic minerals;*
♦ *determination of liberation sizes of economic elements;*
♦ *identification of potential problems that might arise during processing testwork.*

Due to the comparatively small number of gold grains (in ppm range), heavy and light fractions were prepared from the samples using heavy liquid separation. The fractions were examined by ore microscopy, spectrometry and diffractometry.

Ore microscopy

The results indicate that, for the metallurgical evaluation, the ore can be divided into two major groups. The first group comprises of ore minerals - mainly pyrite (FeS_2) with subordinate to minor intergrowth of chalcopyrite ($CuFeS_2$), galena (PbS), sphalerite (ZnS), arsenopyrite (FeAsS), magnetite (Fe_3O_4), hematite (Fe_2O_3), leuxeconite - rutile (TiO_2), hydrated iron oxides (FeOOH) muscovite and gold (Au). Figure 1 shows an example of galena-sphalerite-chalcopyrite aggregate. Pyrite (0.28 - 280 μm) contains inclusions of chalcopyrite, sphalerite, magnetite, rutile, tellurobismuth and galena. The grain size of the tellurobismuth inclusions varies from 1 - 40 μm. Native gold occurs as small grains (< 15 μm) completely occluded within the pyrite. Figure 2 shows an example of tellurobismuth-gold associations in a large pyrite grain; such gold assemblages (9 μm - 15 μm) may not be readily liberated or exposed even by very fine grinding.

Quartz (SiO_2), calcium-magnesium carbonates, chlorite and plagioclase were classified as gangue minerals without intimate association with gold.

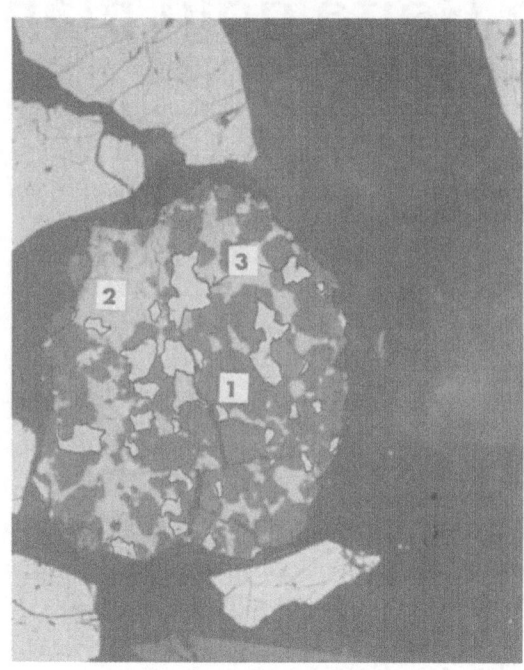

Fig. 1: An intergrowth of ZnS (1), PbS (2) and $CuFeS_2$ (3)

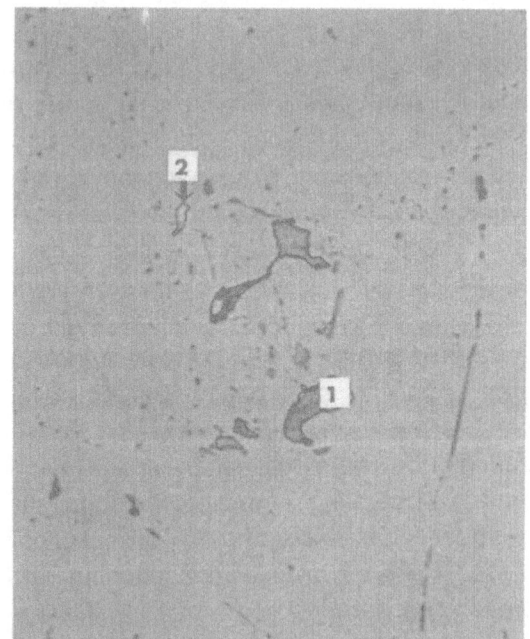

Fig. 2: Inclusions of tellurobismut (1) and native gold (2) in pyrite matrix

Silver content in the gold grains is very low and was not detected by electron microprobe analysis. Liberation of the sulphides from the gangue was estimated to be satisfactory, with a d_{75} between 40 μm and 80 μm. The liberation of locked gold from the pyrite at this grind will be minimal. However, maximum gold recovery can be achieved by pyrite flotation after grinding the ore to a d_{75} below 60 μm.

X-Ray Analysis

The gold assay of the testwork samples (8.2 g/t) were close to the average gold assay (8.0 g/t) in the Buck Reef plant, Geita. For the determination of concentration of ore and gangue minerals AAS, ICP-AES and XRF were used.

The light fraction consists of about 78 - 80 % SiO_2, 10 % iron oxides, 10 - 12 % carbonates. The heavy fraction consists of 0.1 % $CuFeS_2$, 4 % carbonates, 8 % iron oxides, 10 % silica and 78 % FeS_2. The ratio of the ore to gangue minerals is 1:20. For the diffraction work a Siemens D 5000 generator with monocromatic Cu $K_{\alpha 1+2}$ radiation was used. Fig. 3 shows the powder diffraction spectra obtained from the light and heavy fraction samples.

A comparison of the X-ray powder diffaction patterns of the samples with those of data base files gave the following results: The light fraction sample consists mainly of quartz. Most of the crystalline peaks (peaks 2 - 10) obtained from the heavy fraction sample represent pyrite. Small peaks of hematite, (pyrite oxidation) and a peak of quartz were observed. The occurence of SiO_2 in the heavy

Fig. 3: Powder diffraction spectra of light and heavy fractions

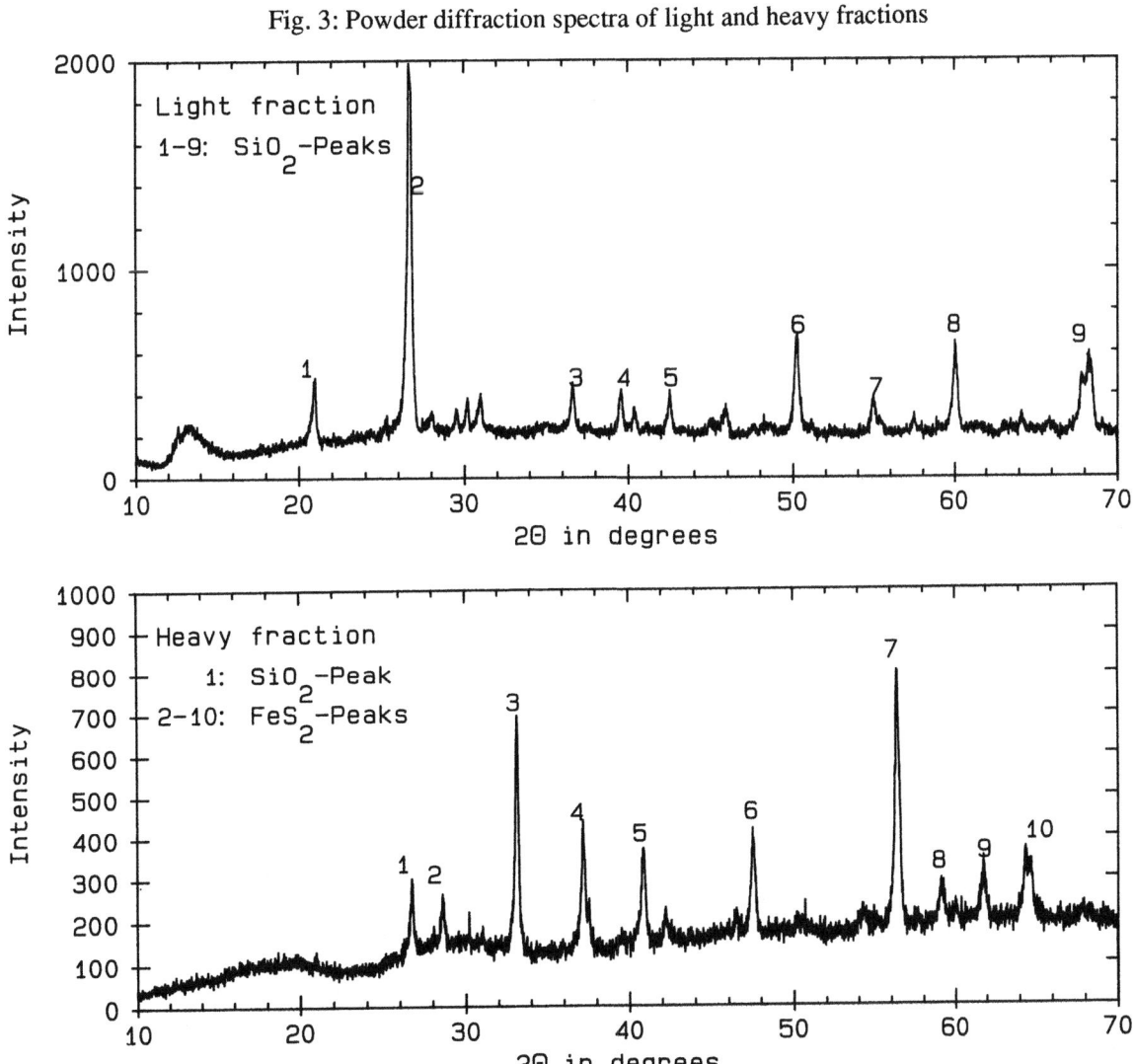

fraction samples is an indication of fine gangue mineral intergrowths within the sulphide phase.

Although these results are considered to be semiquantative, they clearly show pyrite and silica as major minerals in the sulphides and in the gangue respectively.

Mössbauer Spectroscopy

The process mineralogical study confirmed that tellurobismuth occurs as an accessory mineral within pyrite. Such minerals are in many cases chemically associated with gold. The occurence of gold in telluride form may result in poor recoveries during cyanadation, or penalty payments by selling flotation concentrates. For the identification of the

chemical nature of the gold, experiments were conducted using Mössbauer spectroscopy. This method, which is characterized by Isomer shift (IS) and quadrupole splitting (QS) parameters, can identify the chemical state of gold (Au°, Au^{+1} or Au^{+3}) in solids, irrespective of the state of crystallinity[7]. Concentrates with 200 g/t Au were analysed using the method by the Department of Physics, Technical University of Munich, Federal Republic of Germany. The detailed description is explained elsewhere[8,9]. The results of the spectrometic data are summarized in Fig. 4. Most gold compounds have doublet spectra. Native Au has a singlet spectrum. Through matching of IS and QS to corresponding parameters of well characterized crystalline gold compound, an identification is rather simple. The IS of metallic Au

Fig. 4: Mössbauer spectrum of gold concentrate sample (Au-Assay: 200 g/t)

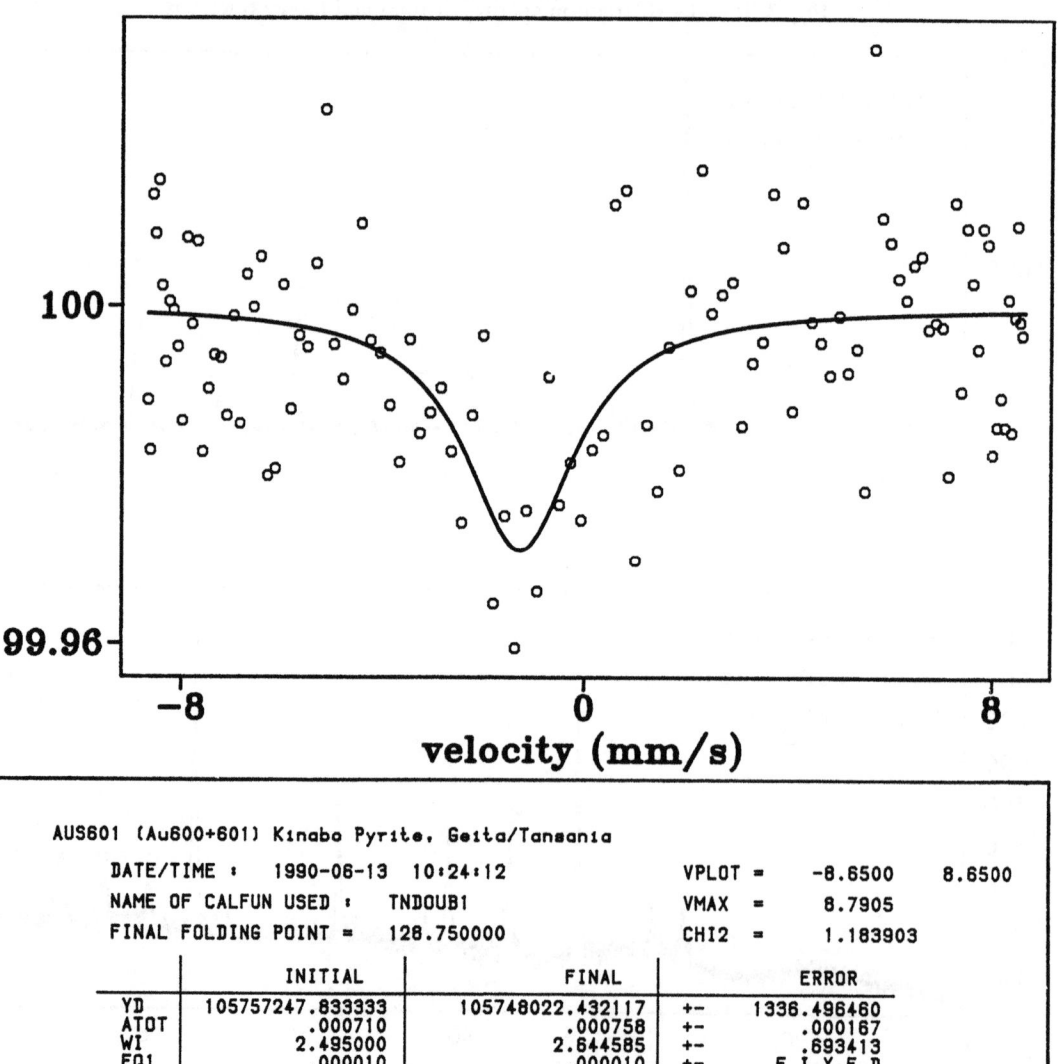

AUS601 (Au600+601) Kinabo Pyrite, Geita/Tansania

DATE/TIME : 1990-06-13 10:24:12			VPLOT =	-8.6500 8.6500
NAME OF CALFUN USED : TNDOUB1			VMAX =	8.7905
FINAL FOLDING POINT = 128.750000			CHI2 =	1.183903

	INITIAL	FINAL		ERROR
YD	105757247.833333	105748022.432117	+-	1336.496460
ATOT	.000710	.000758	+-	.000167
WI	2.495000	2.644585	+-	.693413
EQ1	.000010	.000010	+-	F I X E D
SQ1	-1.204880	-1.275735	+-	.202040

is -1,23 mm/s with QS = 0 mm/s[9]. The IS (-1.27mm/s) and QS (0 mm/s) of Au from Geita agree well with the isomer shift and quadrupole splitting of crystalline Au°. These results reveal that Au in Geita occurs as the native form (Au°) rather than as a telluride.

Metallurgical implications of the mineralogical studies

The refractory nature of the gold may result in poor recoveries by cyanadation[2,4,10,19] In alkaline cyanide circuits, the pyrite surface is oxidized with hydroxyl ions adsorbed on the surface thereby inhibiting any further reaction[11]. Moreover, with the fine nature of the gold grains (Fig. 2), slimes are likely to be formed that might adversely affect thickening and clarification during cyanadation. The successful application of gravity separation is unlikely because gold-carrier minerals are very fine. The association of gold with pyrite means that a high percentage of gold should be recovered to a bulk sulphide flotation concentrate. The recovery of gold from the concentrate will require breaking down of the bulk sulphide matrix by oxidation before cyanadation. Based on these factors, the metallurgical testwork was confined entirely to sulphide (pyrite) flotation. The testwork was conducted at the Institute for Mineral Processing and Chemical Technology, Technical University of Clausthal, F.R.G.

FLOTATION TESTS

Material and reagents

Batch flotation tests were carried out on samples from the Buck Reef plant in Geita, Tanzania. Analytically, pure chemical reagents were used throughout the flotation tests. The pH was regulated with H_2SO_4 and NaOH. Copper sulphate was added for the activation of pyrite. The collectors used were potassium amyl xanthate (KAX) and Hostaflot X-91 from Hoechst (FRG), dithiophosphate from Cyanamid (USA). Baymin F-500 from Bayer

(FRG) was used as a frother. The natural pH of the pulp was 8,5

Apparatus and method

Ore sample batches of 400 g were ground at 66 % solids by mass in a labolatory mill with charge of steel rods. A Denver cell with an effective cell volume of 1,5 liter was used in all batch flotation tests. A pulp density of 28.6 % solids by mass was adopted during flotation. Before commencing flotation, the pulp was conditioned for 5 min. with copper sulphate, followed by pH adjustment and flotation reagents dosage. The collector and frother in the pulp were each conditioned for 3 minutes prior to flotation. Air was pumped in to the cell at a rate of 2.4 $cm^3cm^{-2}min^{-1}$ during flotation. The duration of flotation was 15 to 20 minutes. The assay of gold in the flotation products was determined by AAS and ICP-AES methods.

RESULTS AND DISCUSSION

The effect of particle size

Grinding time has a direct link with the liberation of gold bearing minerals[12,24]. The grain size distributions of five batches of samples ground for different times are given in Fig. 5 and 6. The detailed mathematical representation and evaluation of such analysis data are described elsewhere[13]. The shapes of the distribution curves suggest that the nature of the investigated sample compose of polyminerals that behave different during the grinding. The density distribution curves consist of more than one peak. This is due to fact that quartz is more brittle than carbonates.

The data obtained from the batch flotation tests are shown in Fig 7. The optimum grinding size for gold flotation is considered to be approximately $d_{75} < 53$ μm. At this grind recoveries of about 95 % would be evident for a concentrate grade in excess of 250 g/t Au. Finer grinding is costly and not justified in terms of grade and recovery of gold. Tests with $d_{75} = 69.6$ μm show that both grade and recovery are drastically reduced. At this particle size complete liberation of pyrite was not achieved.

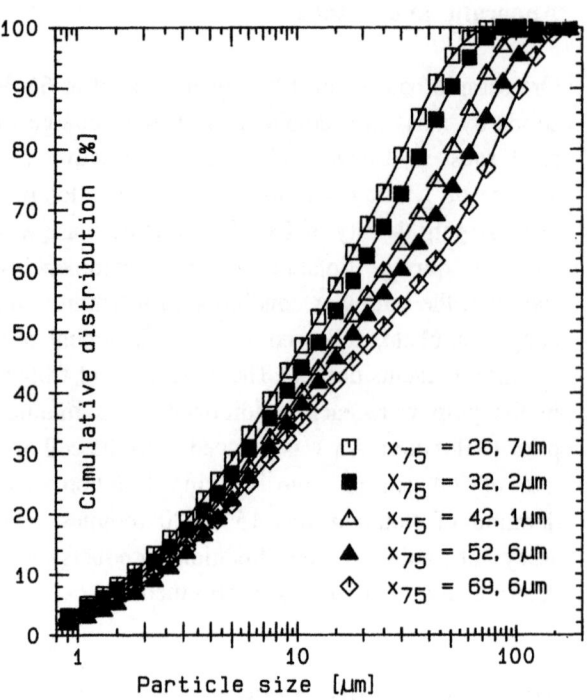

Fig. 5: Particle size distribution

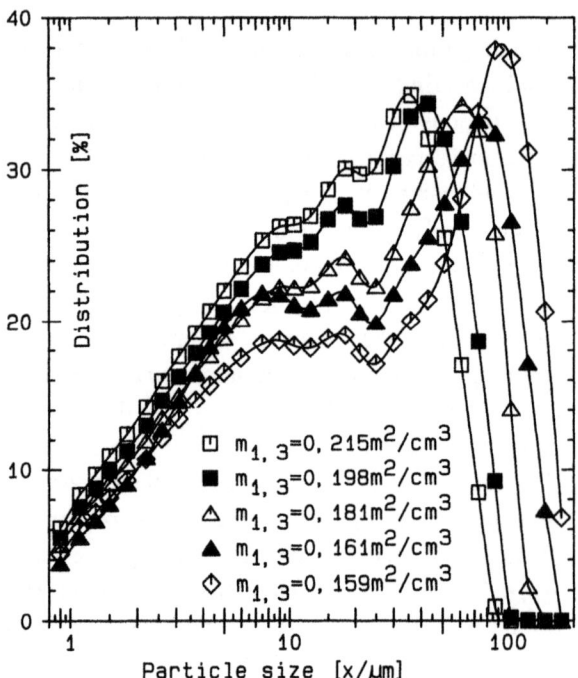

Fig. 6: Density distribution

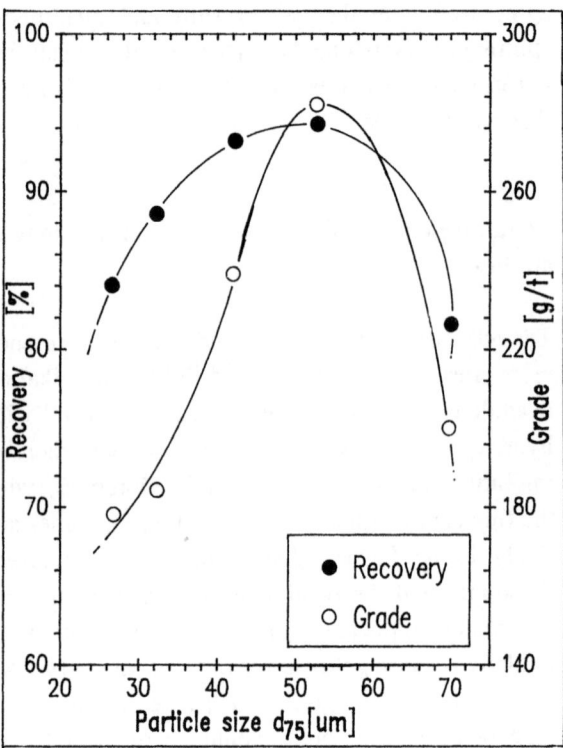

Fig. 7: Gold recovery vs particle size in flotation

The effect of addition of Cu_2SO_4

Previous published reports of the ability to float pyrite with Cu_2SO_4 as an activator indicate that, the activation reactions are usually favoured in a weakly basic pH solutions[14,15]. Tests were therefore conducted by adding Cu_2SO_4 at the natural pH value of the pulp, which was 8,5. Fig. 8 shows the effect of addition of Cu^{2+} ions on the flotability of pyrite at natural pH 8.5 of the ore. The results show that the flotability of gold is significantly increased in the presence of Cu^{2+}. This suggests that the pyrite surface which is the main gold carrier is activated by the precence of copper ions[14-16]. The zeta potential of pyrite at pH 7 is 0; above pH 7 the surface is negatively charged[17]. The activation of pyrite in weakly basic pH solutions can, therefore, be described by the following redox equations:

153

cathodic reduction of Cu²⁺:
$$Cu^{2+}(aq) + e^- = Cu^+(surf)$$

anodic oxidation of pyrite surface:
$$S_2^{-2}(surf) = S_2^{-1}(surf) + e^-$$

The overall reaction is
$$Cu^{2+}(aq)+S_2^{-2}(surf)=CuS_2(surf)$$

Copper adsorbed on the pyrite surface will be in a form of $(Cu^+)(S_2^-)$ that is easly floated with sulfhydryl collectors.

achieved at pH 5 and 9. The sharp decrease in recovery at neutral pH values may be due to formation of insoluble iron or copper hydroxyl xanthates [(Fe(OH)₂X) and (Cu(OH)X)], which are weakly hydrophobic, at the pyrite surface[22]. At acidic or basic pH values xanthate ions (X⁻) and dixanthogen (X₂) dominate. From the flotation tests, it is strongly suggested to run the flotation on the basic pH range; the high carbonate content in the pulp buffers the lowering of the pH to the acidic range.

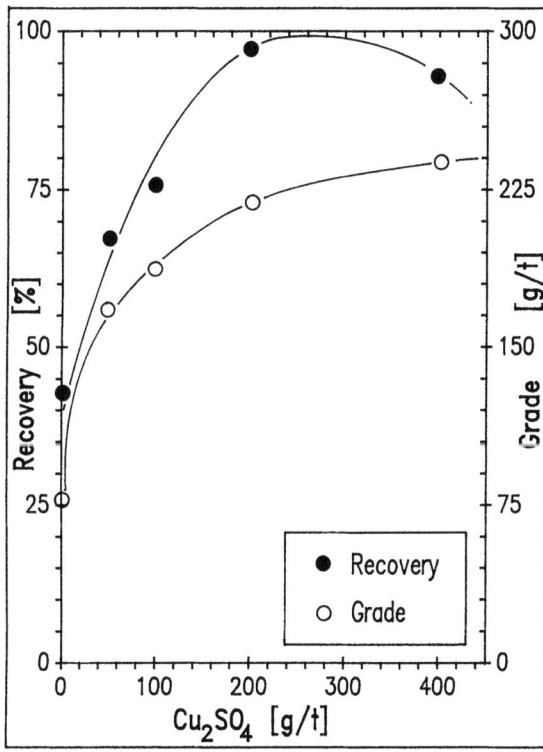

Fig. 8: Effect of Cu_2SO_4 on gold flotation

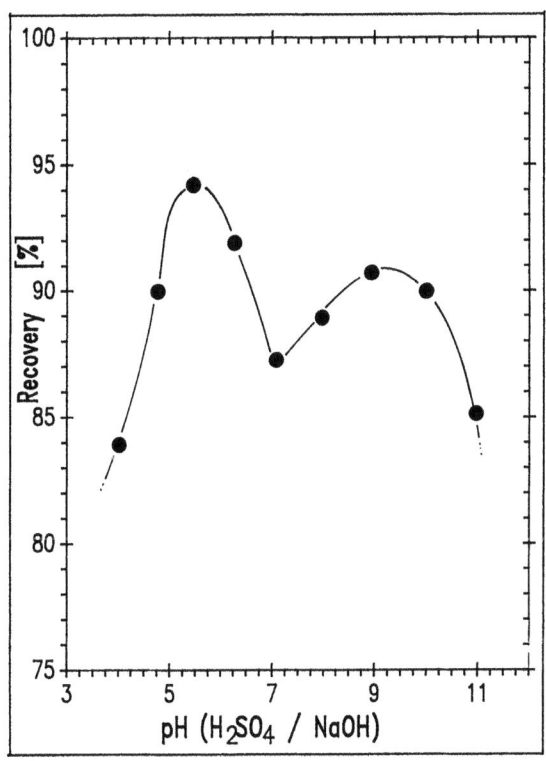

Fig. 9: Effect of pH on the recovery of gold.

The effect of pH

Experiments were conducted to determine the flotability of pyrite using xanthate as a collector with H_2SO_4/NaOH as pH regulator. The collector was maintained at 200 g/t. The flotation of gold as a function of pH is shown in Fig. 9. Gold recovery is characterized by two peaks. Fuerstenau et al.[17] reported similar results with the flotation of pyrite using xathate as collector. Good recovery results are

Effect of collector dosage on recovery and selectivity of gold

One of most common operating parameters of any flotation plant is the collector dosage. A comprehensive list of industrial collectors used in sulphide flotation was presented by Klimpel[18]. A series of experiments were conducted to investigate the effect of three different collector dosages in the flotation of gold. The flotation results are plotted in

Fig. 10 and 11. The xanthate dosage was varied over the range 50 and 400 g/t during flotation tests. Flotation results show that the gold recovery increased from 56 % to 95 %, whereas the gold grade reached maximum at collector dosage of about 200 g/t. An interesting obsevation was that thiourea, which is mainly used for leaching gold and silver[19], can be effectively applied in sulphide flotation circuits in basic medium, with very low dosage. The results of flotation show that thiourea effectively floats pyrite with a maximum dosage of 50 g/t. Above this dosage, for example 75 g/t, both grade and recovery remain constant. The effect of addition of Hostaflot M-91, a mixture of different sulfhydryl collectors, is again presented in Figs. 10 and 11. It is very evident from the plot that increasing the M-91-dosage from 20 to 60 g/t improve both grade and recovery of gold. Above this dosage the recovery remains relatively constant, while the grade significantly decreases. Comparing the three different collectors, it is evident that low

thiourea and M-91 dosage are required for complete flotation of gold. Such results can be achieved with xanthate only by higher dosages, e.g. above 200 g/t.

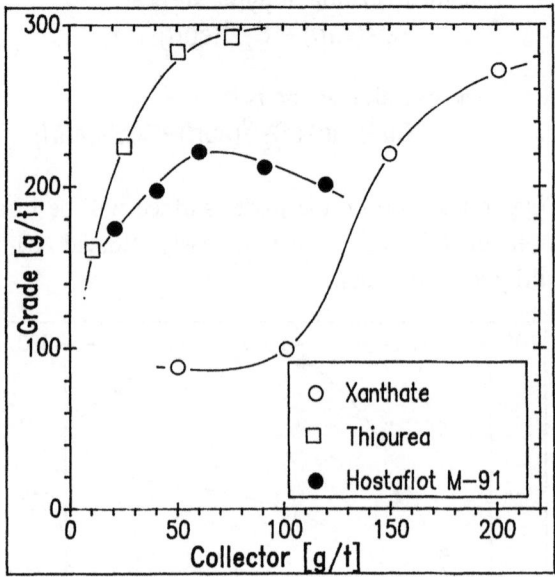

Fig. 11: Effect of different collectors on the grade of gold.

Flotation kinetics

The following kinetic investigation demonstrates whether the collector dosage (M-91) of 60 g/t or 120 g/t is justified. The curves in Fig. 12 show clearly that higher collector dosage (120 g/t) improves recovery up to 5%. With this dosage addition, there is a significant decrease in the grade of the concentrate. Collector addition of 60 g/t led to lower recoveries but the grade was significantly increased. Generally, 1st order flotation rate has been experimentally found to describe flotation processes[21-26]. The rate constant is determined by the formula:

$$f(t) = e^{-kt}$$

where f(t) is the fraction of gold remaining in the pulp at an effective flotation time t; k is the rate constant of Au-flotability. Fig. 13 shows the results of the performance of two collector dosages. It is evident that the flotation of pyrite has two regimes

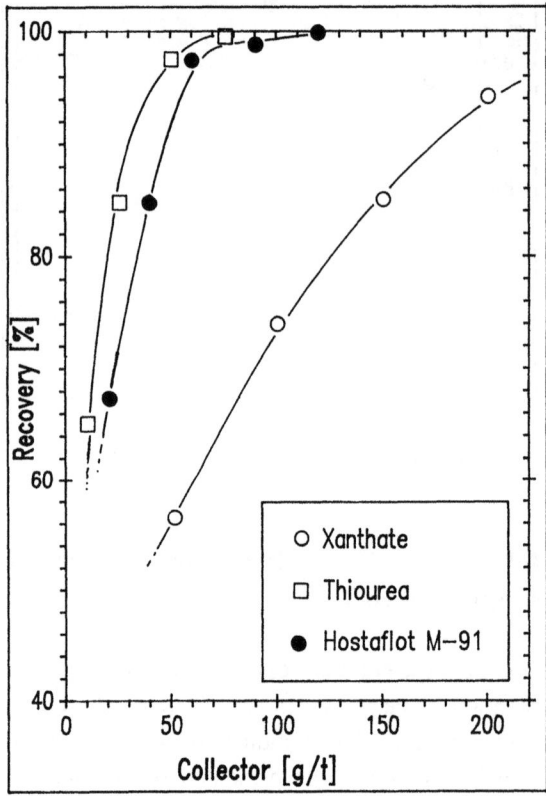

Fig. 10: Effect of different collectors on the gold recovery

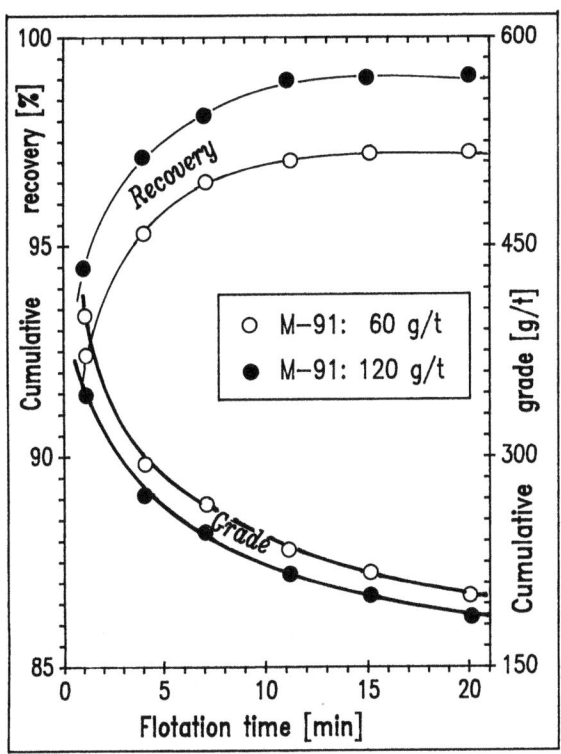

Fig. 12: Kinetic results of flotation of gold.

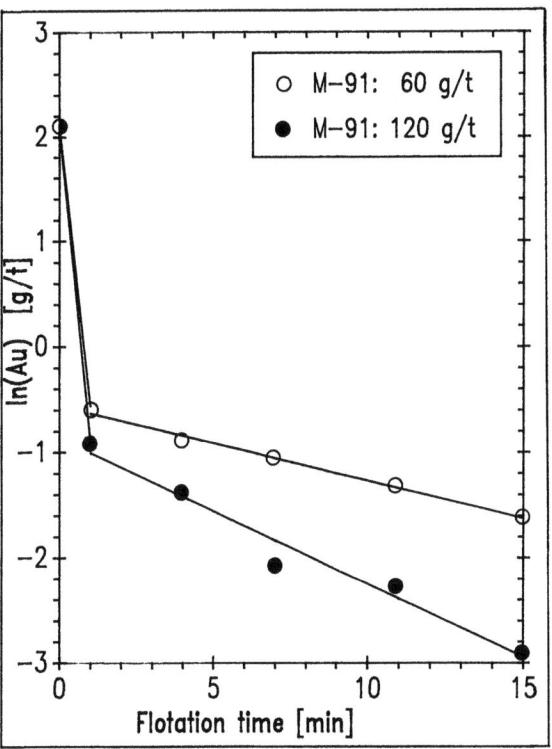

Fig. 13: Effect of collector dosage on the rate of gold flotation.

of rate constants. In tests with collector additions of both 60 g/t and 120 g/t, the first regime lies between flotation time of 0 and 1 minute. The rate constant, k, for the first minute of flotation with dosage of 60 and 120 g/t is 2.68 and 3.01 min^{-1} respectively. Subsequently, as flotation proceeds, the rates diminish. The second regime occurs between 1 and 15 min. Because of the depletion of floatable minerals and/or collector depletion, the rate constant reduces to 0.07 min^{-1} for collector dosage of 60 g/t and 0.14 min^{-1} for 120 g/t. The significantly increase of rate constants with increasing collector dosage indicate that the flotability of gold is enhanced. Higher rate constants resulting from higher collector dosage are, however, accompanied by a loss of selectivity of gold, due to increased chances of flotability of the middlings[24].

Concluding remarks

The mineralogy of Geita gold ore is such that the classical approach of direct cyanadation is neither metallurgically nor economically feasible. Flotation prior to hydro- or pyrometallurgical treatment remains the alternative for the efficient processing of this type of ore. The liberation of gold-carrier pyrite particles prior to flotation is achieved by grinding the ore to $d_{75} < 53$ μm. Under this condition, the response of the ore to flotation is very good. Quality and type of reagents used are also important criteria in determining the economic efficiency of flotation. They determine the flotation rates and the selectivity of mineral separation. More than 96 % of the gold contained in the ore could be recovered by using mixed sulphydryl collector or thiourea as collector, with dosages 60 and 50 g/t respectively. For the further treatment of the flotation concentrates, pressure-, biological- or chemical-oxidation will require to dissolve gold from pyrite particles. Such methods have been discussed in detail elsewhere[19,27-32].

It is hoped that investors or potential gold producers in Tanzania will benefit from this introductory study and invest more in the field of research before taking any risky ventures in the mining and processing industry.

References

[1] Feathers C.E. The significance of the mineralogical and surface characteristics of gold grains in the recovery process. Journal of South African Institute of Mining and Metallurgy, Febr. 1973, p. 223-234

[2] Henley K.J. Gold ore mineralogy and its relation to metallurgical treatment. Minerals Science and Engineering, vol. 7, no. 4, Oct. 1975, p. 289-312

[3] Gasparrini C. The mineralogy of gold and its significance in metal extraction. Bulletin of Canadian Institute of Metallurgy, vol. 76, no. 851, March 1983, p. 144-153

[4] Chryssoulis S.L. and Cabri L.J. Significance of gold mineralogical balances in mineral processing. Transaction, Institution of Mining and Metallurgy, Sec. C: Mineral Process. Extr. Metall., vol. 99 January-April 1990, p. C1-C10

[5] Mineral Resources of South-Central Africa. Pelletier R.A. Oxford University Press, London 1964, 277 pp.

[6] Anhaeusser C.R. The nature of distribution of archean gold mineralization in southern Africa. Minerals Science and Engineering, vol. 8, no. 1, Jan. 1976 p. 46-84

[7] Kongolo K., Bahr A., Friedl J. and Wagner F.E. 197 Au Mössbauer study of the gold species adsorbed on carbon from cyanide solutions. Transactions, Society of Mining Engineers, vol. 21B, April 1990, p. 239-249

[8] Parish R.V. Gold and Mössbauer spectroscopy: the use of Gold-197 Mössbauer spectroscopy to characterise gold compounds. Gold Bulletin, vol. 15, no. 2, 1982, p. 51-63

[9] Untersuchung des Adsorptionsmechanismus von Gold- und Silbercyanidkomplexen aus wäßrigen Cyanidlösungen an Aktivkohle. Kongolo K. Dissertation, Technische Universität Clausthal, BRD, 1989 171pp

[10] Gold Bulletin vol. 21, No. 2, 1988, p. 87

[11] The chemistry of cyanadation. In: Cyanamid Bronchure of American Cyanamid Company, Reprint of Article No. 22, May 1955, p. 48-60

[12] Rickelton W.A. and Dell C.C. Grinding requirements of an ore for flotation. Transactions of the Institution of Mining and Metallurgy, Dec. 1977, p. C207-C213

[13] Leschonki K. Representation and evaluation of particle size analysis data. Particle Characterization, vol. 1, 1984 p 89-95

[14] Wang X. Forssberg K. and Bolin N. Adsorption of copper(II) by pyrite in acidic to neutral pH media. Scandinavian Journal of Metallurgy, vol. 18, 1989, p. 262-270

[15] Ghiani M. Satta F. Barboro M. and Passariell B. Flotation of sphalerite from pyrite by use of copper sulphate and sodium cyanide. In: Reagents in mineral industry. Eds. Jones M. and Oblatt R., Institution of Mining and Metallurgy, London, 1984, p. 89-93

[16] Steininger J. The depression of sphalerite and pyrite by basic complexes of copper and sulfhydryl collectors. Transactions, Society of Mining Engineers, vol. 241 March 1968, p. 34-42

[17] Elgillani D. and Fuerstenau M.C. The role of dixanthogen in xanthate flotation of pyrite. Transactions, Society of Mining Engineers, vol. 241, June 1968, p. 148-156

[18] Klimpel R.R. The industrial practice of sulfide mineral collectors. In: Chemical reagents in the mineral processing industry, eds. Malhotra D. and Riggs W.F., Society of Mining Engineers, Littleton, Colorado, 1986, p. 73-93

[19] Deschênes G. Literature survey on the recovery of gold from thiourea solutions and the comparison with cyanadation. Canadian Institute of Mining and Metallurgy, vol. 79, no. 895, Nov. 1986, p. 76-83

[20] Wang X. Forssberg K. and Bolin N. Thermodynamic calculations on iron containing sulphide mineral flotation systems, I. The stability of iron xanthates. International Journal of Mineral Processing, vol 27, 1989 p. 1-19

[21] Harris C.C. Chakravarti A. Semibatch froth flotation kinetics: species distribution analysis. Transactions, Society of Mining Engineers, vol. 247, 1970, p. 162-172

[22] Harris C.C. Raja A. A modified laboratory flotation cell. Transactions, Society of Mining Engineers, vol. 235, June 1966, p. 150-156

[23] Ackermann P.K., Harris G.H., Klimpel R.R. and Aplan F.F. Evaluation of flotation collectors for copper sulfides and pyrite, I. Common sulfhydryl collectors. International Journal of Mineral Processing, Elsevier Science Publishers B.V., Amsterdam, vol. 21, 1987, p. 105-127

[24] Mehrotra S.P. and Padmanabhan N.P.H. Analysis of flotation kinetics of Malanjkhand copper ore, India, in terms of distributed flotation rate constant. Transactions of the Institution of Mining and Metallurgy, vol. 99, 1990, p. C32-C42

[25] Sousa A.R.B. and Ross V.E. The flotation of pyrite from Buffelsfontein gold mine. MINTEK, Ore Dressing Division, Report no. M320D January 1990 18 pp.

[26] Thorne G.C., Manlapig E.V., Hall J.S. Lynch A.J. Modelling of industrial sulphide flotation circuits. In: Flotation 1976. A.M. Gaudin Memorial Vol. 2, ed. Fuerstenau M.C. American Institute of Mining and Petroleum Engineers, Inc. NY 1976, p. 725-752

[27] Haque K.E. Microwave irradiation pretreatment of a refractory gold ore. Proceedings of the International Symposium on Gold Metallurgy Winnipeg, Canada, August 1987, p. 327-339

[28] Demopoulos G.P. Papangelakis V.G. Acid pressure oxidation of refractory gold ores: Some fundamental and process considerations. Proceedings of the International Symposium on Gold Metallurgy Winnipeg, Canada, August 1987, p. 341-357

[29] Hyden A.S., Mason P.G. and Yen W.T. Refractory gold ore oxidation - Simulation of continuous flow. Proceedings of the International Symposium on Gold Metallurgy Winnipeg, Canada, August 1987, p. 249-258

[30] Fair K.J., Schneider J.C. and van Weert G. Options in the NITROX process. Proceedings of the International Symposium on Gold Metallurgy Winnipeg, Canada, August 1987, p. 279-291

[31] Lawrence, R.W. and Bruynesteyn. A.: Biological preoxidation to enhance gold and silver recovery from refractory pyritic ore concentrates; Canadian Institute of Mining and Metallurgy (CIM), vol. 76, no. 857, Sept. 1983, p. 107-110

[32] Kunter R.S., Honea R.M. and Lear R.D. Effect of mineralogy on the Mclaughlin and other Homestake metallurgical process developments. Proceedings of the International Symposium on Gold Metallurgy Winnipeg, Canada, August 1987, p. 169-170

Effect of particle size in gravity separation processes at Palabora, South Africa

B. R. Brits
Palabora Mining Company, South Africa

SYNOPSIS

Particle size is one of the basic parameters affecting the separation of heavy minerals. The application of gravity separation at Palabora is considered and the effect of particle size in the units used is examined. Factors such as the effect of wide and narrow size distributions in the feed, the recovery by size characteristics and particularly the recovery of fines are embarked upon. Improvements that have been achieved in the circuit, regarding the size effects in the wet gravity separators, are described to show how paying attention to size effects can enhance metallurgical performance.

INTRODUCTION

Gravity separation is one of the oldest fields in minerals processing and yet it remains imperfectly understood. The twentieth century, undoubtedly one of the most inventive periods of all time, brought a decrease in the popularity of gravity separation and favoured the development of processes such as flotation, magnetic separation and leaching. In recent years, however, there has been a resurgence of interest in the field of gravity separation. Gravity separation is regaining popularity because the processes produce comparatively little environmental pollution, require relatively low capital and operating costs and the individual units are generally easy to operate. Furthermore, technological development has resulted in an increase in the areas of applicability of gravity separation.

Gravity separation processes are, in essence, the exploitation of density differences between particles to separate those which are heavy from those which are light. In order to make these separations an environment needs to be engineered where particles, with forces of different magnitudes acting on them, are made to travel in different directions. The force acting on a particle is dependant on specific gravity, size and shape. In order to reduce the size effect and make the relative motion of particles specific gravity dependant, close size control is needed.

In this paper, data obtained from conducting testwork at Palabora Mining Company Ltd. is used to describe the effect of particle size on a number of different gravity separation processing units. Palabora is one of the largest open-pit copper mines in the world. It has a unique orebody in that it contains many of the valuable minerals known, albeit at very low concentrations. The economical recovery of some of these minerals has been achieved by processing the tailings from the copper plant. Two of the bi-product minerals are uranothorianite and baddeleyite which are recovered in the heavy minerals plant.

GRAVITY CONCENTRATION CIRCUIT (Fig. 1)

After recovering 25 000 tpm copper concentrates in the concentrator, 24 000 tpd of magnetite is removed in the magnetic separation plant and the tails from this plant constitute the feed to the heavy minerals plant.

Feed preparation cyclones are utilised to deslime and dewater the plant feed. Approximately 32 000 tpd are treated over Reichert cones where initial recovery and up-grading takes place. Within the cone circuit two double drum magnetic separators are used to remove any residual magnetite.

159

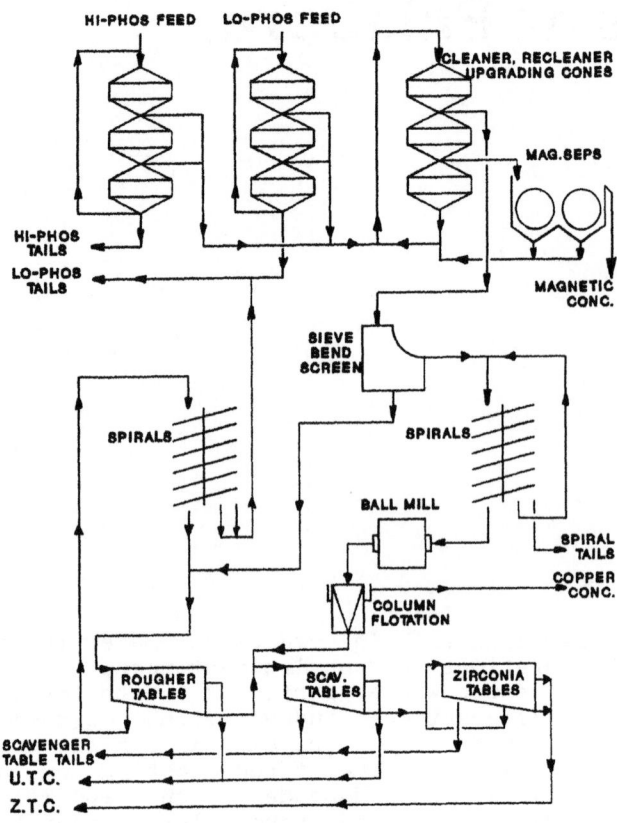

Fig. 1 Simplified gravity section
flowsheet

The final cone concentrate achieved is
classified into a coarse and fine
fraction over a sieve bend screen. The
coarse material is then processed over
spiral concentrators where coarse
particles containing baddeleyite are
recovered. The spiral concentrate is
reground in a small ball mill and
returned to the table circuit.

The sieve bend under size material is fed
to a Stokes hydro- classifier where the
material is separated into six different
size fractions. Each fraction feeds a
triple deck Deister shaking table. The
table circuit is used to further up-grade
the material and two concentrates are
taken off. The first is a
uranothorianite table concentrate (UTC)
which is rich in both uranium and
baddeleyite, and the second is the
zirconia table concentrate (ZTC) which
contains around 75% baddeleyite (ZrO$_2$).
The UTC and ZTC are then processed in
separate chemical plants to produce 98%
U$_3$O$_8$ and ZrO$_2$ final products.

APPLICABILITY OF GRAVITY SEPARATION AT PALABORA

Before considering the influence of
particle size in the heavy minerals plant
it is important to note the applicability
of gravity separation on the basis of
its essential characteristic, namely
density.

In the heavy minerals plant
uranothorianite, a uranium oxide with a
relative density of 9.0, and baddeleyite,
a zirconia oxide with a relative
density of 5.8, are concentrated. The
high densities of these minerals
immediately suggest that gravity
separation should be a viable means of
separation. The Concentration Criterion[1]
provides a widely accepted estimate of
the applicability of gravity
concentration to the separation of a
mineral pair of differing specific
gravities.

$$\text{Concentration criterion} = \frac{SG_H - SG_F}{SG_L - SG_F}$$

where SG_H is the specific gravity of a
heavy mineral, SG_L is the specific
gravity of the light mineral, and SG_F is
the specific gravity of the fluid medium.
In the case at Palabora the light mineral
is the gangue which has specific gravity
of 3.0. The fluid medium is water,
specific gravity = 1.0.

The following criterion are calculated:

Uranothorianite / Gangue   – C.C.= 4.0
Baddeleyite / Gangue   – C.C.= 2.4
Uranothorianite / Baddeleyite – C.C.= 1.7

The size range of applicability of
gravity concentration, assuming that
viscosity effects can be ignored, is then
obtained from Table I.

Table I   Interpretation of concentration
criterion

Criterion	Size Range of Applicability
> 2.50	Separation easy down to fine sand (+ 75 micron)
2.50 - 1.75	Separation effective to + 150 micron
1.75 - 1.25	Separation possible but difficult
< 1.25	Separation only possible in processes where the fluid density is between that of the component ore particles.

This analysis therefore indicates that
the recovery of uranothorianite should be
excellent, the recovery of baddeleyite
good and the separation between the two
heavy minerals possible although
difficult. In the final analysis,
however, the suitability to gravity
concentration, in general, and a
particular device, specifically, is
determined by ore testing.

SIZE PROBLEM

Size effects certainly play a significant
role in gravity separation circuits.
Unfortunately many of the harmful effects
of size in a plant cannot be avoided
economically. The production of fine
particles in comminution operations can
never be eliminated and large losses are
encountered through poor recovery in the
finer sizes.

It is useful to consider the size distribution of particles in three convenient, but arbitrary, regions. Material that is below 75 μm in size comprises the ´fines region.` 75 μm to 300 um is the ´intermediate region` and above 300 um is the ´coarse region`. In the intermediate region recovery of heavy minerals is generally high. In the fine and coarse regions the deleterious effects of size come into play and the recoveries drop off.

The size distribution analysis of solids and Baddeleyite (ZrO2) in the plant feed is given in Fig. 2 with the defined regions shown. The area under the dotted and solid lines indicates the quantity of material by mass and heavy mineral respectively. The plot shows that most of the heavy mineral is in the fine region (31%) and the intermediate region (60%). The high concentration of heavy mineral in the fine region results because of the higher friability of the valuable minerals compared to the gangue particles. Furthermore, when cyclones are used for size classification during comminution, heavy minerals show a higher affinity for reporting to the underflow, for further milling, than the light particles.

The reason for lower recoveries of particles in the coarse region is mainly due to insufficient liberation of the valuable mineral. Because of this fact, coarse particles in the final product will also tend to lower the final grade. Further, in flowing film separations coarse particles are often not able to penetrate through smaller particle suspensions and therefore do not enter the separating environment and do not report to the concentrate.

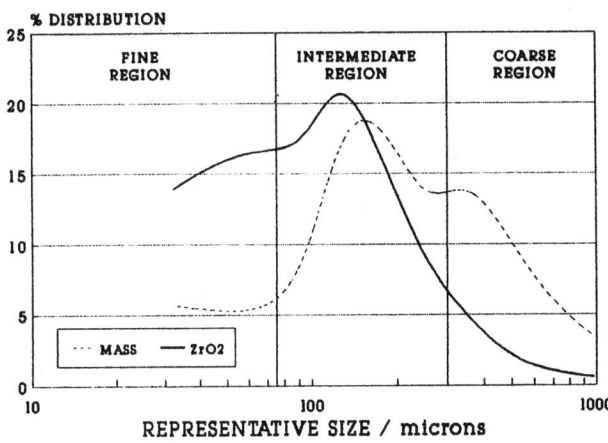

Fig. 2 Size distribution analysis of plant feed

EFFECT OF FINE PARTICLES

In physically based processes the most harmful effect of very fine particles can be that they increase the pulp viscosity changing the flowing film characteristics and inhibiting separation. To ameliorate the situation many gravity separation plants use feed preparation cyclones to remove slimes and excess water from the feed. At Palabora the portion of material in the -45 um size fraction is reduced from over 15% to less than 8% by this stage of the operation.

To ensure that near colloidal particles do not adversely affect the plant operations it is often necessary to use raw water on the units. When the plant water supply is operated in a closed circuit by means of reclamation via thickeners, dams or ponds, the slimes behaviour within the total system can become of crucial importance to the outcome of the process.[2]

RECOVERY OF FINE PARTICLES

A look at the distribution of heavy minerals in the plant feed, (Fig. 2) shows the importance of improving the recovery in the fine region if overall plant recoveries are to be improved. The effective treatment of fines is likely to become increasingly important to the overall viability of the operation.

Fines recovery is inherently less efficient than that of coarser particles. As particles are ground finer, their specific mass decreases and their specific surface area increases. The surface properties of the particles therefore become more relevant. Very fine particles often ´coat` the surface of coarser particles changing the effective surface properties.

In order to address the fines problem a wide range of equipment has been designed for their treatment. Separate circuits for the processing of fines have been, or are being, installed in many gravity concentration plants. The recovery of fines has been avoided for two reasons: firstly, the capacity of a unit treating fine particles is less than that for one treating coarser particles; and secondly, in fine gravity concentration no one device is capable of bringing about a high enrichment at reasonable recovery. Series treatment is therefore almost universally required for optimum plant performance. This results in large, expensive Fines Separation Plants.

SIZE EFFECTS IN UNIT PROCESSES OF GRAVITY CONCENTRATION

There is a wide range of gravity concentration equipment available to today's minerals processing engineer. Some of the key factors in selecting a unit are the throughput and duty required, the size range of particles to be separated and the capital and operating costs of the unit.

At Palabora, Reichert cone concentrators and Deister shaking tables are the two main units used. Spirals and a mineral jig have also been utilised in the circuit. The size effects in these units will be discussed.

REICHERT CONE CONCENTRATORS

Palabora has one of the world's largest installations of Reichert cones utilising 68 cone units to treat some 32 000 tpd of tailings from the main copper circuit.

The Reichert cone is a high capacity flowing film concentrator. Feed pulp is distributed around the periphery of the cone. As it flows towards the centre the heavy particles separate to the bottom of the film. This concentrate is removed by an annular slot, called a tulip, while the tailings flow over the insert. The efficiency of this separation process is relatively low and it is repeated a number of times within a single machine to achieve effective performance. A typical machine contains three or four double single-cone stages in series, each retreating the tailings from the preceding stage and producing a separate concentrate.[3]

The Reichert cone will generally accept feed particles up to 3000 um in size, although the top size which typically gives good recovery is about 600 um. The lower size limit is around 45 um. The proportion of very fine 'slimes' should ideally be below 5% of the weight of solids in the feed to keep pulp viscosity low.

Variations in the feed pulp density and the tulip settings will affect the size of particles being recovered (Fig. 3). At normal pulp densities closing the tulip setting discriminates against the coarser particles. At higher pulp densities the solids recovery generally increases with fine particles benefiting the most. A wide tulip setting is most desirable on the cones to ensure higher recovery in all size ranges. Pulp densities must be closely controlled within specified limits.

The recovery by size achieved on the rougher cones is shown in Fig. 4. The high overall recoveries are achieved with a sacrifice in up-grading ratios. Recoveries are high across a wide range

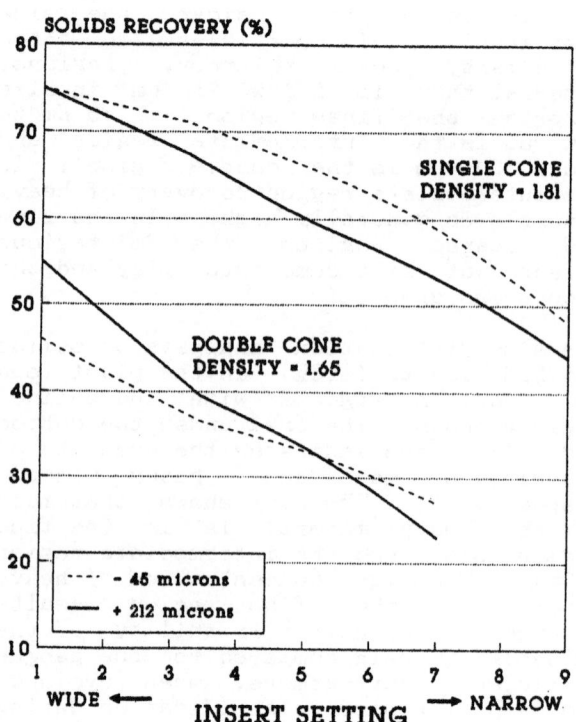

Fig.3 Solids recoveries of coarse and fine particles on the rougher Reichert cones with variations in insert setting and pulp density

but they are best between 40 and 400 um. The 50% recovery above 400 um is reasonable considering the lower grade and poorer liberation of minerals in these particles. Only 3.5% of the ZrO_2 in the plant feed is contained in this size range. Concentrate 3 contains a fairly large portion of the coarse minerals and it should be remembered that when this stream is retreated on Scavenger cones the tulip setting should be sufficiently large to enable coarse particle recovery. The Scavenger cones have been installed with four decks, compared to three on the Rougher cones, to ensure good recovery.

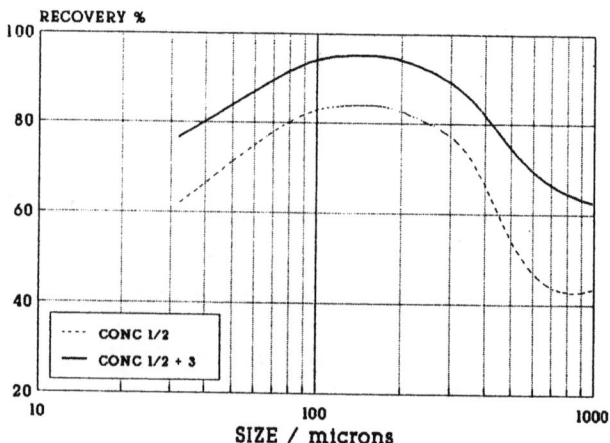

Fig. 4 ZrO_2 Recovery by size on the rougher Reichert cones

DEISTER SHAKING TABLES

12 triple deck shaking tables are used to up-grade material from the Reichert cones to final high grade concentrates. Shaking tables have proved to be the most metallurgically efficient unit to produce finished concentrates. The up-grading ratio achieved is between 5 and 8 for ZrO_2 and between 10 and 20 for U_3O_8.

The range of sizes that can be treated on shaking tables is considerable although the treatment of the full range of sizes on one table is not recommended. As a general rule the efficiency of separation will decrease as the range of sizes in a table feed increases. Fig. 5 shows a theoretical separation. It can be seen that the middlings produced are not only particles of associated mineral to gangue, but often relatively coarse dense particles and fine light particles. If these particles are returned to a grinding circuit, as would normally be necessary with true middlings, then they will be needlessly reground.

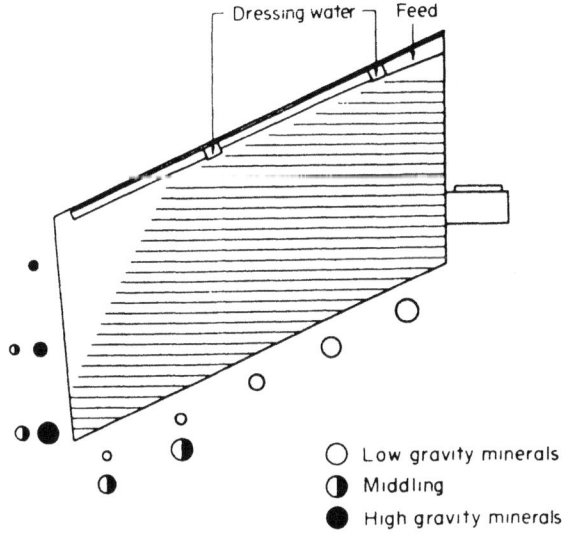

Fig. 5 Distribution of table products

Good fines recovery in the table circuit is necessary for high overall plant recoveries. The finest size that can effectively be treated on tables is a function of the volume of cross water and the movement of the table, since it is essential that particles settle into the bed for them to be collected as concentrates. For fine feed the table should generally be operated with less water and feed, a faster speed and a shorter stroke.

Size classification of feed before the table operation has become common practise to enable a stronger gravity effect on separation than the possible size effect. By narrowing the size range of the feed to a table with the Stoke's hydrosizer, effective separation on the tables has been achieved between 35 and 300 um.

The effect of the hydrosizer on shaking table performance

The Stoke's Hydrosizer is often used to classify material into a number of size fractions and to smooth out the fluctuations in plant running conditions.[4]

A schematic representation of the Stokes hydrosizer is shown in Fig. 6. There is a general flow of material horizontally across the machine. Particles progressively settle out of the flowing stream and are collected in a series of chambers. The collected material in each chamber is discharged via a spigot. Within each chamber an upward current is produced by a supply of water at the base called the teeter water. This water causes three effects: it sharpens classification by reducing the settling rates of the particles in the chamber; it causes the settling particles to form a loose bed where the stratification process can take place; and it scours off some of the fines that may have adhered to the larger particles. It is critically important to maintain the correct pulp density in the Teeter Chamber for optimum performance.

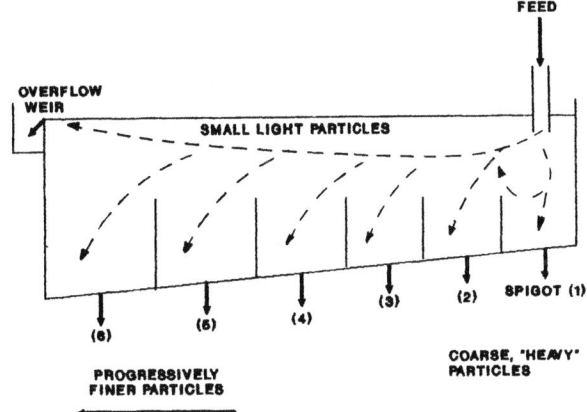

Fig 6 Schematic section of the Stokes
        hydrosizer 4

The teeter water flow rate and the residence time in a chamber affect the d_{50} of a spigot discharge. As the upward current of the water increases the value of the d_{50} also increases. The longer it takes the pulp to flow across a chamber the greater the probability of a particle settling into the chamber. The chambers increase in size away from the feed end resulting in a decreasing upward current and an increasing residence time. These affects tend to decrease the value of the d_{50} down the length of the hydrosizer. An increase in the classification efficiency can be brought about by decreasing the throughput.

Hydrosizers are suitable for preparing the feed to shaking tables because they are capable of producing a number of closely sized products. Within each product the denser particles have a smaller average size than the lighter. This is ideal as shaking tables give maximum separation between fine, dense particles and coarse, light material.

In the Heavy Minerals Plant the feed to the hydrosizer is the -300 um sieve bend undersize material. Fig.7 shows that the machine is effectively sizing the feed into six distinct size fractions. The 'S'-shape of the curves is expected in a settling process and the transport of fines in the coarser size fractions may be due to entrainment in water. The entrainment of fines can generally be decreased by increasing the spigot discharge density. The tonnage of material discharged from each spigot decreases along the unit. This is desirable as the capacity of a shaking table reduces as the particle size of the feed decreases.

To have full understanding of how well a hydrosizer is working it is necessary to study the performance of the hydrosizer and the shaking tables simultaneously.[4] The grade recovery relationship on the shaking tables for U_3O_8 and ZrO_2 before and after the installation of the hydrosizer are shown in Figs. 8 and 9.[5] There is an improvement in the uranium recovery efficiency for a UTC grade of above 1%. Below 1% U_3O_8 there is little difference in the performance. The results do show that with the hydrosizer

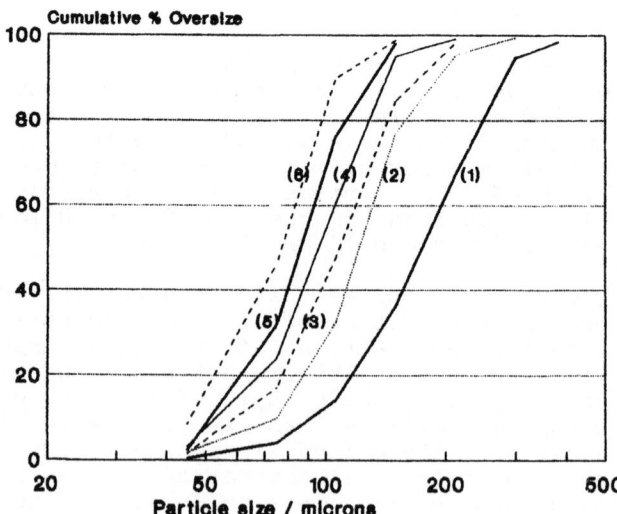

Fig. 7 Size distribution analysis of the six hydrosizer spigot fractions

it is possible to take cleaner UTC cuts which facilitates further processing. The grade recovery relationship for ZrO_2 is far more striking and shows the true value of the hydrosizer. The curve shows that it is now possible to remove a zirconia concentrate of 54% ZrO_2 at a recovery of 89%, compared with a 68% recovery with 54% ZrO_2 grade or a 40% ZrO_2 grade with 89% recovery before installation.

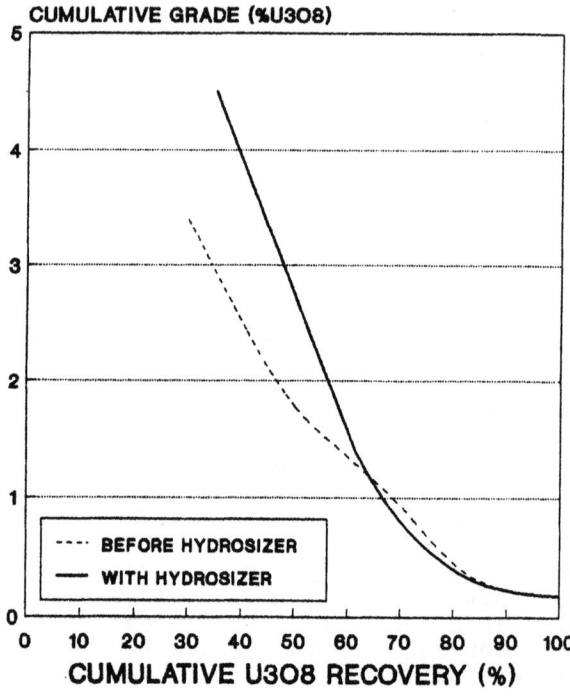

Fig. 8 Cumulative grade recovery curve of U3O8 over the shaking tables, with and without the hydrosizer

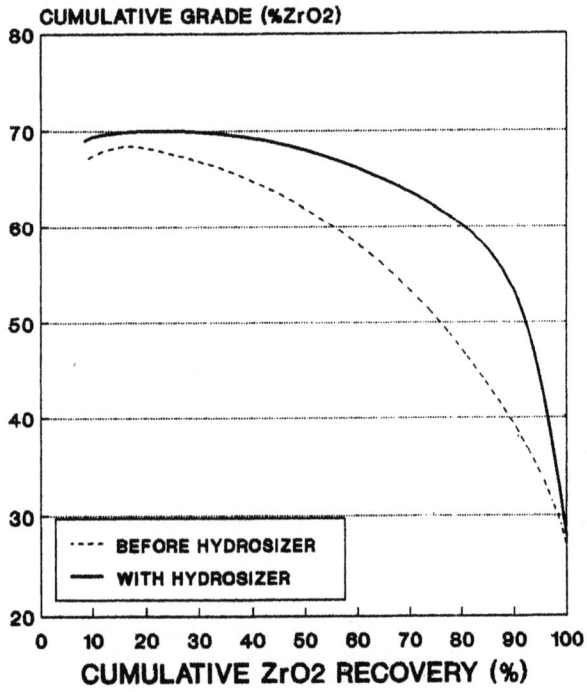

Fig. 9 Cumulative grade recovery curve of ZrO2 over the shaking tables with and without the hydrosizer

SPIRAL CONCENTRATORS

The spiral concentrator is a relatively modern gravity separation device. They have found favour since their inception because, compared to other gravity units, they have low capital and operating costs and they are easy to operate.

The device has been tried on a test scale for a number of applications in the Palabora circuit; from scavenging heavy minerals from the final plant tails to the up-grading of final concentrates. While the spirals are unable to compete with shaking tables for producing high grade concentrates, the spiral has performed extremely well in scavenging and up-grading applications. The most economical use of spirals has proved to be in scavenging heavy minerals from the tails of other units, such as the jig and table tails. Spirals have subsequently been installed for this purpose. From these streams, with feed grades of 3 to 20% ZrO_2, fairly wide size distributions and sizeable concentrations of fine, liberated, high density particles, good recovery and up-grading has been achieved.

The spiral is a unique device in its approach to separation and recovery. The size dynamics occurring in the flowing film of the spiral are a particularly important part of its operation. Feed is introduced through the feed box, which reduces its velocity and establishes the correct pattern of flow. The general flow pattern in a spiral consists of a primary (down trough) flow component with a secondary (transverse) circulation imposed. The secondary flow arises from a greater frictional retardation on the lower layers of the primary flow than on the upper layers. It consists of an outward flow in the upper region and an inward flow in the lower region, the terms 'inward' and 'outward' being defined as motion towards or away from the spiral column, respectively, (Fig. 10).

The spiral trough can be considered as two zones that are governed by different aspects of fluid flow. The behaviour of the inner (near column) zone is controlled by secondary circulation because the rising component of the flow imposes a limit on the size of particle that can just remain in this region. Particles finer than this cut size are lifted into the upper levels of the flow and are transported to the outer zone. At any given size, lower-density particles are more likely to be lifted than higher-density ones, so this zone exerts an influence on concentrate grade.

The outer portion of the trough is considered to constitute a recovery zone because the particles must settle into the lower levels if they are to be transported inwards to the concentrate zone. There is a lower size limit, or cut size, that can just settle out, and this cut size is dictated by the deposition velocity under the prevailing conditions.[6]

Some interesting observations have been made regarding the operation of the spirals with different feed streams. Some of these are discussed below and related to mechanisms that are likely to occurring in the spiral.

Treating material with wide size range

The quinary cone concentrate is the final product of the Reichert cone circuit and it has a wide size distribution. An analysis of the size distribution of the product streams clearly indicates that the mean particle size increases from the inner to outer zones of the spiral trough. Fig. 11 is a plot of the fractional mass percent in a size class against the representative size of that class for each stream. It shows that the spiral concentrate contains most of the particles below 135 um, the middlings contain most of the particles between 135 um and 380 um and the tails dominate above 380 um.

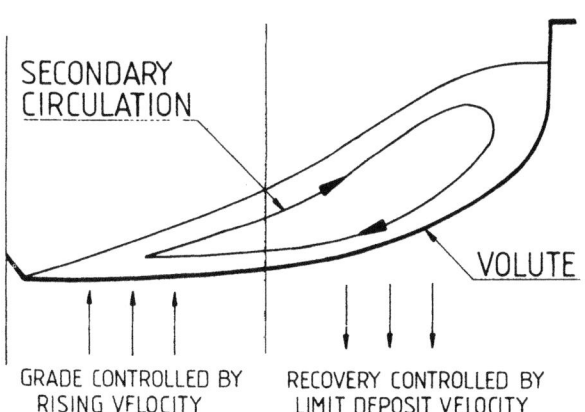

Fig. 10 Spiral flow behaviour

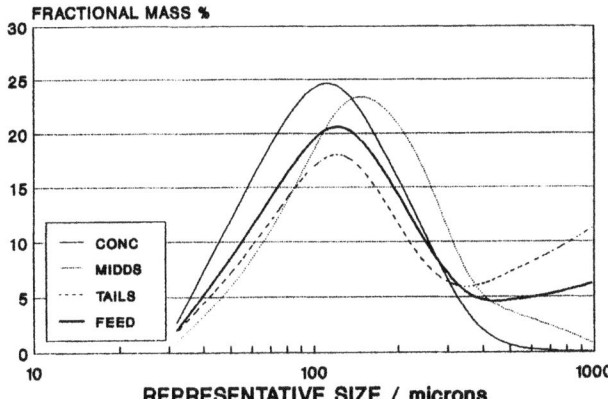

Fig. 11 Size distribution of feed and and product streams when treating the quinary cone concentrate on spirals

The quinary cone concentrate constitutes the feed to the shaking tables where final table concentrates are produced for chemical processing. Before the material is fed to the tables, however, extensive size classification using a sieve bend screen and a hydrosizer is necessary to produce a suitable, narrow size range, feed. Spirals do not require the same pre-treatment of feed before use. Treating the quinary cone concentrates directly over spirals gave comparable results to those achieved on the tables. From a feed of 13% ZrO_2 a concentrate containing 40% ZrO_2 was achieved with a 74% recovery. Taking the recovery to both concentrates and middlings the overall recovery is 94%. The tails grade was reduced to 2% ZrO_2.

Furthermore, the spiral is almost as effective at removing coarse material from the concentrate as the sieve bend. 97% of the concentrate is -300 um and the 80% passing size was 190 um.

Treating coarse material

Currently, the most beneficial application of spirals at Palabora is their use in treating the sieve bend oversize material. This stream consists mainly of coarse material although fine material is present due to the inefficiency of the screening operation. The stream was previously treated over a two-cell mineral jig.

The performance of the two-cell mineral jig varied considerably, but with a large load to the cell performance was generally rather poor. The jig gave a fair zirconia recovery between 150 and 500 um but the overall recovery was only 56%. One of the main reasons for the poor ZrO_2 recovery was the wide size range of the feed. In the size class +150 um to -600 um, containing 65% of the ZrO_2, the recovery was 70%. For the -150 um fraction containing 33% of the ZrO_2 the recovery was only 31%. The low recovery in the -150 um size range resulted because the small, high density, particles did not have sufficient time to settle into the bed of particles and they were therefore washed into the tails.

Spirals were initially introduced to treat the jig tails. It was thought that the jig and spirals working together would enable good recovery from this difficult stream with the jig recovering coarse particles and the spirals scavenging the finer particles. Subsequently it was found that the spirals operating alone performed comparably with the jig and spirals operating together.

Considering the physical characteristics of the sieve bend oversize material shows why the recovery from this stream is difficult. The 80% passing size is 650

um, although only 3% of the ZrO_2 is contained in the +650 um material. 25% of the material is less than 200 um in size. Most of the residual copper and magnetite at this stage of the circuit is contained between 212 and 600 um and is therefore concentrated in this stream. As the relative densities of these minerals are only slightly less than that of the baddeleyite separation is difficult (rel.den.: ZrO_2 = 5.8; Fe_3O_4 = 5.2; Cu minerals = 4.0).

The spiral has proved extremely successful in producing high grade concentrate from this material. From a feed of 16% ZrO_2 a concentrate with 55% ZrO_2 is achieved at a 70% recovery. The middlings recover 18% of the ZrO_2 but also contain most of the copper and magnetite. The recovery by size curve for the concentrate and combined concentrate and middlings is given in Fig. 12. This plot shows that above 130 um the middlings begin to play a more significant role in ZrO_2 recovery. Above 500 um the recovery drops off rapidly.

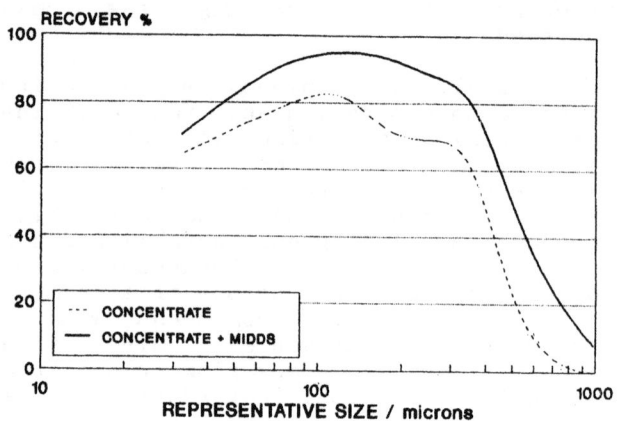

Fig. 12 Recovery by size of the sieve bend oversize material treated on spirals

Treating narrow size range

Another test of interest on the spirals has been the treatment of the discharge from spigot 1 of the hydrosizer. Unlike material used in the other spiral test work, this stream consists of particles in a narrow size range between 75 and 300 um. As the particle size range of the feed coincided closely with the size range of optimum recovery obtained in previous tests an excellent recovery and up-grading was expected. The size distribution analysis did show an increasing average particle size away from the spiral column but the metallurgical performance was not very good. In order to obtain a recovery of over 80% the up-grading ratio was only 2.

Mechanisms at work in spirals

The test results indicate that size dynamics are of crucial importance in spiral concentrators. In most gravity concentration devices a narrow size range enhances performance but on spirals this was not the case. In the outer recovery zone the presence of coarse particles appears to facilitate interstitial trickling of finer particles to the lower layers where they are transported to the inner zone. Hindered settling also plays a role in the separation. The high solids content of the pulp results in particle interaction and the fluid acts as a heavy liquid with the density of the pulp rather than the fluid. This has a profound effect on particle separation by exaggerating the relative density differences between particles. Its effect is seen in the separation of baddeleyite and copper minerals in the sieve bend oversize tests and also explains why the spirals can tolerate a fairly wide size range.

It was also observed that while coarse particles are recovered in the outer region they do not remain in the inner region for long. The coarse particles reach equilibrium more quickly than fine particles but tend to be caught into the upper layers by virtue of their size. In addition the shearing force in the lower layers will result in the Bagnold force playing a significant role in displacing coarse particles. According to Bagnold[7], when a suspension of particles is subjected to continuous shear, pressure develops across the plane of shear and the resulting force acts perpendicular to this plane. The resultant force on a particle from this effect is proportional to the square of particle diameter and the rate of shear. The Bagnold force, relative to the gravitational force, will decrease with increasing particle density and increase with increasing particle size. The sorting thus produced will be vertical stratification; that is, coarse lights on the top followed by fine lights and coarse heavies, with fine heavies on the bottom.

CONCLUSION

The effect of particle size is an inherent part of gravity separation processes. Size always plays a dominant role in the separation mechanism although that role often changes for different processing units. Three types of units in use at Palabora have been considered to make a comparison of these size effects.

Pre-treatment of the feed to the units is important. Reichert cone and spiral concentrators require desliming of the feed as the presence of fines usually increases the pulp viscosity. Both of these units are, however, capable of treating a wide size range with good recovery. The shaking table requires a narrow size range of feed for optimum performance. Once separated into a number of size classes, however, the range of sizes that can be treated is considerable. The Stokes hydrosizer has been successfully used to size the table feed with tremendous benefits to the metallurgical performance.

For the continued viability of gravity separation operations the effective treatment of fines is likely to become increasingly important. Although fines recovery is generally more expensive than the recovery of coarser particles it is likely that the proportion of fines in plant feed will increase rather than decrease as the ore grade and liberation size of minerals decreases within the orebody.

The operation of the units should be carefully controlled to ensure optimum recoveries, particularly of the fine particles. Once lost, fine particles are difficult to recover again. Reichert cones should be operated at pulp densities which are as high as possible without the decks silting up. Shaking tables treating fine particles should be lightly loaded and operated with minimum wash water, fast speeds and short strokes. Minimum quantities of wash water should be used on spiral concentrators.

REFERENCES

1. Taggart A.F. Handbook of Mineral Dressing, Wiley, 1954, Section 11.
2. Holland-Batt A.B. Analysis of slime disposition within a closed system that includes mining, processing and water reclamation. Transactions of the Institution of Mining and Metallurgy, Section C, vol.91, Dec.1982, p. C162-C170.
3. Wills B.A. Minerals Processing Technology, 4th Edition, Pergamon Press, 1988, p. 396.
4. Mackie R.I., Tucker P. and Wells A. Mathematical Model of the Stokes Hydrosizer. Transactions of the Institution of Mining and Metallurgy, Section C, vol.96, Sep.1987, p. C130-C135.
5. Sherring A.J. The Effect of the hydrosizer on rougher table performance, Apr.1986, Unpublished paper.
6. Holland-Batt A.B. Spiral Separation: Theory and Simulation. Transactions of the Institution of Mining and Metallurgy, Section C, vol.98, Apr.1989, p. C46-C60.
7. Bagnold R.A. Experiments in the Gravity-Free Dispersion of Large Solid Spheres in a Newtonian Fluid Under Shear, Proceedings, Royal Society, London, 1954, p.225A.

Influence of some operating parameters on the metallurgy of a column cell around the region of optimum performance

W. T. Musara
Institute of Mining Research, University of Zimbabwe, Zimbabwe
A. L. Mular
Department of Mining and Mineral Process Engineering, University of British Columbia, Vancouver, Canada

ABSTRACT

The influence of the parameters feed rate, feed percent solids, air flowrate, washwater flowrate, frother concentration, collector concentration and froth depth on the metallurgy of a pilot flotation column was investigated based on the efficiency index of coal fines ground to 96 % < 600 μm. The approach taken was to bring the flotation column's performance close to optimum using a first order statistical design followed by the steepest ascent technique. A second order statistical experimental design was then employed to examine the influence of the operating parameters on the efficiency index. The study identified feed rate, feed percent solids and frother concentration in that order as having the greatest influence on the efficiency index. It was observed that there is significant interaction between the feed percent solids and air flowrate, feed percent solids and frother concentration, feed rate and froth depth, feed rate and feed percent solids, air flowrate and froth depth, collector concentration and froth depth, frother concentration and collector concentration, and between the washwater flowrate and collector concentration. Three parameter interactions which were statistically significant were between the feed rate, feed percent solids and air flowrate and between the feed percent solids, air flowrate and froth depth.

INTRODUCTION

The metallurgical performance of a flotation column is influenced by a number of parameters, including the feed rate, feed percent solids, air flowrate, washwater flowrate, reagent dosages and froth depth. The influence of most of the parameters on the metallurgy of column cells has been investigated by various researchers using one technique or another. The approach presented in the literature by most of the researchers is to analyse the influence of the parameters one at a time.

Such an approach does not provide information on the strength of interactions between the different parameters though the interactions ultimately determine the metallurgical efficiency of a flotation system. Somasundaran and Prickett[1,2] and Mular and Bull[3] have pointed out the limitations of such an experimental technique. In order to have more complete information on the performance of a flotation column, it is necessary to design the experiments in such a way that interactions which occur between different parameters and their influence on the column's metallurgy can be identified.

This paper presents the results of a study of the influence of the parameters feed rate, x_1 (l/min); feed percent solids, x_2; air flowrate, x_3 (l/min); washwater flowrate, x_4 (l/min); frother concentration, x_5 (g/t); collector concentration, x_6 (g/t); and froth depth, x_7 (cm); on the efficiency index[4] of coal fines with a size range of approximately 96 % < 600 μm when the column was operating close to the region of optimum performance. The efficiency index is defined as

$$E = 100(\frac{A_f - A_t}{A_o - A_t})(\frac{A_t}{A_c}) \qquad (1)$$

169

where A_f is the feed percent ash content, A_c is the concentrate ash content, and A_t is the refuse ash content.

DESIGN OF EXPERIMENTS

The study was conducted in three steps, as follows:-

(1) Preliminary experiments were carried out in accordance with principles of fractional. factorial experimental designs in order to collect data which were fitted to a first order statistical model of the form

$$E = a_0 + \sum_{i=1}^{n} a_i x_i + e \qquad (2)$$

where E is the efficiency index, ϵ is the residual error variance, a_i is the coefficient of regression and x_i is the i-th parameter.

(2) Using the regression coefficients in the first order model, the steepest ascent technique[3,5] was employed to bring the performance of the column cell to the optimum region, and

(3) A second order statistical model of the form

$$E = a_0 + \sum_{i=1}^{n} a_i x_i + \sum_{i=1}^{n} a_i x_i^2 + \sum_{k} \sum_{j} a_{ij} x_i x_j + e \qquad (3)$$

was utilised to examine the influence of the operating variables on the efficiency index.

More details on how to perform statistical experiments of this kind can be found in Box et al[6] and Mular and Bull[3].

TEST APPARATUS

Test work was conducted in a 6.35 cm internal diameter pilot flotation column with a height of 5.5 m. The pilot flotation column was set up as illustrated in Figure 1. Pulp level control was achieved by means of a system which regulated the opening and closing of a valve on the tailings line in response to changes in the pulp/froth interface. The feed flowrate was regulated by a recirculating line back to the feed conditioning tank. Flowrates of air and washwater were measured by rotameters.

TEST PROCEDURE

The coal used in this test work was initially ground to about 96 % < 600 μm in a laboratory rod mill. The pulp was transferred to a conditioning

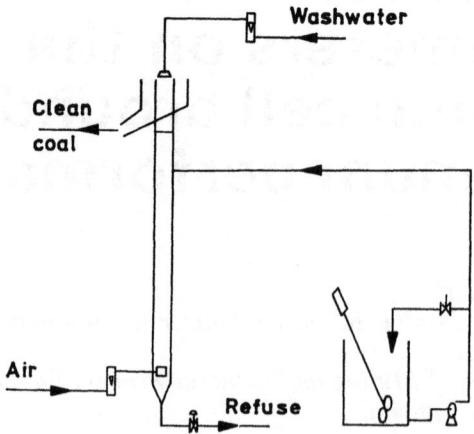

Figure 1: Flotation Column Equipment Setup

tank and mixed for 5 minutes before the necessary dosages of MIBC frother and kerosene collector were added. Feed conditioning was carried out for 5 minutes.

Air was turned on prior to feeding the column and the washwater was turned on as soon as concentrate started overflowing into the concentrate launder. A period of about 3 minutes from the time concentrate started overflowing the overflow lip was allowed so that the flotation column could attain steady state conditions. Samples of the concentrate and tailings were collected and measurements of the pulp densities and mass flowrates of feed, concentrate and refuse were recorded. The samples were filtered, dried and analysed for percent ash content.

The mass flowrates, pulp densities, percent ash contents of the feed, concentrate and refuse constituted the "raw data". Raw data were statistically adjusted using the Simplex direct search routine[7,8,9]. The search routine minimises the objective function

$$O.F. = \sum_{i=1}^{n} \left(\frac{M_i - \hat{M}_i}{M_i} \right)^2 \qquad (4)$$

where M_i is a measured variable (raw data) and $\hat{M}_i$ is the adjusted value of the measured variable.

A computer programme written in Microsoft Basic was used for data adjustment. The output of the program includes adjusted values of the feed ash content, concentrate ash content and refuse ash content, and the efficiency index.

RESULTS

A 2^{7-3} fractional experimental design was utilized to generate the first order model. The conditions employed to generate the first order model are shown in Table 1.

Table 1:Coded Levels For First Order Design

Parameter	Code	Low Level (-1)	Centre Level (0)	High Level (+1)
Feed rate (l/min)	x_1	2	6	10
Percent solids (%)	x_2	5	17.5	30
Air flowrate (l/min)	x_3	6	7	8
Washwater (l/min)	x_4	0.5	1.25	2
Frother conc^n (g/t)	x_5	40	60	80
Collector conc^n (g/t)	x_6	800	1000	1200
Froth depth (cm)	x_7	15	30	45

Table 2: Design Matrix and Responses for First Order Model

Run No.	x_1	x_2	x_3	x_4	x_5	x_6	x_7	E	A_c	A_t	A_f
1	-1	-1	-1	-1	-1	-1	-1	505.6	6.0	45.9	19.4
2	1	-1	-1	-1	1	-1	1	0	0	0*	19.5
3	-1	1	-1	-1	1	1	1	105.2	10.2	29.2	22.3
4	1	1	-1	-1	-1	1	-1	24.3	11.0	18.2	17.1
5	-1	-1	1	1	1	1	-1	676.2	7.7*	59.9	14.8
6	1	-1	1	-1	1	1	1	0	0	0*	18.8
7	-1	1	1	-1	-1	-1	1	99.9	11.4	25.6	19.3
8	1	1	1	-1	1	-1	-1	34.6	8.9	16.2	14.8
9	-1	-1	-1	1	-1	1	1	274.2	6.1	34.5	20.8
10	1	-1	-1	1	1	1	-1	351.6	10.1	40.1	13.5
11	-1	1	-1	1	1	-1	-1	55.9	11.1	20.3	17.5
12	1	1	-1	1	-1	-1	1	13.8	5.5	18.9	18.3
13	-1	-1	1	1	1	-1	1	226.2	6.9	33.3	20.9
14	1	-1	1	1	-1	-1	-1	259.8	8.1	28.8	13.7
15	-1	1	1	1	-1	1	-1	199.1	9.6	35.2	21.3
16	1	1	1	1	1	1	1	26.2	8.7	21.8	20.5
17	0	0	0	0	0	0	0	24.4	5.4	20.5	19.5
18	0	0	0	0	0	0	0	26.9	5.3	20.9	19.8

* No concentrate overflow

The design permits estimation of main effects not confounded with other main effects, or with two factor interaction effects. The results are shown in Table 2. The full first order statistical model for the efficiency index is given by

$$E = 161.3 - 89.5x_1 - 108.4x_2 + 12.0x_3$$

$$- 2.4x_4 + 6.2x_5 + 28.8x_6 - 85.1x_7 \quad (5)$$

By equating the first derivatives of the first order model to zero, the direction of steepest ascent can be predicted. An estimate of the direction of steepest ascent is assumed to follow the vector of coefficient values $[a_i, a_{i+1}, \cdots, a_{n-1}, a_n]^5$, for i = 1 to n. The length of this vector is

$$L = \sqrt{\sum_{i=1}^{n} (\frac{\partial E}{\partial x_i})^2} \quad (6)$$

From equation (6), the length of this vector is

$L = \sqrt{((-89.5)^2 + (108.4)^2 + (12)^2 + (2.4)^2 + (6.2)^2 + (28.8)^2 + (-85.1)^2}$

= 155.4.

A vector of unit length in the direction of steepest ascent has coordinates $[a_1/L, a_2/L, \ldots, a_{n-1}/L, a_n/L]$, or [(-89.5/155.4), (-108.4/155.4), (-2.44/155.4), (6.20/155.4), (28.82/155.4), (-85.09/155.4)] or [-0.58, -0.70, 0.08, - 0.02, 0.04, 0.19, -0.55].

The efficiency index is then measured in steps along the predicted direction until a decrease is observed. In this case, the path of steepest ascent was climbed in steps of 0.25. Table 3 shows the results of the steepest ascent exercise. The highest value of the efficiency index was obtained at the first step of steepest ascent. Therefore, the second order design was constructed around step 1.

Table 3:Responses Along The Path Of Steepest Ascent

Code of Parameter	Step 1	Step 2	Step 3
x_1	- 0.58	- 0.72	- 0.87
x_2	- 0.70	- 0.88	- 1.05
x_3	0.08	0.10	0.12
x_4	- 0.02	- 0.03	- 0.03
x_5	0.04	0.05	0.06
x_6	0.19	0.24	0.29
x_7	- 0.55	- 0.69	- 0.83
E	517.03	438.86	394.48

Second Order Design

The second order design was chosen so that first order, second order, two term and three term interactions could be estimated. A 2^{7-1} design meets these requirements. The levels of factors are shown in Table 4.

Table 4: Coded Levels For Second Order Design

Parameter	Code	Low Level (-1)	Centre Level (0)	High Level (+1)
Feed rate (l/min)	x_1	3.5	4.0	4.5
Percent solids (%)	x_2	5.0	10.0	15.0
Air flowrate (l/min)	x_3	6.0	7.0	8.0
Washwater (l/min)	x_4	1.0	1.25	1.5
Frother conc^n (g/t)	x_5	55.0	60.00	65.0
Collector conc^n (g/t)	x_6	950.0	1000.0	1050.0
Froth depth, cm	x_7	15	18	21

The design matrix and corresponding efficiency indices based on statistically adjusted data is shown in Table 5. The coefficients of the second order model generated from the data in Table 5 are shown in Table 6. The full statistical model for the efficiency index is given by

$$E = 124.96 - 110.67x_1 + 64.16x_2 + 10.02x_3 - 12.23x_4 + 53.37x_5$$
$$- 2.66x_6 - 14.68x_7 + 28.62x_1^2 + 13.98x_2^2 + 6.43x_3^2 + 21.04x_4^2$$
$$+ 30.46x_5^2 - 2.00x_6^2 + 13.14x_7^2 - 35.49x_{12} + 12.88x_{13} + 18.94x_{14}$$
$$- 4.81x_{15} - 3.07x_{16} - 38.20x_{17} - 46.23x_{23} - 1.93x_{24} + 41.93x_{25}$$
$$- 4.05x_{26} + 18.37x_{27} - 11.94x_{34} - 14.14x_{35} - 8.51x_{36} - 34.66x_{37}$$
$$+ 11.14x_{45} - 26.26x_{46} - 11.57x_{47} + 26.29x_{56} - 14.34x_{57}$$
$$+ 26.59x_{67} + 27.78x_{123} + 10.25x_{124} - 16.70x_{125} - 18.37x_{126}$$
$$- 16.26x_{127} + 20.36x_{134} + 1.48x_{135} + 2.59x_{136} - 15.36x_{137}$$
$$- 8.33x_{145} + 0.08x_{146} + 12.02x_{147} - 20.92x_{156} + 22.68x_{157}$$
$$- 7.14x_{167} - 5.43x_{234} + 3.03x_{235} + 4.84x_{236} - 32.32x_{237}$$
$$- 2.37x_{245} + 4.44x_{246} + 6.94x_{247} + 3.35x_{256} + 4.73x_{257}$$
$$+ 7.80x_{267} + 18.37x_{345} + 15.67x_{346} - 5.21x_{347} - 17.24x_{356}$$
$$- 25.08x_{357} + 4.11x_{367} + 13.53x_{456} - 9.68x_{457} - 17.41x_{467}$$
$$+ 11.85x_{567} \qquad (7)$$

The value of the index of multiple determination, R^2 is 0.93, which indicates reasonable agreement between the model and data. Figure 2 is a plot of residuals against the run number. The residuals form a horizonal band parallel to the horizontal axis, confirming that the model does not exhibit serious lack of fit.

Table 5: Design Matrix and Responses for Second Order Model

Run no.	x_1	x_2	x_3	x_4	x_5	x_6	x_7	E	A_c	A_t	A_f
1	-1	-1	-1	-1	-1	-1	1	161.14	6.7	30.74	22.34
2	1	-1	-1	-1	-1	-1	-1	52.59	5.3	18.90	16.89
3	-1	1	-1	-1	-1	-1	-1	361.63	8.0	43.13	19.43
4	1	1	-1	-1	-1	-1	1	37.0	5.8	18.89	17.40
5	-1	-1	1	-1	-1	-1	-1	403.4	9.6	40.68	11.84
6	1	-1	1	-1	-1	-1	1	35.8	6.1	17.90	16.46
7	-1	1	1	-1	-1	-1	1	454.9	7.3	46.86	18.73
8	1	1	1	-1	-1	-1	-1	326.1	9.1	38.24	17.84
9	-1	-1	-1	1	-1	-1	-1	280.7	6.6	36.73	21.51
10	1	-1	-1	1	-1	-1	1	26.6	4.9	21.46	20.46
11	-1	1	-1	1	-1	-1	1	620.1	6.5	56.04	20.43
12	1	1	-1	1	-1	-1	-1	190.7	7.5	29.11	18.46
13	-1	-1	1	1	-1	-1	1	361.9	5.7	40.72	22.87
14	1	-1	1	1	-1	-1	-1	136.4	8.2	26.33	18.61
15	-1	1	1	1	-1	-1	-1	157.2	10.2	30.22	19.58
16	1	1	1	1	-1	-1	1	34.5	5.6	20.67	19.26
17	-1	-1	-1	-1	1	-1	-1	136.1	6.0	26.01	19.76
18	1	-1	-1	-1	1	-1	1	14.5	5.3	16.73	16.20
19	-1	1	-1	-1	1	-1	1	638.1	6.7	56.73	19.07
20	1	1	-1	-1	1	-1	-1	222.7	6.2	30.48	19.51
21	-1	-1	1	-1	1	-1	1	219.9	5.7	34.68	24.16
22	1	-1	1	-1	1	-1	-1	214.4	8.9	31.51	17.81
23	-1	1	1	-1	1	-1	-1	586.9	9.4	65.27	17.89
24	1	1	1	-1	1	-1	1	53.4	6.2	21.95	19.58
25	-1	-1	-1	1	1	-1	1	58.0	5.2	28.28	25.82
26	1	-1	-1	1	1	-1	-1	34.7	6.3	18.83	17.37
27	-1	1	-1	1	1	-1	-1	521.8	7.1	53.03	21.09
28	1	1	-1	1	1	-1	1	310.4	6.2	34.03	18.31
29	-1	-1	1	1	1	-1	-1	381.3	7.9	46.74	21.69
30	1	-1	1	1	1	-1	1	27.6	6.1	16.75	15.68
31	-1	1	1	1	1	-1	1	246.6	6.7	32.94	19.80
32	1	1	1	1	1	-1	-1	444.9	11.6	58.95	17.62
33	-1	-1	-1	-1	-1	1	-1	133.7	5.5	32.95	26.86
34	1	-1	-1	-1	-1	1	1	53.3	6.3	19.00	16.75
35	-1	1	-1	-1	-1	1	1	621.0	7.7	55.16	14.15
36	1	1	-1	-1	-1	1	-1	77.8	10.0	25.08	20.38
37	-1	-1	1	-1	-1	1	1	540.8	6.1	51.48	22.32
38	1	-1	1	-1	-1	1	-1	220.6	8.4	31.83	18.15
39	-1	1	1	-1	-1	1	-1	279.8	10.3	41.30	19.65
40	1	1	1	-1	-1	1	1	41.2	5.9	22.37	20.58
41	-1	-1	-1	1	-1	1	1	9.8	8.6	23.13	22.59
42	1	-1	-1	1	-1	1	-1	25.7	6.0	18.21	17.18
43	-1	1	-1	1	-1	1	-1	181.2	8.6	30.79	19.58
44	1	1	-1	1	-1	1	1	27.5	5.9	20.41	19.26
45	-1	-1	1	1	-1	1	-1	93.6	5.7	26.64	22.43
46	1	-1	1	1	-1	1	1	28.7	5.9	17.95	16.81
47	-1	1	1	1	-1	1	1	184.8	7.1	31.43	21.31
48	1	1	1	1	-1	1	-1	159.6	9.4	30.02	19.71
49	-1	-1	-1	-1	1	1	1	330.1	5.8	37.93	21.62
50	1	-1	-1	-1	1	1	-1	155.9	6.0	23.95	16.97
51	-1	1	-1	-1	1	1	-1	603.5	6.8	48.30	12.92
52	1	1	-1	-1	1	1	1	415.2	6.4	45.12	22.42
53	-1	-1	1	-1	1	1	-1	242.7	6.1	31.26	19.29
54	1	-1	1	-1	1	1	1	66.5	6.3	19.84	16.99
55	-1	1	1	-1	1	1	1	630.6	7.8	66.38	22.88
56	1	1	1	-1	1	1	-1	225.6	10.5	34.76	18.27
57	-1	-1	-1	1	1	1	-1	257.2	5.7	31.17	19.14
58	1	-1	-1	1	1	1	1	37.0	5.5	18.30	16.88
59	-1	1	-1	1	1	1	1	827.6	7.3	69.36	15.16
60	1	1	-1	1	1	1	-1	102.4	7.5	22.20	17.12
61	-1	-1	1	1	1	1	1	164.5	5.9	30.73	22.89
62	1	-1	1	1	1	1	-1	224.4	7.6	31.07	18.15
63	-1	1	1	1	1	1	-1	496.9	9.8	60.67	19.78
64	1	1	1	1	1	1	1	92.5	6.5	23.30	18.97
65	-2.83	0	0	0	0	0	0	658.5	6.4	56.74	19.17
66	2.83	0	0	0	0	0	0	34.9	7.2	21.78	20.10
67	0	-2.83	0	0	0	0	0	210.1	7.9	33.37	20.66
68	0	2.83	0	0	0	0	0	248.9	5.7	31.32	19.65
69	0	0	-2.83	0	0	0	0	64.8	5.7	24.19	18.78
70	0	0	2.83	0	0	0	0	273.3	8.5	35.51	17.91
71	0	0	0	-2.83	0	0	0	146.2	6.7	29.25	21.72
72	0	0	0	2.83	0	0	0	425.8	6.3	44.49	21.62
73	0	0	0	0	-2.83	0	0	84.7	6.1	22.51	18.75
74	0	0	0	0	2.83	0	0	638.2	6.2	52.06	17.27
75	0	0	0	0	0	-2.83	0	110.9	5.8	27.19	22.15
76	0	0	0	0	0	2.83	0	92.2	10.7	25.88	20.10
77	0	0	0	0	0	0	-2.83	338.5	10.4	39.66	13.72
78	0	0	0	0	0	0	2.83	107.0	5.2	26.09	21.66
79	0	0	0	0	0	0	0	217.6	5.9	33.69	23.09
80	0	0	0	0	0	0	0	110.0	6.2	27.16	21.86
81	0	0	0	0	0	0	0	116.5	6.6	29.69	23.70
82	0	0	0	0	0	0	0	143.1	6.2	29.57	22.52
83	0	0	0	0	0	0	0	94.5	6.8	23.32	18.77
84	0	0	0	0	0	0	0	56.0	6.3	23.14	20.58
85	0	0	0	0	0	0	0	122.0	11.7	29.16	20.60

Table 6: Coefficients for Efficiency Index and Corresponding Statistics

Term	F-Ratio	Term	F-Ratio	Term	F-Ratio
$\hat{a}_0$ = 124.9	7.0	$\hat{a}_{46}$ = -26.3	2.7	$\hat{a}_{245}$ = -2.4	0.1
$\hat{a}_1$ = -110.7	60.9	a_{47} = -11.6	0.5	$\hat{a}_{246}$ = 4.4	0.1
$\hat{a}_2$ = 64.2	20.5	$\hat{a}_{56}$ = 26.3	2.8	$\hat{a}_{247}$ = 6.9	0.2
$\hat{a}_3$ = 10.0	0.5	$\hat{a}_{57}$ = 14.3	0.8	$\hat{a}_{256}$ = 3.4	0.1
$\hat{a}_4$ = -12.2	0.7	$\hat{a}_{67}$ = 26.6	2.8	$\hat{a}_{257}$ = 4.7	0.1
$\hat{a}_5$ = 53.4	14.2	$\hat{a}_{123}$ = 27.8	3.1	$\hat{a}_{267}$ = 7.8	0.2
$\hat{a}_6$ = -2.7	0.1	$\hat{a}_{124}$ = 10.3	0.4	$\hat{a}_{345}$ = 18.4	1.3
$\hat{a}_7$ = -14.7	1.1	$\hat{a}_{125}$ = -16.7	1.1	$\hat{a}_{346}$ = 15.7	1.0
$\hat{a}_{12}$ = -35.5	5.0	$\hat{a}_{126}$ = -18.4	1.3	$\hat{a}_{347}$ = -5.2	0.1
$\hat{a}_{13}$ = 12.9	0.7	$\hat{a}_{127}$ = -16.3	1.1	$\hat{a}_{356}$ = -17.2	1.2
$\hat{a}_{14}$ = 18.9	1.4	$\hat{a}_{134}$ = 20.4	1.7	$\hat{a}_{357}$ = -25.1	2.5
$\hat{a}_{15}$ = -4.8	0.1	$\hat{a}_{135}$ = 1.5	0.1	$\hat{a}_{367}$ = 4.1	0.1
$\hat{a}_{16}$ = -3.1	0.1	$\hat{a}_{136}$ = 2.6	0.1	$\hat{a}_{456}$ = 13.5	0.7
$\hat{a}_{17}$ = -38.2	5.8	$\hat{a}_{137}$ = -15.4	0.9	$\hat{a}_{457}$ = -9.7	0.4
$\hat{a}_{23}$ = -46.2	8.5	$\hat{a}_{145}$ = -8.3	0.3	$\hat{a}_{467}$ = -17.4	1.2
$\hat{a}_{24}$ = -1.9	0.1	$\hat{a}_{146}$ = 0.1	0.0	$\hat{a}_{567}$ = 15.9	0.6
$\hat{a}_{25}$ = 41.9	7.0	$\hat{a}_{147}$ = 12.0	0.6	$\hat{a}_{11}$ = 28.6	5.3
$\hat{a}_{26}$ = -4.1	0.1	$\hat{a}_{156}$ = -20.9	1.7	$\hat{a}_{22}$ = 14.0	1.3
$\hat{a}_{27}$ = 18.3	1.3	$\hat{a}_{157}$ = 22.7	2.1	$\hat{a}_{33}$ = 6.4	0.3
$\hat{a}_{34}$ = -11.9	0.6	$\hat{a}_{167}$ = -7.1	0.2	$\hat{a}_{44}$ = 21.0	2.9
$\hat{a}_{35}$ = -14.1	0.8	$\hat{a}_{234}$ = -5.4	0.1	$\hat{a}_{55}$ = 30.5	6.0
$\hat{a}_{36}$ = -8.5	0.3	$\hat{a}_{235}$ = 3.1	0.1	$\hat{a}_{66}$ = -2.0	0.1
$\hat{a}_{37}$ = -34.7	4.8	$\hat{a}_{236}$ = 4.8	0.1	$\hat{a}_{77}$ = 13.1	1.1
$\hat{a}_{45}$ = 11.1	0.5	$\hat{a}_{237}$ = -32.3	4.2		

Analysis of Variance

Source	Sum-of Squares	D.F.	Mean Square	F-Ratio	R^2
Regression	2 975 986.4	70	42 514.1	2.641	0.93
Residual	225 406.6	14	16 100.5		

By comparing the F-ratios obtained by analysis of variance to the F-ratios of individual coefficients, variables which influence the efficiency index significantly can be isolated from those which are of lesser importance. When the F-ratio of an individual term is less than the F-ratio obtained by analysis of variance, that term is statistically insignificant. Only the terms which were statistically significant were included in the discussion which follow.

Within the range studied, the air flowrate, froth depth, washwater addition rate and collector concentration did not have a significant influence on the efficiency index. Table 7 shows the terms which were statistically significant.

DISCUSSION

In base metal flotation circuits where flotation columns are used primarily as cleaners, it has been argued that as well as ensuring that the column operates under positive bias[10], washwater also helps to deepen and stabilize the froth zone[11].

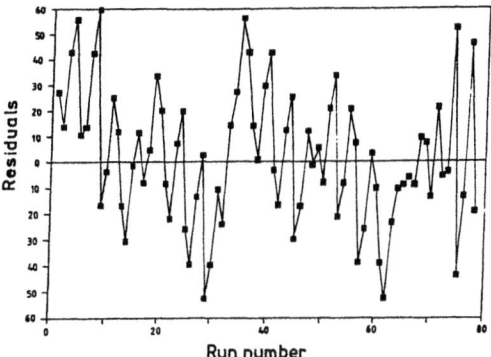

Figure 2: Flotation column residuals

Table 7: Statistically Significant Flotation Column Parameters

a) **First Order Parameters**

feed rate, feed percent solids, frother

b) **Second Order Parameters**

feed rate, frother

c) **Two Parameter Interaction Terms**

feed percent solids/air flowrate, feed % solids/frother, feed rate/froth depth, feed rate/froth depth, collector/froth depth, frother/collector, washwater/collector

d) **Three Parameter Interaction Terms**

feed rate/feed % solids/air flowrate

feed % solids/air flowrate/ froth depth

The results presented here suggest that washwater may not be an important parameter in coal cleaning. Pareth et al[12] have also reported an insignificant effect of washwater on the efficiency index using a Plackett-Burman[13] statistical experimental design. However, they concluded that air flowrate is the most important single factor which affects the efficiency index. The range of air flowrate studied was 1 to 3 l/min, which is below the range of 4 to 8 l/min employed in the experiments carried out here. The insignificant influence of air flowrate observed here suggests that at the upper end of the scale, there is little benefit to be gained from operating a flotation column at high air flowrates. This effect has also been observed by other researchers[14,15].

The froth depth also had a statistically insignificant influence on the efficiency index. This result is in agreement with the results published by Clingan and McGregor[16] who concluded that froth depth does not have a significant influence on the metallurgy of a flotation column. Clingan and Mcgregor's conclusion was drawn from testwork conducted on a 12 m tall column used as a cleaner for a copper concentrate. The froth depth was varied between 0.6 m to 1.0 m.

The first order terms shown in Table 7 suggests that, within the range of variables studied, three variables: (1) feed rate, (2) feed percent solids, and (3) frother concentration have a significant influence on the efficiency index.

The significant influence of feed rate can be explained in terms of the column's residence time and axial dispersion. Residence time and axial dispersion are both negatively affected by increasing feed rate. Increasing the feed rate reduces the column's residence time by increasing the interstitial liquid velocity. Axial dispersion also increases causing the flow regime in the column to deviate from plug flow conditions. The net result is a decrease in the efficiency index.

The results presented here also suggest that high pulp densities have a positive effect on the efficiency index. High feed pulp densities increase apparent hindered settling in the collection zone and increase the residence time, and also reduce axial mixing thereby ensuring that the flotation column behaves more like a plug flow reactor. A similar influence has been observed by other researchers[17,18,19].

High dosages of frother have a positive influence on the efficiency index. Increasing frother concentration increases the gas hold up in the collection zone as smaller bubbles are produced. The small bubbles increase flotation rates by increasing the contact area for bubbles and particles. Hence, an accompanying increase in the efficiency index is observed. However, there is a limit to the amount of frother which can be added, for two reasons:

(1) Smaller bubbles cause a greater percentage of feed water to be entrained to the froth zone[20].

(2) There is little benefit to be gained from generating bubbles less than 0.5 mm in diameter[21].

Two Parameter Interactions

The interaction between feed percent solids and gas rate has also been observed by Laplante et al[22] who proposed that the two parameters are linked to the axial dispersion coefficient by the correlation

$$E = 2.98D^{1.31}EXP(-0.025S) \qquad (8)$$

where S is the feed percent solids. From this correlation, it is apparent that high gas rates increase axial mixing in the collection zone. The negative influence of high gas rate can be counteracted by a high feed percent solids.

There is also a significant interaction between the feed percent solids and frother concentration. The efficiency index can be maximised by adding a high dosage of frother to a feed with a high pulp density.

The interaction between feed rate and froth depth was expected, from the view point of residence time. High feed rates reduce the residence time in the collection zone. For a flotation column of fixed height, the residence time can be increased by reducing the froth zone height.

The interaction between feed percent solids and feed rate was also expected since these two factors are mainly responsible for defining the carrying capacity of the column. High feed rates at high pulp densities will reduce the column's carrying capacity. Thus, there is a maximum amount of solids which can be fed to a column for a fixed feed rate. Beyond this maximum, the carrying capacity falls, together with the efficiency index.

The interaction between gas rate and froth depth is well known, having previously been demonstrated by Yianatos et al[11]. Excessive gas rates can cause increased entrainment into and mixing in the froth zone, with subsequent loss in grade. When a high efficiency index is the objective, a combination of low gas rate and a deep froth depth is desirable.

Significant two parameter interactions between the collector concentration and froth depth, frother concentration and collector concentration, and washwater addition rate and collector concentration were also observed. Since, the collector concentration is rarely incorporated into flotation column statistical experimental designs, the interactions observed here have, to the authors' knowledge, not been reported elsewhere in the relevant literature.

The interaction between the collector dosage and froth depth can be explained in terms of the hydrophobicity of the coal feed particles. A combination of low collector dosage and deep froth has a negative influence on the efficiency index. Under these conditions, weakly attached particles detach from bubbles as a result of the increased residence time in the froth zone. Thus, high efficiency indices are favoured under conditions of high collector dosages and deep froths beds.

The interaction between frother and collector has rarely been discussed in the available literature on column flotation, though some important observations have been made in mechanical flotation cells. Generally, excess collector has the action of inhibiting the functioning of frother. Lynch et al[23] and Rao et al[24] have suggested that the presence of collector in excess of the amount required to render the coal particles hydrophobic diverts the functioning of the frother by causing it to emulsify the excess collector. The joint effect is to reduce the efficiency index. A similar interaction appears to take place in flotation columns.

Though washwater and collector on their own do not have a significant influence on the efficiency index, their interaction with each other is significant. The results suggest that for a fixed collector addition rate, the washwater should be kept to a minimum to ensure good flotation results.

Three Parameter Interactions

Three parameter interactions between the feed percent solids, feed rate and air flowrate, and between the feed percent solids, air flowrate and froth depth were significant.

The interaction between the feed percent solids, feed rate and air flowrate can also be explained in terms of the carrying capacity.

The feed rate and feed percent solids are the primary factors which govern the column's carrying capacity. The influence of the air flowrate on the carrying capacity was first observed by Espinosa-Gomez et al[25]. The three term interaction reported here suggests that gas rate should be taken into account when the column's carrying capacity is determined for flotation column sizing.

The interaction between air flowrate, feed percent solids and froth depth is complex. A possible explanation is that since high air flowrates and shallow froths result in feed water entrainment to the concentrate, high feed percent solids can be used to minimise the amount of feed water entering the froth zone. It is important to note, however, that the amount of solids which can be added to a flotation column is regulated by the column's carrying capacity.

CONCLUSION

The metallurgy of a flotation column is influenced by a number of parameters. Under the conditions employed in this study, parameters which had a statistically significant influence on the efficiency index were ranked in the order feed rate, feed percent solids and frother concentration. Strong two parameter interactions were observed between the feed percent solids and air flowrate, feed percent solids and frother concentration, feed rate and froth depth, collector concentration and froth depth, frother concentration and collector concentration, and between the washwater flowrate and collector concentration. Two three parameter interactions were also observed to be statistically. These were between the feed rate, feed percent solids and air flowrate, and between the feed percent solids, air flowrate and froth depth.

ACKNOWLEDGEMENTS

Financial assistance to conduct this research was provided by the Science Council of British

Columbia. The cooperation of Messrs Bill Fleming and Lumir Bakota of Bullmoose Mining Corporation, Tumbler Ridge, British Columbia was greatly appreciated.

REFERENCES

1. Somasundran, P. and Prickett, G.O., 1967. Response Surface and Computer Methodology for Adsorption Studies, Symposium on Computers in Chemistry of ACS and APS, University of California, San Diego.

2. Somasundran, P. and Prickett, G.O., 1969. Optimisation of a Flotation Operation Using Statistical Methods, Trans. SME, Vol. 244.

3. Mular, A.L. and Bull, W.R., 1976. Mineral Processes: Their Analysis, Optimisation and Control, Department of Mining and Mineral Process Engineering, University of British Columbia.

4. Tsiperovich, M.V. and Evtushenko, V., 1959. Preparation and Carbonization of Coals, Vol. 1, Sverdlovsk, Metallurgizdat, 22.

5. Box, G.P. and Draper, N.R., 1987. Empirical Model Building and Response Surfaces, John Wiley and Sons.

6. Box, G.P., Hunter, W.G. and Hunter, J.S., 1978. Statistics for Experimenters, John Wiley and Sons.

7. Nelder, J.A. and Mead, R.A., 1965. A Simplex Method for Function Minimisation, Computer Journal, 7, pp 308.

8. Mular, A.L., 1976. Optimisation of Flotation Plants, FLOTATION - A Gaudin Memorial Volume (Fuerstenau, M.C., ed.), AIME, pp 895 - 935.

9. Mular, A.L., 1979. Data Adjustment Procedures for Mass Balances, Computer Methods for the 80's (Weiss, A., ed.), AIME, pp 843 - 849.

10. Amelunxen, R.L., 1985. The Mechanics of Operation of Column Flotation Machines, 17 th Canadian Mineral Processors Conference, Ottawa.

11. Yianatos, J.B., Finch, J.A. and Laplante, A.R., 1987. The Cleaning Action in Flotation Column Froths, Trans. IMM, C199 - 205.

12. Pareth, B.K., Bland, A.E. and Groppo, J.G., 1989. A Parametric Study of Column Flotation for Fine Coal Cleaning, Private Communication.

13. Plackett, R.L. and Burman, J.P., 1946. The Design of Optimum Multifactorial Experiments, Biometrika, 33, p305.

14. Huls, B.J., Lachance, C.D. and Dobby, G.S., 1990. Gas Rate and Froth Depth Effects on Performance of a Cu-Ni Separation Flotation Column, Proceedinds of CIM (Dobby, G.S. and Rao, S.R., ed.), Pergamon Press, pp 311-323.

15. Kosick, G.A., Freberg, M. and Kuehn, L.A., 1988. Column Flotation of Galena at the Polaris Concentrator, CIM Bulletin, 81(920), pp 54-60.

16. Clingan, B.V. and McGregor, D.R., 1987. Column Flotation Experience at Magma Copper Company, Minerals and Metallurgical Processing, pp 121 - 125.

17. Column Flotation Company of Canada Ltd., 1982. Column Flotation Separation of High Sulphur Coal by Column Flotation, CANMET Contract Report No. 05Q81 - 00153.

18. Moon, K.S. and Sirios, L.L., 1982. Beneficiation of High Sulphur Coal by Column Flotation, CANMET Division Report ESP/MSL (IR).

19. Bensley, C., Roberts, T., Nicol, S. and Lamb, R., 1987. Column Flotation: The Development of the Tower Flotation Cell, Proceeedings 4 th Australian Coal Preparation Conference, pp 153-169.

20. Finch, J.A., Yianatos, J.B. and Dobby, G.S., 1989. Column Froths, Frothing in Flotation (Laskowski, J.S., ed.), Gordon and Breach Science Publishers, pp 281-305.

21. Dobby, G.S. and Finch, J.A., 1986. <u>Particle Collection in Columns - Gas rate and Bubble Size Effects</u>, Canadian Metallurgical Quarterly, Vol. 25, No. 1, pp 9 - 14.

22. Laplante, A.R., Yianatos, J. and Finch, J.A., 1988. <u>On the Mixing Characteristics of the Collection Zone in Flotation Columns</u>, Column Flotation '88 (Sastri, K.V.S., ed.), AIME Meeting, Phoenix, Arizona, pp 69-79.

23. Lynch, A.J., Johnson, N.W., Manlapig, E.V. and Thorne, C.G., 1981. <u>Mineral and Coal Flotation Circuits: Their Simulation and Control</u>, Elsevier Scientific Publishing Co., New York.

24. Rao, T.C., Pillai, K.J. and Vanangamudi, M., 1982. <u>Statistical Analysis of Coal Flotation - A Prelude to Process Optimisation</u>, IX International Coal Preparation Congress, New Delhi.

25. Espinosa-Gomez, R., Yianatos, J.B., Finch, J.A. and Johnson, N.W., 1988. <u>Carrying Capacity Limitations in Flotation Column Froths</u>, Column Flotation '88 (Sastri, K.V.S., ed.), AIME Meeting, Phoenix, Arizona, pp 143 - 148.

Mining 2

Nonel initiation system trials in ZCCM operations

Benedict C. Chileshe
Weston Mubanga
ZCCM Ltd, Kalulushi, Zambia

SYNOPSIS

In 1989 ZCCM carried out trials at Mufulira, Nkana and Nchanga Open Pit to establish the benefits of using the non-electric delay initiation system compared to the Capped Fuse/Igniter Cord initiation system and electric SPDs in underground and column initiation (using detonating cord) in open pit blasting.

Results of the development trials indicated improvements in linear advance per blast of up to 20% and cost savings of up to K31.00 (using exchange rate at USS1 = K20) per cubic metre of rock blasted compared to ends blasted with capped fuse.

In stoping, though no extensive trials were conducted, the anticipated improvement in fragmentation would in the long run result in reduced secondary blasting, which at the time of the trials accounted for 30% of explosives used in production blasting in ZCCM.

The absence of hard toes and improved fragmentation in open pit blasting would make it possible to expand the blasting pattern which would result in savings in drilling and explosives.

INTRODUCTION

The conventional Capped Fuse/Igniter Cord initiation system has been used, with success, in all of ZCCM tunnelling operations. The system however suffers from the inability to ensure the desired accurate timing of shots. This inability to achieve the desired initiation sequence was partly responsible for the low advance rates and hence high overall development cost.

Since most of the stoping holes are charged with ANFO, it has not been practical, on safety grounds, to achieve toe initiation with the conventional electric detonators, as blow loading ANFO over electric lead wires causes a build up of static electricity and is to be avoided. All holes in stoping are, therefore, collar primed; with subsequent loss of explosive energy resulting in poor blasting results.

The practice of column initiation with detonating cord in open pit blasting results in failure to fully utilize the explosive energy, resulting in poor fragmentation and hard toes, in addition flyrock, which is a common feature with column initiation, poses a safety hazard.

Most of the defficiencies in the Safety Fuse/Igniter Cord System in development blasting can be overcome by the electric delay detonators. However electric detonators are inherently hazardous and so an initiation system was sought that had all the advantages of electric detonators without the disadvantages. A non-electric detonator was seen as the answer and Mining Technical Services ordered some Nonel detonators from Nitro Nobel of Sweden and CXA (ICI) of Canada for trials.

GENERAL DESIGN OF THE DETONATOR

The common design of a detonator is shown in Fig 1. Whether it is ignited with a fuse, electrical fuse head or the shock and hot gas from a Nonel tube, the initiating part of the detonator is similar. Commonly, it is composed of a primary charge of lead azide and a base charge of PETN, RDX/TNT or some other high explosive In front of the primary explosive charge, a pyrotechnical delay element is inserted. In front of this, the electrical fuse head or the Nonel tube is adapted. As can be seen from the figure, the space between the primary explosive and the crimp contains no high explosive components.If no delay element is necessary, there is an open space between the plastic plug in the upper end of the primary charge. In this space, the open end of the Nonel tube or an electrical fuse head is placed.[1]

GENERAL FUNCTION OF THE DETONATOR

The purpose of the detonator is to send a strong shock wave into the blasthole explosive. When the shock wave passes voids like airbubbles, microspheres, or porous grain, "hot spots" are produced and the explosive starts burning violently from these under the high pressure behind the shock wave. As the shock wave propagates through the explosive, further energy is released and the shock is strengthened and accelerates.[2]

THE NON-ELECTRIC INITIATION SYSTEM

The Nonelectric initiation system is based on a plastic tube of 3 mm outside diameter, the inside of which is coated with a reactive substance that maintains the propagation of a shock wave at a rate of 2000 m/sec. This shock wave has sufficient energy to initiate the primary explosive in the delay element of the detonator (Fig 1).[2]

Advantages of the Nonelectric Delay System

- The tube cannot be initiated by electric energy, impact, friction or flame hence it is a safe system.

- Unless brutal force is applied, the Nonel tube will withstand the rough conditions usually found in mining enviroments.

- The inbuilt delays, guarantee sequential initiation.

- The tube has no explosive effect on the explosive charge thus ensuring maximum utilization of explosive power.

- The simple connectors save on charging time.

- In open pit blasting, cut offs are virtually eliminated.

Disadvantages of the Non-Electric Delay System

- During the charging up cycle, great care must be taken to ensure that the numerical sequence of the delay detonators is correct; when delay detonator assemblies are used out of numerical sequence then out of sequence shots will occur.[3]

The Nonelectric system utilizes two connecting up methods in development rounds, namely the bunching method with either detonating cord or the factory made connector block (Bunch Connector) and the harness or clipping method.

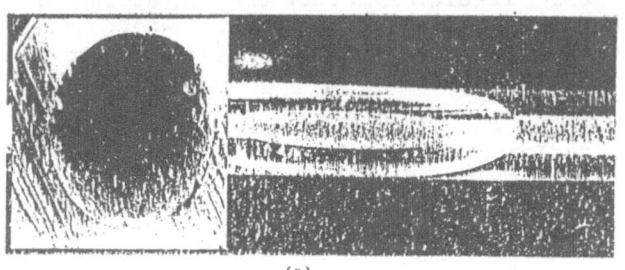

(a)

Section through Nonel tube

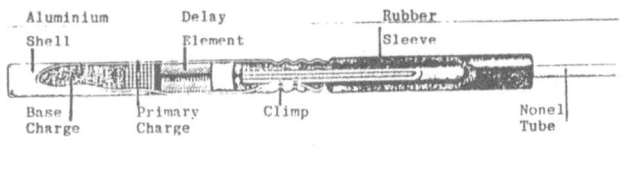

(b)

Section through Nonel Detonator

Fig 1   Nonel Detonator Assembly[2]

In stoping, the system utilizes the bunching method or the Unidet Connector blocks.

In open pit blasting surface detonating cord trunk lines or the Unidet System are used.

OBJECTIVES OF THE TRIALS

- To introduce and determine the applicability of the nonelectric system in ZCCM operations, with a view to adopting it and improving operations.

- To determine the safety aspects of the system.

- To determine the cost benefits of the nonelectric delay system compared to the conventional system.

THE TRIALS

ZCCM's Mining Technical Services, working closely with blasting engineers in divisions directed the trials at Nkana Mine U/G, Nchanga Open Pit and Mufulira U/G sections. These areas were also targetted as the initial implementation sites once trials proved successful.

Advance per round in development, fragmentation in stoping and open pit blasting were measured and assessed. Adaptability, certainty of firing and worker friendliness of the system were also of prime consideration.[4]

DEVELOPMENT

A total of 281 blasts in end sizes of 2.8mx2.6m were recorded, 211 with the bunching method and 70 with the clipping method. An average advance per blast of 1.4m out of 1.5m drilled length and a powder factor of 3.24 kg/t was obtained giving an improvement of over 15% in advance and 13% in powder factor.[4]

A total of 55 blasts in end sizes of 3.5mx3.3m were recorded, 25 with the bunching and 30 with the clipping method. An average advance per blast of 1.86m out of 2.2m drilled length and powder factor of 3.84 kg/t was obtained giving an improvement of over 20% in advance and 19% in powder factor.[4]

Charging in Development

With the exception of lifter holes which were wholly charged with Gelignite (because they were usually wet), every blast hole was toe primed with a 32mmx200mm cartridge of Gelignite 60% and column charged with ANFO.

Two methods of priming were used. In the "Nonel Bunching System" the detonator was fitted with a rubber sleeve which facilitated toe reverse priming.

In the "Nonel Clipping System" the detonator was adapted with an alluminium fastener thereby making it more suited to toe front priming.

The observed general trend was that ends blasted using toe reverse priming gave better advance per round than those blasted using the latter system.

The explanation after assuming other factors to be equal was that the direction of the detonator had some influence on the results obtained. The conclusion was that maximum toe breakage is achieved when the detonator is pointing in the direction of the bulk of the explosives.

Connecting up the Round

Bunch Connection

In this method up to 20 tubes which were at least 2 metres longer than the drill hole were collected together in a bunch. A factory made looped connector block (Bunch Connector) was attached to the bunch and pulled back from the face. When more than one bunch was made in the face then a connector block (UBO) was used to connect up tubes from the Bunch Connectors. A capped fuse was then attached to the connector block (Bunch Connector or UBO) for initiation (Fig 2).

Clipping Connection

In this method a 5 gram detonating cord was run along the perimeter of the face onto which Nonel tubes were connected by means of clips. The cord was then pulled about 2m away from the face and attached to a capped fuse for initiation (Fig 3).

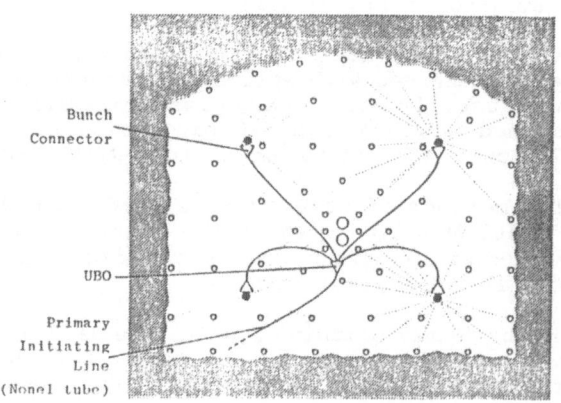

Fig 2   Bunch Connection 2

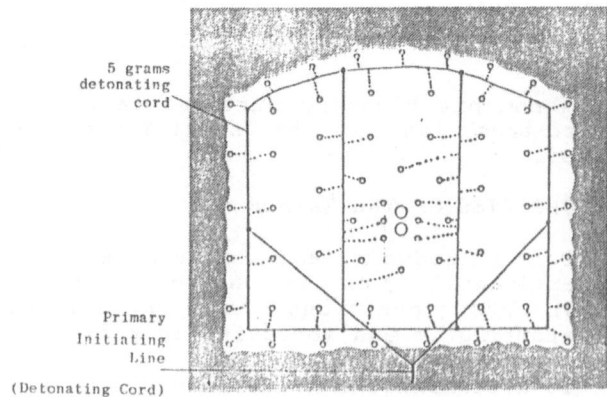

Fig 3   Clipping Connection

Safety

Only eight cases of misfires were encountered during the trials which upon investigation were attributed to human error and not product failure.

Comparison Between Trial Results and Current Performance

Some of the gain in advance of drilled length with the Nonel system can be attributed to the increased supervision in the trial area, as the historical efficiency trends for one of the trial sites indicate.

C/fuse in conventional ZCCM operations -
  1987/88 - 71%
C/fuse with increased supervision   -
  1988/89 - 81%
Nonel with supervision - 26 May - 30 June
1989 - 93%

Nevertheless the considerable improvement during the Nonel test period indicates the value of this system.[4]

STOPE BLASTING

The charging practice with the Nonel system differed from the coventional system in that the detonator was placed at the toe of the holes in the Nonel system as against the collar of the hole in the conventional system. The timing sequence remained the same (Figs 4 and 5).

Trials in stope blasting at one of the sites gave good fragmentation. The cost benefits were however not immediately evident in the short term, but savings resulting from the improved fragmentation are expected to be at least 50% of the current cost of secondary blasting.

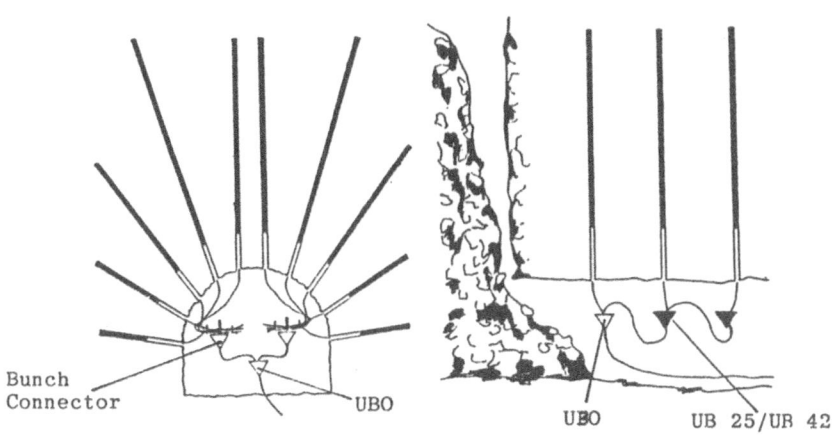

Fig 4   Connecting a Stope
         Ring

Fig 5   Connecting
         Stope Rings

OPEN PIT

Twelve blasts. nine using the 5 gram
detonating cord surface lines and three
using the Unidet System were fired. All
blasts resulted in good fragmentation.
heave and throw. Noise and vibrations were
considerably   lower   than   with   the
conventional method and all blasts broke
to the bottom.

Open Pit Charging

The charging process with the Nonel system
was similar to the conventional system the
only difference being that the Cordtex down
line in the conventional system was replaced
by the Nonel shocktube. Allocation of surface
delays in the Nonel system remained the same
as for the conventional system (Fias 6 and
7).

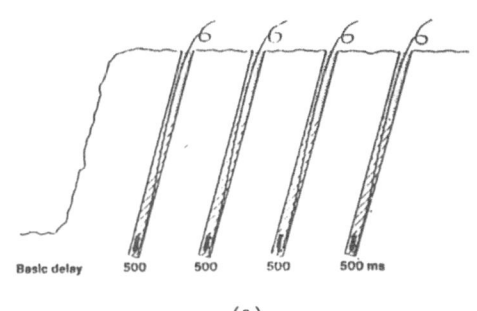

(a)

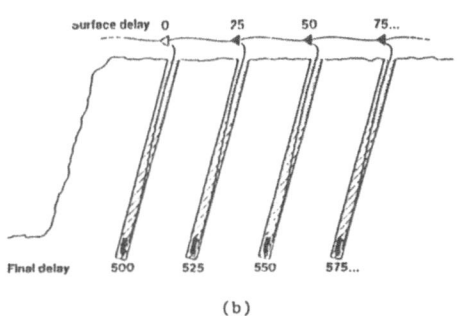

(b)

Fig 6   Unidet Connection with
         Connector Blocks 2

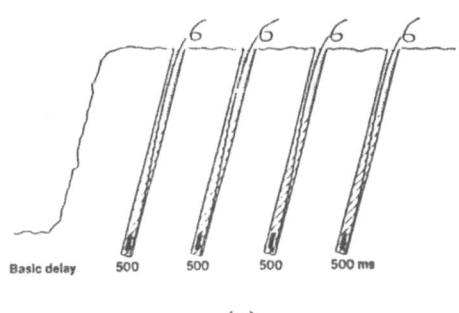

(a)

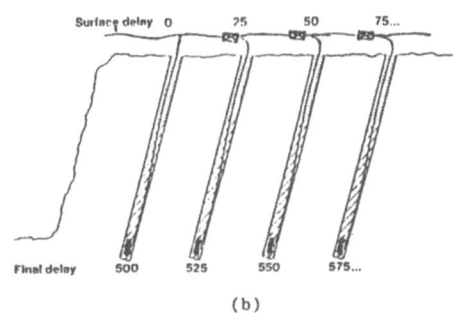

(b)

Fig 7   Detonating Cord Connection
         with Surface Relays

All the holes were toe primed with Gelignite 60% and a 500 milliseconds Unidet, and column charged with ANFO. An inter-row delay interval of 50 milliseconds was chosen. This arrangement allowed all the surface delays to be initiated and advanced about 10 rows before the first hole in the first row detonated: virtually eliminating the risk of cut offs (Figs 8).

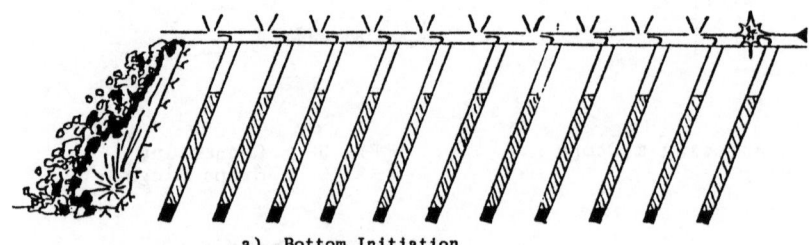

a) Bottom Initiation

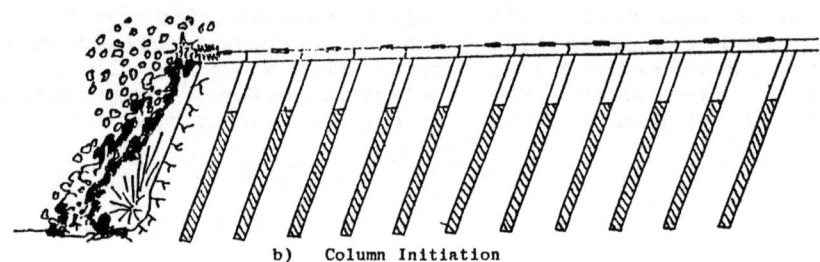

b) Column Initiation

Fig 8   Open Pit Connection/Blasting

(a)   With Unidet System

(b)   With Cordtex Surface and down lines

FINDINGS OF THE TRIALS

- All men adapted well and quickly to the system.

- The nonelectric initiation system is more cost effective than the conventional capped fuse/igniter cord system in development especially in large ends and developed with long drill holes.

- The explosives material input for smaller faces and shorter rounds is higher than for bigger faces and longer rounds making the use of the Nonel system in bigger faces and longer rounds more attractive.

- All rounds blasted with the Nonel system yielded better linear advances, finer fragmentation and fewer sockets than those blasted with the conventional system.

- Tying up a Nonel development round took 25 minutes compared with 30 minutes for the conventional round.

- Better fragmentation was observed in stope blasting though the benefits were not immediately quantifiable.

- All blasts fired with the Nonel system in the open pit exhibited low vibration, less noise, and finer fragmentation. No digging problems were encountered, suggesting clean toes.

- Fragmentation and throw were improved in open pit blasting which resulted in easier loading.

187

As a result of using Nonel, savings of K31.06/m³, K0.93/m³ and K0.18/m³ would be realised in large ends, small ends and open pit respectively (Appendices).

From the studies carried out after the trials, total implementation of the Nonel system in all of ZCCM operations would yield annual savings in excess of K200 million broken down as follows:-

- Development due to improved
  efficiencies   90 000 000
- Development due to price
  difference   -3 400 000
- Stoping due to reduction
  in secondary blasting   50 000 000
- Stoping due to price
  difference   27 300 000
- O/pit due to approx 10%
  expansion in drill pattern   40 000 000
- O/pit due to price
  difference   2 700 000

Total   206 600 000

CONCLUSION

The nonelectric delay initiation system is more cost effective than the conventional system and its implementation in all ZCCMs operations would result in annual savings in excess of up to K200 million.

Implementation of the Nonel system is particularly attractive in large faces where deeper rounds are drilled.

Since the conclusion of the trials ZCCM has imported quantities of Nonel for implementation in all blasting operations and negotiations have been going on between ZCCM and Nitro Nobel AB of Sweden to enter into a Joint Venture to manufacture Nonel detonators in Zambia.

REFERENCES

1. A. Persson, How to achieve performance of the Detonator in the Blasthole. SveDefo, 25 10 1990.

2. Nitro Nobel AB Sweden, Nonel Users Manual.

3. K A Rhodes, The Use of Nonel at Cooke 2 Shaft, Randfontein Estates Gold Mining Company Witwatersrand Limited.

4. ZCCM Limited, Nonel/Shocktube Trials Report, WM/30 Aug 1989.

APPENDIX 1

Trial Results for large size ends - 3.5m x 3.3m

Table 1
Explosives Material Cost Analysis (ZK)

Item	Fuse	Nonel
Total Expl Cost/blast (ZK)	776.46	776.46
Total Acc Cost/blast (ZK)	749.65	825.73
Total Expl+Acc Cost/blast (ZK)	1526.11	1602.19
Drilled depth (m)	2.20	2.20
Advance (m/blast)	1.50	1.86
% eff adv of drilled round	68.18	84.55
Volume blasted (cu m)	18.38	22.79
Powder Factor (kg/t)	4.77	3.84
Cost/m3 (Explosives)	42.26	34.08
Cost/m3 (Accessories)	40.80	36.24
Cost/m3 (Explosives+accessories)	83.05	70.32
Saving/m3 (Explosives)	0.00	8.18
Saving/m3 (Accessories)	0.00	4.56
Saving/m3 (Explosives+Accessories)	0.00	12.74

Table 2
Drilling Cost Analysis

Item	Fuse	Nonel
Target Drill Metres (m)	1608588.00	1608588.00
Target Volume (m3)	547366.75	547366.75
Advance (m/blast)	1.50	1.86
Achieved Drill Metres (m)	2359262.40	1902630.97
Drill Metres/m3	4.31	3.48
Drill Metres/m3 saving	0.00	0.83
Cost/Drill Metre	15.00	15.00
Drilling Cost/m3	64.65	52.14
Drill Cost Saving/m3	0.00	12.51
Labour Cost/m3	30.00	24.19
Labour Cost Savings/m3	0.00	5.81

Table 3
Overall Savings per Cubic Metre

Item	Access Savings	Explos Savings	Drill Savings	Labour Savings	Total Savings
Fuse (Base case)	0.00	0.00	0.00	0.00	0.00
Nonel	4.56	8.18	12.51	5.81	31.06

Trial Results for small size ends - 2.8m x 2.6m

Table 4
Explosives Material Cost Analysis (ZK)

Item	Fuse	Nonel
Total Expl Cost/blast (ZK)	342.80	342.80
Total Acc Cost/blast (ZK)	311.30	493.46
Total Expl+Acc Cost/blast (ZK)	654.10	836.26
Drilled depth (m)	1.50	1.50
Advance (m/blast)	1.22	1.40
% eff adv of drilled round	81.33	93.33
Volume blasted (m3)	8.88	10.19
Powder Factor (kg/t)	3.72	3.24
Cost/m3 (Explosives)	38.60	33.64
Cost/m3 (Accessories)	35.06	48.43
Cost/m3 (Explosives+accessories)	73.66	82.07
Saving/m3 (Explosives)	0.00	4.96
Saving/m3 (Accessories)	0.00	-13.37
Saving/m3 (Explosives+Accessories)	0.00	-8.41

Table 5
Drilling Cost Analysis

Item	Fuse	Nonel
Target Drill Metres (m)	21600.00	21600.00
Target Volume (m3)	6552.00	6552.00
Advance (m/blast)	1.22	1.40
Achieved Drill Metres (m)	26557.38	23142.86
Drill Metres/m3	4.05	3.53
Drill Metres/m3 saving	0.00	0.52
Cost/Drill Metre	10.50	10.50
Drilling Cost/m3	42.56	37.09
Drill Cost Saving/m3	0.00	5.47
Labour Cost/m3	30.00	26.14
Labour Cost Savings/m3	0.00	3.86

Table 6
Overall Savings per Cubic Metre

Item	Access Savings	Explos Savings	Drill Savings	Labour Savings	Total Savings
Fuse (Base case)	0.00	0.00	0.00	0.00	0.00
Nonel	-13.37	4.96	5.47	3.86	0.93

Trial Results in Open Pit

Table 7
Explosives Material Cost Analysis (ZK)

Item	Det Cord	Nonel
Total Expl Cost/blast (ZK)	153205.50	153205.50
Total Acc Cost/blast (ZK)	25221.39	18157.97
Total Expl+Acc Cost/blast (ZK)	178426.89	171363.47
Volume blasted (m3)	38475.00	38475.00
Cost/m3 (Explosives)	3.98	3.98
Cost/m3 (Accessories)	0.66	0.47
Cost/m3 (Explosives+accessories)	4.64	4.45
Saving/m3 (Explosives)	0.00	0.00
Saving/m3 (Accessories)	0.00	0.18

Water management at Rössing uranium mine, Namibia

M. T. R. Smit
C. P. Brent
Rössing Uranium Ltd, Swakopmund, Namibia

ABSTRACT

Water Management at a large uranium mine and leaching plant located in a desert environment is described in respect of reducing water consumption and controlling and containing contaminants. The extent to which water consumption has been reduced by innovative measures to reduce water losses and increase water recycle is described. Although the recycling of untreated solutions generated in the process has had negative effects on plant throughput and recovery, the overall benefit has been significant. Measures employed to ensure that no contamination of local groundwater occurs are described.

INTRODUCTION

Rössing Uranium is located approximately 65 km inland from the West coast of Namibia at latitude 22°25' and longitude 15°01'. The operation consists of an open pit, an acid leach plant and a sulphuric acid plant.

The open pit mines ca. 2,5 million tonnes per month, supplying the plant with ca. 1 million tonnes per month of ore. The circuit consists of four stages of crushing, open circuit rodmills, sulphuric acid leaching, sand and slime separation and washing, continuous ion exchange, solvent extraction and Ammonium diuranate (ADU) precipitation and calcining.

More detailed descriptions of the process are published elsewhere. [1,2,3,4,5].

The Acid Plant produces ca. 720 tpd of concentrated sulphuric acid from pyrite for consumption in the leach process. Tailings are disposed of by pumping the slurry to a disposal dam located nearly 1 km from the plant at a mean elevation of about 50 m above the plant and covers an area of some 650 ha.

The company employs ca. 2300 people, of whom ca. 1000 and their families reside in the town of Arandis, 15 km from the mine site, whilst the rest stay in the coastal town of Swakopmund, 65 km from the mine. The mine is located in a desert environment with a mean rainfall of 30 mm per year, a mean evaporation rate of 7,2 mm per day and temperatures ranging between 4°C and 40°C.

WATER MANAGEMENT

Due to the desert environment of the mine site and the nature of the process, it is necessary to pay considerable attention to management of the available water resources and prevention of groundwater contamination by seepage from the Tailings dam.

Potable water is supplied to the coastal towns of Walvis Bay and Swakopmund as well as Arandis and Omaruru rivers by the Department of Water Affairs. This department is the water controlling authority in Namibia and Rössing has regular and frequent contact with them. The position of the major pipelines and well-fields is shown in Fig. 1.

As a result of population and industry increases on the coast during the past decade (expected to accelerate subsequent to independence), the limited water resources are put under increasing pressure.

Rössing is the largest consumer on the coast (double the consumption of the town of Swakopmund with a population of ca. 18 000) and has a moral as well as commercial interest in ensuring reliability of water supplies and maintenance of water quality. These two objectives have been met by reducing potable water consumption by 50% during the last decade and by establishing efficient control and containment measures to prevent local contamination of groundwater resources.

REDUCTION OF WATER LOSSES

Figure 2 shows water consumption on the mine site since the start of operations in 1977. It can be seen

that fresh water requirements increased to about 28 000m^3/day as the mine worked up to full production in 1980. By this time a tailings pond containing some 10 million m^3 of acidic solution had been built-up. Recycle of this RDS (return dam solution) to the metallurgical plant commenced in 1981. The major components of the Rössing water circuit are shown in Fig.3. It can be seen that not only is water recycled from the tailings dam but also from the seepage dam. This seepage dam is situated about 1 km downstream of the main tailings wall at an elevation some 70 m below that of the tails pond. Also shown on Fig.3 is the utilisation of brack groundwater from the nearby Khan River in the water circuit. The acidic RDS was fed directly to the rodmills[6] as a partial replacement for fresh water.

Later RDS was brought into use as glandseal service on the large tailings slurry pumps. The increasing level of recycle which can be seen in Fig.2 reached a peak in 1985 when about 75% of mill feed water consisted of recycle solution. This allowed a considerable reduction in fresh water consumption but during 1986 and 1987 consumption had to be increased since the tailings pond had been depleted to a very low level. Since the beginning of 1988 new savings measures which are described below, have reduced mine site fresh water consumption to about 12 000m^3/day with confidence that this level should not again be significantly exceeded during the rest of mine life unless production increases.

By setting up a mine site water balance (a simplified version of which is shown in Fig.3 where a balance for the period May to December 1987 is compared to a balance for November 1990), it was possible to identify the major water losses from the Rössing system. From Fig.4 it can be seen that the losses occurring on the Tailings dam are the major component and efforts were thus concentrated in this area. By means of far-reaching changes in tailings deposition method, losses from the dam have been reduced by some 10 000m^3/day over the past three years. These changes fall into two main categories.

Improved Control of Tailings Pumping System

Problems encountered with the original design of the tailings systems, and subsequently resolved, are described elsewhere.[7] Subsequently tailings pumping was controlled by maintaining the slurry velocities in the tailings lines at a level high enough to ensure that no choking of pipelines occurred due to coarse particles settling out. Velocities were typically in the order of 4,2m/s, and required the addition of significant volumes of transport solution to the tailings pulp. Once it was realised that a high proportion of the water losses occur because of the excess transport solution added, efforts were directed at reducing tailings pumping velocities.

Approximately 50% of the transport water is lost to evaporation from the wetted tailings beach, and by changing the velocity control philosophy, typical velocities were reduced to about 2,9m/s, resulting in a reduction in transport water requirement of some 8 000m^3/day and consequent reduction in evaporation losses.

The change in philosophy essentially consists of changing the primary control from tailings velocity measurement and control, to tailings pump sump level measurement and control. Depending on the rheology of the pulp, there is a critical velocity at which settling in the pipeline will start to occur. Whereas in the past velocities were strictly controlled above a fairly conservative minimum level, control is now by simple algorithm which aims to maintain a constant sump level. The onset of settling in the tailings pipeline is manifested by an increase in sump level and is controlled by adjustment of a variable speed pump. In this way tailings velocities are controlled just above a critical minimum, dictated by operating conditions. The control philosophy is depicted in Fig.5.

Reduction in Tailings Dam Wetted Area.

In addition to reducing tailings pumping velocity, significant water savings have been effected by reducing the wetted area on the tailings dam to about 30% of its former size by means of the development of discrete paddock blocks within the larger tailings impoundment.

Because of the exceptionally coarse nature of Rössing tailings (30% + 0,85 mm and 80% + 0,1mm), it has been possible to rapidly develop partition walls with simple open ended pipe deposition of the bulk tailings volume, (40 000 tons/day at 50% solids). Beach slopes of 5-6% are achieved close to the discharge point by natural segregation and no form of cycloning is necessary.

Before 1988 tailings deposition was carried out only from the perimeter of the impoundment which has a total length of some 7 km. (See Fig.6 where the perimeter runs from zone Z5 in the north to zones E and F in the southwest). Discharge from this perimeter crest drained towards the pond in the west across a beach which varied on length between 1 000m and 3 000m. Deposition volumes were kept more or less even at all points along the perimeter such that an overall "smooth bowl" shape was formed, thus maintaining free drainage of all solution towards the pond. Particle segregation down the beach flattens the slope to about 2% within a few hundred metres and to about 0,5% after about 1 300m. By 1988 this had resulted in any material further than about 500m from the perimeter being so fine (and thus wet) as to prevent any access by man or machine.

By extending the deposition pipeline over this zone of fines it was possible to deposit a 3m thick "causeway" of coarse material. This was rapidly developed toward the centre of the impoundment allowing machine access in order to lay permanent tailings pipelines.
Further deposition was then done to raise the elevation of these "causeways" in benches 6m high and 15m wide, effectively forming partition walls and establishing distinct paddock blocks which have sufficient capacity to accommodate all tailings for many months at a time. Once the paddock partitions are established it has proved feasible to control a paddock pool of 5 - 10 ha in size within a paddock block (50 - 70 ha) around a conventional penstock decant point. Clear solution entering the penstock is piped some 800 m directly to the main recycle pond. This main pond has also been reduced in size from about 150 ha in 1981 to about 10 ha at present since only clear solution now enters and no allowance need be made for the rapid silting which formerly occurred. At the time of writing two such paddock blocks have been established and developed to a height of some 12 m above the surrounding tailings beach. The layout of these paddocks within the overall tailings impoundment is shown schematically in Fig.6.

Development of the paddock method took place over a period of two years in association with our tailings dam consultants (Steffen, Robertson and Kirsten of Johannesburg) and in order to facilitate deposition planning, a spreadsheet model has been developed. The model takes into account pumping and maintenance costs and water losses for each paddock due to its size, position and elevation.

The changes in deposition method have led to other major benefits in addition to water savings being realised in the form of reduced power (pumping) costs, reduced pipe abrasion as well as large savings in capital expenditure.
The deferred capital expenditure results from the much slower silting up of the main pond referred to above which has eliminated the need to raise the containment berm which confines the pond in the west. In the long term further benefits are envisaged to result from the steeper overall beach angles which should become possible, thus allowing a much higher proportion of tonnage to be deposited closer to the plant.

A geotechnical investigation similar to reference 8, is currently under way in collaboration with the Rössing Tailings dam consultants and monitors the safe rate of rise constraints. Information from this investigation will allow the benefits of steeper beach angles to be maximised. These constraints exist because the new paddock blocks are constructed on top of many metres of much finer slimes resulting from the former perimeter discharge method practiced during the first twelve years of mine life.

CONTAMINANT CONTROL AND CONTAINMENT.

A series of dewatering wells, cutoff trenches and monitoring wells have been established around the tailings impoundment, in the natural groundwater drainage channels and in the Khan River to ensure the containment of seepage solutions and to monitor the groundwater quality. (Fig. 7).
Due to the large elevation difference between the tailings dam and the Khan River (ca. 250m over 6 km), it was necessary that the cutoff trenches were excavated in stream beds and blasted 3m into bedrock.

French drains were established in the trenches and continuous pumping keeps the water levels below 0,5m in the trenches. Approximately 30 boreholes have been drilled along the western perimeter of the tailings impoundment, siting based on geological and geophysical surveys and input from the Department of Water Affairs geohydrologists and consultants' geohydrologists.

Pumping rates are set to ensure maximum recovery of intercepted groundwater. A further 59 monitoring wells have been established around the property. Samples from the dewatering wells, cutoff trenches and monitoring wells are analysed for contaminants. Table 1 shows typical analyses of potable water, RDS (tailings pond solution), seepage from the tailings dam and Khan River. Table 2 lists comparative analyses outside the mine frontage. No legal standards for water quality exist in Namibia and, therefore, a range of American and Australian standards are also listed in table 2.

Water samples from the various boreholes and cutoff trenches are taken at regular intervals and analysed on site. Duplicate samples are submitted to the Department of Water Affairs for independent control analyses.

Seepage

Intercepted water from the trenches and dewatering wells is returned to the seepage dam. This dam, below the toe of the tailings dam in Pinnacle Gorge (Fig. 7 item 27) is the principal collection point of all non-acidic solution recycled to the process plants.
(On figure 7, items identified as N(14), M, KMW, 1.5, are monitoring borehole codes).
In construction the dam was excavated to bed-rock and equipped with a plastic liner. Cutoff trenches exist below the seepage dam. Tailings dam solution (pH=2) is neutralised (naturally) whilst seeping through the tailings embankment and underlying groundrock to a pH of about 5,5. Due to this pH change, it is not necessary for any treatment of the seepage solution before recycling to the process.

It has been found that a chemical "fingerprint" exists which can be used to indicate the presence of seepage into ground water. The sulphate/chloride ratio in seepage - high in sulphates, low in chlorides - is different from the ratio in natural ground water - low in sulphates, high in chlorides - and can be used to track a seepage front from the tailings impoundment. Due to the fact that this front can be accurately tracked, it is possible to plan corrective measures (i.e. additional cut-off trenches) if this front should move beyond the current cut-off trenches and dewatering wells. It also makes it possible to state with complete confidence that the contaminant front has, up to now, been contained within the mining area.

In addition, a site-specific computer model has been developed for Rössing to allow prediction of contaminant fronts.

The model is calibrated using site data, and used for predicting contaminant front movement to allow pre-emptive planning and action. The model is regularly recalibrated and refined as more data becomes available.

The model is also used in refining and up-dating the decommissioning plan for the mine.

FUTURE WORK

Reliability of Water Supplies

In spite of the major reduction in potable water consumption by Rössing, water resources on the West Coast of Namibia are still under considerable pressure. The Department of Water Affairs is currently in the process of constructing an alluvial dam wall across the Omaruru river to retard flood water and enhance recharge of the Omaruru aquifer. Rössing is investigating a similar concept although much smaller in scale for the enhanced recharge of the Khan aquifer. Abstraction of brack water from the Khan aquifer (situated approximately 6km south of the mine site) currently accounts for about 20% of total water consumed at Rössing. By constructing a 20 m alluvial wall across the river bed (see Fig. 7),

floodwater will be sufficiently retarded to allow decanting of clear water to downstream spreading grounds where rapid infiltration should occur. This scheme will ensure a reliable yield of 0,5 million m^3 per annum from the 8km stretch of Khan aquifer adjacent to the mine frontage. Engineering feasibility studies and an Environmental Impact Assessment will be completed during 1991 and construction is planned for 1992. By establishing additional boreholes upstream from the proposed dam, currently untapped resources in the aquifer will become available at a safe yield of a further 0,5 million m^3 p.a.

The combination of additional boreholes and enhanced recharge of the Khan River will ensure a safe yield of 1,0 million m^3 p.a. for Rössing which would otherwise have to be supplied by the Department of Water Affairs.

In the broader context of the West Coast water supplies, the Department of Water Affairs is pursuing other avenues of investigation in addition to the enhanced recharge of the Omaruru.

Improved Process Solution Quality

At present more than 50% of the total plant water consumption is from tailings solution and seepage, and although considerable benefits have been realised by recycling process effluent, it is of economic importance to improve process solution quality without losing the benefits of recycling.

Due to the nature of the leaching process, the sulphuric acid content and dissolved ferric iron, manganese and uranium in the Return Dam Solution (RDS) recycled to the circuit, constitute considerable cost benefits. However, to date various individual areas of increased costs due to recycling have been identified, and attempts have been made to alleviate these effects. Increased costs are incurred due to rising levels of chlorides in the solution circuit leading to increased stainless steel corrosion; increasing levels of dissolved solids leading to increased erosion of pump glandseal components. In addition precipitation of jarosite and related precipitates in solution pipelines results in solution throughput restrictions, causing uranium recovery inefficiencies. To overcome these negative effects, measures taken to date include poly-urethane coating of some under-solution stainless steel components (i.e. CCD thickener rakes), using stainless steel components in pump glandseal assemblies and mechanical cleaning of pipelines. In spite of the negative effects of recycling, a cost analysis of benefits and drawbacks indicate a favourable ratio of 3:1. This is primarily due to the much reduced need for capital expenditure mentioned earlier as well as the considerable credits obtained by recycling acidic solution high in soluble uranium, manganese and ferric.

However, most of these problems caused by recycling can be solved by treatment of one or more of the recycle streams or new input streams. Therefore, efforts are now being directed to formulating an overall strategy that will take cognisance not only of the economic feasibility of various treatment alternatives, but also of future changes in recycle stream composition and management of effluents generated by treatment of recycle streams.

CONCLUSION

Due to Rössing's moral and economic commitment to an overall water management strategy, it has shown that by developing site specific techniques, modifying accepted methods to suit and utilising technology, it is possible to operate a large mining complex in a desert environment with due consideration for responsible environmental and resource management.

ACKNOWLEDGMENTS

The authors wish to express their thanks to the management of Rössing Uranium Limited for permission to publish this paper and to acknowledge the contribution made by their colleagues.

FIG 1

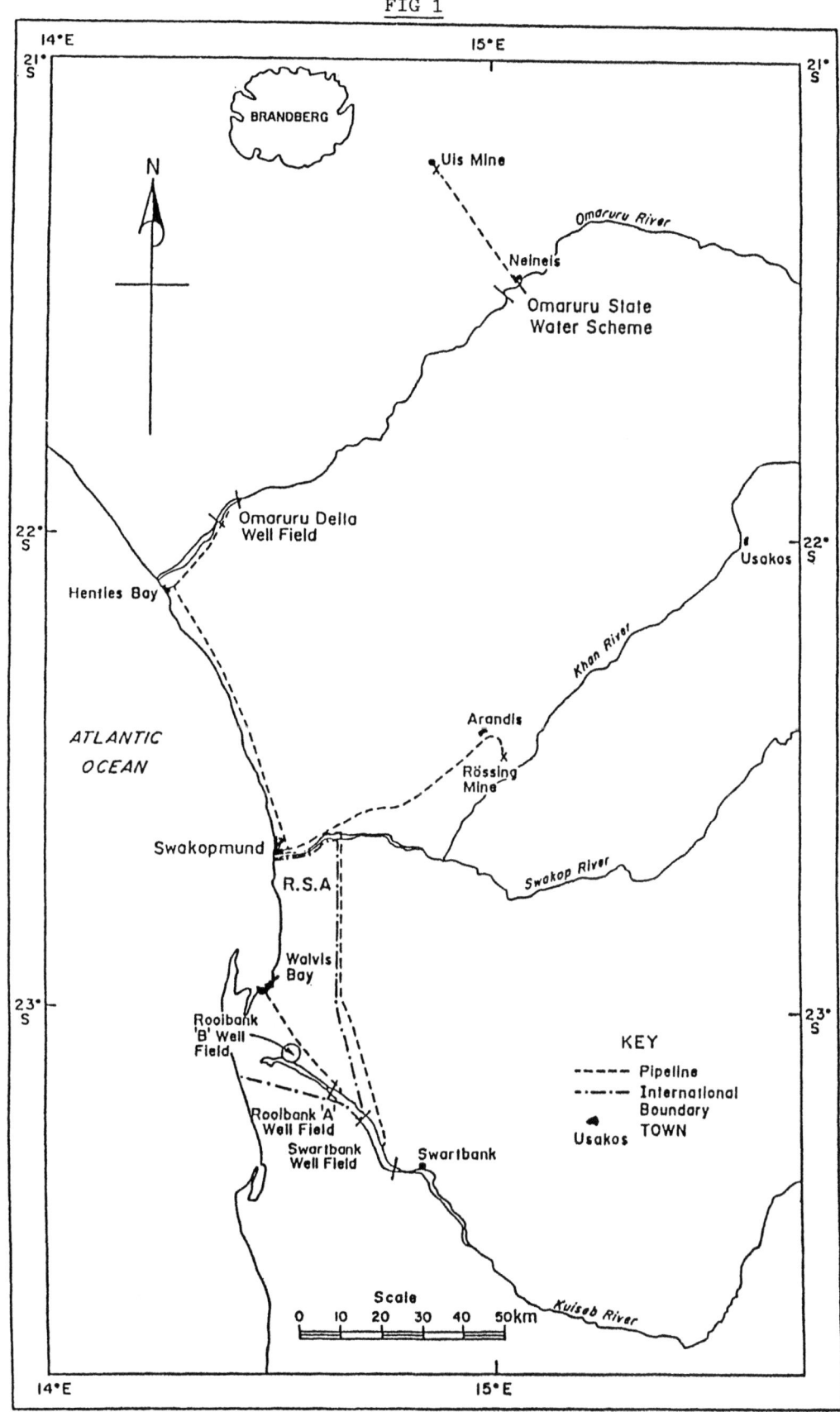

Diagrammatic map of the western portion of the central Namib Desert, showing the locations of water supply pipelines and the positions of water extraction well-fields on the Omaruru, and Kuiseb rivers in relation to nearby towns.

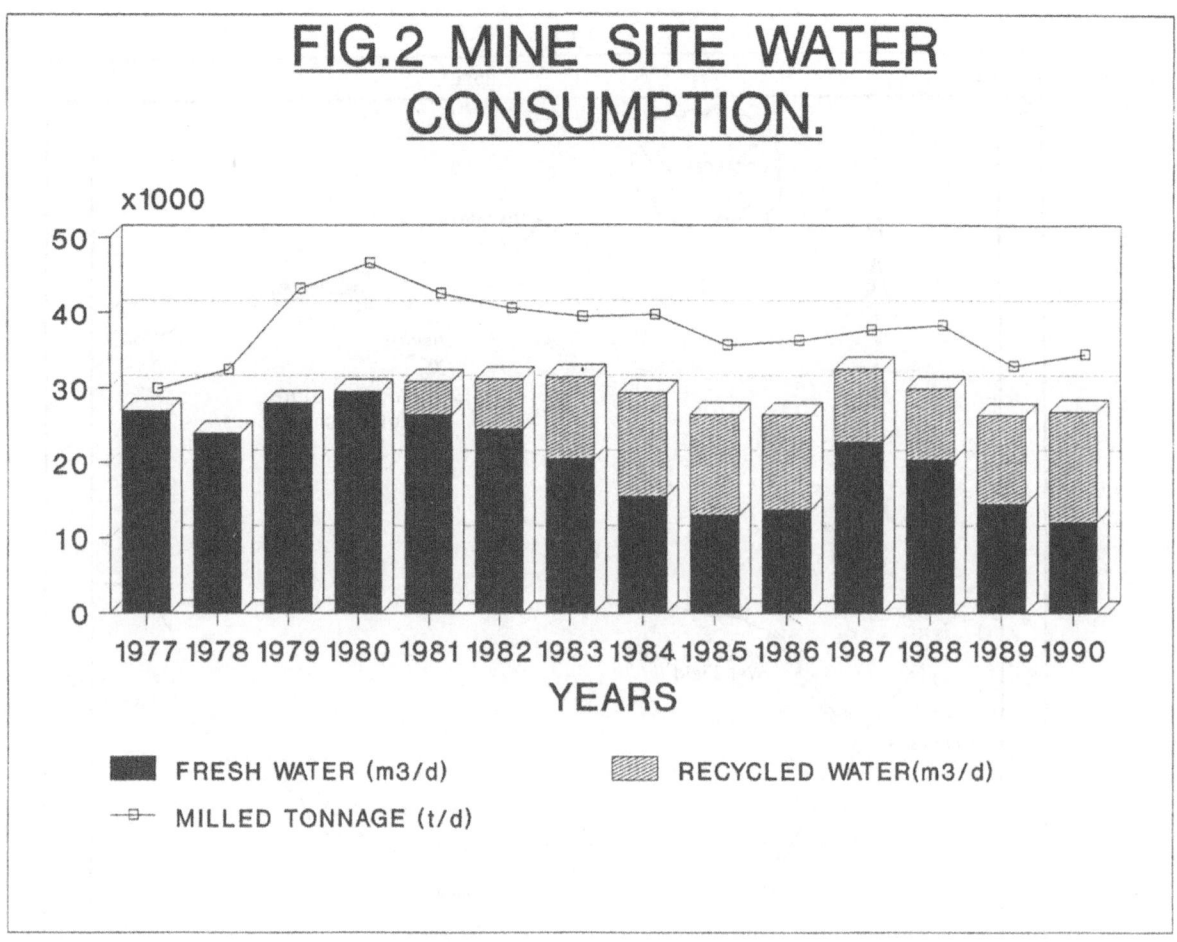

FIG.2 MINE SITE WATER CONSUMPTION.

x1000

- FRESH WATER (m3/d)
- RECYCLED WATER(m3/d)
- MILLED TONNAGE (t/d)

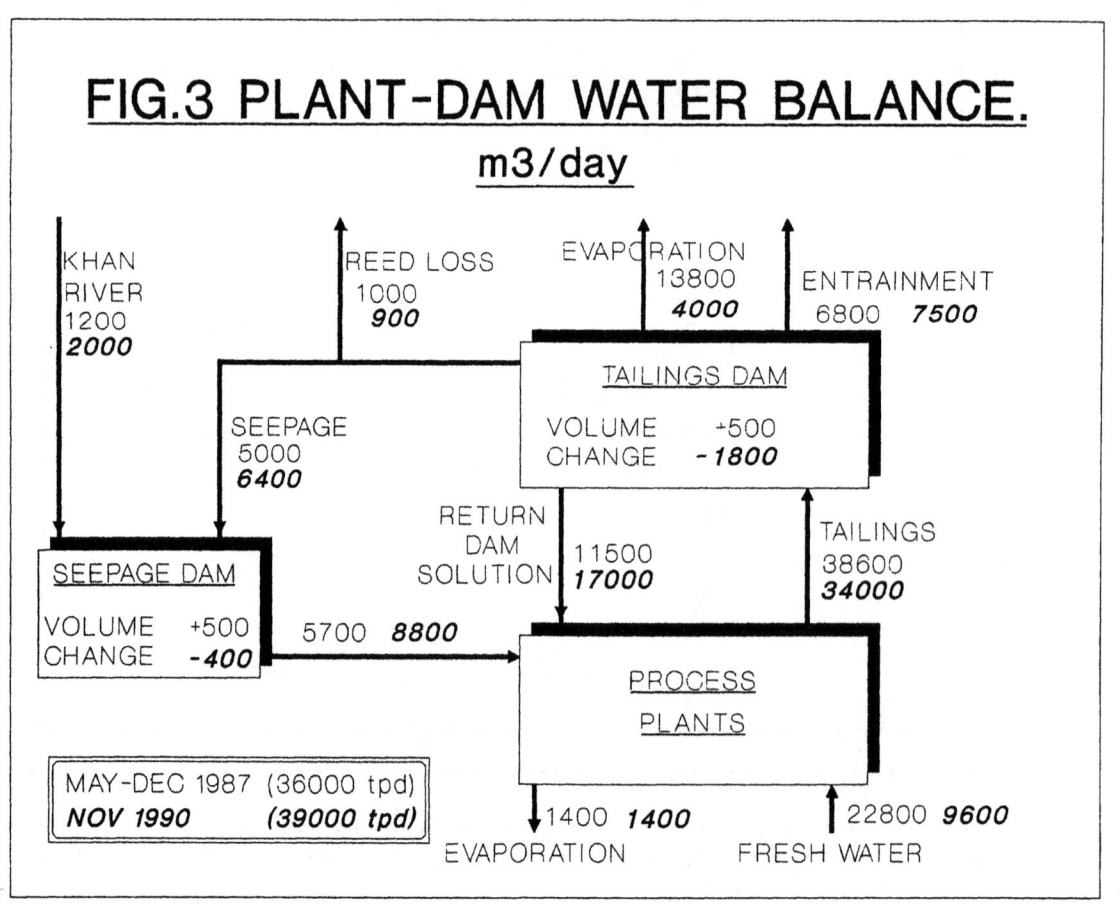

FIG.3 PLANT-DAM WATER BALANCE.
m3/day

KHAN RIVER 1200 *2000*

REED LOSS 1000 *900*

EVAPORATION 13800 *4000*

ENTRAINMENT 6800 *7500*

SEEPAGE 5000 *6400*

TAILINGS DAM

VOLUME CHANGE +500 *-1800*

RETURN DAM SOLUTION 11500 *17000*

TAILINGS 38600 *34000*

SEEPAGE DAM

VOLUME CHANGE +500 *-400*

5700 *8800*

PROCESS PLANTS

MAY-DEC 1987 (36000 tpd)
NOV 1990 (39000 tpd)

1400 *1400*
EVAPORATION

22800 *9600*
FRESH WATER

FIG.4 WATER LOSSES.
m3/day.

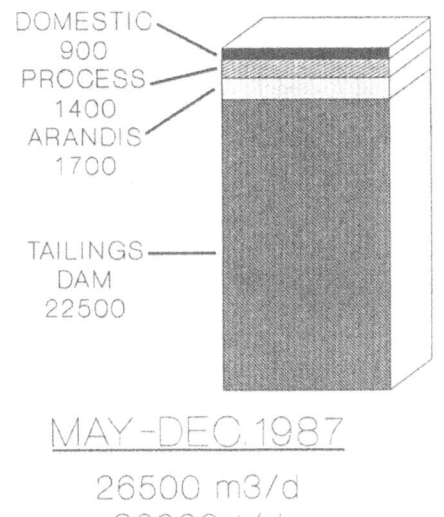

DOMESTIC
900
PROCESS
1400
ARANDIS
1700

TAILINGS
DAM
22500

MAY-DEC.1987
26500 m3/d
36000 t/d

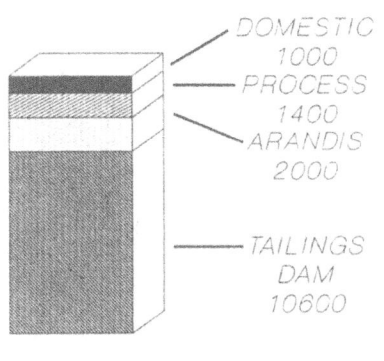

DOMESTIC
1000
PROCESS
1400
ARANDIS
2000

TAILINGS
DAM
10600

NOV.1990
15000 m3/d
39000 t/d

FIG.5 VELOCITY CONTROL.

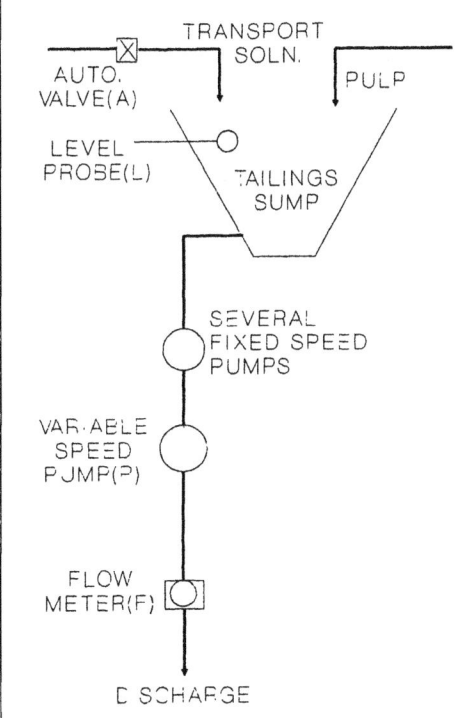

TRANSPORT
SOLN.
AUTO.
VALVE(A)
PULP
LEVEL
PROBE(L)
TAILINGS
SUMP
SEVERAL
FIXED SPEED
PUMPS
VARIABLE
SPEED
PUMP(P)
FLOW
METER(F)
DISCHARGE

ORIGINAL CONTROL

CONTROL LOGIC BASED ON CONTROL
OF VELOCITY(F).
CONTROL VALUE SET FOR FLOW(F),
WHICH CONTROLS PUMP SPEED(P).
LEVEL MEASUREMENT(L) CONTROLS
AUTOMATIC VALVE(A) IN RESPONSE
TO VARIATION IN (P).

CURRENT CONTROL.

CONTROL LOGIC BASED ON CONTROL
OF SUMP LEVEL(L).
CONTROL VALUE SET FOR LEVEL (L)
WHICH CONTROLS PUMP SPEED(P) AS
PRIMARY CONTROL.
AUTOMATIC VALVE(A) RESPONDS TO
CHANGES IN (L) BELOW A SET
MINIMUM LEVEL AS SECONDARY CONTROL.
FLOWMETER(F) HAS NO INFLUENCE
ON CONTROL.

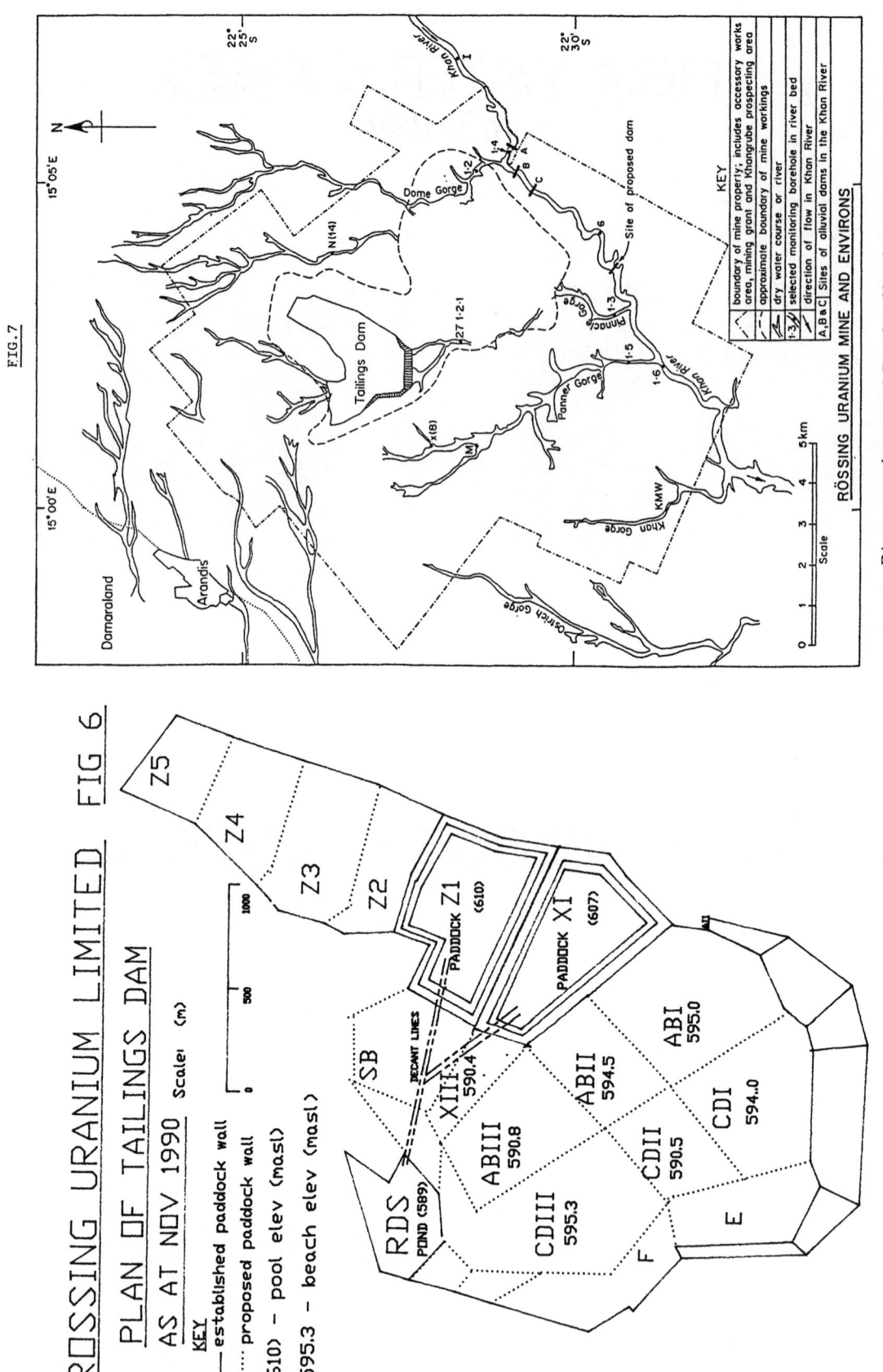

FIG.7

Diagrammatic map of Rössing Uranium Mine and its environs, showing the position of the proposed new dam on the Khan River in relation to the sites of the three alluvial dams and monitoring boreholes.

RÖSSING URANIUM MINE AND ENVIRONS

KEY

—·—·—	boundary of mine property; includes accessory works area, mining grant and Khangube prospecting area
—·—·—	approximate boundary of mine workings
- - -	dry water course or river
———	selected monitoring borehole in river bed
1-3 >	direction of flow in Khan River
A,B & C	Sites of alluvial dams in the Khan River

RÖSSING URANIUM LIMITED FIG 6

PLAN OF TAILINGS DAM

AS AT NOV 1990 Scale: (m)

KEY

——— established paddock wall

········ proposed paddock wall

(610) - pool elev (masl)

595.3 - beach elev (masl)

TABLE 1

TYPICAL WATER ANALYSES

		Potable Water	Return Dam Solution (RDS)	Seepage	Khan River (Mine Frontage)
pH		8,0	1,9	5,5	7,9
TDS	(ppm)	811	42900	14100	4000
Ca	(ppm)	69	550	430	320
Mg	(ppm)	21	1900	1020	130
Cl^-	(ppm)	296	1800	1700	1700
SO_4^{2-}	(ppm)	73	26500	7500	590
U	(ppm)	-	27	0,54	0,15
Na	(ppm)	170	1860	1300	820
Mn	(ppm)	50,1	1400	790	0,1

TABLE 2

COMPARATIVE WATER ANALYSES

		American and Australian Standards	Khan River Upstream	Khan River Downstream
pH		5 - 9,5	7,9	7,9
TDS	(ppm)	50 - 1500	3900	3900
Ca	(ppm)	75 - 200	310	320
Mg	(ppm)	50 - 150	140	130
Cl^-	(ppm)	250 - 2000	1670	1700
SO_4^{2-}	(ppm)	200 - 400	N.A.	590
U	(ppm)	0,02 - 44,3	0,11	0,17

REFERENCES

1. Wyllie, R., Rössing Uranium reaches full production, World Mining, (January - February 1979).

2. Vernon, P.N. and Sylvester, C.W., The Rössing continuous ion exchange plant, presented at a Symposium on Ion Exchange and Solvent Extraction, South African Institution of Mining and Metallurgy, Johannesburg, February 1980.

3. Kesler, S.B., and Fahrbach, D.O.E., Uranium extraction at Rössing, Uranium '82 Symposium, Toronto, 1982.

4. Kesler, S.B., and Knoetze, L., The introduction of central process control for metallurgical optimisation at Rössing , 15th International Congress on Mineral Processing, Cannes, France, June 1985.

5. Vernon, P.N., and Smit, M.T.R., Engineering and metallurgical improvements at the Rössing uranium treatment plant, Hydrometallurgy '87 Symposium, Manchester, England, July 1987.

6. Hamman, P.F., and Vernon, P.N., The re-use of tailings dam water at a large mine in the Namib desert in Namibia, presented at the 16 Congress of the International Water Supply Association, Rome, November 1986.

7. Guzman, A., Beale, C.O., and Vernon, P.N., The design and operation of the Rössing tailings pumping system, Proceedings of the 8th International Conference on the Hydraulic Transport of solids in pipes, Johannesburg, August 1982, British Hydrotransport Research Association.

8. McPhail, G.I., and Dorman, S.A., Increased tailings dam capacity through controlled deposition and pore pressure observation: a case history, Proceedings of the International Conference on Mining and Industrial Waste Management, Johannesburg, 1987.

9. Sharma, D. and Marais, P., A comprehensive mathematical model to predict hydrodynamics and mass transport at a complex mine tailings disposal facility, International Journal of Mine Water, Volume 5, number 1, March 1886.

Black Mountain—the oasis mine

P. E. Bryant
Black Mountain Mineral Development Co. (Pty) Ltd, South Africa

SYNOPSIS

When this mine was established in an arid environment, special attention had to be given to all aspects affecting employees, production, environment and the local community. This paper examines the position now that the mine has been in production for over ten years.

Specific topics discussed are road and rail infrastructure, supply of power and water, the township, and medical, conservation, recreation and socio-economic factors.

INTRODUCTION

To start a mine in a remote arid region entails a great deal more than the raising of finance and the establishment of the mining and metallurgical processes.

Of equal importance are the total infrastructure (road, rail, harbour, power and water, supply points), personnel factors (obtaining and keeping a skilled labour force, housing, schools, medical), and ecological aspects (both the effect of the mine on the environment and the effect of the environment on the mine).

LOCATION & HISTORY

Black Mountain is situated on the farm Aggeneys, in Bushmanland in the North West Cape region of South Africa. Fig. 1 shows the location relative to major and minor centres, infrastructure and geographical features, as it was prior to the establishment of the mine.

Bushmanland lies between the zones of winter and summer rainfall, and gets a little of each. Protracted droughts are a common feature, and in the recent past, some parts did not have any rain for a period of 10 years.

Temperatures at the mine site range between -2° and 45°C, the annual rainfall varies between 50 and 190mm, averaging 90mm, while wind velocities of up to 110km/hr have been experienced. Humidity levels are usually very low.

Bushmanland is generally flat, the topography being broken by stark rocky hills. Fossil red Kalahari sand dunes occur in certain zones and there are several large salt and gypsum pans. Sand depths reach over 100m in places.

The main natural vegetation consists of small perennial bushes and dry grass clumps, although in areas where there has been some rain in winter, spring flowers abound for a few weeks in August/ September.

The population is very scattered, based on a few isolated goat and sheep farms, some mission stations, and now, of course, the mine. In the entire region, twice the size of Wales, there are only three small towns; Pofadder, Kenhardt and Aggeneys, the township built to serve Black Mountain mine. Total population is about 15 000, (1 person per square mile).

The large herds of springbok and other small antelope which were abundant in previous centuries have all but disappeared. With the scant rainfall, over-grazing by domestic farm animals has become a common feature.

The Orange River forms the northern boundary of the region, but, because it runs in a gorge between mountains, does not present the same opportunity for irrigation farming as it does upstream in the Gordonia area.

Exploration

In contrast to the neighbouring Namaqualand, which has a long history of copper and diamond mining, there was in Bushmanland only one small, and now defunct, mine exploiting a sillimanite deposit at Swartkoppies, near Pella, and a small barytes prospect on top of Gamsberg.

Figure 1 : North-west Cape region : 1970

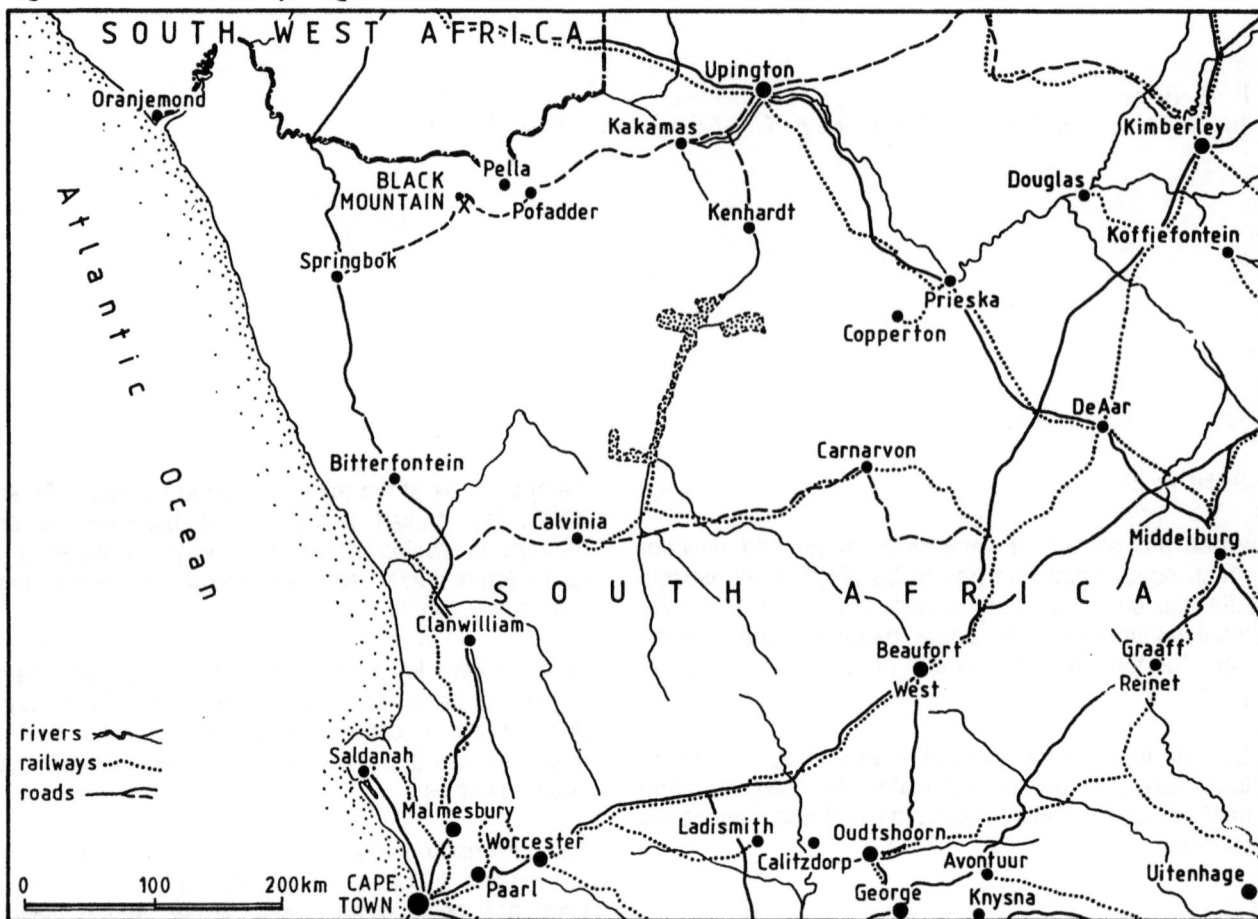

Only sporadic base mineral exploration had been done prior to 1970, when Phelps Dodge commenced exploration at Black Mountain, discovering three mineralised zones, of which the highest grade Broken Hill deposit was the obvious initial target for mining. Here surface diamond drilling, together with underground drilling from an exploration adit developed in 1974, outlined two sub-conformable orebodies, with an initial combined reserve of 33Mt at 0,45%Cu, 6,35%Pb, 2,87%Zn and 82g/t Ag.

Exploitation

Once viability had been proved by a feasibility study, sinking of the Broken Hill main access decline from surface was commenced in 1976.

In 1977, Phelps Dodge Corporation (PD) entered into a substantive agreement with Gold Fields of South Africa (GFSA) for the development of the deposits, in terms of which GFSA subscribed for a 51 per cent interest in Black Mountain Mineral Development (BMMD) and consequently PD interest was reduced to 49 per cent. GFSA was, thereafter, responsible for the management of Black Mountain.

The mine was brought into production in October, 1979.

Underground mining is being conducted on the Broken Hill deposit, initially by post-filled blast hole open stoping, but now entirely by mechanised cut and fill stoping. The mine is served by a vertical hoisting shaft, an access decline, and a conveyor sub-decline feeding ore below shaft bottom elevation up to the shaft.

An average of 120 000t/mth is mined, of which 100 000t is ore, treated on site by multi-stage flotation to produce three concentrates, copper/silver; lead/silver and zinc.

While the Cu and Pb concentrates must be shipped overseas, the Zn concentrate can be treated in South Africa by Zincor, in the Witwatersrand area.

Initial Infrastructure

At the start of exploration in 1970, the only communication link available to the property was a dirt road linking Springbok and Kakamas, via Pofadder.

The nearest railheads were at Kakamas, 200km to the east, and Bitterfontein, 300km to the south. The Sishen/ Saldanha railway, dedicated to the transport of iron ore, had been approved and the proposed route passed the property some 150km to the south, but construction of this line had only just begun.

Saldanha harbour, some 600km away, was still to be adapted for ore carriers. Cape Town harbour, the next possible port which could handle large ships, was over 700km away by road.

The nearest airport was at Springbok, 120km to the west. This was also the nearest potential power supply point, although at that time Springbok was not connected to the national grid, but drew power from a local power station, which could not be expanded.

Water could be obtained only in relatively small quantities from poor quality boreholes. The only perennial river, the Orange, flowed some 50km to the north of the property.

A daunting picture, indeed.

Exploration camp

During the exploration phase, a base camp was established beside the Black Mountain itself. This camp grew and eventually comprised an airstrip, mess, store, transport workshop, electrical workshop, temporary mine offices, and prefabricated housing. Power was supplied by diesel alternators while water was obtained from 16 separate boreholes, interlinked with 45km of piping to three water tanks.

This camp was used until the permanent township, Aggeneys, was established, and, even today, still has a few families living in it.

INFRASTRUCTURE

Fig. 2 shows the completed infrastructure.

Pella pump station

The first priority for a mining operation in this sort of climate is water - an adequate, continuous supply of water.

The Water Act makes provision for the establishment of Water Boards to design, construct and operate schemes for the supply of water to communities and enterprises and, with the consent of the Treasury, to raise funds on the open market to finance the scheme. Since it is a statutory board, any borrowings are regarded as approved securities, enjoying a lower interest rate, and also attracting a 20 per cent subsidy from the Treasury. This mixture of state and private enterprise results in a considerable reduction in the eventual cost per cubic metre consumed.

After discussions with the Department of Water Affairs and all other interested parties, the Pella Water Board was established in November 1974, in terms of the Water Act, whose area of jurisdiction was eventually to include Black Mountain, Pella Mission Station, Pella Refractory Ores at Swartkoppies, Pofadder, and Gamsberg.

Following comprehensive investigations, the decision was made to establish a waterworks and pump station at Pelladrift, on the Orange River, and work began on this project in April, 1978.

The scheme consists of the following:

Intake tower - situated in the river with the low lift pumps positioned 13,5m above the river bed, just above the assumed "100-year flood level". Water is pumped at a maximum rate of 432m³/hour through a total head of 21m to the:

Clarification works - the turbid river water is prechlorinated and then sent to a 23,6m diameter clarifier. The clarified water is sent to a 2000m³ balancing reservoir, which feeds the:

High lift pump station - Two 1218kW and two 746kW five stage pumps operate as required to meet the demand. The total pipe length to the mine is 52km long and is made up as follows:

Rising main - A 30km long 400mm ID steel rising main carries the water from the pump station to the 2000m³ Horseshoe reservoir, with take-offs for supply lines to the Pella mission station, Pofadder municipality, and Swartkoppies. With a static head of 640m and a friction head of 160m at the designed capacity of 12 000m³/day, the total head is 800m.

Gravity feed - From the Horseshoe reservoir, water is gravity fed in a 13km 400mm ID asbestos cement pipe to the 1200m³ domestic water Kokerboom reservoir, situated 4km north of Aggeneys township, and then through a 9km long 360mm ID pipe to the 20 000 m³ Saddleback reservoir, which feeds the mine complex. Nett head over this section is 8m.

The first Orange River water reached Aggeneys on 29 March, 1979. Current average daily water (m³) consumption is:

	Summer	Winter
Mine and plant	4220	4706
Township	7785	4379

Two points of interest should be noted:

Flood

In February, 1988, abnormally heavy rains fell in the Orange River and the tributary Vaal River catchment areas, causing extensive flooding and damage. On 23 February, initial information forecast a 13m rise in the river level at Pelladrift, which would flood both the low

Figure 2 : North-west Cape region : 1990

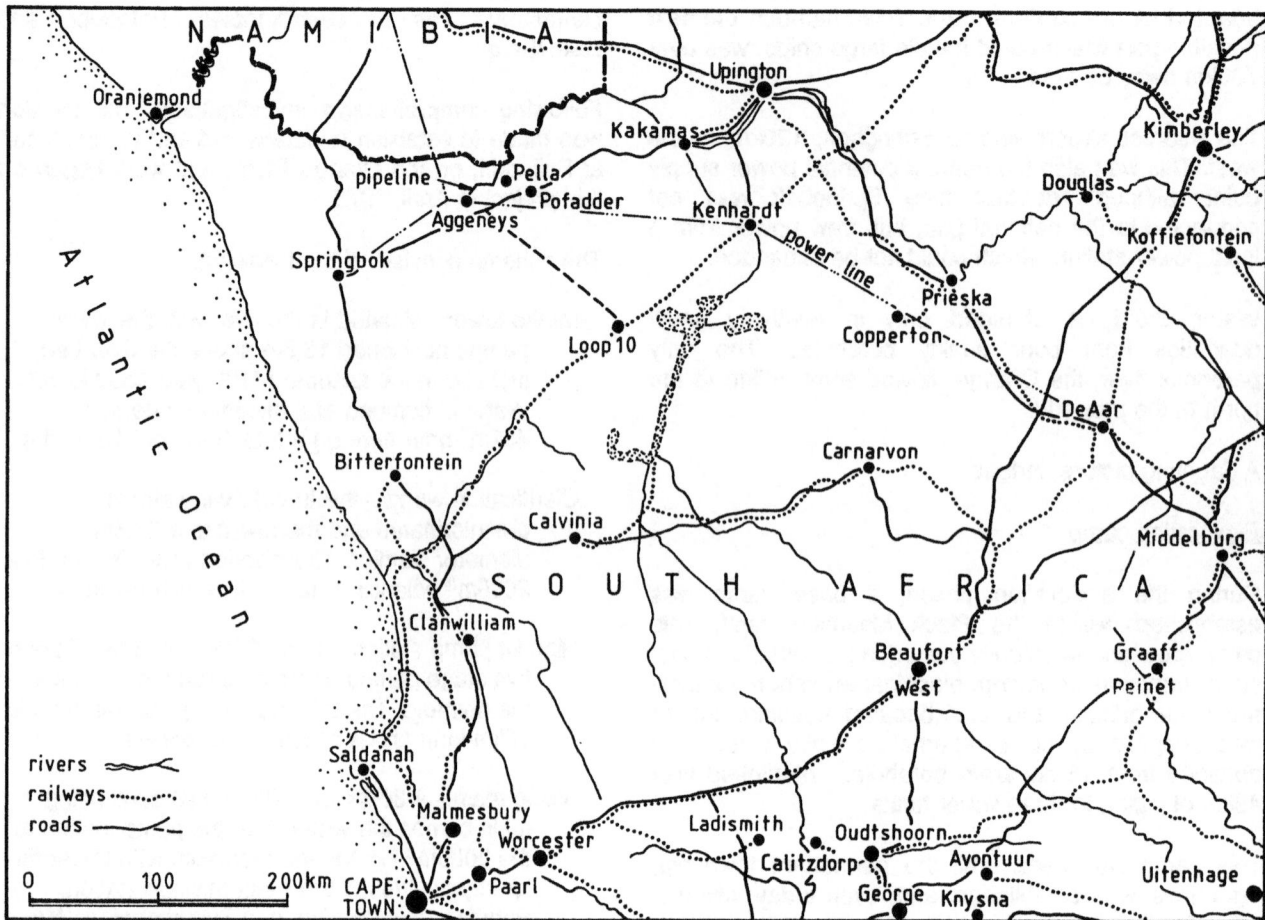

and high level pumps.

had not seen rain for months!

The decision was immediately made to remove the equipment from both pump stations, since delivery periods for the replacement motors were estimated to be 6 - 9 months, and a loss of water to the mine for this protracted a period would have been disastrous.

By 28 February, it had become apparent that, due to the river having burst its banks in so many places upstream, and the subsequent flooded areas acting as accumulators to slow the river down, the flood level would be not be as high as had been predicted, whereupon, with the river still rising, 50% of the equipment was reinstalled. The river reached its maximum height, just over 1m below the level of the low lift pumps, at 04h00 on the following day. By 16h00 on the same day, the Pella pump station was again delivering water to the mine. Some 20 000t of production had been lost.

Either the prediction of the 100yr flood level was remarkably accurate, or else that flood was not the 100yr one. Whichever, in the final analysis, the company could not afford to take the chance of leaving the pumps in position.

As an aside, the irony of the situation was that we were having to battle with a flood when we ourselves

Capacity

The bottleneck in the system is the rising main, and this means that any future expansion to the system, if required, will involve major costs. With hindsight, it would have been preferable to install a larger column, allowing the bottleneck to be the pump delivery, since pumps can be added relatively easily, whereas a parallel rising main would be very expensive.

Concentrate transport

Railway

The railway line from Sishen to Saldanha, which was completed in 1976, is 861km long and, from Sishen, some 1200m above sea level, slopes gradually downwards to meet the coastline at Strandfontein, from where it runs 100km along the West Coast to Saldanha.

There are no conventional stations, but crossing places with long loops are installed at intervals along the line. Loop 10, which is 447km from Saldanha, is the halfway crossing point, where train crews are rested after 8-9 hour shifts. This loop was the nearest one to Black Mountain, and so a 2,5km long siding was installed there for BMMD concentrate loading. Concentrate storage sheds, with separate bays for the three concentrates, were erected, to which concentrate could be delivered by trucking, and from which the railway trucks could be loaded. A workshop facility, shower/toilet block, kitchenette, weighbridge, road loop and accommodation were also required.

Cu and Pb concentrates are sent to Saldanha in special 50t trucks which are also used for the transport of iron ore, and are dedicated to this line. These trucks are not suitable for use in the country wide railway system over which the Zn concentrate must pass on its way to Zincor. Thus standard 40t railway trucks must be brought in to load this latter concentrate.

Of importance are the precautions taken against lead poisoning. Apart from the compulsory use of respirators, men working in the sheds are allowed to eat only at a fixed time during the shift, and must shower and change into clean overalls prior to eating. Overalls are laundered on the mine.

Concentrate haul road

Early in 1979, the Cape Provincial Roads Department awarded a contract to build a heavy duty gravel road from the Springbok/

Pofadder road, at a point about 14km from Aggeneys, to Loop 10 on the Saldanha/Sishen railway line. This road was 155km long and was completed in the same year. A contractor is responsible for the concentrate trucking and loading operation.

Harbour

The iron ore loading facilities at Saldanha harbour could not be utilised by BMMD, and so, in 1978, South African Railways and Harbours was commissioned to design and build a ship-loading facility for lead and copper concentrates. This consisted of a storage/reclaim area, approximately 600m north of the main quay and serviced by suitable shunting trucks, and a wharf for medium tonnage bulk carriers, approximately 1km down the length of the main quay.

The concentrate shed has an upper level, where eight 50t rail trucks at a time can be offloaded through grizzleys onto a lower storage floor. This storage floor can hold a total of about 30 000t of lead and copper concentrates.

Reclamation is done by means of front end loading into a travelling hopper which traverses a conveyor running the length of the concentrate shed. This conveyor feeds a bucket type conveyor to a surge bin, from where the concentrate discharges into two 12t measuring flasks mounted on load cells.

Concentrates are conveyed from the reclaim area to the wharf by means of tractor-drawn trailers, each conveying two 12t overturning skips. The skips are lifted into the ship's holds by means of high speed quayside cranes, where they discharge automatically once they come to rest in the hold.

The reclaim conveyor and batch weighing system are controlled by computer, and the system is capable of handling 300t/hr.

Power

Power is supplied from the Electricity Supply Commission network at the Hydra sub-station, De Aar(some 650km away), via two independent lines, to the Aries substation at Kenhardt. From here a 400kV single line supplies the terminal sub-station at Aggeneys, whence three separate feeders supply Springbok, Namibia and Black Mountain.

Average daily power consumption in MVA is as follows:

	Summer	Winter
Mine and plant	22,9	23,5
Village	1,2	1,6
Pella pumps	2,4	2,0

Material supplies

At Springbok, a certain amount of secondary industry is based on the Namaqualand mining activities, and several mining equipment agencies are established there, but the main centres from which supplies can be obtained are chiefly Johannesburg (1200km away), and Cape Town. From both these centres, supplies can be railed only part of the way, and then have to be trucked by road.

When the Black Mountain project started, there was an urgent need for the dirt road to be upgraded, and, by 1981, the section between Springbok and Aggeneys had been tarred, followed later by the complete road between Aggeneys and Kakamas.

Despite this, transport costs of goods to Aggeneys are still extremely high. In certain instances, the transport cost of an item will exceed its own cost.

Air field

Initially, the mine used the original air strip established for the base camp. Owing to the proximity to several hills, and the distance of this old strip from the newly established township, a new surfaced runway was eventually established closer to the township.

A commercial air line runs a regular service, connecting Aggeneys to Johannesburg.

General

I cannot leave this section without mentioning the co-operation given by the oft-maligned State-run organisations, who gave every assistance possible with the project, and who ensured that the various infrastructural needs were provided on time.

PERSONNEL

The current labour requirement for the mine is listed in Table I, (Patterson ratings are shown).

Sourcing the labour

The crushing socio-economic conditions experienced by the resident population of Bushmanland are hard to picture with First World eyes.

Generations have grown up on a subsistence basis, eking out a living by occasional work for mission stations and seasonal crop picking in the Upington/ Kakamas area. Education levels are generally low, children often having to leave school early to seek some form of employment to help their families.

This population group could not therefore without lengthy training provide the skills required in modern mining, though some immediate use could be made of it for the supply of unskilled labour. In any event, the Bushmanland population was too small to provide all the labour requirement, and so some unskilled migrant labour had to be obtained from the homelands

The majority of the miners and u/g equipment oper- ators could be sourced from the Springbok area, where certain copper mining operations had been curtailed at that time.

Only a few persons with the necessary supervisory, technical, or management skills could be obtained from either of these areas, and most of this category was sourced from the Witwatersrand.

Thus it was important from the outset that living and working conditions were made as attractive as possible at Black Mountain so that all those persons who were out of their normal environment could feel contented. It was equally important that suitable members of the local population be trained to partake more fully in the employment advantages offered by the mine.

Training

The mine runs a training centre which trains personnel not only in the functional aspects of the various jobs up to, and including, miner, but also in loss control, induction, literacy, and language. The Namaqualand Technical Training Institute in O'okiep, run jointly by the mines of Namaqualand and Bushmanland, is used for the training of apprentices.

Further training facilities are available at Gold Fields Training Services at Kloof Mine, and use also is made of various tertiary institutions.

Township

Housing

As there is no established community within a reasonable distance of the mine, accommodation has to be supplied for all categories of employees mentioned above. Additionally, accommodation must also be made available to persons who are not in the employ of the company, but are required in the township, such as post office staff, police, doctors, chemist, teachers, shop owners, and that most important person in the community - the hairdresser.

Since housing is a very expensive item, it has not been possible to supply married housing to all staff, and houses are currently supplied to all labour in categories B4 and above, with some of the more senior B3 persons being given housing if available.

To reduce the pressure on housing, it is mine policy that as many posts as possible should be filled by wives of male employees, and two nursery school/ creches have been established. Apart from the normal clerical and nursing jobs, females are employed in the plant control room and assay laboratory. Our requests to the Inspectorate to be allowed to employ females as hoist drivers and backfill control room operators have not, as yet, met with success.

Management and skilled persons are housed mainly in the North Village, which has 235 houses, a single

Table I: Labour requirement at BMMD		
Management: D band :- down to mine overseer level		33
Skilled: C band :- staff and clerical		156
C band :- miners and artisans		145
Semi-skilled B4-B6 :- heavy m/c operators/drivers		180
Unskilled B3 down:- labourers, light m/c drivers		874
Total operating labour		1388
Others various:- apprentices & sundry		40
Total labour		1428

Figure 3 : Typical Aggeneys house

quarters containing 12 rooms, 8 flats for senior single staff and visitors, and 28 mobile homes, (ex the original base camp).

All semi-skilled persons, some skilled persons, and the more senior B3 persons, are housed in the South village, which comprises 254 houses, a male single quarters for 110 men, and a female single quarters for 30 women.

The remaining employees are housed in single quarter accommodation in the hostel complex, in which are included 12 married quarters for visiting wives.

The size of a house allocated to an employee is dependent on his job category. Every house is designed to be cool. Fig. 3 shows a typical house design having a double flat roof with an air gap between the roofs to prevent heat build-up. All houses are supplied with evaporative desert coolers, which are very effective in the Bushmanland climate due to the extremely low humidity levels. Cooking stoves and electric geysers are supplied in all houses.

The rooms in the hostel complex are fitted with air conditioners, for both cooling and heating.

Shopping centre

A shopping centre, hospital, childrens' playground and a multi-denominational church lie between the North and South Villages.

The small shopping complex contains a post office, cafe/vegetable shop, supermarket/butcher, hairdresser, public library, chemist, garage, two banking agencies, clothing shop, video outlet and a needlework shop. A bus service provides transport to Springbok three times a week.

Initially, the mine ran a vegetable garden, piggery, bakery and a sheep herd to ensure the availability of basic foodstuffs to the inhabitants. As the lines of supply have become established over the years, all of these have become redundant.

Gardens

From the beginning, residents have been encouraged to establish and maintain gardens, and a nursery is run by the mine to facilitate this. Great emphasis has been placed on tree planting in the gardens and streets, both for aesthetic reasons and for natural cooling, and this has paid off handsomely.

Water is supplied free of charge for both domestic and for garden use. To avoid excess use, the larger gardens and the public park areas have automatic timer operated pop-up sprays, which allow watering at night, when evaporation rates are lowest, and which can be adjusted during the year to suit climatic conditions. Trees along the roads are watered with a drip irrigation system.

It is the sight of the lush grass lawns, the thousands of trees, and the flower-filled gardens that have led to visitors calling Black Mountain "The Oasis Mine"

Schooling

There are two primary schools in Aggeneys, but no secondary schools. The company compensates parents who must send their children to schools away from the township by paying the boarding fees and providing free transport to and from the schools.

Recreation

So far away from canned entertainment, recreational activities are an important factor of life in the township, and each of the two villages has its own recreation club, supporting all the popular sports such as squash, bowls, tennis, badminton, rugby, soccer, cricket, running, shooting, swimming, etc. The golf and riding clubs each have separate club houses. The 9-hole golf course is fully grassed.

The hostels have beer gardens, tribal dancing areas, swimming pool, video show rooms, and soccer fields.

Medical

A 19 bed hospital, run by the mine, and two private doctors cater for the normal medical needs of the community. A dentist and eye specialist visit Aggeneys at regular intervals.

Until recently, a plane was kept on site to cater for emergency medical needs and to provide transport 2 - 3 times a week to assist patients to see specialists in Cape Town. With the improvement in emergency medical flights in the province, this service is no longer felt to be necessary and has been discontinued.

A kennel is run by the mine, and a veterinarian visits at least once a month.

ENVIRONMENT & ECOLOGY

An arid region is more vulnerable to ecological damage than others, firstly because the eco-system is much more specialised and thus vulnerable to pollutants and physical damage, and secondly because the recovery of the veld takes so much longer, which can lead to the permanent loss of fauna and flora.

Because of this, extra care has to be taken by mining companies in this sort of environment, and a responsible management outlook is essential.

Fig. 4 shows the surface rights owned by the mine, just under 25 000ha in extent, together with the main features referred to in this section.

The property is crossed by the Springbok/Pofadder road, and the zone to the south of this road, some 9000ha with extensive sand dunes, is treated as a wilderness reserve. The centre portion of the property is allocated to operations, housing and recreation, while the northern hilly section is largely untouched, but might be required for mining operations at some future time.

Waste disposal

An important aspect of the environmental/mining interaction is concerned with the disposal of waste products. Waste products are disposed of as follows:

Waste rock

About $2/3$ of the waste rock mined is packed into the stopes as fill and does not reach surface - the remainder is hoisted and stacked in a dump resting on the lower slopes of a hill.

Tailings

About $1/6$ of the tailings produced is used underground as hydraulic backfill, the remainder being stored in a conventional slimes dam. Oxidation of the slimes in the sides of the dam has caused crusting, and this eliminates wind-blown dust. A major portion of the slimes is composed of a particularly pure form of magnetite, which may be of future economic value.

Waste water

Ageing pond - water from the slimes dam is caught by drainage trenches around the foot of the dam and is led to an ageing pond, from where it is pumped for re-use in the plant and for hydraulic backfill.

Evaporation pond - surplus water from the ageing pond runs to an evaporation pond at Plaatjiesvlei where it is permanently retained. Although the level rises and falls with varying climatic conditions, overall the system is in balance, with no water leaving the property. One advantage of an extremely dry climate is that disposal of water by evaporation is quite effective. Supposedly this water should be a real chemical soup, since evaporation concentrates the dissolved chemicals, but

Figure 4 : Surface area
         Black Mountain Mine

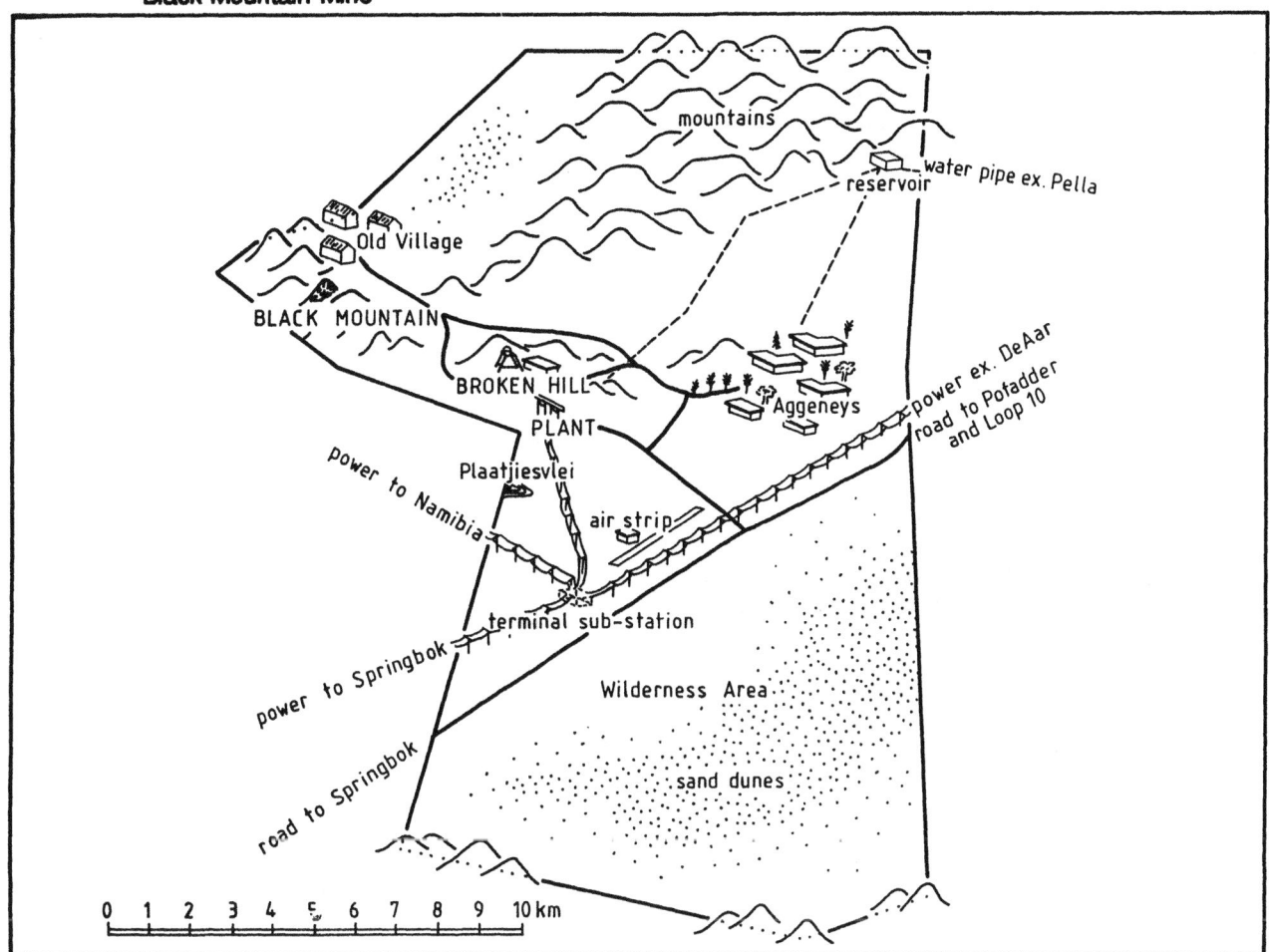

natural processes are at work to break them up, and this pond has become the habitat of various sorts of water fowls. It would appear that unplanned "wetlands" purification is happening here.

The mine has a herd of sheep grazing around the area of the pond- to monitor any problems with leakage, should it occur!

Due to an impervious calcrete layer below surface there is little or no possibility of pollution of any sub surface water, but sampling of water contained in a borehole drilled at the lowest point of the property has been done to confirm this.

Underground water - Very little natural water is encountered underground. Orange River water is used for service and drinking, and also in the backfill plant when cemented fill toppings are being thrown. The bulk of the water being pumped from underground originates from backfill drainage. Dirty water pumps deliver unsettled water from underground to a water clarifier on surface. The underflow from this clarifier is sent to the slimes dam, while the clarified water is mainly used for backfilling operations, with any surplus being sent to the concentrator.

Sewage

Sewage is treated in sewage evaporation ponds, water emanating from these ponds being used for irrigation of lucerne and for watering the golf course. In the period that the mine has been in operation, it has been necessary to clean out the solid residue once only, and, on this occasion, the residue was buried in trenches.

Rubbish

Two rubbish pits are used. Rubbish is burnt and buried.

Air pollution

All crushing and allied plant operations that create dust have exhaust systems that lead the dust-laden air to filter banks.

Fortunately from the point of view of pollution, the mine does not have a smelter, but it does run a small plant producing sulphurous acid for the process, from which a small amount of sulphur fumes enters the atmosphere through a stack. As far as can be ascertained, the quantity is too small to have had any deleterious effect on any aspect of the local ecosystems.

Fauna and flora

The sand dune region in the south of the property and the hill region in the north of the property are both being allowed to return to their natural state. In the period that the mine has been in operation a significant recovery from the original overgrazed condition is evident.

Dune region

In the dune region to the south, natural dune grasses are again stabilising the dunes, and the increased production of grass seed has allowed the numbers of small seed-eating birds, insects, rodents and other small mammals to increase, which, in turn, has enabled species higher up the food chain to augment their numbers. Conspicuous among these are the raptors, ranging from the Martial Eagle to goshawks, harriers and the smaller kestrels. Probably the most interesting feature of these dunes is that they contain the only known nesting place of the Red Lark in the world, a sobering thought.

A herd of about 800 springbok also lives in this area.

Central region

The flat stony plain on which the mine and township are situated is a strange mixture of ecosystems. Various types of large mammals, including springbok, eland, and camels have been introduced into, and are allowed to roam in, the areas not being used. Some ostriches also exist here.

At the sewage and ageing ponds, various species of duck, geese and waders live permanently, and these ponds also provide temporary stops for migrating birds.

In the township, the gardens and trees host an increasingly wide variety of birds. Unfortunately, certain insect species, notably crickets and gnats, also flourish, probably because their natural enemies have not become sufficiently well established in this artificial environment.

Hill Region

This zone has nesting Black Eagles, Lanner Falcon and Pigmy Falcon, but is mainly noted for the plant life. Apart from Kokerboom, Halfmens, Stone Flower, and various other succulents and aloes, the most remarkable is a conophyte known as Burger's Onion (Conophytum burgeri)

This plant is found nowhere else in the world, and is confined to two small parts of this region. Its main bulbous stem is concealed in areas of opaque to translucent quartz pebbles in scree slopes, and while being shielded from the harsh sun is still able to photosynthesise in the light filtering through the pebbles. People from all over the world have come to see this tiny plant, but only restricted access is allowed- It would be too easy to eradicate this plant totally by irresponsible actions.

SUMMARY

The establishment of Black Mountain Mine is an achievement of which all those involved can justifiably be proud. Not only has a flourishing mine been created in this desolate region, but a high degree of responsibility to the people and the environment has been shown, and this is certainly one mine that has improved, and not degraded, the area that it is in.

ACKNOWLEDGEMENTS

This paper is presented with the kind permission of the management of Gold Fields of South Africa, Limited.

The author has gathered some of his information from a symposium of 15 papers entitled 'The Black Mountain Project', compiled and edited by D.A. Blair Hook and presented to the Association of Mine Managers of South Africa on 28 November, 1980. Anyone wanting more details about the establishment of this mine would be well advised to refer to this comprehensive set of papers.

Further information was obtained from an article entitled 'Mining- an ally of conservation' written by D.J. Mourant and published in the 1989 Goldfields Review.

Metallurgy 3

Application of Codelco–Chile pyrometallurgical techniques at the Nkana smelter, Zambia

C. Salazar R.
Codelco–Chile, El Teniente Division, Chile
F. Hernandez A.
Codelco–Chile, El Teniente Division, Chile
M. D. Strachan
Techpro Mining and Metallurgy, Ashford, Kent, England
A. Cross
ZCCM Ltd, Nkana Division, Zambia

SYNOPSIS

The Zambian copper producer, Zambia Consolidated Copper Mines Ltd. (ZCCM) is presently carrying out a modernisation and optimisation project at its Nkana Smelter. As part of the implementation, ZCCM have entered into a contract with CODELCO-Chile for the provision of front end engineering services at the design stage, for detailing by TECHPRO Mining and Metallurgy, as part of an overall retrofit strategy. There is also an Agreement covering the transfer of operating know-how relating to the pyrometallurgical techniques both for oxygen/fuel-oil burners in reverberatory furnaces and for the autogenous smelting of concentrate with the simultaneous conversion of reverberatory furnace matte using oxygen enrichment of the blowing air in a Teniente Converter. These techniques have been developed by El Teniente Division, and have been in use for over a decade at the Division's Caletones Smelter, as well as in other smelters owned by CODELCO-Chile and ENAMI, within Chile.

This paper sets out the process parameters for various capacity options considered for the Nkana Smelter, including mass and energy balances, and sizing of basic equipment, as well as some aspects of the general design of the main equipment.

INTRODUCTION

Nkana Smelter, one of the three copper smelters of Zambia Consolidated Copper Mines Ltd. (ZCCM), is typical of the old fashioned copper production process. The smelting of green charge copper concentrates, together with limerock flux and recycled reverts, is carried out in reverberatory furnaces using fossil fuels in end-wall burners. The only recent modification of note was to introduce partial enrichment of the preheated combustion air with oxygen. Primary smelting is then followed by the conversion of matte to blister copper using standard Peirce-Smith converters. The copper is conditioned in rotary furnaces to produce anodes, which are then electro-refined. The complete complex also includes facilities for producing cobalt[1], and four plants producing sulphuric acid[2].

Concentrates are supplied from various mines owned by the company; the main concentrators being at Nchanga, Luanshya, Nkana and Konkola.

To implement a far-reaching modernisation and optimisation programme of the Nkana Smelter, ZCCM decided to introduce new technology with the aim of attaining four main objectives:-

- improving the energy balance of the smelting circuits.

- optimising the use of available oxygen.

- reducing operating and maintenance costs.

- improving the quality, and so the potential for collection, of the process off-gases for the production of sulphuric acid.

To achieve these objectives, ZCCM carried out a number of cost and feasibility studies to determine the best technical and economic option. TECHPRO Mining and Metallurgy were contracted to investigate in detail modern pyrometallurgical technologies, such as electric smelting and flash smelting, as well as continuous smelting-conversion processes of the "bath smelting" type. The main criterion was that the process must have been successfully operated on an industrial scale in other copper smelters. Also of importance were aspects related to slag chemistry, impurities, gas treatment, operability, and use of available oxygen, as well as ease of retrofit into an existing operating plant. The result was that Teniente Technology was selected[3].

PRESENT EQUIPMENT AND PROCESSES

The Nkana Smelter was first operated in the 1930's. Present scheduled production of primary copper is in the region of 240,000 tpy, although historically production in excess of 330,000 tpy has been attained.

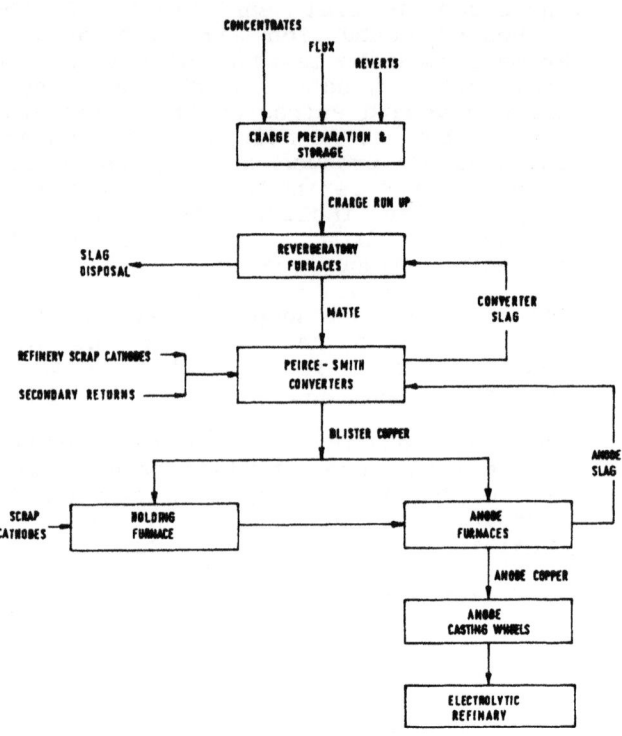

Figure 1: Original Flow Diagram.

Figure 1 sets out a general flow diagram of the facilities at the Nkana Smelter before the current retrofit.

A detailed description of the smelter has been published[4] previously. However, a brief summary of the original operation is described below.

Reverberatory furnaces

The reverberatory furnace section has five identical units 29.6 m long, 9.1 m wide and 3.3 m high; three or four of these units are usually in operation together at one time, depending on production planning.

For slag extraction, each furnace has a taphole with launder in the rear wall, from which the slag is removed in pots and taken to a disposal site. The matte is removed in ladles via three tap holes situated in the side wall at the settling zone.

Each furnace has two boilers for waste-heat recovery, two air heaters, four induced-draught fans and two forced-draught fans.

The firing facilities enable the furnaces to operate either with heavy fuel oil, pulverised coal or a mixture of both . The front wall has two pairs of lateral burners and one central burner. This can be seen schematically in figure 2. The maximum heat load in the furnace does not exceed the thermal equivalent of 150 tonnes per day of pulverised coal i.e. 168 GJ/h.

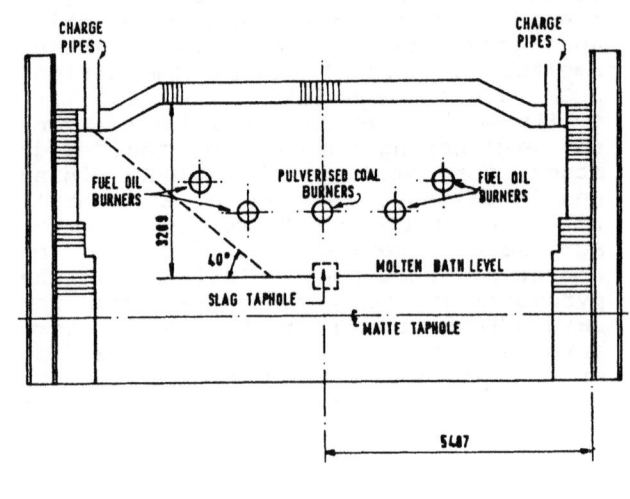

Figure 2: Section through original furnace burner end-wall.

Combustion air is supplied to the burners as preheated primary and secondary air, together with oxygen that is introduced for enrichment purposes at the burners.

Converter slag is returned to each furnace through a launder in the upper part of the burner wall to recover copper and for control of the bath slag chemistry.

Peirce-Smith Converters

The conversion of matte from the reverberatory furnaces is carried out using six Peirce-Smith converters, each of which is 4.0 m in diameter and 9.1 m long. Normally four or five of these are in operation at any one time.

Silica flux is charged using a garr-gun system, and air is introduced via fifty-one manually punched tuyeres fitted to each converter. Facilities are available to introduce oxygen enrichment to the blowing air. However, this is not used in normal practice. The offgas circuit of each converter consists of an air cooled cast steel collection hood and a waste-heat recovery boiler. The latter's sole function is to cool the gases. The steam produced is subsequently recondensed. The gas, at a temperature of 350°C, is then sent either via electrostatic precipitators for dust recovery before going to the sulphuric acid plants or vented to the atmosphere via the stacks.

DESIGN BASIS FOR NKANA

The project at the Nkana Smelter comprises the conversion to oxygen/fuel-oil burners of one reverberatory furnace and installation of a Teniente Converter as an integrated process. The remaining reverberatory furnaces will continue to operate on the current firing scheme. However, the number of operating units will be reduced to only those needed to provide sufficient seed matte to drive the Teniente Converter reactions whilst balancing the material flow. Full implementation involves the conversion of a second reverberatory furnace to oxygen/fuel oil burners. This will be carried out only when sufficient oxygen is available. The Teniente Converter will replace an existing Peirce-Smith converter.

Three levels of concentrate supply were specified by ZCCM, designated as low, normal and high. Table I shows the mineralogical and chemical composition of the concentrates for the throughput requirements when Nkana Smelter is operating after project completion.

OXYGEN/FUEL-OIL BURNER TECHNIQUE

The technique of smelting in reverberatory furnaces using oxygen/fuel-oil burners (OFB) has been used at Caletones Smelter since 1974[5]. As is already known, this technology aims at the partial or total replacement of conventional fuel-air burners (AFB) by a new type of burner, which is sited in the roof of the furnace and which uses oxygen alone for combustion. This leads to a more efficient use of fuel in the "green-charge" smelting process, and thus to a significant reduction in operating costs, also enabling a significant increase in each unit capacity to be achieved.

Table I Concentrate Data.

Level		Tonnes per year (dry)
Low	(L)	695,000
Normal	(N)	746,000
High	(H)	800,000

Mineralogy		(%)
Chalcopyrite	$CuFeS_2$	33.2
Bornite	Cu_5FeS_4	8.9
Covellite	CuS	0.2
Chalcocite	Cu_2S	15.1
Malachite	$Cu_2(OH)_2CO_3$	4.8
Carrolite	Co_2CuS_4	1.3
Pyrite	FeS_2	5.2
Silica	SiO_2	15.7
Alumina	Al_2O_3	2.7

Chemical Analysis		(%)
Copper	Cu	34.3
Iron	Fe	14.9
Sulphur	S	19.4

Figure 3 indicates the different amounts of heat available for smelting green charge in reverberatory furnaces, as a function of the oxygen content of the combustion air, between firing using normal air (21% O_2) and theoretically using industrial-grade oxygen (95% O_2).

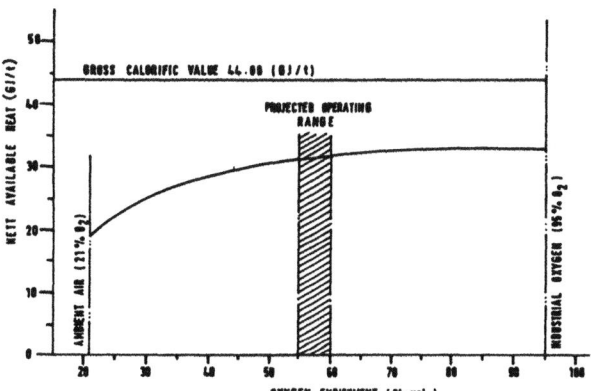

Figure 3: Effect of the Oxygen Enrichment on the Nett Available Heat from HFO.

In practice, overall enrichment within the furnace reaches values between 40% and 60% due to natural or forced ingress of air. The increase in the efficiency of the use of combustion heat is due basically to the replacement of a significant proportion of the nitrogen associated with normal air by reactive oxygen.

The fraction of heat generated by combustion which leaves the furnace with the nitrogen associated with the combustion air decreases from 35% with normal air, to approximately 0.5% when using industrial oxygen of 95% purity. This can be seen in table II.

Table II: Heat Loss by Nitrogen.

Calorific value HFO: 41,644 MJ/t

O_2 Enrichment %	Heat Lost by N_2 MJ/t	%
21	14,828	35.6
70	1,691	4.1
95	210	0.5

Additionally, there is an increase in the residence time of the gases at a higher temperature in the furnace, due to the formation of a smaller volume of gas. This gives rise to an improvement in the heat-transfer mechanisms to the charge bank due to part of the reverberation effect being replaced by an increase in the heat transferred by convection. Another important advantage of the use of oxygen is the greater concentration of sulphur dioxide in the gases leaving the furnace when compared with the conventional process. This increase, when linked to the advances made in the production techniques of sulphuric acid, make the recovery of sulphur economically feasible.

Nkana Engineering Design

Engineering design of furnace n° 4 of the Nkana Smelter took into account changes in the height and the profile of the roof. After initial analysis, it was determined that the height of the inner roof should be 706 mm lower than its original position over the first 18.9 m of the furnace ie in the charging zone. It was also considered necessary to change the existing trapezoidal profile to a uniplanar one. These changes resulted in a decrease in the distance from the surface of the molten bath to the inside of the roof, from approximately 3.3 m to 2.6 m. This can be seen in figure 4. The design changes were confirmed using a model for heat transfer by zones, under steady-state conditions[6]. On the basis of this model it was determined that the optimum height range of the roof for a maximum flow of heat transfer to the charge banks would be between 2.3m to 2.75m , with a balance being struck between thermal and flow-dynamic requirements.

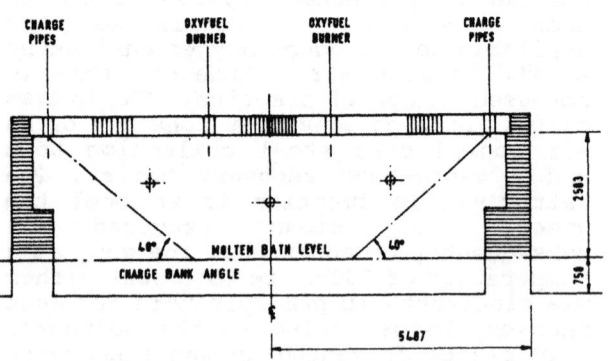

Figure 4: New Furnace Cross-Section.

Another important aspect of the design is the layout of the new burners in the furnace roof. For the project, up to eighteen burners were distributed lengthwise within the furnace; six OFBs in the centre, and the remaining twelve in two groups of six on either side of the central group. Their placement is determined by the position of the base of the charge banks towards which the flames are directed, as well as such considerations as the distance between the burners and the charge pipes, the molten bath level and the position of the burners relative to the old burner wall. This can be seen in figure 5.

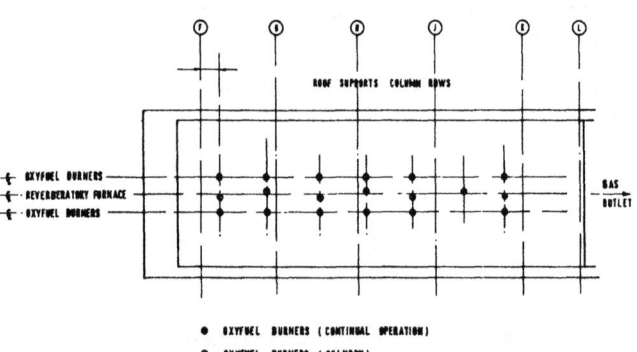

Figure 5: Position of New Burners in the Furnace Roof.

One important factor for Nkana was that introduction of this technique would not mean substantial changes in the design of other related systems. However, there was a need to redesign the piping systems for the supplies required for the process, as well as the instrumentation and control systems and operating requirements such as working platforms, access points, etc. The engineering for this was carried out by TECHPRO Mining & Metallurgy.

Smelting Characteristics

In general terms, the concentrate smelting process in reverberatory furnaces does not differ greatly whether by using OFBs or conventional methods. The oxygen-using process applied in practice follows the same reactions in respect of concentrate decomposition, oxidation of pyritic sulphur and the formation of the matte, slag and gas phases. These reactions and their corresponding heats of reaction for minerals present at Nkana Smelter are as follows:

Chalcopyrite:

$$CuFeS_2 = \tfrac{1}{2}Cu_2S + FeS + \tfrac{1}{4}S_2$$
$$\Delta H° = 448.23 \text{ MJ/t } CuFeS_2$$

Covellite:

$$CuS = \tfrac{1}{2}Cu_2S + \tfrac{1}{4}S_2$$
$$\Delta H° = 462.6 \text{ MJ/t } CuS$$

Bornite:

$$Cu_5FeS_4 = (^5/_2)Cu_2S + FeS + \tfrac{1}{4}S_2$$
$$\Delta H° = 219.0 \text{ MJ/t } Cu_5FeS_4$$

Carrolite:

$$Co_2CuS_4 = \tfrac{1}{2}Cu_2S + (^2/_6)Co_9S_8 + (^{31}/_{26})S_2$$
$$\Delta H° = 837.4 \text{ MJ/t } Co_2CuS_4$$

Pyrite:

$$FeS_2 = FeS + \tfrac{1}{2}S_2$$
$$\Delta H° = 1124.5 \text{ MJ/t } FeS_2$$

Oxidation of pyritic sulphur with oxygen from air infiltration:

$$\tfrac{1}{2}S_2 + O_2 = SO_2$$
$$\Delta H° = -11,259.1 \text{ MJ/t } S_2$$

Reduction of the magnetite introduced into the furnace from converter slag return:

$$FeS + 3Fe_3O_4 = 10FeO + 5SO_2$$
$$\Delta H° = 723.8 \text{ MJ/t } FeO$$

The formation of the products of the process, namely matte and slag, obey the following considerations:

Matte: The magnetite content of the matte is determined by the following equation:

$$Fe_3O_4 (\%) = (0.16-(0.0025 \times Cu \text{ matte } (\%)) \times 100$$

Slag: Its formation from gangue constituents in the charge (SiO_2, Al_2O_3, CaO, etc.) and iron oxides in the slag returned from the converters, follows the Modified Viscosity Ratio (MVR) parameter, as defined by ZCCM, which consists of the relationship between basic and acidic oxides. For ZCCM concentrates, based on existing theoretical and practical data, a value of 4.0 has been taken for the charge bank, and 2.5 for overall slag. This MVR parameter enables a calculation of the required amounts of flux and returned slag needed from a strictly metallurgical point of view.

The following figures and tables show graphically the main characteristics of the process with OFB's.

Figure 6 indicates the unit heat requirements of the individual and bulk concentrates at Nkana Smelter, as a function of the moisture content of the concentrate. This also highlights the refractory nature of the concentrates and is characterised by a low pyritic-sulphur content and a high level of gangue components giving rise to a high melting point slag system.

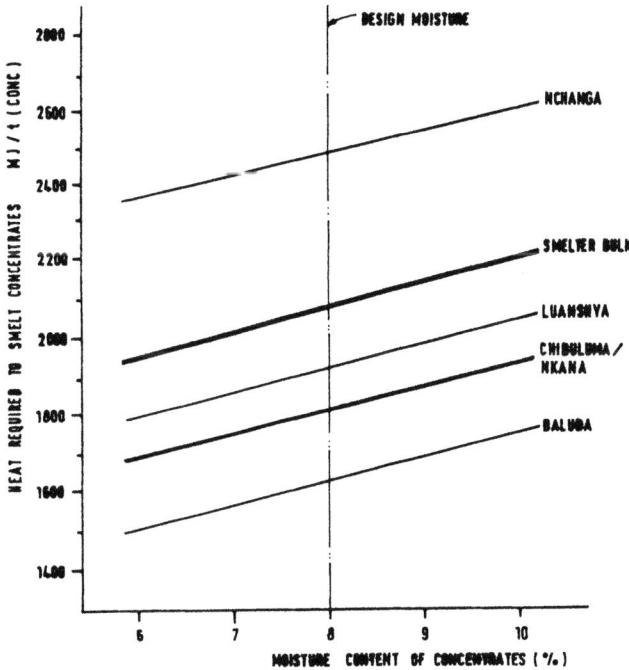

Figure 6: Smelting Effect of Moisture.

Figure 7 sets out the fuel-oil and industrial-oxygen requirements of the reverberatory furnace with OFBs, for various capacities for smelting the bulk concentrates at Nkana, taking into consideration the oxygen content of the exit gases and a fuel-oil flow rate per burner varying between 4.5 kg/min and 5.0 kg/min.

Air ingress into the furnace is a parameter which is difficult to control and quantify. Figure 8 sets out the number of OFBs required, at a firing rate of 4.5 kg/min per OFB, for smelting

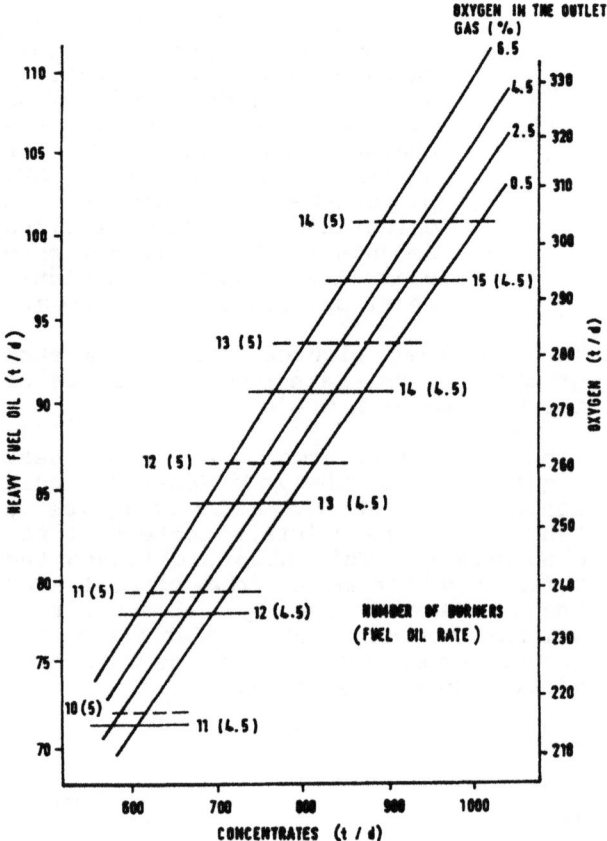

Figure 7: Smelting Rate as a Function of Fuel/Oxygen/Air Ingress.

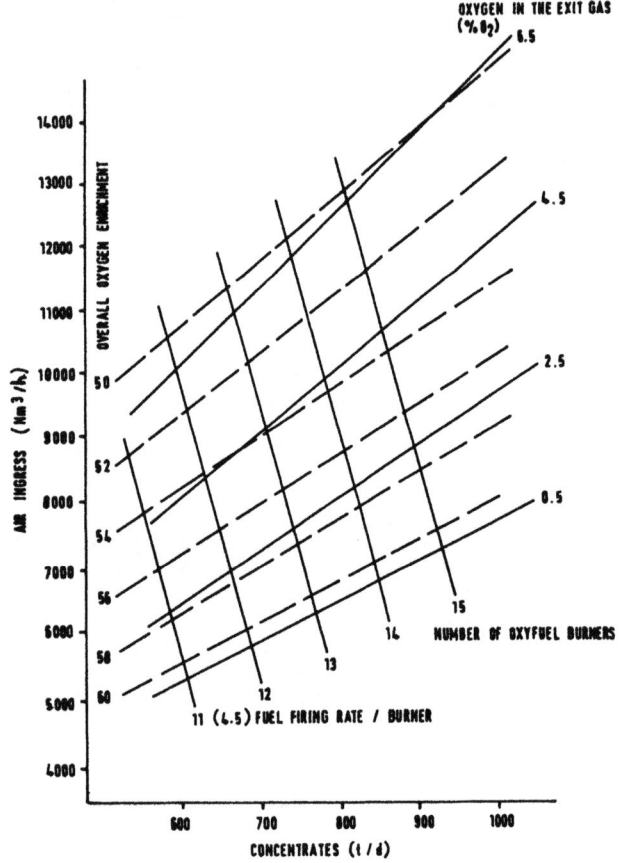

Figure 8: Effect of Air Ingress.

varying quantities of ZCCM bulk concentrate, depending on the volume of air ingress into the furnace. Also indicated is the percentage of oxygen in the exit gases in a range from 0.5% to 6.5%. Additionally, lines of overall oxygen iso-enrichment are shown, which, according to practical values determined at the Caletones facility, should be in the region of 50% to 60%.

Finally, Table III sets out a comparison of the main current operating parameters and those forecast once the OFBs in n° 4 reverberatory furnace at the Nkana Smelter are in operation.

Table III   Operating Parameters for Reverb No 4.

Parameter	Conventional	Future
Total smelting capacity (t/d)	870	1066
Combustion:		
Heavy fuel oil (t/d)	40	86
Pulverised coal (t/d)	90	–
Oxygen (t/d)	110	256
Oxygen enrichment (%)	30	55-60
Fuel-Ratio		
(MJ/t charge)	5000	3600
(MJ/t Conc)	6500	4100
Gas flow (Nm³/h)	50000	29000
Conc. SO_2 (%)	2.5	5.5

TENIENTE CONVERTER TECHNIQUE

The smelting-conversion technology in Teniente Converters has been used on an industrial scale at Caletones since 1977, and constitutes the basis for the increase in capacity and the low operating costs recorded in the recent past.

The Teniente Converter process consists of the simultaneous conversion of matte from the reverberatory furnaces and the autogenous smelting of copper concentrate, by oxygen-enriched air introduced through tuyeres, to produce "white metal" (73% - 75% Cu), slag and gases containing a high proportion of sulphur dioxide.

The autogenous smelting capacity of a Teniente Converter depends, amongst other factors, on the size of the equipment (design characteristics), the operating conditions under which the process is carried out, and the chemical and mineralogical composition of the materials to be processed, as well as the products obtained, mainly the copper content of the white metal.

Chemical reactions involved in the process are basically those already mentioned for the reverberatory smelting process, which are complemented by oxidation reactions of iron sulphide and the formation of slag:

$$FeS + (^3/_2)O_2 = FeO + SO_2$$
$$\Delta H° = -5,276.0 \text{ MJ/t FeS}$$

$$3FeS + 5O_2 = Fe_3O_4 + 3SO_2$$
$$\Delta H° = -6,497.8 \text{ MJ/t FeS}$$

$$2FeO + SiO_2 = 2FeO.SiO_2 \text{ (fayalite)}$$
$$\Delta H° = -252.4 \text{ MJ/t FeO}$$

From the mass and energy balances for the process carried out in the reactor, a "concentrate/matte" ratio, R, can be defined, expressed as:

$$R = \frac{\emptyset_1 \times \emptyset_2 \times q_m + f_m}{\emptyset_1 \times \emptyset_2 \times q_c + f_c},$$

where:

$\emptyset_1$ parameter principally connected with the duty factor (i.e. blowing time per day).

$\emptyset_2$ parameter principally connected with the design characteristics of the reactor. It includes the heat-loss factor due to radiation and convection; this value depends on the design of the equipment and the flow of air used, the latter being linked to the number and availability of the tuyeres fitted.

q_m heat released (MJ/t matte) by the conversion of matte into "white metal".

q_c heat required (MJ/t concentrate) for smelting the concentrate and converting the matte produced into white metal. This value varies in line with the amount of recirculated materials per tonne of new concentrate fed into the reactor.

f_m, air or oxygen-enriched air (Nm^3)
f_c required for the oxidation of one tonne of matte and one tonne of concentrate, respectively, to "white metal", slag and gas.

The values of q_m, q_c, f_m and f_c depend on the chemical composition of the materials in the feed and mineralogical composition of the concentrate as well as of the products obtained, the level of oxygen-enrichment used in the blowing air and the oxygen efficiency of the process. The value of q_c is also affected by the moisture content of the solids processed.

Nkana Engineering Design

The Teniente Converter to be installed at Nkana Smelter will be 4.5 m in diameter and 18.2 m long, and will replace the Peirce-Smith converter n° 1.

The sizing of the vessel was based on a comparative analysis between 4m and 4.5m diameter vessels. This analysis reconciled aspects related to the requirements for concentrate smelting at Nkana with matters of a strictly operational character. Many of these favoured the larger diameter vessel. Namely;-

- a greater distance between the surface of the bath and the garr gun, which enables better concentrate feed with fewer blockages due to splashing of molten material.

- a slower gas velocity inside the reactor for a fixed blowing flow rate, which affects the residence time of the concentrate particles in the vessel and thus allows such particles to enter the molten bath rather than exit via the mouth.

- a larger settling area in the slag-removal zone with less turbulence, which helps reduce the copper content of the slag.

Table IV sets out the general characteristics of the Teniente Converter as designed for the Nkana Smelter.

Table IV Design Characteristics of Nkana Teniente Converter

Principle dimensions		
- diameter	(m)	4.5
- length	(m)	18.2
Other characteristics		
- garr-gun	(n°)	1
- mouth	(m²)	7
- removable port	(n°)	4
- tuyeres	(n°)	55
- "White metal" tap hole	(n°)	1
- slag tap hole	(n°)	1
Drive mechanism		
- electric motor	(HP)	200

For the vessel design, an analysis was carried out of aspects related to the feed system for wet charge material (garr-gun type), water cooled tapholes for white metal and slag, mouth size for charging matte and removing gases, and the number and size of tuyeres. The shell also includes special doors in the tuyere-line area, to enable use of the "hot repair system", as used at the Caletones Smelter, to be adopted. Additionally, consideration was given to the basic design of the furnace refractories and of the foundations for the vessel and its drive system.

Figure 9 is a schematic diagram of the Teniente vessel with outline dimensions.

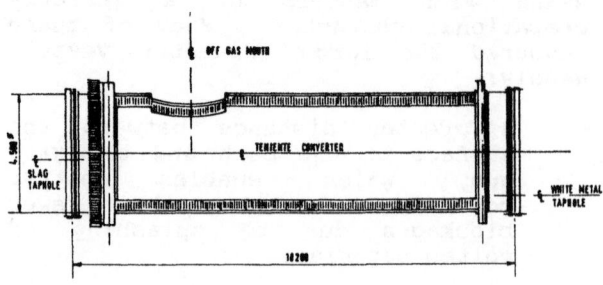

Figure 9: Nkana Teniente Converter.

The low pressure air duct and the corresponding wind box were designed for a maximum of 55 tuyeres. Oxygen, at 95% purity, will be injected into the duct to a maximum oxygen enrichment level of 34% O_2 (by volume). Normal operation will be at a 32% oxygen enrichment level.

For operational control of the reactor and ancillary equipment, consideration was given to local and main control panels. Functions included are:

- reactor-rotation drive system.
- starting up and stopping the concentrate and flux feeders.
- control of air-blast flow rate and pressure
- control of oxygen enrichment
- measuring the flow of concentrate and flux.
- measuring the internal temperature of the reactor.
- interlocks

Smelting-converting characteristics

As has already been mentioned, the characteristics of the process carried out in the Teniente Converter can be viewed in terms of the "concentrate / matte" ratio, R, which is determined from the mass and energy balances of the process. The following figures and tables graphically show the main parameters of the process.

Figure 10 gives the values of q_c for the various individual concentrates at Nkana Smelter to give white metal with a 75% Cu content and a fayalite slag with an Fe/SiO$_2$ ratio of 1.8. The notable differences in the values of q_c are

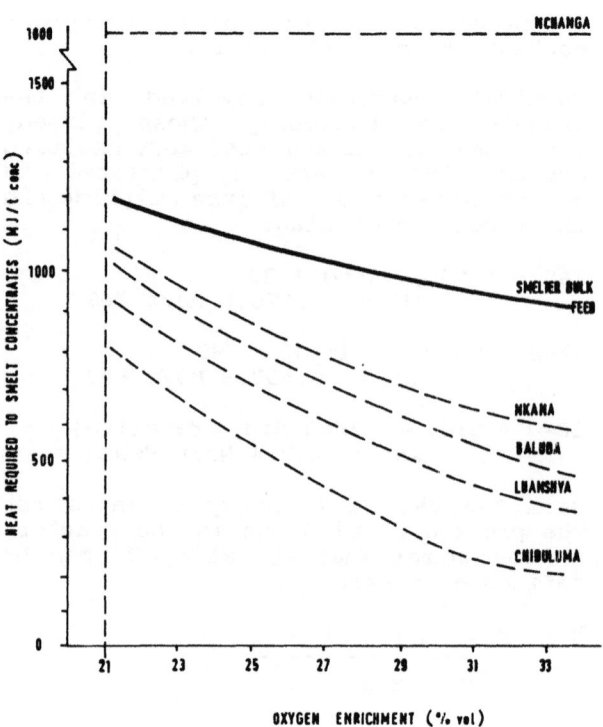

Figure 10: Heat Required for Conversion of ZCCM Concentrates.

basically explained by the different levels of sulphur and iron which produce heat energy during their oxidation to sulphur dioxide and oxides of iron, as well as the levels of gangue and/or refractory constituents in the materials. In the same way, the conversion of the reverberatory furnace matte also has low energy potential due to its high copper and low iron and sulphur content.

Figure 11 sets out the heat released in the oxidation of a range of seed mattes from the reverberatory furnaces, for two levels of oxygen enrichment, 21% and 32%, according to the purity of the white metal obtained. For project design, seed matte grade will be around 55% Cu. The corresponding enriched-air requirements are given in figure 12 for the bulk concentrate mixture (f_c) and the resulting mattes (f_m).

The iron and sulphur content of the concentrates and of the seed matte, as well as oxygen enrichment of the blast air, have a direct influence on the unit consumption of air f_c and f_m required for the oxidation of the said constituents. This can be seen in figure 13.

Figure 13 shows the daily smelting and conversion capacities for concentrate and seed matte respectively, for varying oxygen contents in the blowing air and the corresponding variation in the consumption of industrial oxygen. This figure also indicates the parameter, R, which does not exceed 0.6; this compares

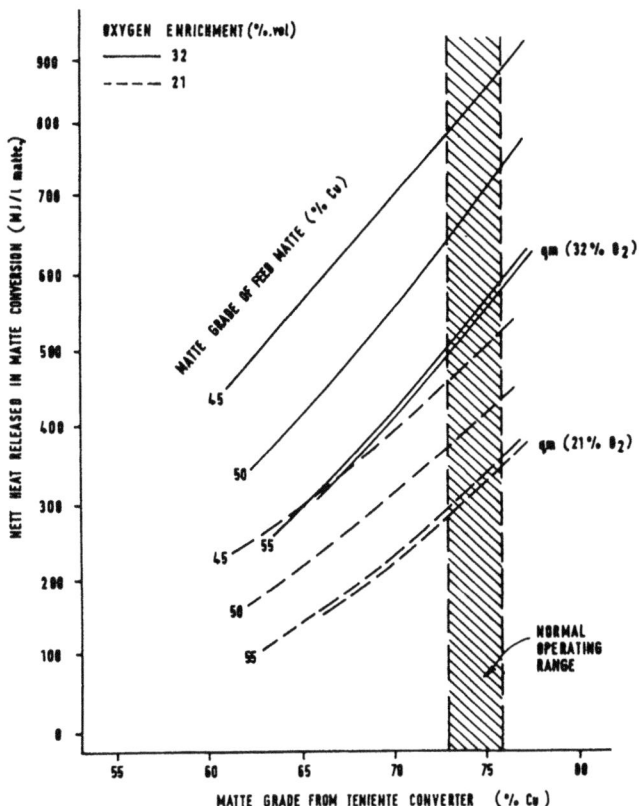

Figure 11: Effect of Oxygen Enrichment on Heat Released in Matte Conversion.

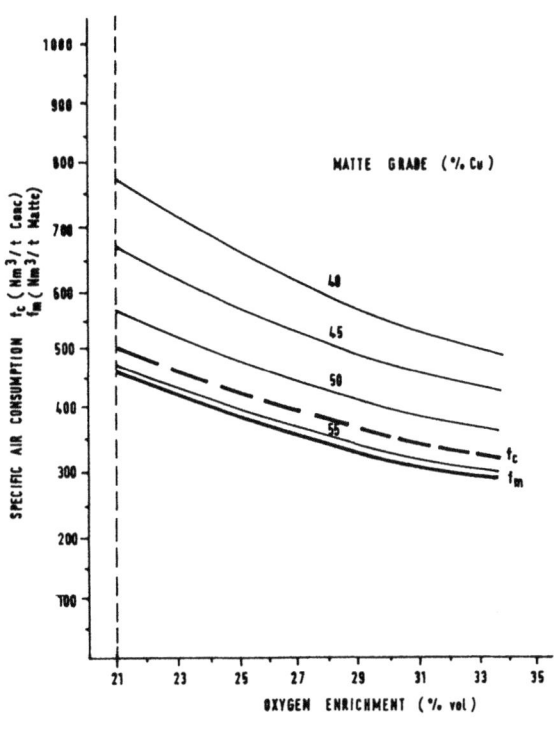

Figure 12: Specific Air Consumption & Oxygen Enrichment for varying Reverb Furnace Matte Grades.

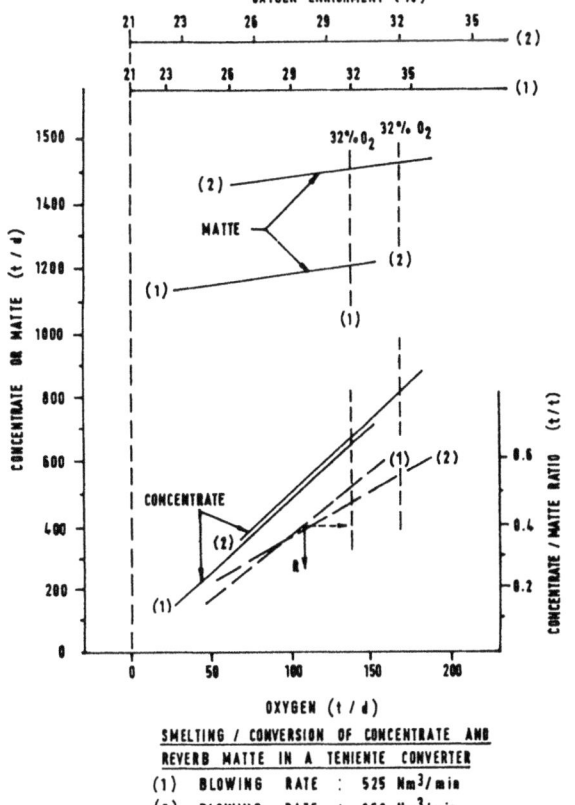

Figure 13: Smelting/Conversion of Concentrate & Reverb Matte in a Teniente Converter.

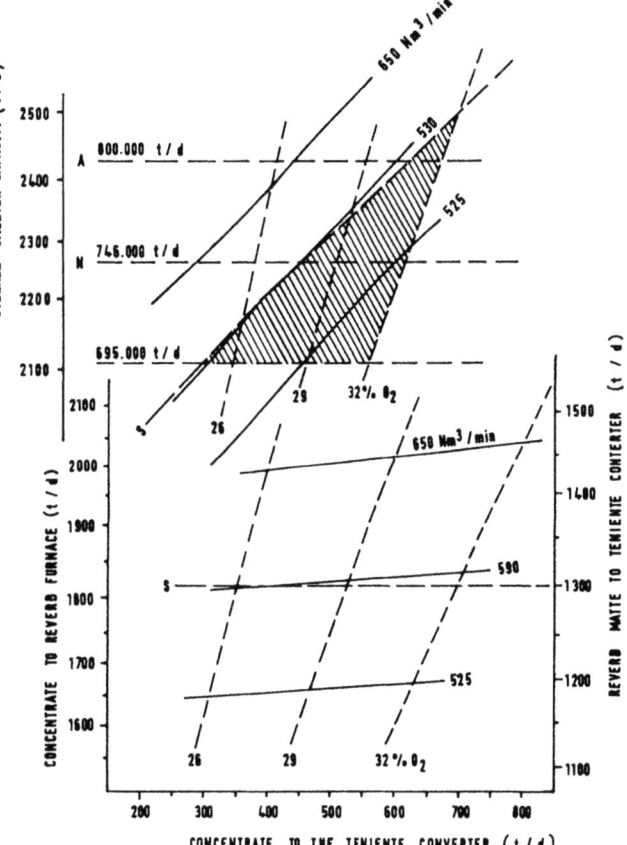

Figure 14: Overall Smelter Capacity.

Table V Basic Teniente Converter Mass Balance.
(excluding gases)

Item	t/d	%Cu	%Fe	%S	%SiO$_2$
Feed					
Concentrate	632	34.3	14.9	19.4	15.7
Flux	120	-	-	-	68.0
Feed matte	1206	54.9	20.2	24.4	-
Total	1958	879	338	417	181
Products					
"White metal"	1103	75.0	3.9	20.8	-
Slag	734	6.0	40.0	1.5	24.4
Dust	16	34.1	17.5	14.9	16.2
Total	1853	879	338	243	181

Air blowing rate	:	525 Nm^3/min
Oxygen enrichment	:	32%
Concentrate moisture	:	8%
Gas flow	:	32,980 Nm^3/h

Gas Comp:			
	SO$_2$	:	18.9 %
	O$_2$	:	2.1 %
	CO$_2$	:	0.5 %
	N$_2$	:	67.1 %
	H$_2$O	:	11.4 %

unfavourably with values in excess of 1.0 at Caletones, and is directly attributable to the levels of iron and sulphur in the base system.

Table V shows the overall mass balance of the system based on the main constituents of the charge, for smelting bulk concentrate, subject to the following operational conditions: blowing air flow = 525 Nm^3/min and 32% O$_2$ enrichment. The corresponding energy balance is set out in table VI.

Finally, table VII sets out the principal operating parameters for the Nkana Smelter Teniente Converter.

Table VII Teniente Converter Operating Conditions.

Blowing air (Nm^3/min) :	400 - 650
Oxygen Enrichment (%) :	26 - 32
Blowing availability (%) :	85
Concentrate moisture (%) :	8 - 10
SO$_2$ conc in offgas (%) :	17 - 19

Table VI Teniente Converter Energy Balance.

Item		GJ/d	%
Input			
Conversion reactions	(25°C)	2538	70.6
Feed matte	(1130°C)	1055	29.4
Total		3593	100.0
Output			
White metal	(1220°C)	887	24.7
Slag	(1210°C)	1101	30.6
Dust	(1200°C)	17	0.5
Gases	(1200°C)	1358	37.8
Heat loss (Convection & Radiation)		230	6.4
Total		3593	100.0

TENIENTE TECHNIQUES INTEGRATED INTO THE NKANA SMELTER

The ZCCM project has proceeded with the installation of oxygen/fuel-oil burners in one reverberatory furnace as a first phase. A Teniente Converter will be installed and commissioned as a second phase in 1993.

For the process to operate effectively, it is vitally important to establish an adequate balance between the rate of extraction of the matte from the reverberatory furnaces and the rate at which it is required in the Teniente Converter, so as to satisfy the energy balance of the system. There is a high degree of flexibility when using the Teniente Process. By rationalising the use of available oxygen and modifying the operating conditions in either of the two units, it is possible to satisfy the overall smelting and conversion requirements in different ways. A decrease in seed matte grade increases the tonnage of concentrate that can be treated in the Teniente Converter but at an increased blowing rate. However, the parameter, R, improves. Should the matte grade be fairly constant, then it is possible adjust the blowing rate and oxygen enrichment to balance the system.

At Nkana, advantage has been taken of the smelting characteristics of one of the concentrates. The Nchanga concentrate is low in iron and sulphur but high in refractory components. By removing the bulk of this material from the reverberatory furnaces, the grade of the seed matte drops. Then, by feeding this material exclusively to the reactor, it is possible to convert with a low blowing rate and also reduce the fluxing requirements by utilising the elements present in the concentrate. It should be noted here that the normal operation in Chile is to treat the same material in both furnaces.

Figure 14 shows the smelting conversion capacity for concentrate and matte in the Teniente Converter at Nkana for air blown at rates varying between 525 Nm^3/min and 630 Nm^3/min, as well as for levels of oxygen enrichment varying between 26% and 32% O_2. The vertical axis (left-hand side) shows the level of concentrate-smelting required in the reverberatory furnaces, to balance the matte between the two types of furnace. It can be noted that for Nkana Smelter an increase in oxygen enrichment in the Teniente Converter, for a constant blowing rate, leads to a greater demand for seed matte. By comparison, at Caletones Smelter, this situation is exactly the opposite; the explanation for this lies in the different levels of iron, sulphur and gangue in the respective concentrates.

The S line in the figure indicates the maximum amount of matte, 1,300 tonnes/day, which can be supplied to the Teniente Converter, and corresponds to a blowing time equal to or better than 85%.

The top half of the figure shows the maximum daily smelting capacity of the reverberatory furnace/Teniente Converter system, which enables the annual smelting requirements forecast by ZCCM to be reached. From this figure, it is clear that it is possible to treat the maximum forecast level of 800,000 tonnes/year of concentrate, using an air-blast flow rate of approximately 560 Nm^3/min with 32% O_2 in the Teniente Converter. For each level of capacity required, the operating conditions of the Teniente Converter can be adjusted.

Facilities already exist at Nkana Smelter to feed heavy fuel oil and oxygen to the reverberatory furnaces. Conveyor systems also exist to enable concentrate to be routed to the converter side of the operating aisle. In addition there is an oxygen plant with a capacity of 540 tpd. This has meant that integration of Teniente Technology into the smelter is fairly straightforward. Apart from the design related purely to the technology, certain additional work has been undertaken by TECHPRO Mining & Metallurgy to gain the maximum benefits of the new process.

On the reverberatory furnace, additional matte tapholes were placed in the old burner wall to allow matte to be tapped directly to the converter aisle. A second slag taphole was also installed. Both were done to cater for the increased smelting throughput from 600 tpd to 1100 tpd of material. The original waste heat boilers were also replaced by an evaporative cooling system. Dust collection cyclones were installed to increase copper recovery and a new exhaust stack was provided to improve the environment.

On the Teniente Converter side, new low pressure blowers will be provided to supply the blast air. The opportunity has also been taken to redesign the gas collection hood and replace the waste heat boiler system with an evaporative cooling system using in-house engineering designs for both. This is additional to the provision of a material storage and feed system to the reactor. The tuyere punching duty was felt onerous enough to require mechanical punching.

CONCLUSIONS

A conventional existing copper smelter can be optimised and modernised by incorporating each of the Teniente Technology techniques, either independently or in a single integrated process. The decisive factor in installation being the availability of industrial oxygen at the plant in question. Oxygen/fuel-oil burner and Teniente Converter Technologies can be employed at the ZCCM Nkana Smelter to efficiently increase its capacity to smelt concentrate and reduce the overall use of energy in the pyrometallurgical process. Only a minimum level of modification to the existing infrastructure is required to adapt to the new technology. This means that the capital cost can be kept low.

Adoption of both techniques at Nkana Smelter will reduce the number of operating units needed to meet production targets. It will be possible to achieve the target production rate of 240,000 tpy new copper with a maximum of three reverberatory furnaces (1 OFB, 2 AFB) and three converters (1 TTE, 2 PS); a reduction of one unit on each side of the aisle. It should be noted that with full implementation only two oxy-fuel fired reverberatory furnaces would be needed to generate the feed matte requirements to drive the process.

The consumption of energy, excluding that used for making industrial oxygen, should fall from a current level of 6,500 MJ/t of concentrate to values in the range of 3,900 to 4,200 MJ/t of concentrate, assuming the incorporation of both techniques and the range of smelting capacity forecast by ZCCM. This, and the reduction in unit operations, will result in decreased operating and maintenance costs.

The higher concentration of SO_2 in the reverberatory furnace gases and in converters, resulting from the incorporation of the Teniente Converter, together with the greater level of continuity of operation, make it possible to recover the gas efficiently to optimise the production of sulphuric acid.

One significant advantage of undertaking a retrofit of this nature is that the unit operations remain essentially the same. This means that the learning curve can be controlled and confidence of both operating and maintenance staff retained. Retraining should also not be onerous.

This is the first time that Teniente Technology has been installed outside of Chile. The first phase of Nkana Smelter modernisation was completed in August 1990, with the commissioning of the oxy-fuel fired reverberatory furnace. Initial indications of furnace operation have been encouraging with better than predicted smelting performance being realised.

In the final analysis, it can be confirmed that implementation of Teniente Technologies for the rehabilitation and modernisation of existing copper smelters by better utilisation of oxygen is feasible. It offers a serious alternative to more sophisticated approaches particularly where capital investment is restricted and the basic changes to infrastructure are too costly or disruptive.

Acknowledgements

The authors would like to thank the managements of CODELCO-Chile, El Teniente Division, and Zambia Consolidated Copper Mines Ltd.,for permission to publish this paper.

References

1 Hansen P., Cross A., Chibuye G. & Willis G., The new Nkana Cobalt Plant for ZCCM, 115th AIME, New Orleans 1986.

2 Strachan M., Production of Sulphuric Acid in Zambia, Sulphur 90, Cancun, Mexico.

3 Strachan M., Bradburn J., Hanschar L. & Cross A., Various Study Reports for ZCCM.

4 Hansen P., Hanschar L. & Morgan M., Reverberatory Furnace Firing & Operating Practices on the Zambian Copperbelt, 109th AIME, Las Vegas 1980.

5 Vera G. & Campos R., CODELCO-Chile Copper Concentrate Smelting Technologies, Extraction Metallurgy 85, I.M.M. London.

6 Otero A., Application of a Heat Transfer Zonal Model to Pyrometallurgical Furnaces, CIMM Internal Report, Santiago 1985.

Process options for modification of the Outokumpu refinery circuit at BSR Ltd to improve base and platinum group metal recoveries

J. Schwarz
M. R. Richardson
Anglo American Corporation Services Ltd, Harare, Zimbabwe

SYNOPSIS

With the probable increase in the 1990's of the treatment of nickel/copper concentrates containing PGMs at the BSR Ltd, it is necessary to modify the refinery. The conventional Outokumpu closed circuit process treating a low sulphur (6%) matte, as practised at the BSR, cannot achieve the desired recoveries of base, and particularly precious metals because of the high circulating load of leach residue through the smelter. Within the constraint of maximising the use of existing equipment, this paper considers those options available to achieve acceptable recoveries.

Many current operations produce a high (20%) sulphur matte from which the PGMs are recovered by pressure leaching and separation of base metal values - sometimes with an intermediate magnetic separation step. These processes have the inherent problem of disposal of the sulphur from the refinery in the form of sodium sulphate.

Testwork has indicated that blowing the matte to 6% sulphur does not significantly contribute to the loss of precious metals. Thus, by reducing the sulphur, the conventional atmospheric leaching circuit can be retained but followed by a pressure leach step in two possible modes:

1) Oxidative pressure leach - leaving a PGM-rich residue the copper being electrowon.

2) Non-oxidative pressure leach - followed by conversion of copper/PGM rich residue to copper anodes for subsequent electrorefining.

In totally removing the high circulating load of copper leach residue containing precious metals, by adding a pressure leach step, acceptable PGM recoveries and improved base metal recoveries will be realised.

1.0 INTRODUCTION

Bindura Nickel Corporation Limited (BNC) currently operates four nickel/copper mines within Zimbabwe - Epoch, Madziwa, Shangani and Trojan. It also owns a smelting and refining complex - BSR Ltd (BSR) - located at Bindura adjacent to Trojan Mine, where it treats the concentrates produced from these mines. The location of these operations is shown in figure 1. In addition BSR undertakes toll treatment to utilise the spare capacity available. In the foreseeable future BNC is faced with the closure of Madziwa (1993) and Epoch (1996) and the termination of one of its current toll treatment contracts.

Exploration activity, within Zimbabwe, to locate new sources of nickel has to take into account the low grade nature of most deposits. The emphasis is therefore focussed on those with additional revenue sources viz. the platiniferous nickel deposits of the Great Dyke Igneous Complex. However it has not precluded low grade nickel deposits, such as the Damba orebody close to Shangani, being examined. Thus by 1996 the possible mining activity providing feed sources for the BSR is shown in figure 2.

To remain viable and attract these possible feed sources to BSR, improvements have to be made to the current process to:

1) Improve recovery of Au/Ag and PGMs

2) Improve recovery of nickel and copper.

3) Improve quality of copper cathode produced.

Figure 1 : 1990 Operating Mines, Smelters and Refineries

ZIMBABWE

KEY

✱ SMELTER AND REFINERY

▣ NICKEL MINES

····· RAILWAYS

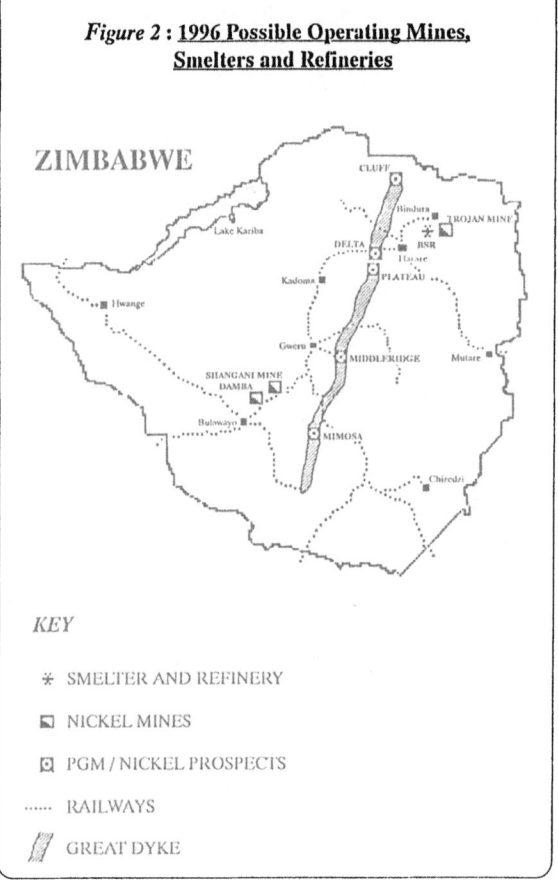

Figure 2 : 1996 Possible Operating Mines, Smelters and Refineries

ZIMBABWE

KEY

✱ SMELTER AND REFINERY

▣ NICKEL MINES

▣ PGM / NICKEL PROSPECTS

····· RAILWAYS

▨ GREAT DYKE

2.0 PRESENT OPERATION AND ITS LIMITATIONS

BNC comprises five operating divisions - four mines, each equipped with a concentrator, and the BSR which treats these concentrates to produce nickel and copper in cathode form and cobalt as a 40-50% Co cake. Additionally, some toll treatment is undertaken in the form of a nickel/copper matte from BCL in Botswana and nickel sulphate from Western Platinum. Some precious metals are also recovered periodically as a low grade leach residue from the refinery. Typical monthly production tonnages are shown in Tables 1 and 2.

Table 1. Monthly Production BNC Group Mines

		Trojan	Shangani	Madziwa	Epoch	Total
Ore milled,	t	89 000	83 000	26 000	42 000	233 000
Head grade,	%Ni	0.69	0.55	0.80	0.58	0.63
Cons. grade,	%Ni	10.5	13.0	8.5	17.0	12.0
Contained Ni,	t	401	351	167	163	1 082
Cu,	t	28	35	46	9	118
Co,	t	10.2	9.7	3.7	3.4	27.0

Table 2. Monthly Production - BSR Ltd

	Smelter		Refinery
	Tonnes	Grade	Tonnes
Leach Alloy	1 680		
Contained NI	1 148	68 %	1 153
Cu	379	23 %	193
Co	10.6	0.6 %	8.9

In the BSR circuit, the smelter treats the concentrates, blended with leach residue returned from the refinery, to produce a granulated blown matte (leach alloy) containing 6% S. This matte, together with toll inputs from BCL and Western Platinum, is treated in the refinery by atmospheric leaching, cobalt removal and nickel/copper electrowinning. Unleached residue, which carries the precious metals, is returned to the smelter as an inherent circulating load which is subjected again to smelter dust and slag losses. Thus the precious metals circulate and build up in concentration until equilibrium is reached where losses equate to input from new concentrate. On an annual basis a fixed tonnage of residue is isolated, dried, crushed, screened and bagged for export.

The overall BSR mass balance is shown in figure 3.

*Figure 3 :*BSR - Overall Mass Balance (tonnes per month)

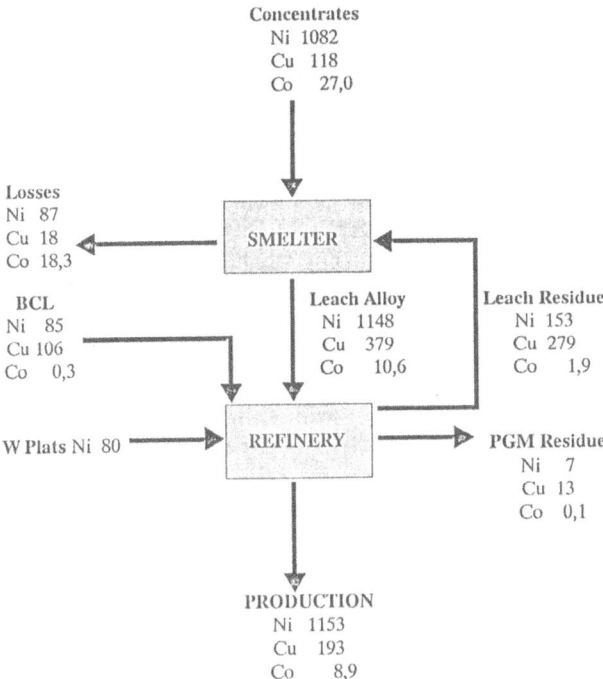

Recoveries typically applying to this balance are shown in Table 3.

Table 3. Current BSR Recoveries (%)

	Smelter	Refinery	BSR Overall
Ni	94	87	93
Cu	95	39	92
Co	37	82	33

The effect on the overall recovery of the circulating load, as determined by the refinery recovery (or plated : input ratio) is evident. The effect on precious metals recoveries is even more severe as the circulating load to build up to reasonable concentrations is many times higher, and recoveries can be estimated as shown in table 4.

Not only are the precious metals recoveries extremely poor, but the low grades make for high transport and treatment costs as well as locking up considerable tonnages of nickel and copper with this residue. In addition contractural recoveries available of only 65/70% of metal values further reduce the income derived from the sale of the residue. On an annual basis, the composition of the PGM residue removed is shown in table 5.

Table 5. Typical Saleable PGM Residue Composition

	Tonnes	%
PGM Residue	600	
Nickel	84	14
Copper	156	26
Cobalt	1.2	0.2
Precious Metals	600 kg	1000 g/t

The main reason for the relatively low base metal recoveries and even lower precious metals recoveries can be found in the limitations imposed by the conventional Outokumpu refinery circuit as currently employed at BSR. This circuit relies purely on atmospheric leaching of both nickel and copper, as a consequence of which sulphur in the feed must be carefully controlled at 5-6% by removing the excess sulphur in the converter. Even so, leach efficiencies achievable for nickel and copper, respectively 92% and 47%, result in high metal values remaining in residue thus necessitating the closed circuit configuration with the smelter.

The basic reactions by which atmospheric leaching is effected in the Outokumpu system are as follows:-

$$Ni^o + Cu^{2+} \longrightarrow Ni^{2+} + Cu^o \qquad \text{........} \qquad (1)$$

$$2Cu^o + {}^1\!/_2 O_2 \longrightarrow Cu_2O \qquad \text{........} \qquad (2)$$

$$Cu_2O + 2H^+ \longrightarrow 2Cu^{2+} + H_2O \text{........} \qquad (3)$$

Table 4. Typical PGM Residue Balance

	Tonnes / Annum	Au		Ag		Pt		Pd		Rh	
		g/t	kg	g/t	kg	g/t	kg	g/t	kg	g/t	kg
Inputs											
Concentrates	106 000	0.77	82	11.4	1209	0.85	90	2.0	215	0.26	28
BCL Matte	2 500	2.0	5	137	342	2.0	5	2.8	7	0.17	
Total			87		1551		95		222		28
Output											
PGM Residues	600	45	27	600	360	80	48	245	147	30	18
Recovery %			31		23		50		66		64

Details relating to these and other side reactions have been described by Groom et al.[1]

This double leaching/electrowin circuit (excluding cobalt for convenience) can be represented by a mass balance as shown in figure 4.

Figure 4. :Mass Balance for BSR Leach / Electrowin Circuit

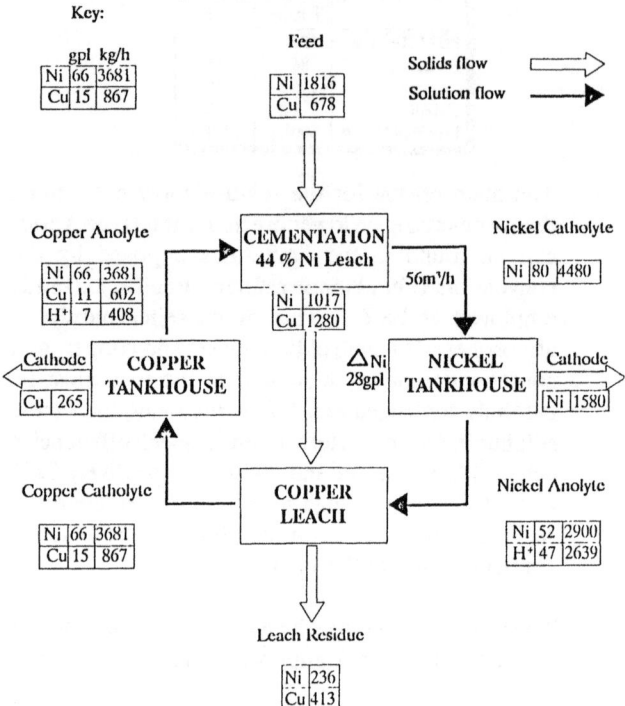

In addition to the limitations in respect of recoveries as previously described, this circuit also imposes a limit on the quality of the copper cathode product. It can be readily appreciated that to electrowin copper from a catholyte containing only 15 gpl Cu against a background nickel tenor of 66 gpl is difficult and inefficient. A current efficiency in excess of 75% is rarely achieved under these conditions, and a significant proportion of the copper product is recovered as a non-coherent sludge.

3.0 PHILOSOPHY OF PROPOSED PROCESS CHANGES FOR EFFICIENT PRECIOUS METALS RECOVERY

The fundamental pre-requisite to achieving acceptable precious metals recoveries is to open the BSR circuit such that all base and precious metals values are subjected to but one pass through the smelter where all the losses occur. This can be achieved by introducing a pressure leach stage to dissolve selectively either, the remaining nickel in residue, or both the nickel and copper, thus providing a continuous extraction route for precious metals at a high grade. Not only would the objectives in terms of precious metals recoveries and grades be met, but significant improvements in nickel and copper recoveries of 2-3% would also be achieved, together with an improvement in copper cathode quality.

As an alternative, the well known slow cooling and magnetic separation method has also been considered.

To maximise recoveries, consideration must also be given to reducing smelter losses. These involve conventional procedures such as improvements to the dust handling system and possibly the installation of a slag cleaning furnace.

In considering the refinery options, however, a considerable number of variables need to be taken into account, and inevitably a "best compromise" solution is sought. The most important of these variables are listed below:

   Precious metals recoveries and point of extraction.
   Point and method of sulphur removal.
   Retention of existing plant.
   Capital cost of new plant.
   Flexibility in accepting a variable feedstock.

4.0 PROCESS OPTIONS FOR PGM RECOVERY

4.1 Early Removal of PGMs by Magnetic Separation

It has been shown that by carefully controlling the sulphur and iron contents during the converter blow, followed by subsequent slow cooling, a PGM rich fraction can be produced from the Ni/Cu matte by magnetic separation. A simplified flow sheet is given in figure 5.

Figure 5 :Simplified Flowsheet for Magnetic Separation of PGMs

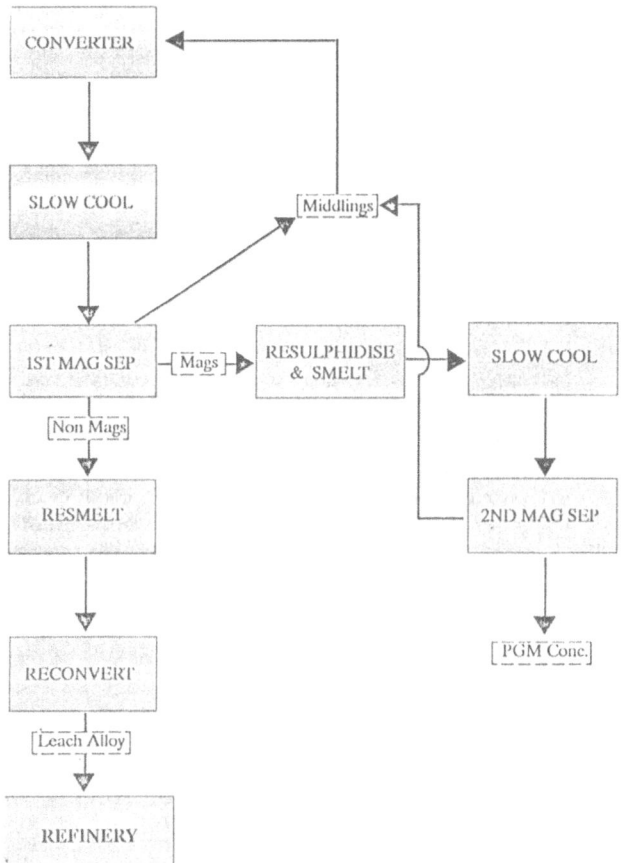

The process is complicated and therefore expensive in capital. Materials are recycled and the bulk of the nickel and copper has to be re-smelted and converted to be acceptable to the refinery.

Advantages

i. Early removal of the bulk of PGMs, i.e. improved cash flow.
ii. Minimum alteration to refinery required.

Disadvantages

i. Multiple steps each with possible loss of PGMs.
ii. Recycle of middlings.
iii. Additional smelting/converting required with additional dust/gas handling.
iv. Unless the refinery is altered no improvement in nickel/copper recoveries.
v. PGM recovery lower than for other options.
vi. Expensive in capital.

4.2 High Sulphur Pressure Leach

This option gives the best possible PGM recovery. However it is expensive in capital as larger autoclaves are required than those for blown matte (low sulphur 6%). Also disposal of sulphur in the form of Na_2SO_4 is expensive due to high cost of imported NaOH.

The typical flowsheet is given in figure 6.

Figure 6 :High Sulphur (20%) Leach Circuit

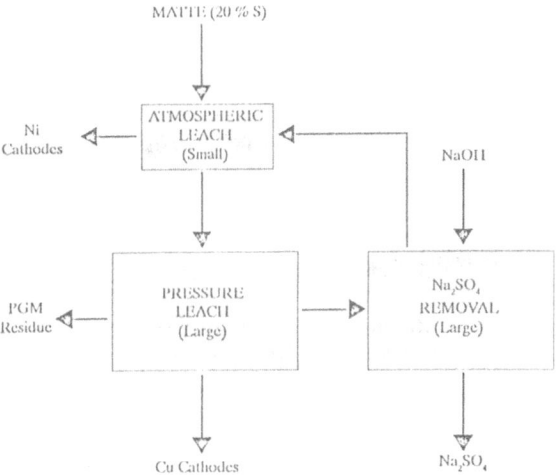

Advantages

i. Highest possible recovery of PGMs. Improved Cu/Ni recoveries. Ni/Cu ratio flexible.
ii. High Co recovery.
iii. Lower refractory consumption in the converters.

Disadvantages

i. Higher capital cost of autoclaves.
ii. Late removal of PGMs.
iii. Sulphur removal problems in Refinery.

4.3 Low Sulphur (Blown Matte) Pressure Leach

Preliminary testwork has indicated no significant loss of PGMs when blowing from high (20%) to low sulphur (6%) in the converter[2]. Results obtained in current testwork[3] are shown in table 6.

Table 6. **Analyses of BSR Matte**

	Furnace Matte	High S Matte	Blown Matte
Ni %	21.7	59.6	64.9
Cu %	8.5	19.3	25.0
Co %	1.1	1.1	0.6
Fe %	26.3	0.3	0.5
S %	26.1	17.3	5.3
Pt g/t	13	29	38
Pd g/t	39	84	109
Au g/t	8	15	16

This has the advantage of removing the bulk of the sulphur at the smelter which reduces the complications of its removal in the refinery. A simplified flow-sheet given in figure 7 shows the relative size of the high S and low S options when compared with figure 6.

Advantages

i. Minimal sulphur removal from refinery.
ii. Improved Cu/Ni recoveries.
iii. Least changes to refinery.

Disadvantages

i. Slightly lower PGM recovery.
ii. Ni/Cu ratio limited to max. 4:1 (due to S blow)
iii. Late removal of PGMs.
iv. Poor Co recovery.
v. Higher refractory consumption in the converters.

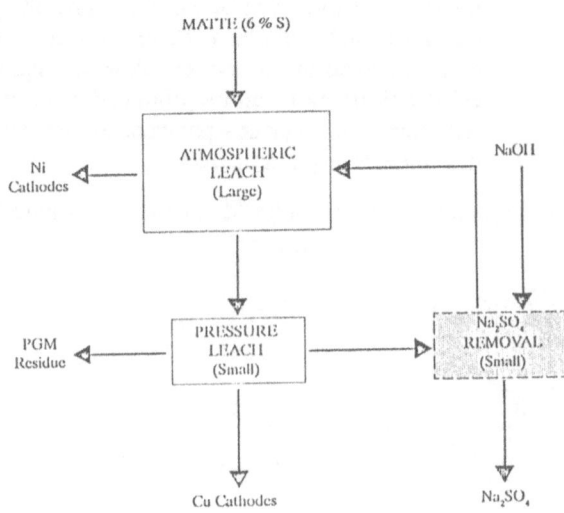

*Figure 7 :*Low Sulphur 6% Leaching Circuit

The low sulphur pressure leach route offers the most attractive option. It fits best into the current circuit and does not introduce the problem of massive sodium sulphate disposal - without significantly jeopardising PGM recoveries. The poor cobalt recovery is not considered a true disadvantage as the need for a major uprate of the cobalt plant is precluded.

5.0 OXIDATIVE AND NON-OXIDATIVE PRESSURE LEACH

Having established that a low sulphur pressure leach appears to be the most suitable option for the BSR refinery, and assuming that a suitable Ni : Cu ratio feed will be available, the fundamental question of how the copper, hence precious metals, is to be treated remains.

The pressure leaching regime chosen determines the way in which the copper, and precious metals are handled. The non-oxidative pressure leach selectively solubilises nickel leaving a copper-rich sulphide residue containing the precious metals, while the oxidative pressure leach results in the dissolution of both nickel and copper from the atmospheric leach residue to leave a high grade precious metals residue.

The BSR atmospheric leach residue has proved to be very amenable to both pressure leaching techniques which can be represented by the following reactions -

Non-oxidative pressure leach:

$$2NiS + Cu_2S + CuSO_4 + H_2SO_4 + \tfrac{1}{2}O_2$$

$$= 2NiSO_4 + 3CuS + H_2O$$

Oxidative pressure leach:

$$NiS + CuS + Cu_2S + H_2SO_4 + 6O_2 = NiSO_4 + 3CuSO_4$$

Preliminary testwork conducted by Sherritt Gordon[3] on BSR atmospheric leach residue samples to investigate both pressure leach options indicated the results shown in table 7.
These two pressure leaching techniques in turn lead to two fundamentally different flowsheets, shown in figure 8 and figure 9.

5.1 Non- Oxidative Pressure Leach

In this flow sheet a two-stage counter-current atmospheric nickel leach is employed. Anolyte from nickel electrowinning is fed to the second stage (as it does presently to the copper leach) and the resultant leach liquor is returned to the first stage where conventional cementation takes place to produce a nickel catholyte for electrowinning. The focal point of this system is that copper electrowinning is eliminated, and all incoming copper reports to atmospheric leach residue which is fed to the non-oxidative pressure leach stage. Here nickel is selectively extracted and returned to the atmospheric nickel leach circuit, while the residue, containing 0,7% Ni and 71% Cu and the precious metals, reports to a copper smelting step to produce a top quality copper cathode and the precious metal values collected as a high grade anode slime.

Advantages

i. Existing Cu leach adapts to 2° atmospheric nickel leach .
ii. Sulphur removed in Cu smelting.
iii. Higher PGM grades in anode slime.

iv. Selenium removal not necessary.
v. O_2 not required.
vi. Top grade Cu produced.
vii. High Pb in copper anolyte, reporting to nickel circuit, eliminated.

Disadvantages

i. Capital cost of Cu smelting and anode casting.
ii. Cu tankhouse to be adapted to electro-refining.
iii. Possible PGM losses in Cu smelting.
iv. Late PGM removal.

Table 7: Sherritt Gordon Pressure Leach Tests on BSR Leach Residue

			Non Oxidative	Oxidative
Temperature,°C			150	150
Leach Time, Minutes			180	180
Oxygen overpressure, kPa			-	345
		Head	Tail	Tail
Solution, gpl	Co	-	0.51	0.28
	Cu	10	0.12	67
	Fe	-	4.9	0.4
	Ni	-	34	19
	H_2SO_4	50/85	32	30
Solids, %	Co	0.27	0.05	0.19
	Cu	45.5	70.8	3.1
	Fe	2.5	0.19	17.2
	Ni	13.7	0.72	2.8
	S	17.6	21.0	2.8
Weight Ratio		100	84	6.35
Extraction, %	Ni		96	99
	Cu		-	99
	Co		85	96
	Fe		94	56
	S		-	99
Precious Metals, g/t	Au	34	40	540
	Pt	71	85	1120
	Pd	248	295	3900

Figure 8 :Possible BSR Refinery Circuit Incorporating Non-Oxidative Pressure Leach Step

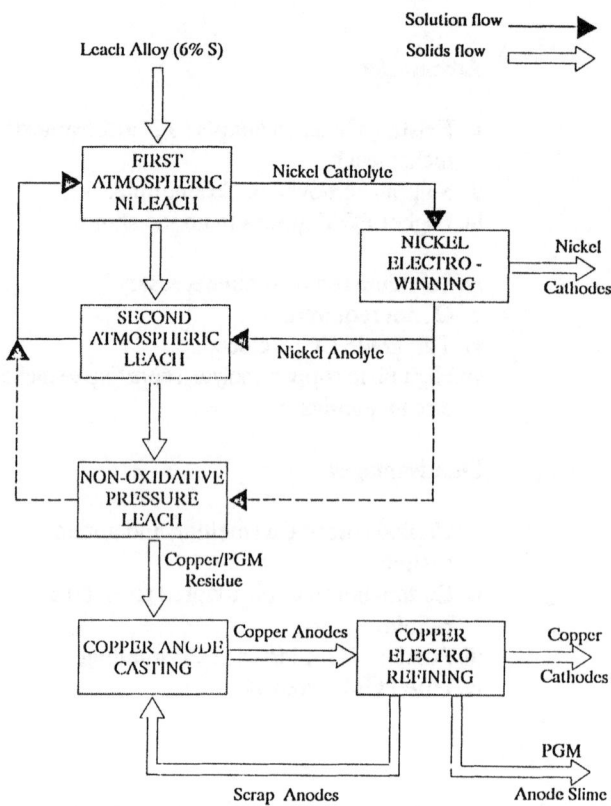

5.2 Oxidative Pressure Leach

In this scheme nickel catholyte can be produced utilising a similar two-stage counter-current atmospheric leach, but with anolyte received from copper electrowinning. The atmospheric leach residue reports to the oxidative pressure leach stage where both nickel and copper are solubilised leaving a PGM residue. The discharge solution, after selenium removal, constitutes the copper catholyte from which cathode copper is electrowon. As all the incoming sulphur is solubilised in the pressure leach step, the resultant acid imbalance calls for the inclusion of a neutralisation step.

Advantages

i. Cu smelting and anode casting step avoided.
ii. No further PGM losses.
iii. Earlier collection of PGM values.
iv. No changes necessary to Cu electrowinning (capacity allowing).

Disadvantages

i. Capital cost of selenium removal.
ii. All incoming sulphur dissolved requiring a neutralisation step.
iii. Additional reagent costs for O_2, SO_2 and NaOH.
iv. Copper quality not as good as electrorefined.
v. High Pb in copper anolyte reports to nickel circuit.

Figure 9 :Possible BSR Refinery Circuit Incorporating Oxidative Pressure Leach Step

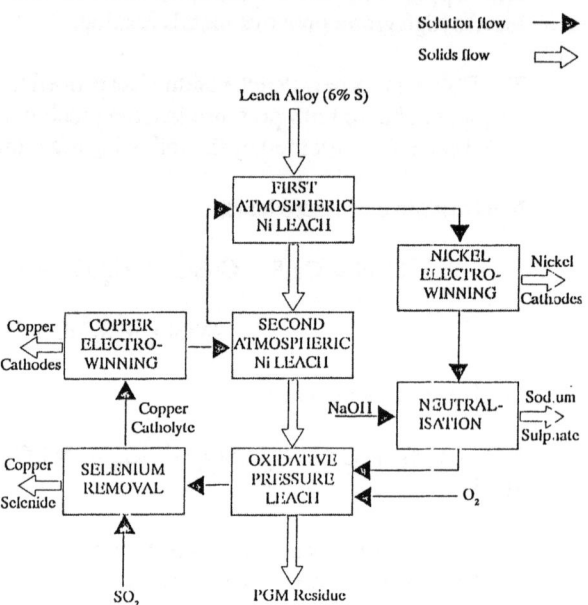

5.3 Process Selection

From the pre-feasibility study completed by Sherritt Gordon [4], it was possible to make a final process selection based on capital and operating costs, and financial benefits. These, in turn, reflect and satisfy the original parameters listed in section 3.0.

It was immediately evident that the low sulphur oxidative leach option brought with it the additional problems of the high sulphur option, but no additional benefits, and therefore was rejected at the outset.

Thus, the high sulphur oxidative leach compares financially with the low sulphur non-oxidative leach as shown in table 8.

As the additional operating cost for the high sulphur oxidative leach option exceed the additional revenue, the return on capital investment is negative, and therefore clearly the low sulphur non-oxidative leach option, with a 4 year payback period, is the only choice.

The large divergence in additional operating costs is accounted for mainly by the large extra quantity of caustic soda needed for precipitation of additional cobalt, and particularly for neutralisation of the vastly increased sulphur input to the refinery. (In the case of a Zimbabwe based process, the caustic soda is an imported item and with taxes/duties added is extremely expensive).

Table 8. :Financial Comparison of Oxidative and Non-oxidative Pressure Leash Options

	High Sulphur Oxidative Leach	Low Sulphur Non-oxidative Leach
Capital Cost, Z$ million	62	45
Additional Operating Cost, Z$ million/annum	38	4
Additional Revenue, Z$ million/annum	30	15
Payback Period, years	Negative	4

6.0 CONCLUSION

With the present operation, base metal recoveries are restricted by the circulating load and PGM recoveries are extremely poor.

For BNC to survive in times of low metal prices, the BSR refinery circuit must be modified to improve recoveries of both base and platinum group metals. This can be achieved by opening the present Outokumpu circuit by the addition of a pressure leach stage. This will eliminate all circulating loads returning to the smelter. Expected recoveries are compared with present recoveries in table 9. This improvement will realise an additional Z$15 million annually from current low grade PGM concentrates alone.

Table 9. Metal Recoveries (%)

	Ni	Cu	Au	Ag	Pt	Pd	Rh
Present	93	92	31	23	50	60	64
Forecast	95	95	>90	>90	>90	>90	>90

Investigations carried out to date have indicated that acceptable PGM recoveries, associated with improved base metal recoveries, can be readily achieved by further treating the low sulphur matte residue in a pressure leach stage, without introducing major problems such as sodium sulphate disposal.

From preliminary investigations it has been concluded that the only viable process, both technically and financially, is the low sulphur non-oxidative leach. By adopting this opinion it can be confidently predicted that the BSR will become an important PGM producer by the mid - 1990's.

7.0 ACKNOWLEDGEMENTS

The authors wish to thank the Chairman and Managing Director of Bindura Nickel Corporation Ltd for permission to publish this paper.

8.0 REFERENCES

1. The Development of the Outokumpu Nickel Refinery Process in Zimbabwe. Groom J D G et al, Proceedings of the International Conference on Mineral Processing and Extractive Metallurgy, Editor, M J Jones and P Gill, Kunming - People's Republic of China, I.M.M., 1984 pp 381 - 396.

2. High Temperature Behaviour of Platinum Group Metals in Oxidising Atmospheres. Jehn H. Journal of the Less Common Metals, 100 (1984) pp 321 - 339.

3. Pressure Leaching and tests on BSR Nickel/Copper Mattes and Leach Residue, Sherritt Gordon, Confidential Test Report, 1989.

4. Prefeasibility Engineering Study on Bindura Smelter and Refinery Process Options, Sherritt Gordon Ltd, Confidential Report, 1990.

Challenges associated with electrowinning nickel at BSR Ltd, Zimbabwe

D. C. Alexander
G. Chiwara
BSR Ltd, Zimbabwe

ABSTRACT

The BSR Refinery was designed to extract nickel, copper and cobalt from a nickel copper matte utilising Outokumpu technology. Since the refinery was commissioned in 1968 process modifications have had to be made in response to the economic climate, market demands on quality and the need to cater for toll treatment arrangements. The more important modifications on the leach plant and improvements in leach plant control philosophy are described. The more recent process improvements have centred around the tankhouse where cathode quality and current efficiencies have been the focus of attention. Changes in operating parameters leading to improved products both physically and chemically are described. The major challenges being faced are in the context of competing with western world producers from a third world base while remaining financially viable.

INTRODUCTION

The BSR Refinery, situated in Bindura has since 1968 treated a nickel/copper/ cobalt/sulphur matte utilising the Outokumpu process developed in the 1960's to produce cathode nickel. The Refinery has seen mixed fortunes over recent years but none more devastating than the depressed world nickel prices leading up to the sudden price surge in 1987.

The economic climate over this period necessitated a number of innovative modifications and additions to the refinery circuit to increase
competitiveness on the international markets. This was initiated with the commencement of toll treating external feedstocks resulting in:-

a)   An in-depth review of the leaching philosophy.

b)   A streamlining of the cobalt removal circuit.

c)   Improvements to cathode chemical quality particularly with respect to lead.

Emphasis on tankhouse operations has been to regain lost ground to produce a cathode nickel of similar high purity to that obtained in the earlier 1980's. The effects of austerity measures implemented in 1987/88 had eroded the ability of the refinery to maintain the high standards set and it was imperative that these standards be regained to maintain Bindura Nickel Corporation's reputation as a producer of high quality nickel cathode.

PROCESS DESCRIPTION

The Outokumpu Process as applied to the BSR Refinery operation has previously been described in detail [1], but a brief description is given here. The refinery treats a nickel/copper/cobalt/sulphur matte to produce electrowon nickel and copper and a cobalt hydroxide cake by-product. The matte is initially milled in two ball mills (2,5m x 3m) to a fineness of 90% -53 microns (270 mesh) before undergoing a first stage atmospheric "cementation leach" where 44% of the contained nickel is leached to produce an "impure" nickel catholyte solution. Copper anolyte from the electrowinning cells is used as the leach liquor.

In order to produce the purified nickel catholyte for electrowinning it is then necessary to remove cobalt from the impure nickel catholyte since electrochemically cobalt and nickel are very similar. This is effected by reacting the cobalt (II) sulphate with nickel (IV) hydroxide produced electrolytically for that purpose. The cobalt (II) sulphate is precipitated as the cobalt (III) hydroxide and removed by filtration. This precipitation and filtration step has the added benefit of scavenging impurities, such as lead, zinc, selenium, manganese etc from the nickel catholyte and a typical assay for the purified nickel catholyte has copper and cobalt as the main impurities in concentrations of less than 0,5 parts per million.

Nickel of 99,99% purity is electrowon from the final catholyte solution and the anolyte is returned to a second stage "copper leach" to leach the remaining soluble nickel and copper from the residue of the cementation leach to form a "copper catholyte". Copper is electrowon from this catholyte and the anolyte returns to the cementation reactors to close the circuit.

The process is illustrated in Figure 1.

Fig.1 : A Simplified Flow Diagram of the BSR Refinery Process

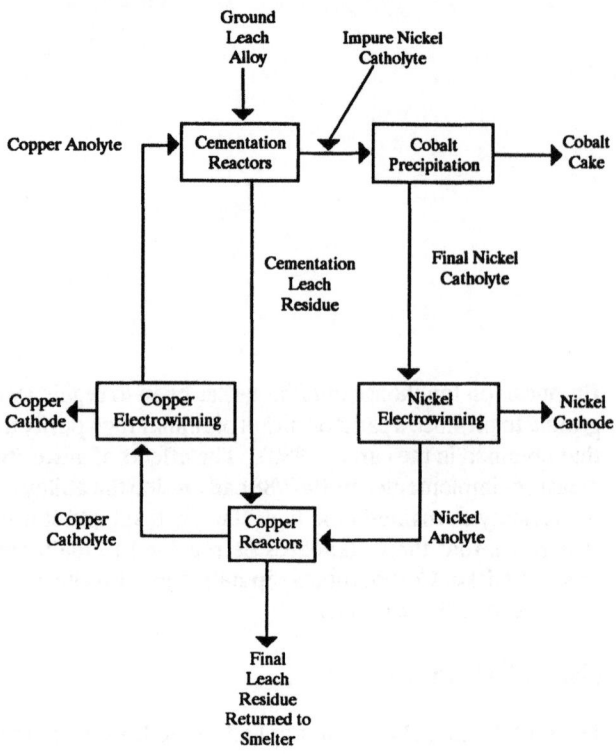

Typical single pass recoveries from the refinery are:-

Nickel - 87%
Copper - 39%
Cobalt - 82%

MAJOR DEVELOPMENTS

The treatment of external feedstocks utilising spare capacity at the refinery commenced with the introduction of Bamangwato Concessions Limited (BCL) matte in late 1985 followed by significant quantities of nickel sulphates in early 1986. Treating these materials required an in depth reivew of the refinery operating parameters and subsequent developments can be traced back to this period.

A. Cementation Leach

In order to produce the nickel catholyte, copper anolyte typically analysing 8 gpl and 10 gpl sulphuric acid is reacted with the ground leach alloy in the cementation reactors. The resulting reactions can be represented simply as:

$$Ni^0 + Cu^2 \rightarrow Ni^{2+} + Cu^0 \quad ----- (1)$$

$$2Cu^0 + 1/2 O_2 \rightarrow Cu_2O \quad ----- (2)$$

$$Cu_2O + 2H^+ \rightarrow 2Cu^{2+} + H_2O \quad ----- (3)$$

Reaction (1), the "cementation reaction", is the main nickel leach reaction and by its nature produces the copper free solution for electrowinning. It is a fast reaction as long as metallic nickel surface is exposed to the solution.

Reaction (2) is a slow reaction involving the oxidation of metallic copper and is generally the rate controlling step.

Reaction (3) is the dissolution of the oxidised copper in sulphuric acid thus providing more copper ions for reaction (1).

As can be appreciated the above reactions, being diffusion controlled, occur on the surface of the alloy particles and consequently the available surface area, or fineness of the grind is an important factor in the overall kinetics.

Typical performance of the cementation reactors is shown in Table 1 as extracted from present operating data.

Table 1: Typical Reactor Conditions Under Present Operation (44% Leach)

	First Reactor	Intermediate Reactor	Terminal Reactor
Soluble copper	100ppm	10ppm	2ppm
H_2SO_4	4 gpl	6,0 pH	6,4 pH

Following the introduction of the BCL matte to the refinery circuit soluble copper control became very difficult while filtration performance deteriorated to the extent that nickel plating capacity was restricted.

Prior to the introduction of the BCL matte the refinery treated a matte with a nickel to copper ratio of 2,50 - 3,00 leaching 58% of the nickel in the cementation reactors. The BCL matte ratio averaged 0,85 which effectively reduced the ratio of the blended matte being treated in the refinery to 2,00. The effect of this is best illustrated by considering an individual particle.

The matte to the cementation reactors is first ground to +90% -53 microns to maximise the surface area of metal available for reaction. As the reaction progresses on the leach alloy particle the nickel on the outer surface becomes depleted leaving a rim of unreacted and cemented copper. This continues until a point is reached when no nickel is available for reaction at which point a fourth rate controlling reaction appears (reaction (4)) at pH >4,5.

$$3Cu^{2+} + 3SO_4^{2-} + 4H_2O \rightarrow CuSO_4 \cdot 2Cu(OH)_2 + 2HSO_4^- + 2H^+ ..(4)$$

The basic copper sulphate (antlerite) so formed is a particularly fine precipitate which is extremely difficult to filter. Although the existence of this reaction was known its effects on plant performance had not previously been quantified.

It was evident therefore that the amount of nickel available for initial leaching was heavily dependent on the nickel to copper ratio of the matte being treated and consequently the leach was dropped to 44% as being the optimum compromise. The soluble copper and filtration problems virtually disappeared overnight and the refinery performance improved significantly as illustrated in Table 2.

Table 2: Comparison of Operating Data on Reducing The Cementation Leach

	58% Leach Soluble Cu	44% Leach Soluble Cu
First Reactor	300	140
Intermediate Reactor	30	10
Terminal Reactor	4	2
Filtration cycle time, hrs	9	19
Filter Aid Consumption	10,2	7,5
% of Downtime due to Cementation and Filtration Problems: Leach Plant	55%	3%
T/House	33%	-

B. Cobalt Precipitation

In the cementation reactors a copper free nickel catholyte solution is produced. However this catholyte does contain 300 ppm cobalt (II) sulphate, and it is necessary to remove this from the nickel catholyte prior to electrowinning in order to produce high purity nickel.

The cobalt is removed from the solution using electrolytically generated nickel (IV) hydroxide in the following sequence of reactions (3).

Nickel (II) hydroxide is generated

$$2NaOH + NiSO_4 \longrightarrow Ni(OH)_2 + Na_2SO_4 \quad ..(5)$$

Nickel (II) hydroxide is electrolytically oxidised.

$$Ni(OH)_2 + 2H_2O \longrightarrow Ni(OH)_4 + 2H^+ + 2e^- \quad ..(6)$$

Cobalt is precipitated

$$Ni(OH)_4 + 2CoSO_4 + 2H_2O \longrightarrow$$
$$2Co(OH)_3 + NiSO_4 + H_2SO_4 \quad ..(7)$$

Acid generated in reaction (7) is neutralised

$$H_2SO_4 + 2NaOH \longrightarrow Na_2SO_4 + H_2O \quad ..(8)$$

Prior to the introduction of the BCL matte the cobalt was removed in a two-stage counter-current process with intermediate filter presses. The cobalt cake was then treated with acid to redissolve unreacted nickel to upgrade the cake.

$$Ni(OH)_2 + H_2SO_4 \longrightarrow NiSO_4 + 2H_2O \quad ..(9)$$

At high plant throughput rates with high cobalt, and during periods of poor filtration, filtration capacity in this configuration was often exceeded and cathode nickel production was affected. In addition the batchwise pH control system (reaction 7) was insensitive to excessive caustic soda addition when nickel was unnecessarily precipitated without a further increase in pH being evident.

Again a major review of the operating philosophy was necessary and the bold step of altering the counter-current two-stage precipitation to a once through co-current system was taken to release the intermediate filter presses for other duties. In addition a system for stoichiometric caustic addition was implemented with major cost benefits as the optimum usage of caustic was easier to attain. This effectively alleviated the filtration problem, and an added spin-off of the tighter control over the cobalt precipitation was a reduction in acid consumption.

With the release of the intermediate filter presses for other duties it became possible to treat the cobalt cake with nickel anolyte (typicallly 48 gpl acid), utilising the acid component of the nickel sulphates now being toll-treated, rather than to treat with new concentrated acid. The additional presses so released were quite capable then of handling the increased load.

The benefits arising from the adjustments to the cobalt precipitation circuit are illustrated below:-

Table 3: Benefits Arising from Changes to the Cobalt Precipitation Circuit and Reducing the Cementation Leach

Cementation leach	58%	44%
Filter Aid Consumption, kg/tNi	10,2	7,5
Sulphuric Acid Consumption, kg/tNi	96,3	58,8
Caustic Soda Consumption, kg/tNi	77,9	41,8

C. Leach Efficiency/Leach Residue

In the refinery process typical nickel and copper leach efficiencies are:-

Nickel - 93%
Copper - 45%

The operating philosophy with regard to leach efficiencies and nickel to copper ratios has been discussed in a previous paper [1] where it was postulated that the maximum copper leach achievable depends on the nickel to copper ratio and percentage of sulphur in leach alloy.

Prior to 1986 the maximum copper leach attainable was 35%. This increased as expected (to 45%) after BCL matte was introduced to the circuit since the proportion of copper to sulphur increased through the reduction in the overall Ni:Cu ratio. However with the termination of BCL toll treatment it has been the refinery's experience that, despite the increase in the nickel to copper ratio, the copper leach has remained at 45%, and this appears to be another spin-off from the reduction in the cementation leach from 58% to 44%. This has still to be quantified but can be correlated to the reduced copper circulating load through the copper leach and electrowinning circuit.

As a result of the leach efficiencies mentioned, a significant proportion of the matte treated is returned to the smelter for further processing. Prior to 1986 this leach residue was taken out on a rotary vacuum filter which produced a cake containing about 45% moisture and this served as a bleed in controlling the Na_2SO_4 generated in the cobalt precipitation circuit. (Reactions (5) and (8))

However, with the treatment of the BCL matte with a significantly higher copper content, the proportion of leach residue increased significantly. This resulted in unacceptably high losses of nickel in entrained solution to the smelter and hence a drop in the overall nickel recovery. To circumvent this the drum filter was replaced by a plate and frame press filter and again the improvement was immediately evident. Cake moisture contents dropped to 25% and a system of partial water washing and air blowing was implemented for greater nickel recovery. As a further improvement the plate and frame press will shortly be replaced by a centre feed chamber plate press with installed wash ports to maximise washing efficiency, and nickel recovery. This dovetails with another proposed development in the refinery process, being the introduction of pressure leaching as described by Schwarz et al[5].

D.   **Sodium Sulphate Removal**

A negative effect of the improved washing of refinery residues has been an increase in the concentration of Na_2SO_4 in the refinery circuit and it was therefore necessary to find an alternative means of removal. Over this period the Na_2SO_4 concentration increased from 130 gpl to over 180 gpl, and, though no ill-effects of the higher concentration were evident, a potential problem was that of the saturation level being reached with disastrous consequences.

A centrifuge was subsequently acquired to test the separation of Na_2SO_4 from the nickel (IV) hydroxide prior to addition to the cobalt precipitation tanks, and this proved very successful.

A centrifuge has now been installed in the circuit, and follow on developments will see the installation of a crystallizer plant to produce a saleable product of anhydrous sodium sulphate for the local detergent industry, thus substituting presently imported material.

Fig. 2 : Centrifuge Operation

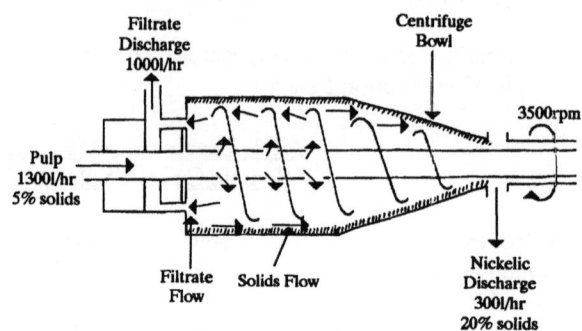

E.   **Lead Removal**

In order to produce high quality nickel cathodes, emphasis is placed on maintaining the lowest possible concentration of contaminants in the nickel catholyte solution, lead being a particularly important one.

In the refinery two main sources of lead have been identified, these being leach alloy (100 - 200 ppm Pb) and the lead anodes in the electrowinning cells which slowly corrode, and resultant soluble Pb in process streams is:-

        Nickel anolyte - 2 ppm Pb
        Copper anolyte - 10 ppm Pb

Lead removal is effected by co-precipitation with barium sulphate and prior to 1988 barium was added through the mills as a solid carbonate. By this method lead was removed from the solution in the cementation reactors and left the refinery with the leach residue. This system was inefficient in that the barium carbonate addition was effectively batchwise, and, since $BaCO_3$ and $BaSO_4$ are both insoluble, only the surfaces of the $BaCO_3$ particles are available for reaction. Testwork indicated that only one-third utilisation is possible i.e. a diffusion reaction becoming progressively more difficult as it proceeds.

Fig. 3 : Schematic Representation of the BaSO₄ Formation

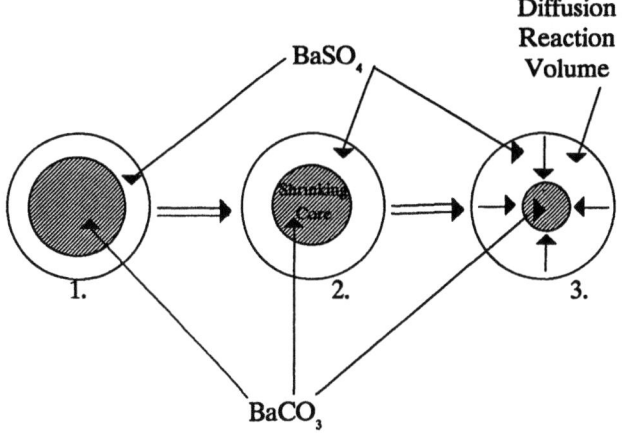

In 1988 the barium carbonate was replaced by barium hydroxide as the lead precipitant since it is water soluble and hence could be added continuously in solution at varying flowrates to multiple dosage points within the circuit (reaction (10)). Its solubility also means that 100% utilisation is possible as illustrated in Table 5.

$$Ba(OH)_2 + NiSO_4 \longrightarrow BaSO_4 + Ni(OH)_2 \quad ..(10)$$

Reaction (10) is very rapid and to obtain the maximum benefit it is necessary to provide for maximum dispersion of the barium hydroxide solution in the smallest practical volume of process solution. In addition this should be followed by a period of relative calm to allow for the precipitate to consolidate. This second point has been difficult to achieve but the best compromise has been the addition of barium hydroxide into pumpboxes as far upstream as possible from agitated tanks which has proved successful as illustrated in Table 4.

Table 4: Comparison of Lead Removal Performance - Barium Carbonate (1987) vs Barium Hydroxide (1988 - 1990)

	1987	1988	1989	1990
Lead in copper anolyte, ppm Pb	9,89	13,62	10,79	9,20
Lead in nickel catholyte, ppm Pb	0,33	0,10	0,15	0,10
Lead in cathode, ppm Pb	21	11	10	4

Reaction (10) has been found to be independent of the acid concentration at the addition point and approximately 80% of the lead is removed at each addition point at acid levels ranging from 10 gpl to 6,4 pH.

Reagent consumption has also reduced accordingly. (See Table 5).

Table 5: Comparison of Barium Carbonate and Barium Hydroxide Consumption Rates (kg Ba²⁺/tNi)

	1987	1988	1989	1990
Barium as Carbonate	18,1	11,5	-	-
Barium as Hydroxide	-	4,8	5,0	4,6

F. Tankhouse Operations

Nickel Tankhouse operations have been described by Saarinen and Seilo[7]. The BSR refinery tankhouse configuration is illustrated in Table 6. A typical anode/cathode pair is shown in figure 4.

Fig. 4 : Schematic Cross Section of an Anode / Cathode Pair in the Nickel Electrowinning Cell

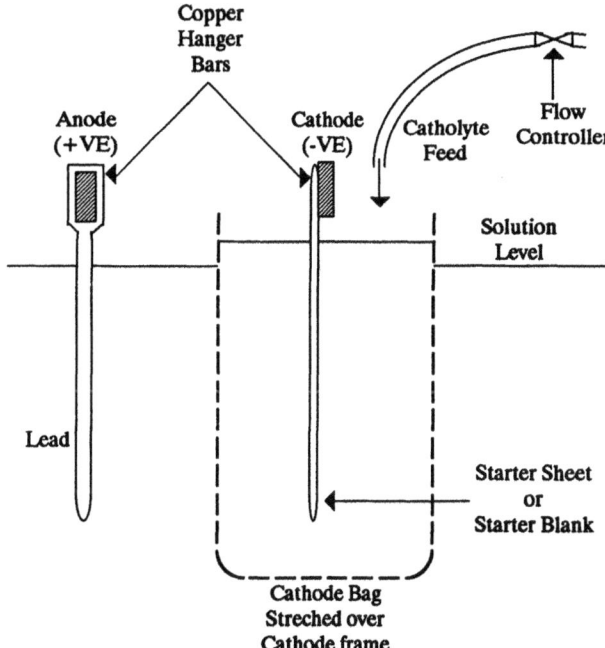

In the period of low nickel prices leading up to 1987, severe austerity measures were necessary which ultimately had an adverse effect on the tankhouse performance. The run-down condition of plant and equipment manifested itself in several areas:-

(a) The general condition of cathode bags deteriorated, these bags being patched as much as possible before being replaced. This made pH control within the bags difficult as a result of anolyte back diffusion with a consequent increase in lead levels in nickel cathodes.

(b) Recasting of antimonial lead anodes with minimal new lead make-up resulted in a deterioration in anode quality through greater corrosion rates and cracking. Testwork carried out at the time indicated a reduction in antimony content in recast anodes by as much as 30%.

(c)   Current distribution in the electrowinning cells was aggravated by the poor anode quality and breaking down of the anode to hanger bar contact. This was difficult to pick up through voltage measurement with the instruments then in use. At that time the problem of "burning loops" manifested itself. The problems being encountered in the tankhouse culminated in a drop in current efficiency from the normal 96% in 1986 to 87% in 1987 and 1988.

Table 6: Tankhouse Configuration

Cell Construction		Electrodes	
Shell	Concrete	Anodes	Lead
Lining	Fibre Glass	No. per cell	40
Wall thickness	155mm	Cathodes	Nickel
Length	6600mm	No. per cell	39
Width	1140mm	Starter blanks	Titanium
Depth	1150mm	Length	980mm
Live volume	8,6m^3	Width	880mm

Electrolyte		Operating Parameters	
Flowrate	13 l/hr/cathode	Cell Current	13,6 kA
Catholyte	80 gpl nickel	Cell voltage	3,5 V
pH	3,0	Current density	180A/m^2
Temperature	55°C		
Anolyte	52 gpl nickel		
pH	48 gpl H$_2$SO$_4$		

Following the nickel price surge in 1987 the task of regaining lost ground began in earnest.

1.   **Cathode Bags**

The problems associated with cathode bags arose as knock-on effects from the deterioration of the anodes and hence current distribution. Higher electrical current flows through individual cathodes resulted in excessive nodular growths which would tear the bags during pulling. pH control within the bag therefore became difficult as a result of anolyte back diffusion which resulted in an increase in the proportion of cathodes splitting along the starter sheet interface when cutting the cathodes on guillotines.

As a result of the problems listed a system of flushing cathode bags before and after loading starter sheets was developed in order to ensure the correct bag pH at the start of the cycle. This has proved very successful and has now replaced etching as a means of minimising the inherent weakness of the starter sheet interface in the nickel cathodes. Table 7 illustrates the success of the flushing procedure.

Table 7: Results of Testwork on Minimising Cathode Splitting

Procedure	% Split
Untreated Starters	44
Etched Starters	25
Flushed cells	14
Flushed cells/etched starters	12

2.   **Lead Anodes**

Lead Anodes have found widespread use in the electrowinning industry. When in use, the anode surface layer is oxidised to lead dioxide (PbO$_2$) which is stable, and being semi-conductive will pass current. However lead is weak structurally and tends to creep which results in cracking and peeling of the oxide layer which is the mechanism by which the anode corrodes. Various alloying elements are added to the lead in order to counter this, antimony being one such element.

The problem of corrosion is magnified when power to the cells is switched off and the lead dioxide surface layer discharges to form lead sulphate. This accelerated anodic corrosion increases the risk of lead contamination. To counter the effects of this passivation a system of "trickle charging" is now in practise whereby during any shutdown a current of 10 A/M^2 is maintained on the anodes. Testwork has indicated this to be the minimum current necessary to prevent discharge of the PbO$_2$ layer.

With the problems being encountered with the poor distribution of antimony in recasting anodes, testwork was conducted comparing the performance of pure-lead, 4% and 6% antimony lead anodes with recast anodes. Of these, pure-lead performance was significantly better with corrosion rate of less than 20% of the antimonial lead anodes. The refinery tankhouse has now converted completely to pure-lead anodes.

The problem of maintaining a good contact between the lead and copper hanger bar has now eased with all hanger bars being tin-coated prior to anode casting.

In order to maintain an even current distribution the practice adopted by the refinery is to measure current flow through individual hanger bars, moving away from voltage drop as a means of control.

All of the measures adopted above have had a significant impact on the tankhouse performance. In addition improvements on the leach plant inventory control have enabled all aspects of housekeeping to be fully implemented and the effects are best illustrated in Table 8.

Table 8   Comparison of Tankhouse Operation
           1987/88 versus 1989/90

	1987/88	1989/90
Anodes	Pb/Sb	Pb
Current Efficiency	87%	96%
Bag Consumption	120/month	240/month

Surface Quality

	1987/88	1989/90
Smooth surfaced cathodes	28%	37%
Noduled	72%	63%

3. Further Development

Having regained ground lost, the refinery continues to move ahead. In order to improve operating performance and cathode quality further, present testwork revolves around replacement of fibre glass cathode frames and a review of cathode spacing within the electrowinning cells.

a) For some time fibre glass cathode frames have been viewed as unsatisfactory for use in the electrowinning cells since fibre glass particles, released either through physical damage or natural degradation, can cause nodulation on the plating surface. Studies have shown that these nodules generally originate on the surface of the starter sheets [6] from inclusions, predominantly silica, thus reinforcing this view. A second source of silica also arises from the filter aid presently in use but this is easier to contain.

Polypropylene frames are presently under test as possible replacements for the fibre glass frames, the increased cost being more than offset by their longer expected life. These frames are more robust and of a narrower width and hence this raises the possibility of closer spacing of electrodes within the cells resulting in a reduced current density on the electrodes. Present spacing of electrodes is 160mm and this has been reduced to 135mm with a significant effect on cathode surface quality as illustrated below.

Table 9   Comparison of Operational Data for Polypropylene Frames versus Fibre Glass Frames

Cathode Grading	Fibre Glass (160mm spacing)	Polyprop (160mm spacing)	Polyprop (135mm spacing)
Smooth	39%	41%	46%
Noduled	61%	59%	54%

Results to date indicate that current density is of prime importance in producing even plating on the cathodes, and that fibre glass particles may account for only a small proportion of nodulation. Testwork in this regard is continuing.

b) With the closer spacing of cathode frames now possible the design of the cell busbars and insulators has also come under scrutiny with added emphasis on sealing all joints to ensure that they are airtight, thus preventing oxidation of the conductor surfaces and maintaining good contacts.

c) A further development has been closer packing of electrodes within the nickelic oxidation cells through the purchase of Outokumpu technology. In oxidising the nickel (II) hydroxide to nickel (IV) hydroxide the reaction occurs at the anode surface. Therefore by increasing the anode surface area the efficiency of oxidation will improve accordingly. Initial results have been encouraging and testwork is continuing.

With all of the above mentioned improvements the BSR refinery again produces a nickel cathode of the highest quality, both physically and chemically. (Table 10).

Table 10:   Typical Nickel Cathode Analyses

		ppm			
Year	% Smooth	Cu	Co	Fe	Pb
1986	32	11	24	<2	15
1987	45	14	19	<2	21
1988	28	20	20	<2	11
1989	19	16	22	<2	10
1990	37	12	24	<2	4

CONCLUSIONS

Over the years BSR Limited has experienced mixed fortunes, the low world nickel price leading up to 1987 threatening its very existence. The refinery has profited from this experience and from the toll treatment of external feedstocks while the development of a pressure leaching circuit for enhanced base metals and PGM recoveries presently under way maintains the momentum into the foreseeable future. This paper serves to describe from a practical point of view the problems encountered in developments in the refinery process and how they were overcome, thus maintaining BSR Limited's reputation as a producer of quality electrolytic nickel for the world market.

ACKNOWLEDGEMENTS

The authors wish to record their thanks to the Managing Director of Bindura Nickel Corporation and management and staff of Anglo American Corporation Services for permission to publish this paper.

REFERENCES

(1) GROOM J D G, STEWART R J E, NIXON J L, SAARINEN H. The Development of the Outokumpu Nickel Refining Process in Zimbabwe Proceedings of the International Conference on Mineral Processing and Extractive Metallurgy, Peoples Republic of China, 1984.

(2) SCHWARZ J. The effects of reducing the Cementation Nickel Leach on Refinery Performance. An Internal Memorandum, 1986.

(3) SCHWARZ J. Reduction in the consumption of Caustic Soda and Sulphuric Acid. An Internal Memorandum, 1987.

(4) SCHWARZ J, CHANNON W P. BSR - Recovery Investigation - 1986. AACS Memorandum.

(5) SCHWARZ J, RICHARDSON M R. Process Options for Modifying the Outokumpu Refinery Circuit at BSR Limited. Proceedings of the Anglo American and De Beers Symposium, January 1990.

(6) FERNANDES T R C. Nodular growths on Nickel Cathodes from Bindura. I M R Confidential Report C308, August 1983.

(7) SAARINEN H, SEILO M. Production of the Cathode Nickel in the Outokumpu Process.

Production of ferrosilicon in Zimbabwe

K. E. Mudzimu
Zimbabwe Alloys Ltd, Refinery Division, Zimbabwe

SYNOPSIS

The history of ferrosilicon production at
Zimbabwe Alloys is described scanning a period
from October 1967 to March 1990 albeit with a
long break between 1969 and 1986[1]. The theory
and metallurgy of the operation is explained in
order to provide insight into the operation. An
account of the exciting challenges and
experiences of the early campaigns is given
culminating in the current successes. Reasons
are postulated for the failure of the early
campaigns and the main factors leading to the
successful production of FeSi75 from 1987 are
described.

INTRODUCTION

Zimbabwe Alloys is the sole producer of FeSi
alloys in Zimbabwe and production is geared to
satisfying local requirements in line with the
import substitution policy of the State. The
first attempts to produce FeSi75 were made
between October 1967 and September 1969 in four
campaigns. The trials were in a 17.5 MVA
furnace and operating data pertaining to these
trial campaigns are presented in Table 1. The
first attempt failed due to an accumulation of
very hard silicon carbide inside the furnace and
extensive damage to electrode assemblies was
incurred due to overheating of the surface. The
campaign lasted one and one-half months after
which the furnace was dug out and rebuilt for
the second campaign. The second campaign carried
on for five and one-half months before it was
aborted in June 1968. The digout revealed signs
of overcoking but the exact cause of failure
could not be pinpointed with accuracy.

Expert advice was sought during the third
campaign which lasted just over four months but
by November of the same year there was growing
evidence of a build-up inside the furnace. The
furnace was successfully washed down using iron
ore three times in a space of two weeks. A
premature shutdown was effected in January 1969
due to poor performance.

Poor electrode penetration necessitated
frequent burndowns during the fourth and final
campaign which lasted seven months before it too
was aborted.

The need for import substitution as one of
the strategic management policies of Zimbabwe
Alloys management revived the idea of producing
FeSi75 in late 1986. The alloy would be
produced mainly for the domestic steel producer,
ZISCO, and local foundries in campaigns
alternating with ferromanganese, also for ZISCO,
and FeSiCr for the international market. The
first attempts were made in September/October
1986 in a 700 kVA furnace. These proved
unsuccessful due to the excessive heat above the
furnace bath which led to the failure of
ancillary equipment. It was, however, noted
during these trials that operations were more
manageable at lower silicon levels. A decision
was taken to produce FeSi50 in February 1987 and
operations at this level were successful. In
April of the same year production was
transferred to a 7.5 MVA furnace.

METALLURGY

Chemistry of Silicon Reduction Process

Ferrosilicon was first obtained in 1808 by
common reduction of silicon and iron oxides with
carbon in a crucible furnace and subsequent
developments led to the production of the alloy
in electric smelting furnaces[1,7].

The main raw materials are quartzite and
carbonaceous reductants with steel scrap or iron
ore.

The overall simplistic reaction is given as
follows[2]:

(i) $SiO_2 + 2C \rightarrow Si + 2CO$ The initial
temperature of this reaction is
approximately 1255°C.

In the upper layers of the burden the SiO
gas from the crucible reacts with carbon in the
descending charge to form SiC as follows:

(ii) $SiO + 2C \rightarrow SiC + CO$

(iii) $3SiO + CO \rightarrow 2SiO_2 + SiC$

or condense according to the equation

(iv) $2SiO \longrightarrow SiO_2 + Si$

From theoretical thermodynamic considerations the temperature of the beginning of reduction of silica is 1487°C for 75% FeSi and 1377°C for FeSi45.

The temperatures are derived from calculations of changes in free energy of the reaction of reduction of silica by carbon and the change in free energy during dissolution of silicon in iron.

$$SiO_{2(liquid)} + 2C_{(graphite)} \longrightarrow Si_{(liquid)} + 2CO_{(gas)} \qquad Z° = 159230 - 87.17T$$

Assuming the constant of the considered reaction to be $K_d = P^2_{CO}$ and considering that the reaction starts at $P = 1,01325N/m^2$, we obtain

$$\ln P_{CO} = O; \quad Z° = O \text{ and } T = \frac{159230}{87.17} = 1827°K$$
$(1554°C)$

The presence of iron in the charge makes the reduction of silica easier, as silicon, dissolving in iron, is continuously removed from the reaction.

The equilibrium constant of the reduction reaction of silica by carbon in the presence of iron has the following form: $K = a_{Si} P_{CO}$ where a_{Si} is the activity of dissolved silicon.

In practice there will always be some SiO gas which passes through the burden without reacting and becomes oxidized by air:

(v) $2SiO + O_2 \longrightarrow 2SiO_2$

Although the very fine submicron SiO_2 dust is amorphous and so far not suspected to be hazardous in most countries[4], the gas has to be cleaned before discharging into the atmosphere. To this end attempts by ZARD to obtain foreign currency for dust cleaning equipment in 1990 were unsuccessful. This is the main reason for silicon loss resulting in typical silicon recoveries of 80 to 90 per cent for FeSi75.

In the crucible the SiO_2 in the burden reacts with SiC formed earlier according to the equation:

(vi) $3SiO_2 + SiC \longrightarrow Si + 2SiO + CO$

This reaction develops at approximately 2100°C, near the zone of electric arcs. A deep reaction zone and a cool furnace top are essential for the aforementioned reactions to take place with minimum silicon and heat loss. Fig.1 summarizes reactions taking place in different zones inside the furnace[3].

Alloy silicon content was gradually increased but at concentrations[1] around 60% the alloy was crumbling and therefore it was decided to go direct to 72-78%Si.

Some grades of ferrosilicon especially those with silicon in the range 50-60% are known to crumble[2]. The mechanism of crumbling is due to the fact that phosphides of aluminium, calcium and other elements disposed along the grain boundaries react with moisture in the air to give gaseous phosphine(PH_3) and the grain surface is oxidized.

Transformation of the supercooled solid solution to a eutectic which is accompanied by a reduction of the specific gravity thereby increasing the volume of the alloy is also believed to explain the crumbling phenomenon[2].

OPERATING EXPERIENCE AT ZARD

Raw Materials and Preparation

Batching

A typical charge employed at ZARD would be made up of:

200kg Quartz   : 12-100mm
 95kg Coal   : 6-25mm
 16kg Steel Scrap : 30mm x 100mm
 80kg Wood Chips : 75mm x 100mm

A batch weighing machine weighs components in a weighing hopper which discharges into a holding hopper. Raw materials are weighed starting with the hardest (quartz) through to the most friable (coal) to avoid crushing and fines generation. When the batch is complete the holding hopper discharges into the charging pan which is picked up by a charging machine and discharged onto the furnace top. After six charges have been so discharged the machine picks up a spreader and levels the charge. A charge height of one foot above sill level is aimed for. The weighing system is checked at least once per shift for accuracy.

Quartz is obtained from Zimbabwe Alloys owned Broadside Quarries 20km outside Gweru; coal comes from Hwange. Steel scrap is supplied locally and wood chips come in as poles which are chipped at the refinery. Charcoal is the optimum source of carbon because of its low ash, high reactive capacity and a low electrical conductivity. The only known source in the country is fully committed to exports. Locally sourced raw materials have on the one hand advantages of being readily available and cheap but on the other hand may be of inferior quality compared to those used by foreign competitors. The materials used at Zimbabwe Alloys Limited have given satisfactory results.

Raw materials are charged to furnace silos by hoist skip.

Typical raw material analyses are given in Table 2.

It is important to have uniform distribution in the upper layers of the burden to enhance the possibility of reactions taking place; the charge should therefore have high and uniform porosity. For this reason the quartzite and coal are screened to remove the minus 12mm and 6mm sizes respectively. The quartzite must be of a quality that does not decrepitate as temperature increases during the descent of the charge.

Figure 1

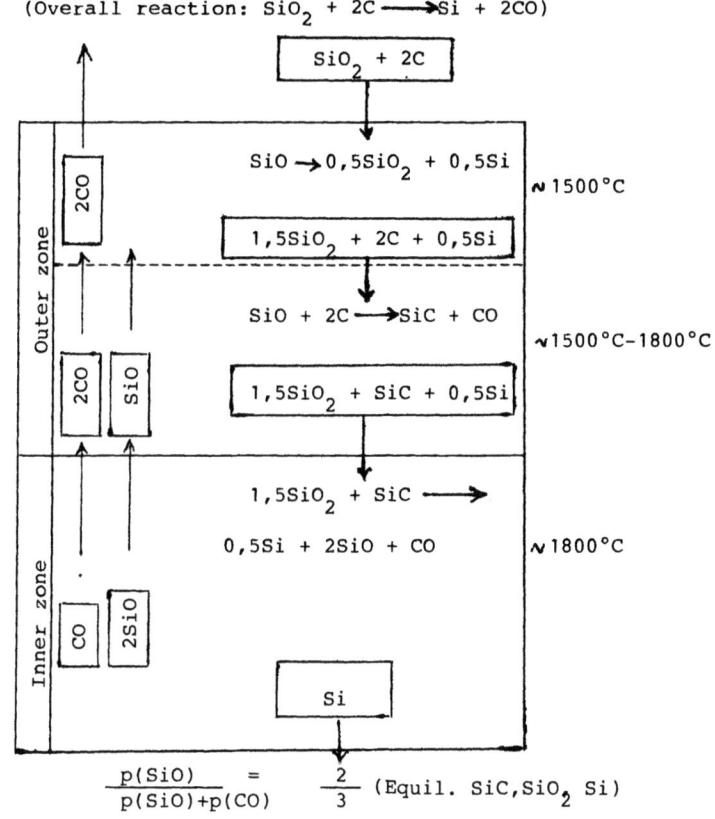

Zone Partition of the Furnace
Idealized Conditions

(Overall reaction: $SiO_2 + 2C \longrightarrow Si + 2CO$)

Table 1

METALLURGICAL DATA FOR FeSi PRODUCTION CAMPAIGN BETWEEN 1967 and 1969

	FIRST CAMPAIGN 26.10.67-19.12.67	SECOND CAMPAIGN 17.1.68-29.6.68	THIRD CAMPAIGN 4.8.68-12.1.69	FOURTH CAMPAIGN 5.2.69-27.9.69
RAW MATERIALS:				
Quartz)	3,014	1,9577	2,110	1,800
Quartzite)	–	–	–	0,205
Steel Scrap) (t/t of alloy)	0.284	0.275	0.190	0.205
Mill Scale)	–	0.008	–	–
Iron Ore)	–	–	0.034	–
REDUCTANTS:				
Coke)	1,565	–	–	0,121
Coke Fines)	0,155	–	0,114	0,006
Coal) (t/t of alloy)	–	–	0,013	–
RC Char)	–	1,110	0,931	0,893
RC Char Fines)	–	0,007	0,087	–
Wood Chips)	–	–	0,1927	0,320
CONSUMABLES:				
RC Paste)	0,178	0,087	–	0,054
French Paste) (t/t of alloy)	–	–	0,093	0,045
Steel Sheets)	12,091	6,460	5,539	5,483
Bars)	0,004	0,004	0,008	0,006
Production (t)	715	4270	4429	6303
Production Rate (t/d)	13.3	25.4	26.3	26.8
Specific Power Consumption (kWh/t)	16383	9515	9114	9200

Table 2   TYPICAL RAW MATERIAL ANALYSES

	SiO$_2$	FeO	Al$_2$O$_3$	MgO	CaO	Fe	Si	Mn	P	C	S
Quartz	97.0	1.24	0.19	0.53	0.21						
Quartz Pebbles	93.8	1.58	0.80	0.74	0.27						
Steel Scrap						86.15	0.96	0.54	0.023	0.233	0.006
Pig Iron						76.70	3.62	1.16	0.026	1.550	0.059
	FC	VM	ASH								
Coal Nuts	63.0	24.5	12.5								
Coal Duff											
Coke	86.5	0.9	12.6								
Wood Chips	74.4	22.7	2.9								

Factors influencing decrepitation include poor mechanical strength, chemical composition, grain size and capacity to absorb moisture.

A basic decrepitation test involves heating up a sample of the material in a muffle furnace after which it is allowed to cool and is subjected to impact in a tumble drum. The material is then screened and the results analysed.

Silicon Containing Materials

Silicon occurs in nature in the form of many minerals among which the most widely distributed are quartzites, quartz, chalcedony, sandstones and sands. Not all of these minerals are suitable for the production of silicon alloys. Sand, for example, makes the charge impervious to gas and one finds that generally, quartz, quartzite and chalcedony are mostly used for the smelting of ferrosilicon.

Reducing Agent

Reductants must have high reactivity and porosity, low ash content typically less than 13% and low volatile matter, typically less than 25% for ZARD. Mechanical strength at high and low temperatures is essential as is high electrical resistivity and low moisture content. Coal is beneficial due to its high resistivity compared to coke and promotes deep electrode immersion; char formed due to coking of coal during descent results in a reductant with high reactivity and porosity. The low density of wood chips results in a light mix with high porosity and relatively high electrical resistance. The wood chips may provide a supplementary source of carbon, although this is not taken into account for charge calculation purposes, neither is carbon from electrodes. No account is taken of the possible effects of hydrogen in the operation.

Iron Containing Substances

Iron is usually introduced into the furnace in the form of steel scrap which prevents the compacting of the charge and promotes the increase of its permeability to gases. The scrap should be in short pieces (100mm); use of long spiral shavings hampers the operation and impairs the full mechanization of charge loading. This is because ZARD does not have a swarf cutter which would make such shavings usable. Iron ore is not recommended because it causes a considerable increase of slag and a deterioration of product quality by introducing unwanted elements such as phosphorus, calcium and aluminium.

OPERATIONS

Furnace Design and Operating Parameters

The general layout of the furnace is given in Fig.2 where the main feature is the arrangement of three electrodes at the corners of an equilateral triangle in an essentially cylindrical furnace. In this case the eletrode diameter is 900mm, the inter-electrode distance is 2149mm, the shell diameter is 6400mm and the furnace depth is 2200mm. The complete set of data is given in Tables 1 to 5. The furnace is lined with firebrick and the hearth is essentially of carbon blocks and rammed paste.

Power Control

The electrodes are Soderberg type and consist of mild steel cylindrical casings into which carbon paste is loaded. The I^2R heating resulting from the passage of current leads to baking of the electrode paste so that with the melting away and oxidation of the casing sheet, a baked rigid carbon electrode exists in the furnace. Heat radiated and conducted from the furnace also assists the baking process. As the electrode is consumed inside the furnace by arcing and oxidation, the electrode has to be slipped down with respect to the electrode holder; this function is carried out on load.

The furnace itself may be viewed as a high temperature reaction vessel with an input of raw materials and an output of metal, gas and sometimes slag. The highest temperature zone exists in a small region immediately surrounding the electrode tips where temperatures of the order of 2000 - 3000°C[3] are encountered. The furnace is essentially a resistance furnace despite being referred to as a submerged arc furnace; the arcing itself is not crucial to the generation of heat energy. The active power per electrode is given by

$$W = VI \cos \phi$$

Where W is the active power per electrode V and I are transformer secondary voltage and current respectively and $\cos \phi$ is the power factor decided by the raito of power utilized (MW) and power input (MVA). This can also be written as

$$W = I^2R \cos \phi \text{ where R = operating resistance}$$

Table 3
FeSi PRODUCTION DATA

	Actual Tpd	Budget	% Avail	%Si Rec	%Si	%C	kWh/t	Furnace	Load MW	%P	%S
February 1987	1.63	2.10	88.8	55.7	46.5	0.034,0	10,509	M1	0.81		
March 1987	2.19	2.10	92.1	76.6	51.1		8,104	M1	0.80		
April 1987	11.2	14.0	93.0	55.6	48.6	.022,0	10,604	S1	5.3	.022	.005
May 1987	9.7	13.0	82.9	62.8	50.2	.027,0	10,904	S1	5.3	.022	.005
June 1987	12.6	13.0	63.7	101.8	55.5	.036,0	7,169	S1	5.9	.021	.005
July 1987	17.98	18.0	86.8	92.3	59.9	.036,0	7,273	S1	6.3	.020	.005
August 1987	20.20	18.0	95.9	85.9	54.8	.030,0	6,419	S1	5.6	.021	.005
September 1987	16.2	18.0	94.8	90.4	67.9	.040,0	8,188	S1	5.8	.020	.005
October 1987	13.3	14.0	95.7	80.2	73.7	.036,0	10,638	S1	6.2	.018	.005
November 1987	12.8	14.0	95.1	81.1	76.8	.029,0	11,287	S1	6.3	.019	.005
August 1988	9.2	11.0	85.5	73.5	71.9	.043,0	13,577	S1	6.1	.019	.004
November 1988	13.1	13.0	96.6	90.4	75.1	.043,0	10,854	S1	6.1	.020	.004
December 1988	14.0	13.0	94.9	95.6	75.0	.046,0	10,044	S1	6.2	.020	.005
February 1989	10.6		86.4	73.0	71.2	.074,0	11,713	S1	6.0	.019	.007
March 1989	12.3		92.5	79.7	77.9	.050,0	11,020	S1	6.1	.019	.007
October 1989	10.1	13.0	90.0	66.9	65.9	.045,0	13,152	S1	6.2	.020	.006
November 1989	10.4	13.0	91.9	82.5	70.3	.047,0	12,270	S1	5.8	.019	.005
December 1989	14.1	13.0	94.8	91.7	75.1	.045,0	9,857	S1	6.1	.020	.005
January 1990	13.1	13.0	94.4	83.4	75.0	.041,0	10,662	S1	6.2	.020	.005
February 1990	15.0	13.0	95.5	90.2	72.2	.048,0	9,535	S1	6.2	.019	.005
March 1990	14.1	13.0	94.0	92.9	74.1	.069,0	9,936	S1	6.2	.019	.005

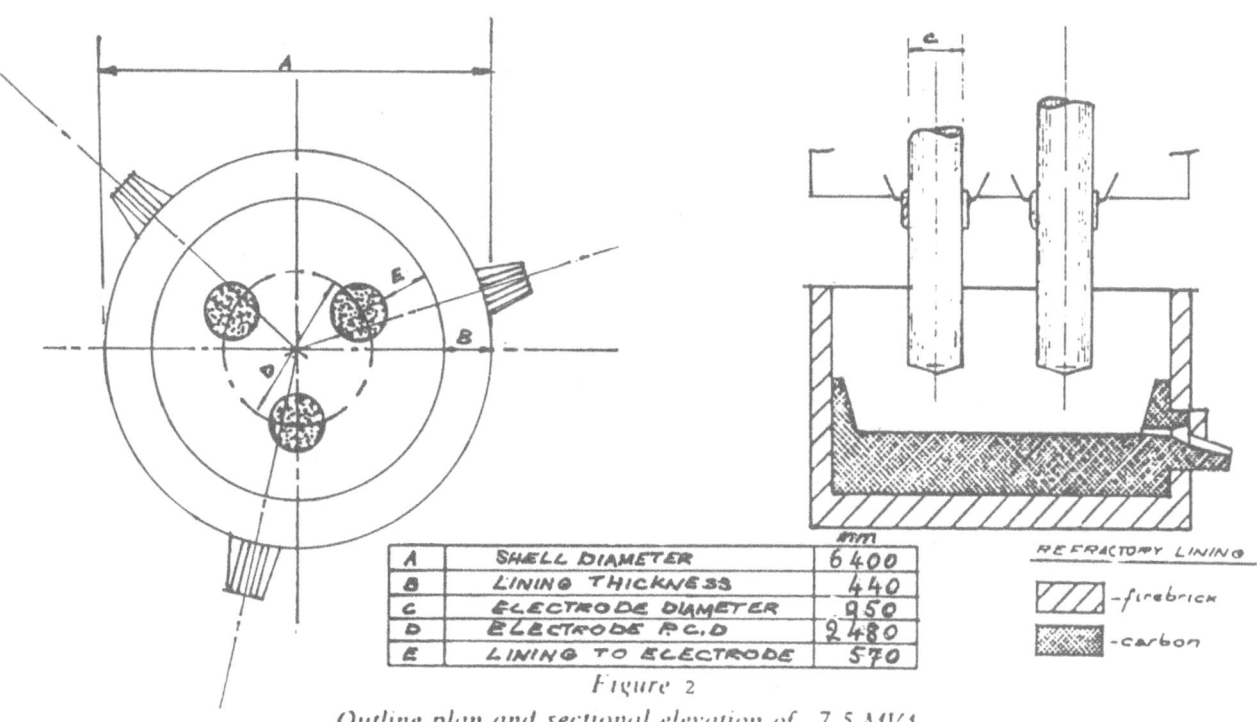

A	SHELL DIAMETER	6400
B	LINING THICKNESS	440
C	ELECTRODE DIAMETER	950
D	ELECTRODE P.C.D	2480
E	LINING TO ELECTRODE	570

REFRACTORY LINING

-firebrick

-carbon

Figure 2
Outline plan and sectional elevation of 7,5 MVA
furnace

Table 4
SPECIFIC CONSUMPTIONS OF RAW MATERIALS

	Quartz	Pebbles	Coal	Coal Duff	Coke	Steel Scrap	Pig Iron	Chips	Paste
April 1987	1.684		0.263		1.000	0.526		0.316	0.105
May 1987	1.446	0.283	0.603		0.497	0.720		0.373	0.080
June 1987		1.199	0.346		0.285	0.342		0.425	0.086
July 1987	1.095	0.313	0.393		0.320	0.346		0.261	0.065
August 1987	1.382		0.400		0.332	0.641		0.153	0.054
September 1987	1.591		0.461		0.383	0.282		0.191	0.082
October 1987	1.989		0.577		0.477	0.121		0.402	0.087
November 1987	2.057		0.595		0.494	0.112		0.411	0.099
August 1988	2.099		0.610		0.505	0.128		0.420	0.210
November 1988	1.457	0.347	0.896		0.140	0.094		0.696	0.126
December 1988	1.601		0.890		0.089	0.107		0.739	0.080
February 1989	2.107		0.877		0.256	0.173		0.841	0.126
March 1989	1.974		0.966		0.097	0.163		0.652	0.136
October 1989	2.119		1.030	.062	0.106	0.073		0.826	0.124
November 1989	1.764		0.850	.018	0.080	0.153		0.700	0.134
December 1989	1.776		0.938		−	0.115		0.711	0.140
January 1990	1.891		1.027		−	0.111	0.042	0.757	0.184
February 1990	1.725		1.000		−	0.129	0.017	0.690	0.103
March 1990	1.731		1.011		−	0.105	0.041	0.692	0.092

Table 5
FURNACE OPERATING PARAMETERS

		S1	A1	M1
Furnace Type		SECEMAEU	TAGLIAFERI	
Year Installed		1963	1967	1958
Transformer Capacity		7.5 MVA	17.5 MVA	0.9 MVA
Primary Voltage		11kV	33kV	11kV
Frequency		220Hz	220Hz	220Hz
Primary Current	Minimum	304A	182A	
	Maximum	385A	306A	46.4A
Secondary Current	Minimum	40900A	6300A	7220A
	Maximum	31800A	54000A	4450A
	Operating	35290A	63000A	7000A
Secondary Voltage	Minimum	85V	95V	40-65V
	Maximum	136V	262V	80-130V
Bus Bar Connection Operating		120V	165V	70V
Voltage Tap Positions		7	18	4
Maximum Ratio Sec. Amps to Sec. Volts		481.2	663.1	90.3
Operating Ratio Sec. " " " "		294.1	381.8	100
Electrode Type		Soderberg	Soderberg	Soderberg
Electrode Disposition		Regular triangle's apex		
Electrode Diameter		950mm	1100mm	290mm
Electrode Stroke		1200mm	1240mm	330mm
Electrode Cross Sectional Area		$7084cm^2$	$9500cm^2$	$660cm^2$
Electrode Current Density		$5.77A/cm^2$	$6.63A/cm^2$	$10.94A/cm^2$
Operating Current Density		$4.98A/cm^2$	$6.63Acm^2$	$10.61A/cm^2$
Shell Diameter		6400mm	8000mm	3050mm
Brick Diameter		5520mm	7210mm	2590mm
Depth		2200mm	2675mm	1229mm
Electrode Centre to Centre dist.		2149mm	1980mm	535mm
Electrode Centre to Shell dist.		1485mm	2355mm	1146mm
Pitch Circle Diameter		2480+172mm	2800+250mm	1100+50mm

The shape of the electrode may be compared to a cylinder with a hemispherical tip. The current distribution in the zone below the tip of the electrode is given approximately by the solution of the Laplace equation for the hemispherically-tipped-cylinder-to-plane problem where the plane is the metal pool at the bottom of the furnace[4].

The zone described is referred to, in electric smelting parlance, as the coke bed[5,6].

The metallurgical reactions thermodynamically permitted in the coke bed are decided by the temperature attained which in turn is a function of the I^2R power dissipated per unit volume in the coke bed depending upon the specific heats and thermal conductivities of the materials present.

If the regulation of the electrode tip position in the furnace is based on constant current it follows that the metallurgy is controlled by the total resistance of the coke bed which is determined by the volume of the coke bed. The extent of electrode immersion inside the furnace is estimated using power factor figures and electrode slip rates are also aligned to the calculated power factor.

When the power factor rises towards unity this generally indicates that the electrode tip is coming up towards the surface due to either shortening or upward movement of the electrode.

The rate of slipping is increased when the power factor increases or vice versa.

The furnace operator will control the operating resistance by selecting the voltage and allowing automatic raising and lowering of electrodes, i.e. the operator actually selects the resistance.

Tapping

Alloy is tapped straight into chill moulds mounted three to a bogie which is moved by electric winch.

The taphole is drilled and burnt through with oxygen; gumpoles and steel bars are used to maintain flow. A mould fills up with 600kgs of alloy and a standard tap weighs about 1 800kgs. The furnace is tapped eight times a day.

Handling

The solid ingots are stripped with ingot dogs by an overhead crane and weighed before being put into a pan labelled with the cast number. A sample is taken from the stream of metal during tapping for analysis of silicon, manganese, phosphorus, carbon, sulphur and aluminium. The results of the analysis may be cross checked with those from a sample taken from the solid ingot.

DEVELOPMENTS

The first major development was the change in the size of wood chips in May 1987 from thin flakes measuring 25mm x 50mm x 10mm to much bigger cylinders measuring 100mm in length and 75mm diameter. Later during the same month the consumption of wood chips was significantly increased and the chips were soaked in water for about 24 hours before use. These changes resulted in improved electrode penetration, improved burden porosity and improved tapping conditions.

During the same month tapping and casting of alloy was modified to allow tapping directly into chill moulds. Prior to this alloy was tapped into a ladle and then teemed into chill moulds; this caused loss of alloy through skulling in the ladles. Another significant development was the raising of the furnace roof and electrode assemblies by one meter in June 1987 in order to minimize the adverse effects of heat on water circuits.

Between the 21st and the 28th of July 1987, alloy silicon content was around 65% to 70% and it was noticeable that the alloy tended to crumble to fines less than 3mm. Fines generation during crushing was excessive. A decision was taken to go direct to 72-80%Si alloy to avoid the crumbling phase.

Product changes in the furnace invariably resulted in contamination of the incoming alloy by elements from the outgoing alloy. In November 1988 a product changeover from high carbon ferromanganese to ferrosilicon 75 was facilitated by introduction of a charge for ferrosilicon 50 so that the iron would mop up manganese units to form an intermediate alloy, thus reducing manganese contamination of the ferrosilicon 75. In a major step the ratio of coal to coke was progressively increased from 0.9:1 to 8:1 over a period of two months from November 1988. Benefits accruing from this development included deeper electrode immersion and improved alloy output and quality. Coke was eventually withdrawn from the charge a year later.

Electrodes frequently overbaked resulting in breaks, the problem was solved by installing additional electrode cooling fans so that more air could be blown when necessary.

The installation of heat shields and subsequently chain curtains around the furnace in December 1989 protected the charger from the adverse effects of heat.

SUMMARY AND CONCLUSIONS

The first campaigns to produce FeSi75 in the 17.5MVA furnace were fraught with problems largely related to inexperience in the operation. Slagging up of the furnace was a common feature and the control of electrode length was extremely difficult. The operation was becoming a very costly exercise and it was not surprising that it was abandoned in 1969.

The successful production of FeSi75 alloy at Zimbabwe Alloys Refinery was attributed to perseverance in the face of difficulties, lessons that the team learnt from past mistakes and a number of new developments that were implemented. Fig.3 shows graphically the production performance during the period under review. The main reasons for success were as follows:

1) The size of wood chips was increased from approximately 25mmx50mmx10mm flakes to approximately 100mmx75mm cylinders.

2) The specific consumption of wood chips per charge was doubled from about 0.40 to 0.80 tonnes per tonne of alloy thereby increasing burden porosity and resistance.

3) The furnace roof and electrode assembly were raised by one metre in June 1988 as protection against heat.

4) The electrode mantle cooling fans were doubled to prevent overbaking by providing more air.

5) Heat shields and subsequently chain curtains were installed around the furnace to protect the charger from the adverse effects of heat.

6) Increased use of coal caused improved electrode immersion due to increased burden resistance.

7) Tapping directly into moulds as opposed to ladles increased alloy recovery.

FIGURE 3

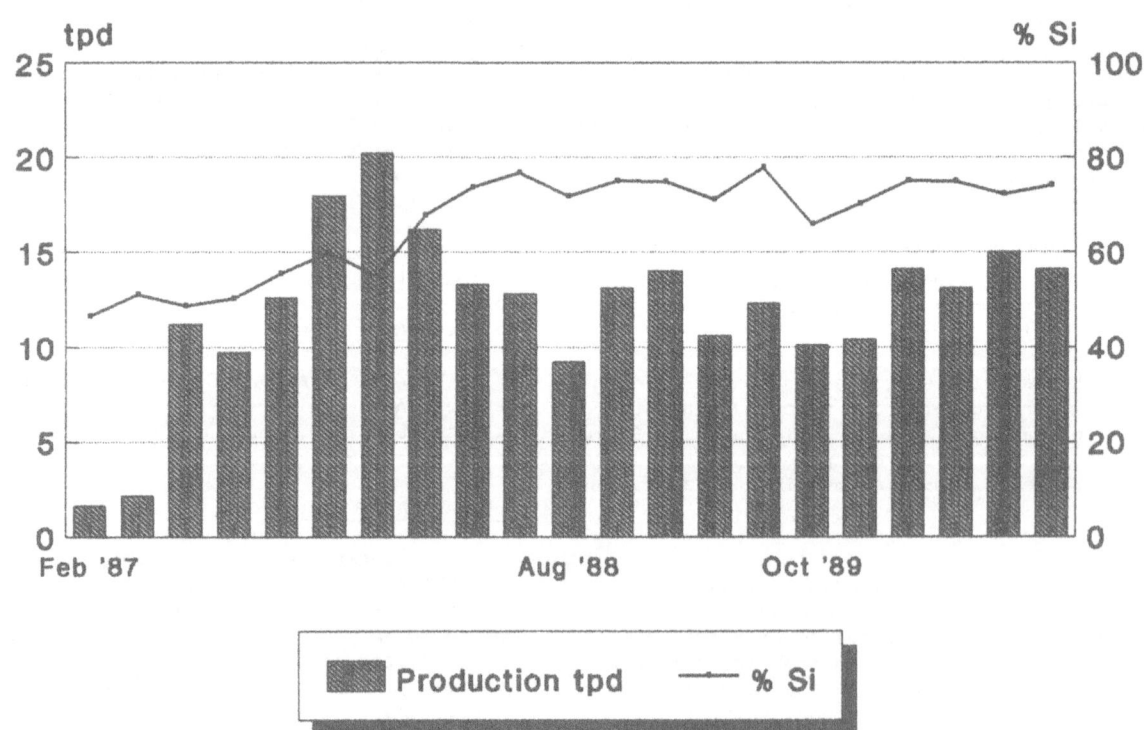

FeSi PRODUCTION DATA
February 1987 - March 1990

APPENDIX

Diary of Events

The following diary of events summarises the various attempts to produce 75% silicon alloy.

August 1986 : Test production of FeSi50 in a 700 kVA furnace. Running time was low (88.8%) due mainly to electrode failures. The campaign was stopped on March 23rd and production was transferred to a 7.5 MVA furnace.

April 1987 : Operations in the 7.5 MVA furnace were erratic characterized by poor electrode penetration, extremely hot surface conditions and difficult tapping.

May 1987 : Two pressure rings, 19 contact pads and 15 loop pipes were renewed in a month with extremely poor furnace availability. The size and consumption rate of wood chips was increased.

June 1987 : Operating load was increased from 5.3MW to 6.0MW on the 2nd and this allowed the bath to open up. A dramatic improvement in electrode penetration and tapping was noted before a transformer failure on the 19th disrupted operations.

A bigger 8.5 MVA transformer was installed and during the shutdown the furnace roof and electrode assemblies were raised by one metre in order to minimize the adverse effects of heat on water circuits.

July 1987 : Operations resumed on the 2nd and the furnace was on full load on the 5th. Between the 21st and 28th alloy silicon content was between 65% and 70% and it was noticeable that the alloy tended to crumble to fines less than 3mm. Fines generation during crushing was excessive.

August 1987 : Operations were satisfactory due to increased power input and benefits accruing from structural modifications in terms of running time. The value of experience gained in previous months was beginning to be felt.

September 1987: The first attempts to produce 75% silicon alloy were made.

October 1987 : The first successes in producing FeSi75 were scored.

November 1987 : Alloy output was rather low due to periods when very high silicon levels were achieved.

December 1987 to July 1988 : The furnace produced FeSiCr and FeMn alloys.

August 1988 : Production of FeSi75 resumed and the month was characterized by extremely poor operations due to poor running time and the adverse effects of adding too much residue material. Extended stoppages became necessary to replace ancillary equipment which had gradually succumbed to heat from previous campaigns. The temporary 8.5 MVA transformer was replaced by the original 7.5 MVA unit during a shutdown from 16th to 19th August. A new roof was cast on water cooled beams.

Electrode breaks and water leaks were prevalent in the period after the shutdown.

September 1988 to October 1988 : The furnace produced high carbon ferromanganese alloy mainly for Zimbabwe Iron and Steel company at twice the monthly consumption rate.

November 1988 : Production changed over from high carbon ferromanganese to ferrosilicon 75 via ferrosilicon 50 during the first four days. The ratio of coal to coke was progressively increased.

December 1988 : Ferrosilicon 75 was produced up to the 8th after which the furnace changed over to ferrosilicon chromium production.

February 1989 : Formation of a layer of silicon carbide during the changeover at the beginning of the month adversely affected tapping conditions in particular and furnace operations in general. Poor movement caused electrodes to overbake and oxidize leading to breakages. Electrode casing failures and water leaks caused prolonged stoppages and agravated already poor furnace operations.

March 1989 : Metal broke out through a thermocouple site in the hearth on the 21st and the furnace was shut down for a reline.

April 1989 : Ferrosilicon chromium alloy

was produced.

May 1989 to September 1989	:	High carbon ferromanganese alloy was produced. Severe power cuts during this period were such as to prevent the more sensitive ferrosilicon operation.

October 1989 : Poor furnace availability and poor electrode penetration affected operations. On the 9th, metal broke out through three thermocouple sites and power was switched off to allow the melt to freeze back. The thermocouple sites were gouged out and grouted as a precaution against further breakouts.

Three electrode breaks occurred and four contact pads were changed. The performance of the charging car was unsatisfactory and contributed towards the poor performance.

November 1989 : Running time was low. An intermittent fault within the batching system was identified and rectified on the 23rd and a dramatic improvement in operating efficiency was noted thereafter. Furnace conditions were difficult during the first three weeks and electrode penetration was poor. Coke was with-drawn from the charge so as to improve electrode penetration.

December 1989 : Operations were extremely satisfactory and the previous highest daily production rate of 14.03 tonnes achieved in December 1989 was surpassed by a production rate of 14.06 tonnes achieved in the month under review.

Chain curtains were installed in order to protect the charging car from radiant heat from the furnace.

January 1990: : Satisfactory operations prevailed despite slight metallurgical difficulties when pig iron was used as replacement for steel scrap.

February 1990 : A new record production rate of 15 tonnes per day was achieved.

March 1990 : The campaign ended on March 21st with successful performance.

ACKNOWLEDGEMENTS

The author would like to thank the Managing Director and General Manager ZARD for their permission to publish this paper. The tireless efforts of the Manager, (Technical Services) AAC Services, who is also the chairman of IMM Zimbabwe Section in vetting the paper are acknowledged.

REFERENCES

1. ZARD Production Manager's reports

2. Production of Ferroalloys, Second Edition by V P Elyutin, Yu A Pavlov, B E Levin and E M Alekseev

3. The Theory and Design of Electric Smelting Furnaces by A K N Reddy

4. Ferrosilicon and the Steel Industry, International Iron & Steel Institute, Committee on Technology, Brussels 1987.

5. Resistance and Heat distribution in Submerged Arc Furnaces by Jens Westly

6. Critical Parameters In Design and Operation of Submerged Arc Furnaces by Jens Westly

7. Electric Smelting Processes by A G E Robiette.

Mining 3

Use of trucks in the underground false footwall block caving mining method of King mine, Zimbabwe

N. J. W. Bell
African Associated Mines (Pvt) Ltd, Bulawayo, Zimbabwe

Synopsis

The False Footwall Cave mining method was described in African Mining '87 and subsequently investigations were made into how this method could be improved further with resultant reduction in costs. This paper describes how this is expected to be achieved.

Currently the ore is loaded and hauled by LHDs from the drawpoint crosscuts to the ore passes situated in the collecting drives. The ore is then trammed by loco to the crusher situated near the shaft. Analysis showed that by moving the crusher closer to the orebody, it would be possible to haul the ore in trucks, loaded by LHDs at the entrance to the drawpoint crosscuts on each sub-level. This would result in a reduction in haulage and orepass development with all its associated support and construction. It would also eliminate the loco tramming for the production phase as the crushed ore would be transported to the shaft bins by conveyor. As the ore would be fed directly to the crusher the grizzly apertures could be increased with resultant improvement in productivity.

The costing of all the operations involved in the extraction of ore and its delivery to the surface mill stock piles showed that it was a viable possibility. Therefore, the decision was taken to follow this path : planning is at an advanced stage and mining has commenced for the new layout.

Introduction

Shabanie and Mashaba Mines (Pvt) Ltd. has a mining lease at Mashava covering a fairly large number of chrysotile asbestos deposits. Most of these have been worked at one time or another although, at the time of writing, only King Section of Gath's Mine is still in production.

All fibre produced is processed on site at King Mill where it is cleaned, graded, blended, checked for quality, packed and transported by road to the railhead at Zvishavane, some 60km to the west of Mashava, for onward despatch to the coast.

Geology

All the chrysotile asbestos deposits on the property are associated with the Mashaba ultramafic complex which is intrusive into altered sediments, volcanics and the ancient gneiss basement, representing Bulawayan, Sebakwian and pre-Sebakwian systems.

King Section is in the footwall of an embayment of serpentine bounded by steeply dipping metasediments. There are two contiguous orebodies: the Main orebody which strikes east-west and the Western Section orebody which strikes north-south. Both zones are separated from the sediments by a sheared footwall which contains talc carbonate and carbonated serpentine rocks.

Prominent shear zones run parallel to the orebody strikes and there are associated secondary shears and slip planes which combine to form a complex jointing system. The geomechanics rock classification ranges from Class 2 chert to Class 5B shear zones, with the majority of the orebody rocks being Class 3A to 3B.

Geomechanics Classification

The Group uses a geomechanics classification[1,2], which rates the rock according to set parameters. Using these, rock mass classification ratings are generated from borehole core and development mapping information. They range from 1 (very good) to 5 (very poor) with A and B being upper and lower sub-divisions.

False Footwall Mining Method

The drawpoints for the false footwall cave mining method[5,6] are mined in an echelon up-dip along a false footwall to give a strong structure that is readily reinforced. The drawpoint spacing is 10m along strike, 12m horizontally up-dip and 10m vertically between levels. (Figs. 1 & 2.). The actual drawpoint positions can be adjusted to allow for the geomechanics rating in the brow area. The drawpoint crosscut directions are as near to right angles as possible to the slot but can be adjusted to cross prominent shears at a better angle. The size of the drifts is 3,2m x 3,2m, supported according to the ground conditions[2], their purpose and the anticipated mining stress to which they will be subjected (Fig 3.). The services are located in the footwall and are well protected from damage.

Possibilities Considered for Improving the Method

During 1987 the idea of improving the false footwall mining method was mooted. Several possibilities were considered, namely -

1   The use of larger LHDs.
2   The use of conveyors to a central pass system.
3   The use of trucks to an orepass system on each of the main legs of the orebodies
4   The use of trucks to a central pass system direct to the crusher.

Briefly, the first three possibilities were rejected for the following reasons:-

1   Larger LHDs would have required larger excavations than the current 3,2m x 3,2m ends; while this situation could be tolerated in the main section of the orebody, it would be untenable in the West Central body or the North East Extension where the predominantly 3B and 4 ground conditions would

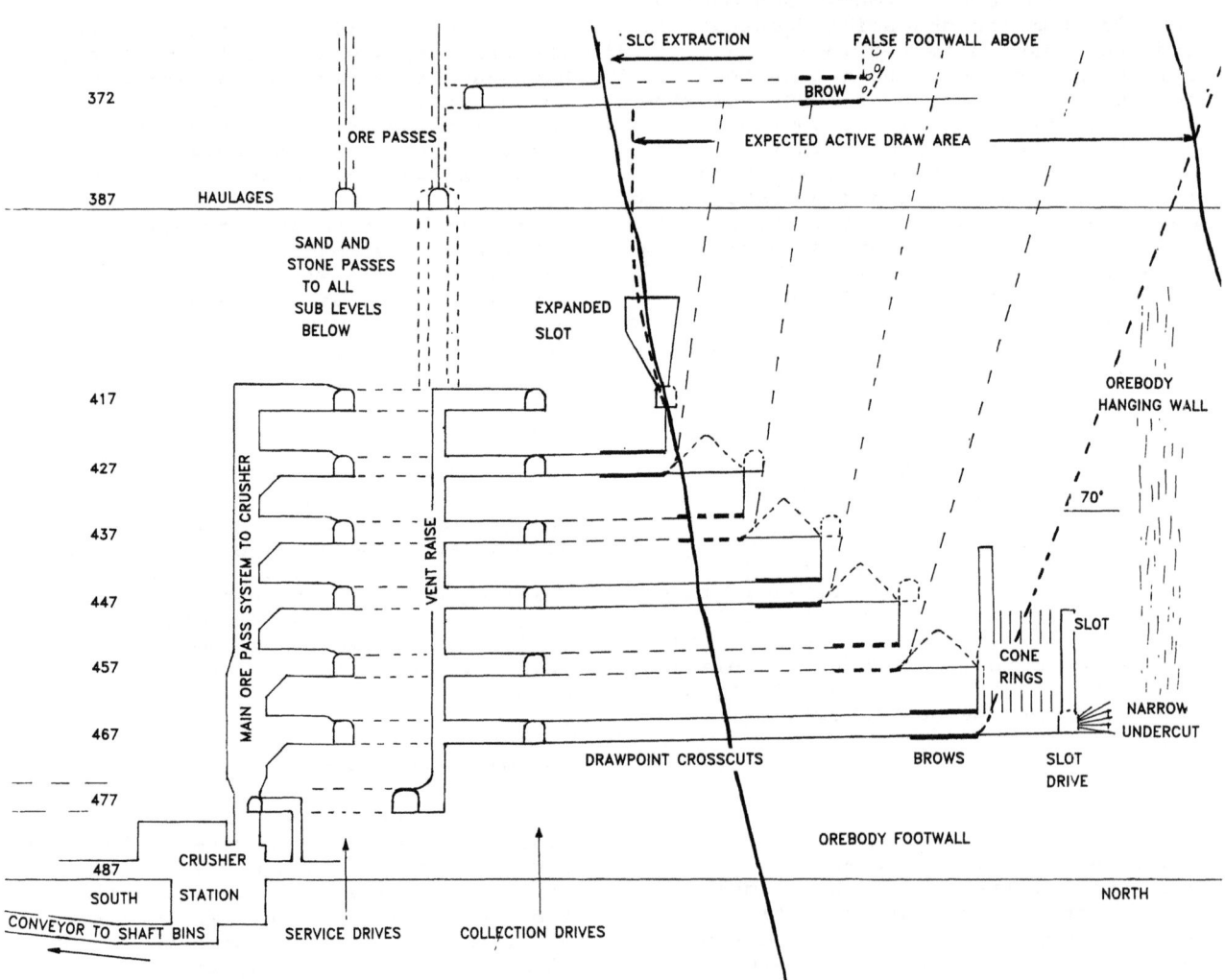

Fig. 1   Section showing modified "False Footwall" mining method with the production ore being tipped direct into passes to the crusher and after crushing conveyed to the shaft.

necessitate extensive and costly additional support during the initial excavations. There would also have been loss of flexibility, with fewer LHDs across the five levels of production.

2   The idea of using conveyors to transport the ore to a central pass was abandoned owing to the high capital cost: four separate conveyors with transfer chutes on each of the five levels would have been involved and as un-crushed, poorly fragmented rock was to be handled, the operating costs and potential for breakdowns were considered to be unacceptably high.

3   The third possibility was to use trucks, loaded by LHDs and trammed to two sets of passes, one for the Main and East and the other for the West Central orebodies. Thereafter the ore would be hauled with conventional locos and granbies to the crusher situated near the shaft. The retention of the loco haulage system and its accompanying costs nullified this idea.

The final possibility was to prove the viable option, namely:-

4   2 cu.yd. LHDs would be used, loading into small dump trucks of 6 to 8 tonne capacity, which would tram along the sub-levels to a central orepass system and deliver directly into the crusher which would be moved from the shaft closer to the orebodies. The production ore would then be transported by conveyors to the shaft bins with a transfer arrangement so that the ore could

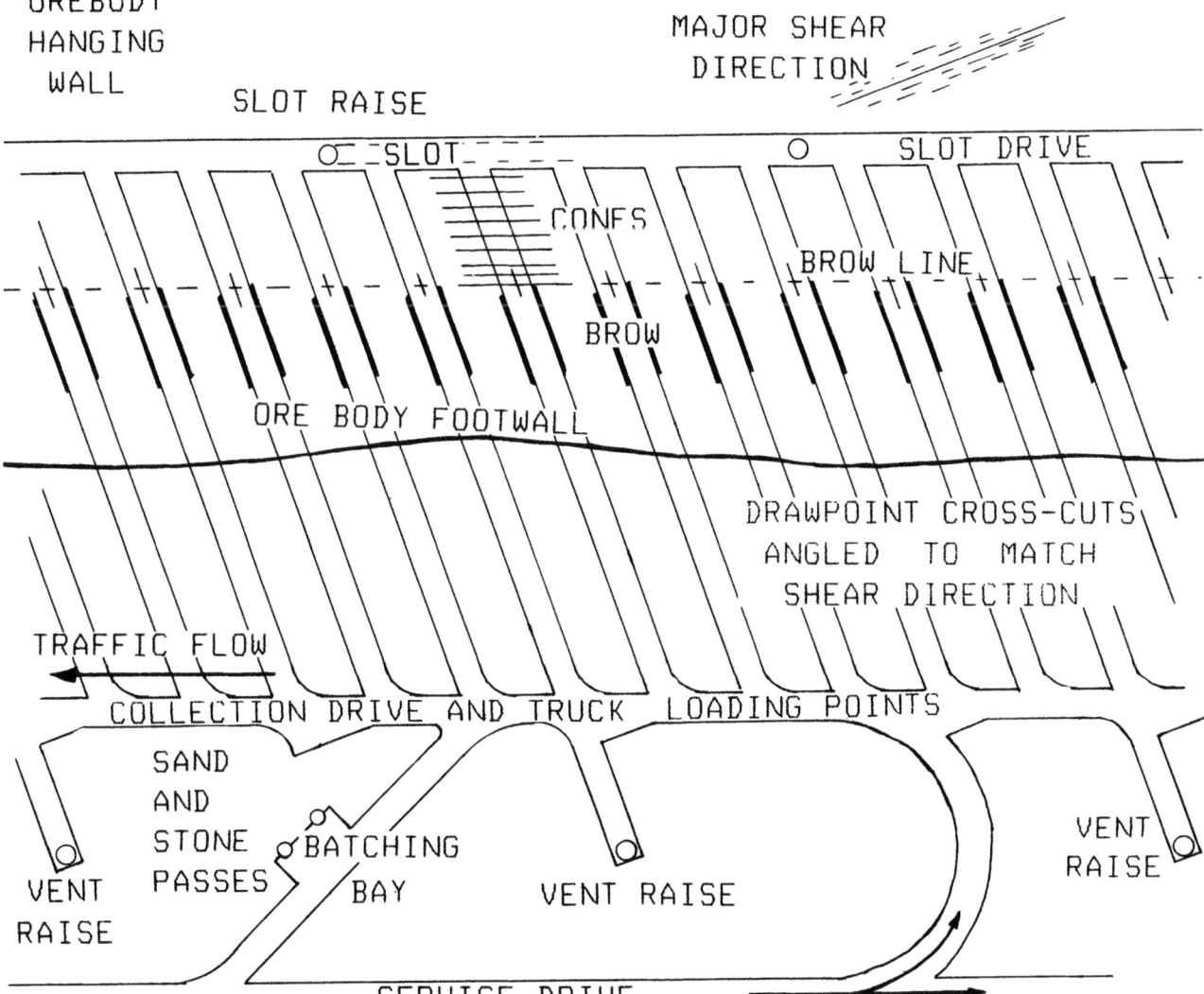

Fig. 2   Plan (447 level) showing the various connections between the service drive and the collection drive.

be directed to the bins at either No.1 or No.2 shaft. A few locos would still be needed for all the development waste and the initial development ore until the crusher was commissioned.

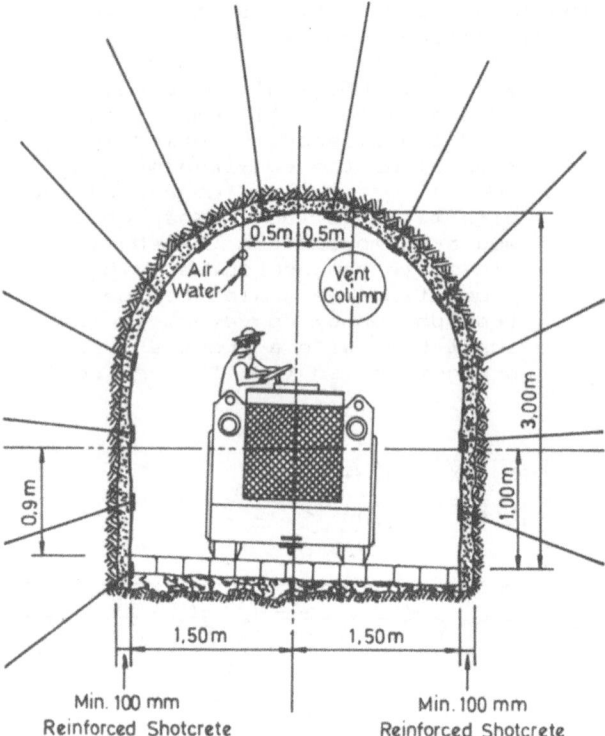

Fig. 3   Typical section through drawpoint cross-cut showing the relative size of a 2cu.yd. LHD and the normal support.

Modifications to the Layout

In March 1989 it was agreed that the new method would be introduced to extract the ore (some 16 million tonne) above the next major haulage level (487 m), which had already been established from the shaft.

The development not required, namely the haulages and ore passes required for the loco system, was stopped immediately, with resultant cost savings.

The development layouts were then adapted to suit the proposals. The tips, ore passes, crusher station, and conveyor distribution designs were commenced. The existing crushers were tested with large rocks, plus 1m diameter, purposely trammed to them and it was shown that a square grizzly with 0,95m square spacings would be suitable.

It was planned that the collection drives would be used for the loading (Fig 4) and tramming of full trucks from the loading point to the crusher tip; thereafter the trucks would return to the loading point via the service

drives. The service drives which normally stopped some 50m from the ends of the orebodies east and north have been extended to within 20m from the ends of the orebodies to allow the trucks to enter the collection drives for loading when mining is at the extremities of the orebodies. The connections between the service and collection drives have also been modified from all 'S' configurations to a combination of 'S's and semi circles to facilitate the traffic flow.

Currently the service drives are equipped with tracks to facilitate the movement of materials in rail cars along the sub-level from the sub-vertical shaft to the working area. The possibility of eliminating the tracks from the service drives is being considered if a suitable system to transfer materials from the tracked haulage supply system to the trackless production area can be developed.

The workshop system for trackless equipment still has to be extended. Greater emphasis is to be placed on suitably located satellite workshops where maintenance up to the 250 hour service which takes 3 hours can be carried out and minor breakdowns (estimated time to repair 2 hours or less) handled.

The modifications to the layout reduced the overall metres for a lift by some 4 500.

Other Considerations

Intake Airways

With the production ore being transported by conveyor to the shaft bins there would be a reduction in the number of connections, which also served as the intake airways, from the shaft to the workings. It is imperative that sufficient drifts into the area are developed to allow for the free flow of fresh air and its distribution along strike and up into the mining areas. This will be a consideration for new lifts.

Trucks

A general enquiry for a suitable truck which would fit the development size constraints was circulated and there were several options open to us. However, it was considered that local support was the most important factor and therefore the initial trucks have been the Wagner MT 406s which have worked exceptionally well; their drivers being LHD trained. As the fleet of trucks expands we envisage the drivers of the trucks, being those most recently trained, graduating to LHDs as vacancies become available.

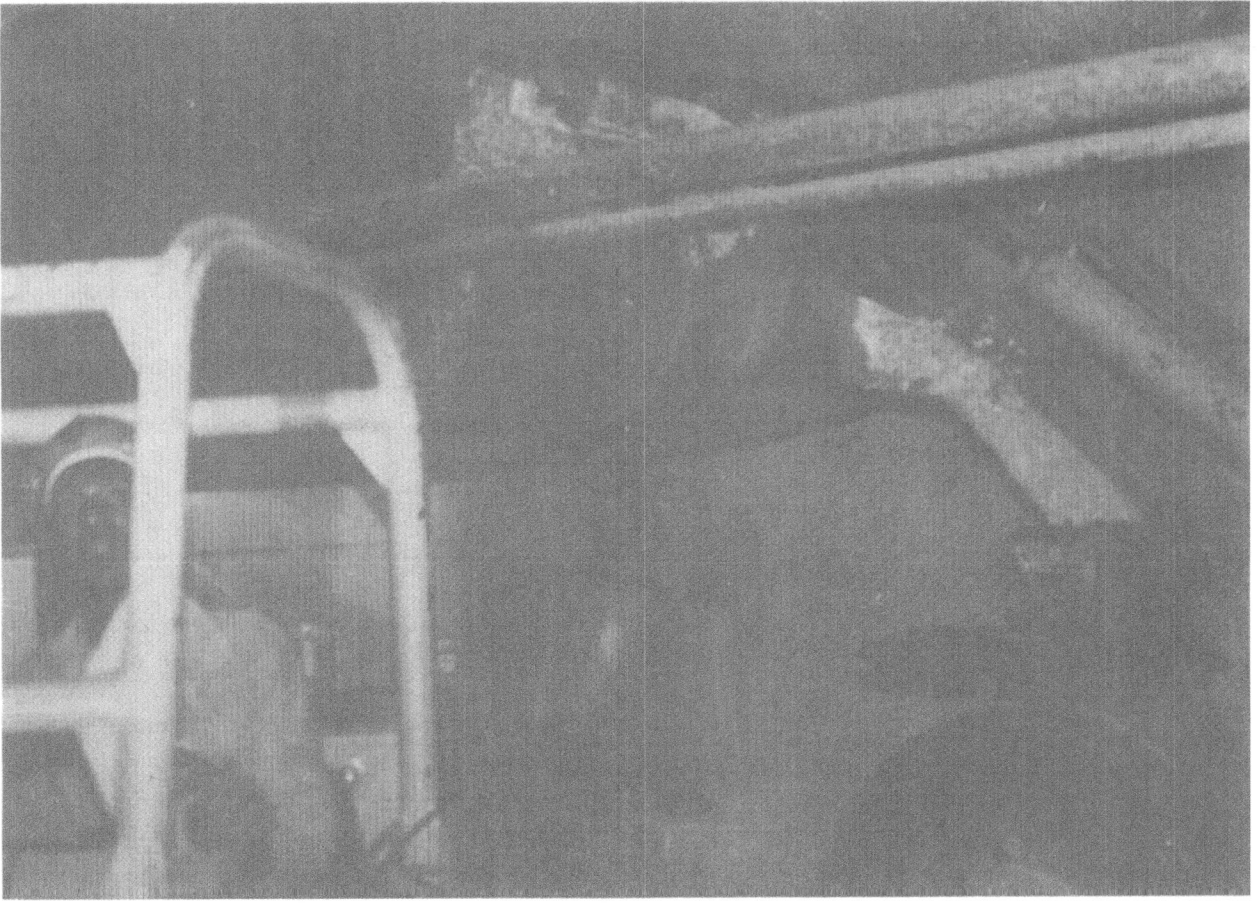

Fig. 4   LHD loading a dump truck in the collection drive.

The number of trucks per LHD is
dependent upon the tramming distance
from the drawpoints to the truck loading
point and thence to the tip. This is
depicted graphically tonne/shift
(Graph 1) and cost/tonne (Graph 2)
assuming a 50m run by the LHD to load
the trucks. The scale of cost in graph
2 has deliberately been left blank as
the costs will depend on individual
operations and the life of the units.
The graphs show that it is better from
both a productivity and a cost view
point to anticipate the need for an
additional truck rather than waiting for
the need to manifest itself: this is
owing to the higher operating costs of
the LHD. The LHD must therefore be kept
operating continuously without waiting
for a truck to load.

Tips

The tipping areas will have to be very
well lit with suitable guide markers for
the reversing of the trucks to the
tipping points. It is noted that this
is one of the drawbacks to the use of
trucks.

**Capital Cost Savings with the Chosen
Method**

See Table 1

Development

For the purposes of this paper all the
development is regarded as capital.

At present the small capacity of the
trains involved, namely, 30 tonne per
trip, necessitates a large fleet all
operating on the same level and
requiring twin tracks or double-width
haulages to allow the passage of "fulls"
and "empties". There will be the
additional costs for the conveyor race -
mining, support and equipping but these
will be more than offset by the saving
effected by the reduction in haulage
development of some 2 500m. There will
be an ongoing saving for every lift,
which will increase for subsequent
lifts, as these will be planned from the
outset.

Further, a considerable number of
orepasses are required in order to
provide the tipping capacity for the
LHDs, especially on the lower levels
where the highest tonnage has to be
produced. These, too, will be
eliminated in the new method with a

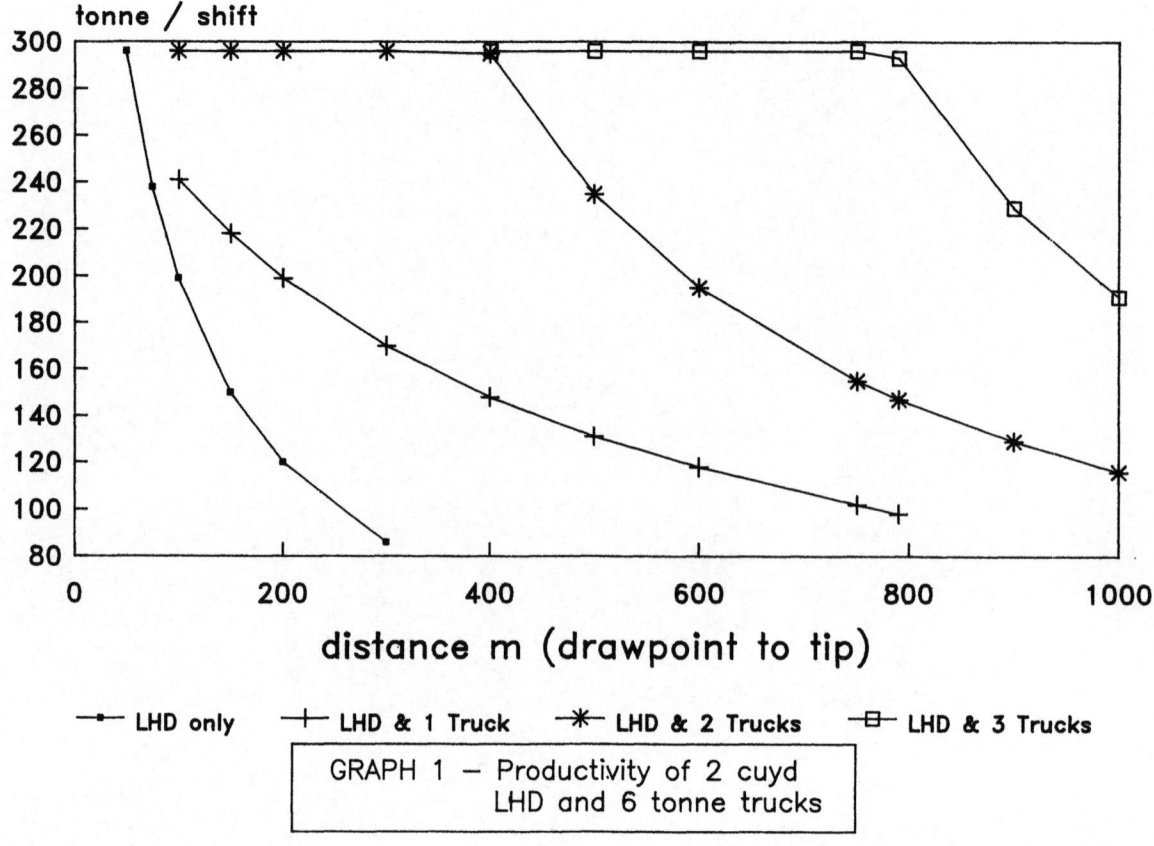

tonne / shift

distance m (drawpoint to tip)

→ LHD only + LHD & 1 Truck ✳ LHD & 2 Trucks ⊟ LHD & 3 Trucks

GRAPH 1 — Productivity of 2 cuyd
LHD and 6 tonne trucks

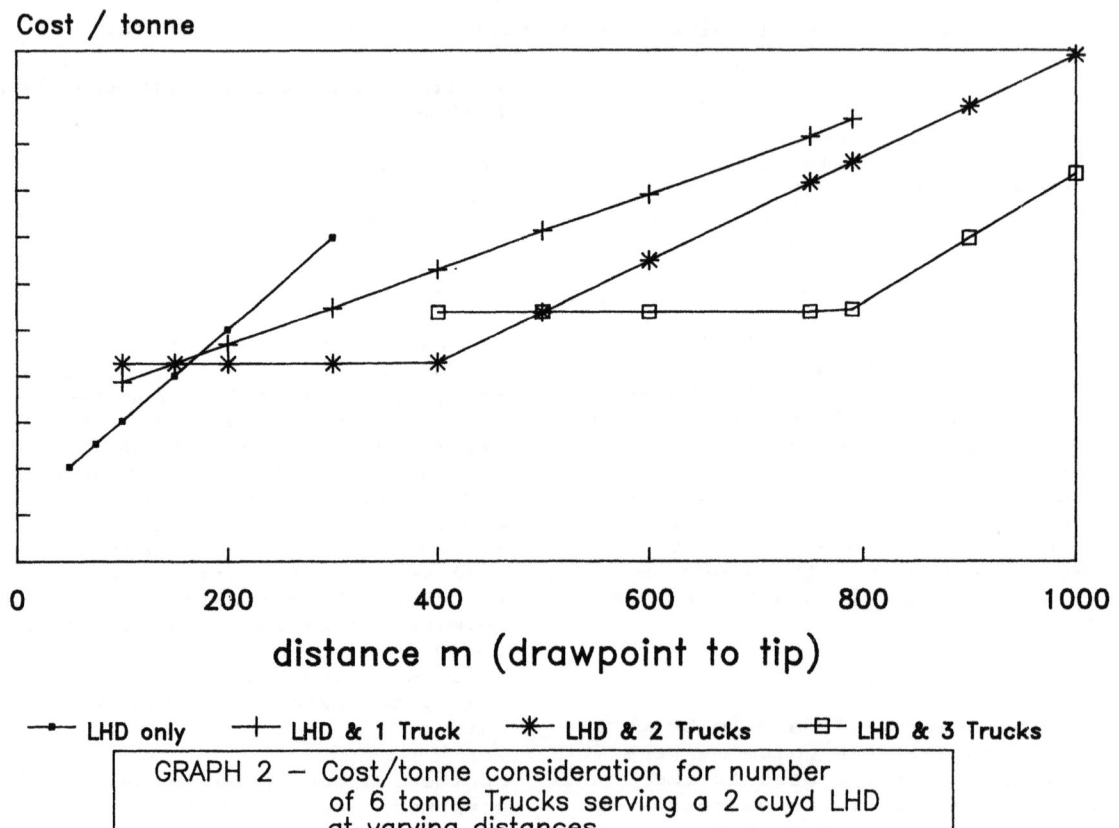

Cost / tonne

distance m (drawpoint to tip)

→ LHD only + LHD & 1 Truck ✳ LHD & 2 Trucks ⊟ LHD & 3 Trucks

GRAPH 2 — Cost/tonne consideration for number
of 6 tonne Trucks serving a 2 cuyd LHD
at varying distances.

reduction of some 2 200m of raising which is always particularly hazardous and, hence, a most welcome spinoff.

The removal of the tips along the collection drives will result in reduction in their span and, consequently, in their support requirements. This is particularly advantageous in those drives that have to be situated in less competent ground.

Boxes and Grizzlies

The orepasses each required a centre loading box and grizzlies for the tips; all of which could be saved under the proposed system.

Loco Replacement

If the method had not been changed there would have been an ongoing capital charge for replacement of locos and their batteries. As the new mining method supersedes the old, the locos currently in use at King Mine will be transferred to the sister mine, Shabanie and this will offer an overall saving in replacement costs to the company.

LHD Replacement

Fewer LHDs will be required and hence replacement costs will be less.

Additional Capital Costs with the Chosen Method

See Table 1

Crusher Station

The extended conveyor race from the new crusher site to the bins is an additional cost.

Furthermore, the orepass feeding the crusher had to be larger to carry the tonnage and anticipated rock size and the tips had to be more robustly constructed, although, as the rock was moving directly to the crusher, the apertures could be enlarged considerably.

The apron feeder to be reclaimed from the original crusher station on 296 level will have to be extended and two large control chutes are to be installed to regulate the flow of rock.

Table 1

Additional "Capital" costs for a lift in 1990 Zimbabwe dollars

Cost Centre	Conventional Haulages	Haulage On Sub-levels with Trucks	Difference	Remarks
Development : including Support, Construction and Equipping				
Conveyor Race :		4353000	4353000	Includes the conveyor belt from the crusher to the shaft.
Haulages	11824000		-11824000	Includes Boxes and Tips
Additional Costs for the Crusher Station		600000	600000	Additional excavation work and the ore control arrangements.
Sub Total	11824000	4953000	-6871000	
Equipment : Purchase and Replacement				
Rolling Stock	1105000		-1105000	Mainly Locos and batteries
Trucks	1750000	10850000	9100000	10 Year life assumed
LHDs	19360000	12760000	-6600000	5 Year life assumed
Sub Total	22215000	23610000	1395000	
Grand Total	34039000	28563000	-5476000	$$$$$$$$$$$

Production tonne in One Lift   15500000   equivalent to 9.6 years
                                                         production life.

therefore cost saving per tonne   -0.35
                                            $$$$$$$$$$$

Trackless Equipment

Calculations showed that while the fleet of trucks will increase considerably, there will be a reduction in the number of LHDs operating. It is anticipated that the final fleet will be 18 LHDs and 31 trucks.

The Capital Expenditure for a Lift

The capital expenditure related to the proposals for a lift with the life expectancy of 9,6 years is summarised in Table 1 as 1990 Zimbabwe dollars. The saving in ongoing development and construction costs will provide the necessary finance required for the method; the main saving being realised in greater efficiency and lower operating costs.

Operating Costs

Table 2 shows these as 1990 Zimbabwe dollars in annual figures.

There will be a reduction in the maintenance costs for the boxes, the tips and the tracks. Finally, the loco operations will cease to be required.

The removal of the loco tramming, which has caused a bottleneck in the production system for a long time and necessitated the grizzlying of all rock being fed to them, means that less secondary blasting will be required with a substantial cost saving. Further, LHDs will not have to tram rocks for blasting from one drawpoint to another and a more continuous flow from priority

drawpoints will be possible thus improving the draw control. As a consequence it is anticipated that LHD operations will become even more efficient. Offset against this is the cost of operating and maintaining the trucks which, it is considered, will be approximately one-quarter of that for LHDs. The use of the first trucks indicates this to be an accurate prediction.

With the reduction of secondary blasting, it is also considered that maintenance in the drawpoint areas will be reduced thus giving a further cost saving.

It is anticipated that the operating costs for the overall mining operation will decrease by some 5 to 8 percent – a worthwhile investment against inflation.

Conclusions

Trucks which have been purchased are proving to be as effective as was hoped, both from a cost and an operational point of view.

The mining of the modified layout is already well under way and production is scheduled to commence in the first quarter of 1994.

Table 2

Additional "Operating" costs on an Annual Basis in 1990 Zimbabwe Dollars

Cost Centre	Conventional Haulage	Haulage on Sub-levels with Trucks	Difference	Remarks
Boxes and Tips	358000	117000	-241000	
Loco Operations	1474000		-1474000	including track maintenance
LHDs and Trucks Development and Transport	666000	493000	-173000	
Production	2705000	3791000	1086000	
Reduced secondary breaking	130000		-130000	explosives usage * 172 g/t to 151 g/t
Reduced Maintenance of Drawpoint Brows	81000		-81000	*
Total	5414000	4401000	-1013000	
Reduced Working Costs / tonne			0.63 $$$$$$$$$$	* areas in which further cost savings are anticipated

Acknowledgements

The author wishes to thank the Directors
and management of Shabanie and Mashaba
Mines (Private) Limited for their
permission to present this paper. He is
also grateful to G.F.Buchanan and all
his staff at Gath's Mine for their
assistance in developing the method
proposed.

References

All from Institution of Mining and
Metallurgy, Zimbabwe Section,
"Mining and Metallurgical Operations
in Zimbabwe", 1983.

1. **Wilson, A.D.**
    The Geological and Geomechanics
    Requirements to Define the Mining
    Environment.
    Vol. II   pp   173 - 197
2. **Marano, G.**
    Rock Mechanics Requirements to
    Define the Mining Environment.
    Vol. II   pp   198 - 235
3. **Laubscher, D.H.**
    The Design and Effectiveness of
    Support Systems in Different Mining
    Environments.
    Vol. II   pp   265 - 289

                    and

4. **Pautz, P.N.** and **Whewell, B.W.**
    (1985) "False Footwall" Drawpoint
    System as Adopted by King Section.
    Association of Mine Managers
    of Zimbabwe.
5. **Buchanan, G.F. et al**
    The "False Footwall" Mechanised
    Caving Method, King Section,
    Gath's Mine, Zimbabwe.
    African Mining 1987   pp 53 - 62

Problems of mining around a sill intersecting the Shangani orebody, Zimbabwe

L. Chimimba
Shangani Mine, Shangani, Zimbabwe
J. E. Smiles
Anglo-American Corporation Services Ltd, Harare, Zimbabwe

SYNOPSIS

A gabbro sill cuts across the Shangani orebody at around 490 metres below surface. The change in the mining method from mechanised sublevel stoping to sublevel caving in order to achieve maximum economic extraction from immediately above and below the sill is discussed.

The change in mining method involved accessing the ore-body via the hangingwall in the area above the sill. Severe ground problems have been encountered in the hanging-wall which is weak sheared talc carbonate. The various factors influencing the design of the mining layout around the sill, including provision for wrecking the sill if required, are discussed. Problems encountered to date with mining above the sill are also described.

INTRODUCTION

Shangani Nickel Mine is situated 90 kilometres north-east of Bulawayo (Fig 1). Mining commenced in 1975 from two open pits at a rate of about 75 000 tonnes per month. Underground production started in 1979[1] from 160 metres below surface when surface mining ceased. It is currently 85 000 tonnes per month from stopes between 440 and 490 metres below surface.

Early underground production was from sublevel open stoping with rib pillars beween stopes being recovered later. This method was replaced by the current sublevel stoping under choke conditions which provided better extraction and recovery. The presence of a barren sill traversing the orebody around 490 metres below surface rendered sublevel stoping unattractive around the sill area. This paper explains the need for a change in mining method to sublevel caving to extract the ore immediately above and below the sill. Production from sublevel caving has now been in progress for just under a year and the experiences and problems encountered so far using this method above the sill area and their influence on plans for mining below the sill are discussed. The original production plan from sublevel caving was scheduled to take three and a half years.

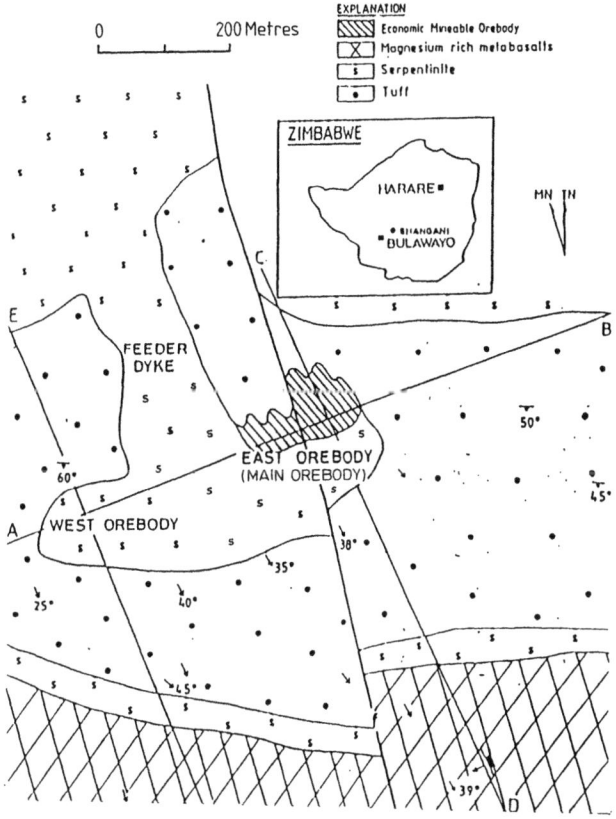

Fig 1   Plan showing geology of Shangani Mine

GEOLOGY OF SHANGANI OREBODY

The nickel orebody is situated in the northern segment of the Shangani greenstone belt[2] which is part for the Archean craton of Zimbabwe. The nickel sulphide mineralisation occurs within a mushroom shaped structure (Fig 1) termed the Shangani Ultramafic complex. This is situated within and appears to intrude mafic and felsic volcanics. The lobes of this structure broadly strike east-west and dip at about 40° to the south. Th nickel sulphide distribution suggests typical gravity control of sulphide mineralisation in komatiites. Narrow pods (⁺ 2m) of massive sulphide ore

267

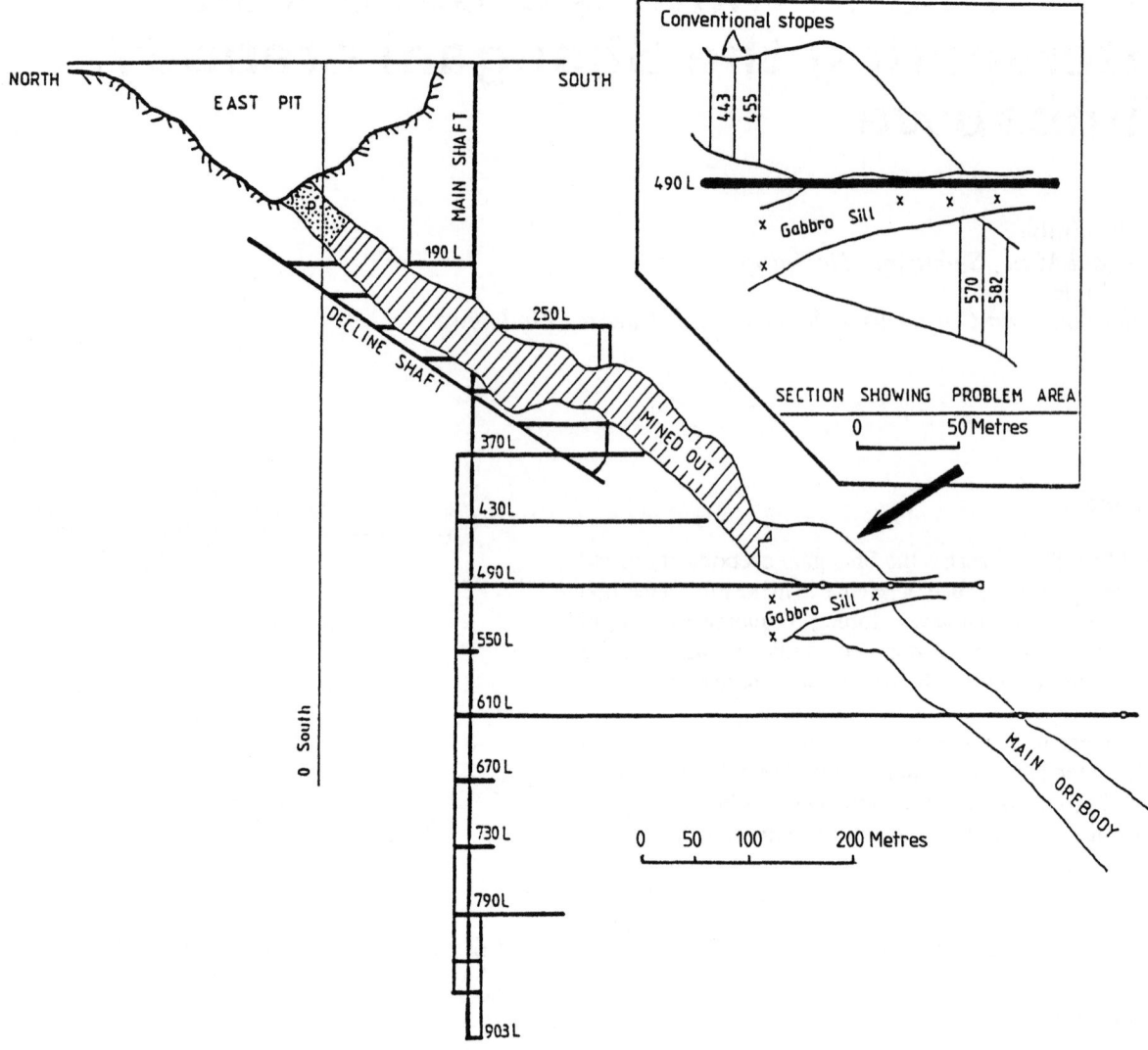

Fig 2   Dip section of Shangani Mine

occur along the contact of the eastern lobe and the footwall tuffs. Disseminated sulphide mineralisation occurs within the serpentinites in both lobes immediately above the footwall contact, with concentration of nickel sulphide decreasing towards the hangingwall or upper side of the lobes. The orebodies are thus defined by a geological footwall and an assay hangingwall of 0,4% total nickel. There is however, evidence of local remobilisation of sulphides during metamorphism and deformation.

Orebodies in the two lobes have been mined in two separate pits on surface. Only the orebody in the eastern lobe which is larger and better mineralised is being mined underground. This orebody (main orebody) is about 120 metres long and 40 metres thick.

A sub-horizontal gabbro sill intrudes the main orebody about 490 metres below surface (Fig 2). The sill, which is wedge-shaped, is about 40 metres thick around the footwall of the orebody and tapers towards the hangingwall. It dips towards the footwall of the orebody.at about 15°.

Metamorphism and deformation have transformed the ultramafic rocks hosting the orebody to serpentinite and talc carbonate schists. Generally serpentinites occur along the footwall and eastern portion of the orebody and the weaker talc carbonate schist occurs towards the hangingwall and western part of the orebody. The rock quality designation for these rocks is considered to be 2A to 2B and 3A to 4A respectively (Fig 3), following Laubscher's classification[3].

THE CONVENTIONAL MINING METHOD

The strike retreat sublevel stoping (choke) method has been employed at Shangani since 1979. This is referred to as the conventional method in this paper and is shown in Figure 4. The stopes in the conventional method have a width of 11,5 metres and are contiguous down dip. Stopes are mined along strike for the full height (width of orebody). Troughs and drawpoints are set into the footwall waste to recove as much of the footwall massive ore as possible and due to the undulation of the foootwall along strike, troughs for any one stope will occur o ' several levels.

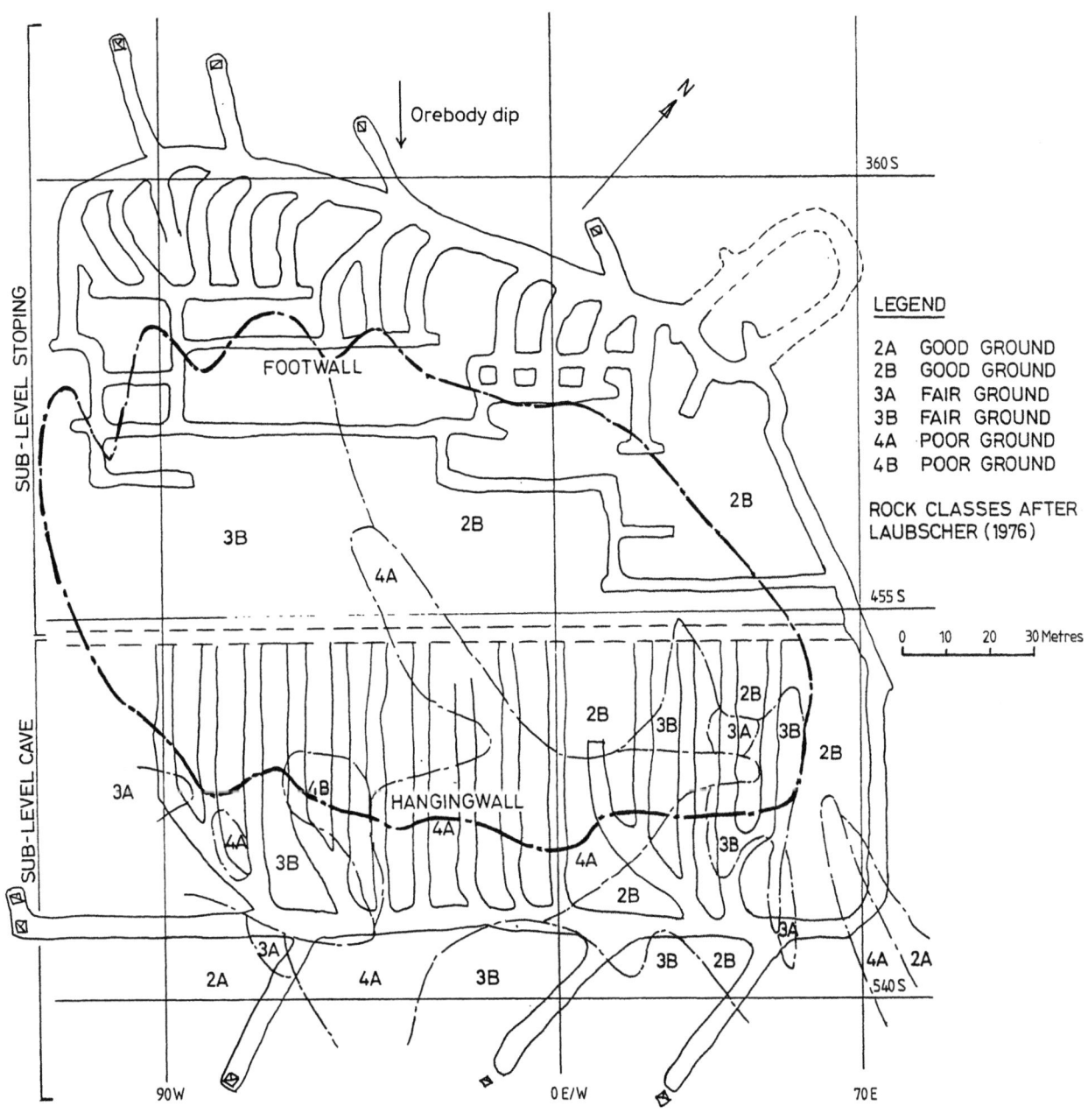

Fig 3   PLan view of 440L showing rock classification

Sublevels are spaced at 10m vertical intervals and drawpoints at 10 metre centres along strike in each stope block, a drawpoint thus ideally serving an 11,5m x 10m column of ore. Drawpoints are accessed by a gathering drive in the footwall serving the whole of the sublevel with lashing being carried out by Wagner ST5 and ST6 'N' and GHH 7,8FL scooptrams. Tip raises at about 30m intervals along strike transfer ore from gathering drives down to the main haulage. The haulages are at 60m vertical intervals. Tramming is by granby cars and battery locomotives to the central ore and waste passes next to the shaft. Sublevels are interconnected by a spiral ramp in the footwall to the east of the Main orebody.

Stope drill drives at 10 and 20 metre vertical intervals are developed within subsequent stopes leaving a 3 metre pillar between stope and drill drives (Fig 4). The situation allows

access back into the drill drive after blasting, the collar remnants being blasted at the same time as the subsequent stope. East and West access crosscuts off the ends of the orebody connect drill drives with the gathering drive.

Slot raises are mined from the troughs to the first sublevel. Rings are drilled in a vertical plane at 1,5 metre burden and toe spacing, this design allows blasting to be carried out in a vertical or horizontal direction. After cutting the initial slot in the trough using the slot raise, it is retreated vertically (3 metre wide) by timing the appropriate blast holes to break downwards. If cut-offs occur it is a relatively simple matter to "step" sideways and then continue blasting upwards. Once the slot is complete to the top of the stope, blasting retreats horizontally. Initially blasting is limited to 5 rings either side of the slot to the full height of the stope. At least 30% of this freshly broken ground is drawn prior to

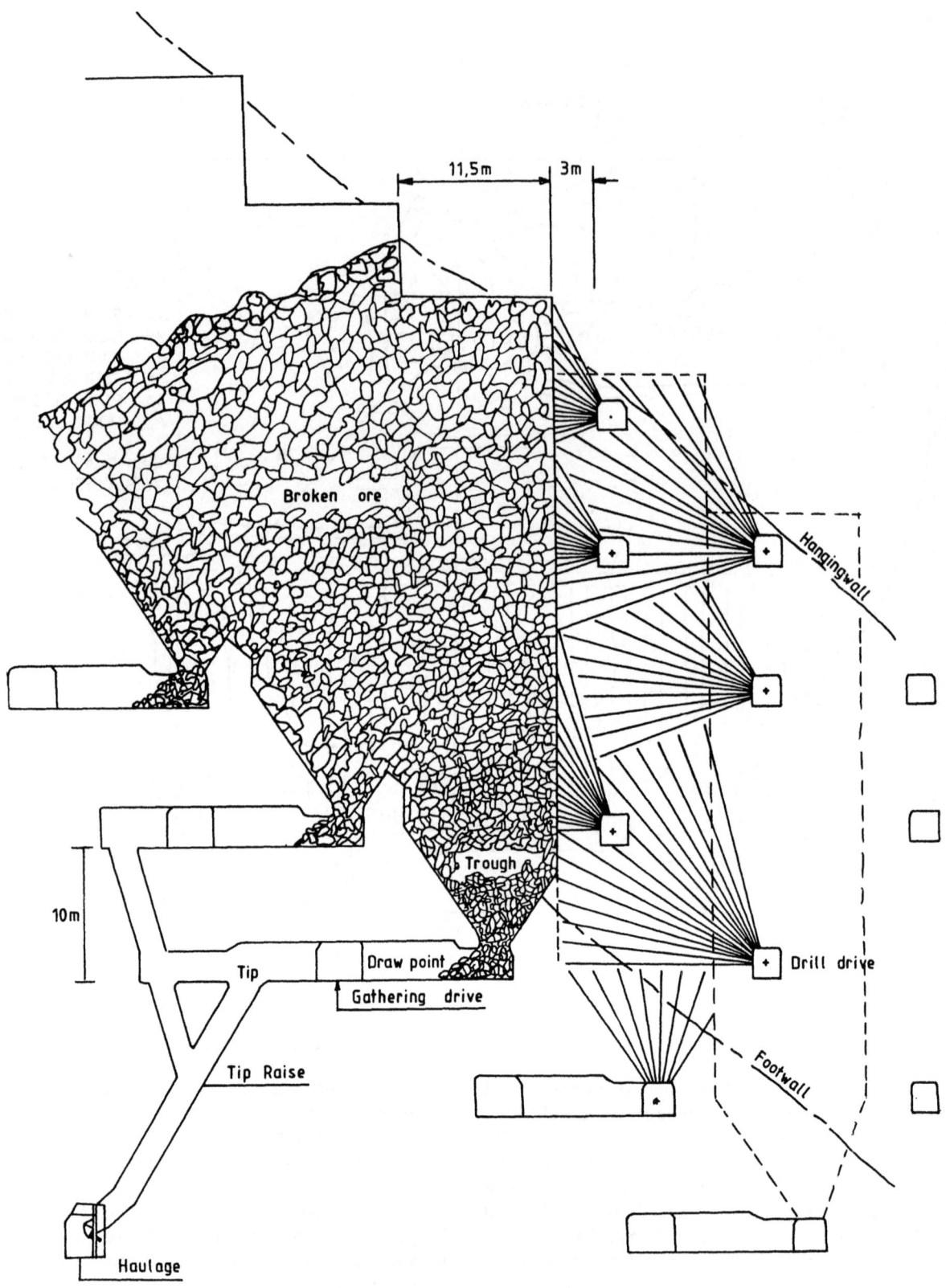

Fig 4   **Dip section showing conventional mining method (sublevel stoping)**

subsequent large blasts of up to 80 000 tonnes of ore.

Since the hangingwall of the orebody is an assay cutoff, typical extraction (rock drawn/ore broken) and recovery (nickel drawn/nickel in ore broken) from the sublevel stoping mining method have been 111% and 115% respectively. Development efficiencies, excluding the ramp, have been about 128 tonnes per metre.

ORE EXTRACTION PROBLEMS PRESENTED BY THE GABBRO SILL

The barren gabbro sill traversing the orebody is shown in Figure 2. The ore presenting the extraction problems are the wedges of orebody immediately above and below the sill.
The effect of the sill on the wedge of ore above it is to flatten the footwall of the orebody. If the conventional mining mehtod were to be used in this area, it would be necessary to have almost all draw points on the same elvation. This would not be acceptable for both production and safety reasons. Mining the wedge of orebody below the sill using the conventional method would require virtually standard development for very little ore extracted. This would increase unit costs and be uneconomic.

The other problems associated with the presence of the sill pertain to safety and dilution. Assuming the wedges of ore above and below the sill were extracted efficiently, the sill would remain as a thin (30m) bridge undercut over a span of about 120 metres by 100 metres. This sill would almost certainly fail sooner or later presenting an air blast hazard. Potential dilution of ore being drawn below it would be significant as there is about 850 000 tonnes of sill material in the orebody area. Leaving the wedges of ore above and below the sill intact to improve the pillar would be unacceptable. This area contains 22% of total current mine ore reserves.

The challenges associated with the presence of the sill therefore encompassed efficient mining of substantial ore reserves immediately above and below it, and special treatment of the sill itself to minimise potential dilution of ore with waste and avoiding the risk of an airblast underground.

SELECTION OF AN ALTERNATIVE MINING METHOD

The mining method for the area around the sill had to be capable of meeting the following requirements:-

- high extraction and recovery
- development requirements particularly of waste per tonne of ore mined similar or better than the conventional method
- unit cost of mining similar or lower than conventional method
- good layout for ventilation, tips, etc.

- minimum support
- amenable to mechanised loading.

Another important consideration in selecting a mining method was the number of drawpoints available for draw. Experience with the conventional method determined that to achieve the mine's targets, production has to be from at least two levels. This requirement narrowed options down to the sublevel caving method.

Some evidence of hangingwall caving had been indicated by the high extraction (+100%) in the conventional stopes and inspections of the hangingwall on upper levels, but the rate of caving was unclear. An additional attraction of the sublevel caving method was that the talc carbonate schists in the hangingwall for the area above the sill would not present major dilution problems as it contains marginally sub-economic nickel grades. The only decisions remaining were how the method was to be implemented and the details of spacing, ring patterns etc.

Sublevel caving was new on the mine. Besides possible production problems presented by the method itself, there had to be a significant period of learning. It was necessary therefore to have production sources shared between the conventional and sublevel caving mining methods. Accurate scheduling of production fron the two sources was therefore vital.

Figure 5 shows the areas to be mined by sublevel caving above the sill (SLC1) and below the sill (SLC2). The last conventional stope mined above the sill is 455 stope and the first conventional stope planned below the sill is 570 stope. It is planned to commence production from conventional stopes below the sill before SLC2 production commences to avoid a possible production hiatus. This timing makes it possible to mine the first two conventional stopes together as a double block (570 -582 block) with drill drives on either side of the double block. The added benefit that would arise from mining a double block would be the rapid build up of production sources below the sill.

THE SUBLEVEL CAVING LAYOUT

The following three possible layouts were designed for the area above the sill. They are illustrated in Figure 6 and their main features are summarised in Tables 1 and 2.
a)   A retreat from the last conventional stope towards the hangingwall.
b)   A retreat from the hangingwall towards the last conventional stope.
c)   A retreat along strike from West to East.

The SLC layout A (hangingwall retreat) was selected ahead of B and C. The main benefits of this layout were the possibility of mining 455 stope in sequence, the possible high rate of extraction and a good tipping plan. The final layout had 10 metre sublevel intervals and drawpoints at 10 metre centres. From experience with the conventional mining method, this was considered to provide the

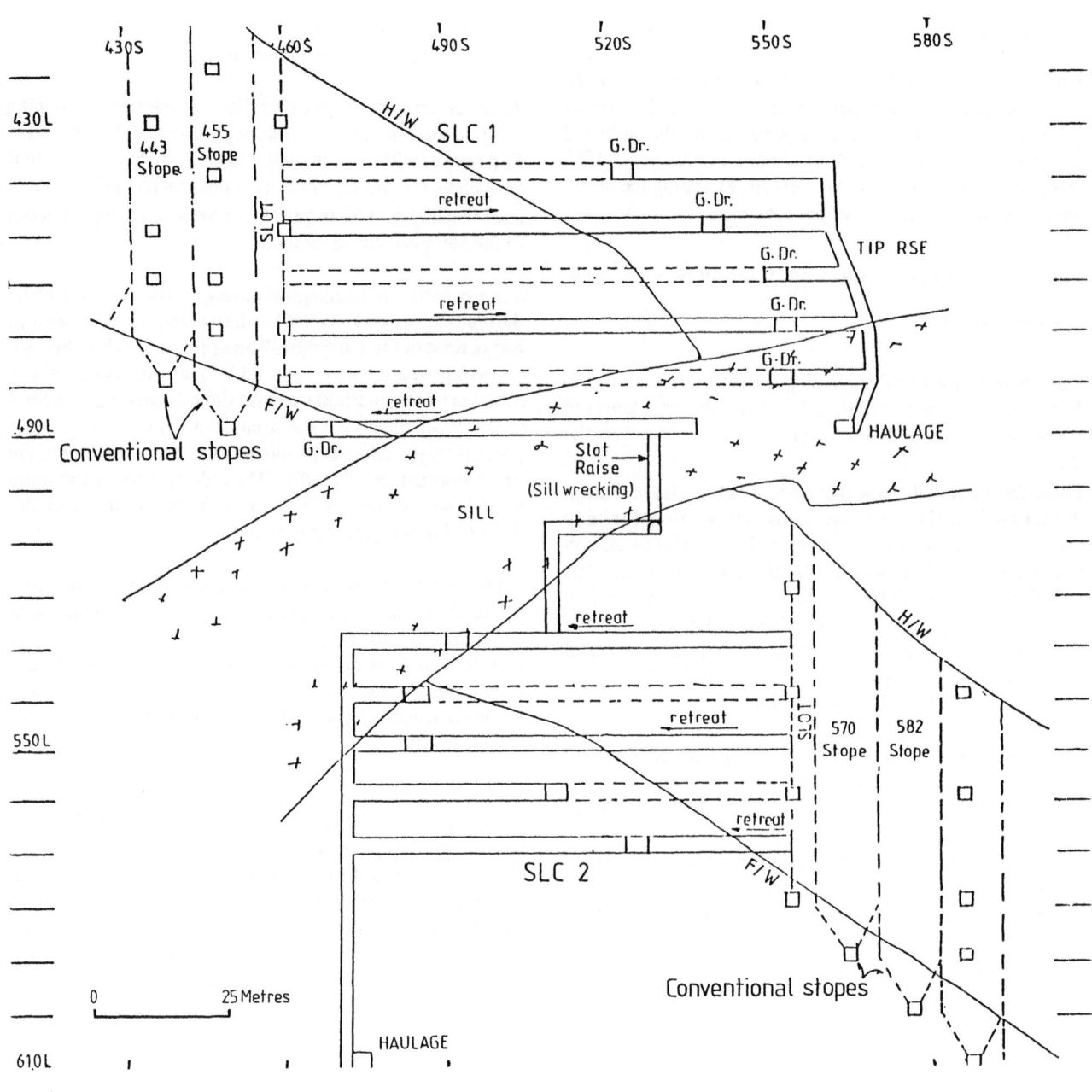

Fig 5   Dip section showing original SLC mining layout above and below the sill

Table 1: Comparison of SLC layout alternatives of SLC 1 and a conventional stope.

	Development in ore (metres)	Development in waste (metres)	Total metres	Tonnage	t/m	No. of tips	No. of production cross-cuts
Layout A F/W to H/W retreat -450L	909	515	1424	184300	129,4	4	14
Layout B H/W to F/W retreat -450L	1091	158	1249	184300	147,6	3	14
Layout C Strike SLC, West to East retreat - 450L	1109	243	1352	184300	136,3	4	8
Conventional Stope 294	818	632	1450	185190	127,7	4	13

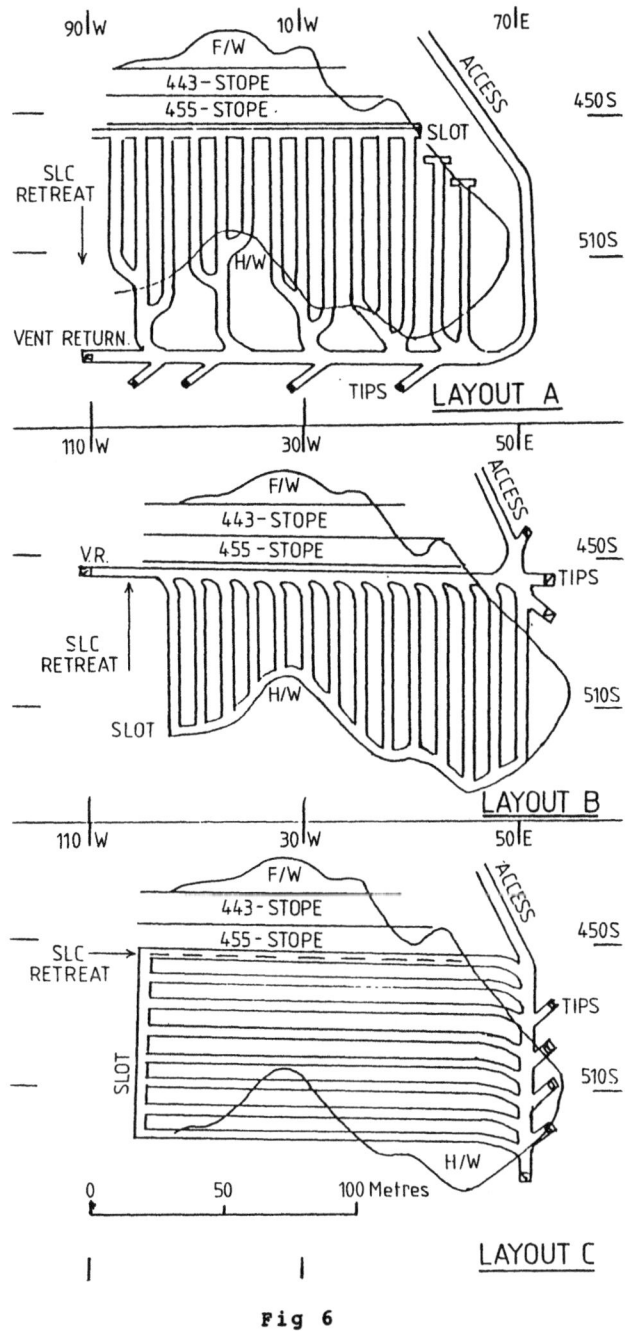

Fig 6

Plan views showing sublevel cave layout alternatives on 450 level

optimum draw geometry. Sublevels were to be established on the 440, 45, 460, 470, 480 amd 490 levels (Fig 5). The drill drives for the 455 stopes were to act as slot drives, with the normal stope ring drilling being carried up to the vertical so as to create a vertical slot face for the SLC.

An important factor in the design of the SLC layout was the size of development excavations in relation to poor ground conditions which layout A would present. It was accepted that smaller LHD units would be ideal in the SLC area, making it possible to mine smaller excavations. However, the company had just come out of a period of severe financial constraints and could not afford purchasing smaller loaders just for this production area, when the larger loaders (5 cu. yard) would remain more efficient in the conventional stoping area. Tunnels were therefore to remain at 3m high x 4m wide.

The blast ring for the SLC layout was a standard fan fore-shortened as the hangingwall was reached. Again drawing on experience from the conventional mining method, ring burden was set at 1,5 metres and a toe burden of 1,5 metres was accepted. In order to preserve the brow in the production crosscut and drawing on experience elsewhere[7] the blast rings were inclined forward at 70°. From the size of the production face and the size of bucket of the scooptrams in use at the mine, it was felt that SLC blast size would be limited to 2 rings to achieve optimum extraction.

After the layout for the area above the sill (SLC 1) had been fixed, the same logic was used to develop future layouts for the area below the sill (SLC 2). Since the first conventional stope below the sill had to be mine ahead of SLC 2 to avoid a production hiatus, the SLC 2 would retreat towards the footwall (Fig 5). This was an attractive layout as accesses and tips for the SLC would be more competent ground in the footwall. The one complication that arose from this layout was that the production crosscuts and gathering drives in SLC 1 are situated above the first stope to be mined below the sill. By reversing the retreat on the lowest level of SLC 1 (490L), it would be expected that production could continue on this level while stoping could commence below the sill without under-cutting the upper levels while they were still in use.

Since the rate of caving of the hangingwall was unclear, it was decided to initially pull only 40% of rings on the upper most SLC level, 440L, the balance of the ground would

Table 2: Advantages and disadvantages of SLC layout options A, B and C.

SLC Layout	Advantages	Disadvantages
A	- high extraction rate - good tipping layout - last conventional (455) stope provides slot - mined in sequence with conventional stope - standard ring layout - gathering drive and tips are outside orebody. - vent return close to tips	- long tramming distance - poor ventilation at face because of long production crosscuts.. - high waste development. - gathering drive and tips in potentially bad ground.

| B | - low waste development
- short length of drawpoint cross-
  cuts improves ventilation at face
- high extraction rate | - ventilation return too far
  from tips.
- Last conventional stope mined
  out of sequence (after SLC)
  giving potential ground
  problems in this stope
- Conventional stope extraction
  can be poor.
- difficulty slotting in
  hangingwall with likelihood of
  cut-back.
- Last conventional (455) stope
  "rib pillar" may unhibit caving
  of hangingwall. |
| C | - good tip layout
- little waste development
- conventional stopes can be mined
  ahead of SLC. | - low extraction rate expected
  (fewer prod. x/c available)
- long tramming distance in
  production crosscut
- poor ventilation at face.
- slot development required
- development parallels shear
  direction.
- production x/c in hangingwall
  side pass through patches of
  waste which may need blasting
- production x/c adjacent to
  last conventional stope (455)
  could deteriorate as the stope |

provide a cushion during mining of subsequent levels. Target extraction on the lower levels is illustrated in Figure 7. Extraction on these levels would be 100% provided that the area under draw was beneath an area already drawn on the level above. On 480 and 490 levels drawpoints would be pulled beyond 100% extraction until waste started reporting to these drawpoints. It was expected that part of the ore left behind on the upper levels would be recovered from these two levels.

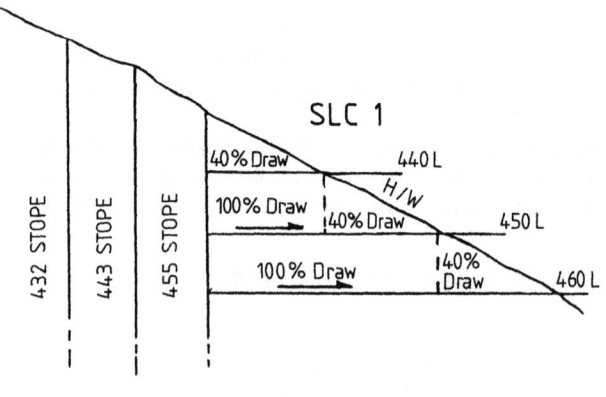

Fig 7

Schematic dip section showing planned extraction of SLC 1

MINING PROBLEMS EXPERIENCED IN SLC 1

Production from SLC 1 commenced in April 1990. By the end of the year about 40% of SLC 1 had been mined. A review of the experience and problems encountered during the commissioning of this mining method is presented.

Poor ground conditions placed the greatest demands on the SLC layout. Although the presence of poor ground was known at the planning stage, the magnitude of possible problems was under-estimated. It became evident early in the development stage that the split set and grouted two-metre long twist steel support used in the conventional mining method was inadequate in the SLC. New support methods had to be adopted and substantial remedial support work done in the areas already developed.

After some initial trials, the following support standard was adopted:

| Production crosscuts, Tip and ventilation crosscuts | - grouted 2,8m long bolt with a 125mm square washer, bolts being placed at 1m square grid and kept within 6m of face. Additional tendon strap and mesh support in very bad areas as required. |
| Gathering drives and breakaways | - tendon straps in combination with wire mesh and grouted 2,8m bolt. Additional 6m long rope anchors placed in bad |

areas as required. Initially mechanical anchors were used, but these tended to slip in the soft rock. Resin anchors are now in use.

When some pillars between production crosscuts on 440L began to fail, it became clear that the severe ground conditions were being exarcerbated by the configuration of the SLC layout. Consequently both sublevel and production crosscut spacing were increased from 10m to 15m. At the time this decision was made commitment on development had progressed such that it was only possible to effect the change in crosscut spacing from 450L and sublevel spacing below 460L. The 475L replaced the original 470 and 480 levels.

Problems of poor ground conditions were also manifest in ring drilling and blasting. Ground movement cut-off blast holes. Holes close to the stope face were very frequently damaged by the previous blast. Modifying charge patterns and trials of a different explosive did not significantly reduce this damage. The net result was that almost all rings on 440, 450 and 460L which had been drilled well ahead of blasting had to be re-drilled. A drill-blast-load cycle has now been adopted.

Ground movement and blast hole cut-off gave rise to severe slotting problems on 440 and 450L. The drill drive for 455 stope was used as the slot drive for both these levels. A large amount of re-drilling was necessary before each slot blast, severely delaying the slot cutting. On 460 and 475L new slot drives independent of the 455 stope were mined. This speeded up the slot cutting significantly.

Roof cut-backs have been experienced on 440L and 460L. These are considered to be a result of both poor ground and the ring design. On 440L, empty stopes were reported in two production crosscuts after only 30% draw following a blast. Inspection confirmed cut backs in the roof (Fig 8). Raises were mined further back in the production crosscuts and semi-horizontal holes drilled and blasted to obtain the necessary stope height. The stope cut-backs on 460L (Fig 8) illustrate the weakness of the initial ring layout, which required very long holes (up to 32m) to cover the pillar between production crosscuts on the level above. This pillar becomes destressed when the production crosscuts have been blasted and pulled, and hole cut off can be expected. These cut-backs have been treated the same was as those on 440L. The ring design has been modified on the lower levels (Fig 9) to avoid similar cut-backs[8].

Poor ground conditions necessitated a modification of tonnage accounting for draw control. Although the standard blast in the SLC production crosscut was set at two rings, it became necessary to blast more than this number if the brow condition was poor or if there was a slip on the third ring which would come down even if it was not blasted. When more than two rings are blasted, some tonnage from the furthest ring in this blast would not be collected on this

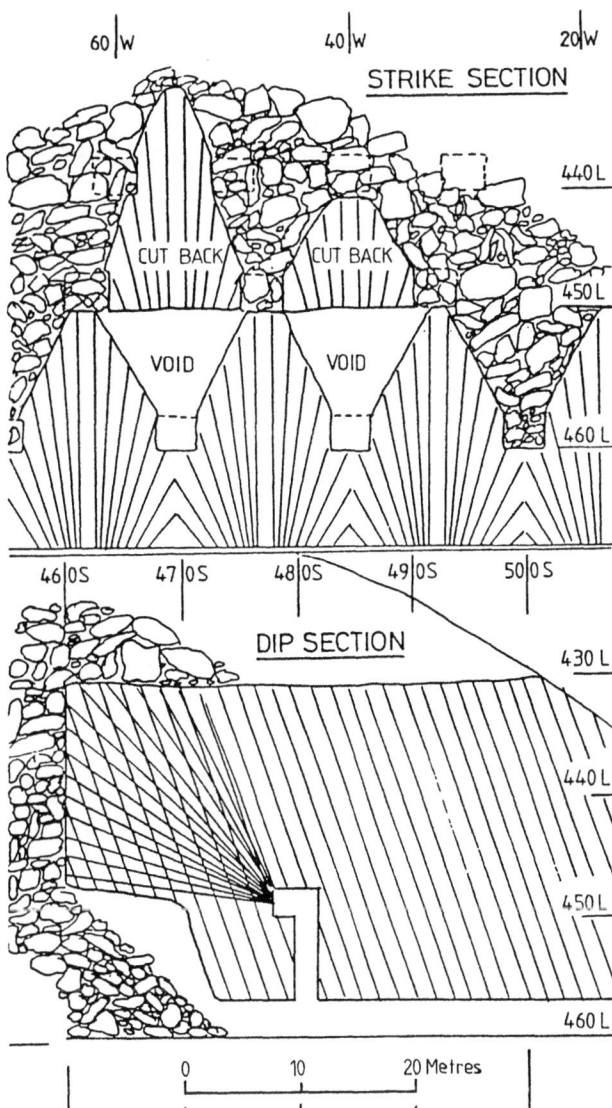

Fig 8 Section showing cut backs in SLC 1

level as the scooptram bucket would not reach it. Adjustments to the tonnage of a corresponding ring below (taking note of ring inclination) would be made so that this tonnage left behind is accounted for on the lower level.

At the end of December 1990, production on 440L was complete. This level was overdrawn. An extraction and recovery of 58% and 53% respectively was achieved against a target extraction of 40%. On 450L, 95% of the available tonnage had been blasted. An extraction and recovery of 89% and 93% had been achieved on this tonnage. On 460L, 14% of the available tonnage had been blasted. On 475L slot-cutting was still in progress.

The SLC mining method requires tight control of the distribution of ventilation air. Problems encountered in the SLC 1 related to personnel training and establishing a control system.

When the decision to commence with the SLC mining was made, it was necessary to communicate the reasons for the new mining method to the workforce. General literature on

SLC mining ana the main SLC layouts for the mine were made available to mine officials. It was important that this level of supervisors understood the basic principles of SLC mining early during the change. New standard procedures covering support, ring blasting and retreating, draw control and ventilation were prepared and communicated to the crews involved in each function.

Draw control and the provision of forced ventilation in production crosscuts required particular co-ordination and communication between departments in the early days of SLC production. Frequent inspections by senior officials were necessary to check that objectives were being met. The workforce responded satisfactorily and routines have now been established.

TREATMENT OF THE SILL

Geologist mapping and some geotechnical work has defined the sill as a medium grained gabbro of geomechanics class 2A[4,5]. The sill contains four main joint sets. Three of these joint sets are steeply dipping and the fourth is shallow dipping (about 15°). The steeply dipping joints strike at about 115°, 172° and 60°. The joints are widely spaced (8m to 13m), however the set that strikes at 115° forms a eries of major shear zones (up to 10m wide) arranged en echelon. The shear zones are of class 3A to 3B. The flat dipping joints are closely spaced (0,5m).

The joint sets contained in the sill, if considered in conjunction with the thickness of the sill itself, suggest that the sill will cave readily. When mining immediately below the sill is complete, the sill will be undercut over a span of 120 metres x 100 metres. The questions to be addressed however are how soon the sill will cave after mining below it commences, and at what rate will it cave? These are important for safety and for controlling dilution. Since the behaviour of the sill could not be deduced with any confidence, a decision was made to induce its caving by cutting a slot through it. The slot would cover the entire strike length of the orebody.

The important objectives considered in laying out a slot for the sill were: connecting the stopes above and below the sill to minimise possible air blasts, cutting the slot as soon after commencement of production below the sill as possible, and low costs. The final layout adopted is shown in Figure 5. The slot, which is close to the hangingwall of the orebody, is expected to provide a zone of weakness from which the rest of the sill would cave outwards across the orebody as mining progresses beneath it. Some waste lashing would be necessary to clear swell from the initial slot blasts and ensure the rest of the slot breaks. The actual rate of caving of the sill would be monitored by use of extensometers. Dilution in the conventional stopes would be minimised by strict draw control that would ensure an even ore/waste interface over the stope. In the SLC 2, strict adherence to 2-ring blasts which are currently the target in the SLC 1, would also minimise the loss of ore through dilution.

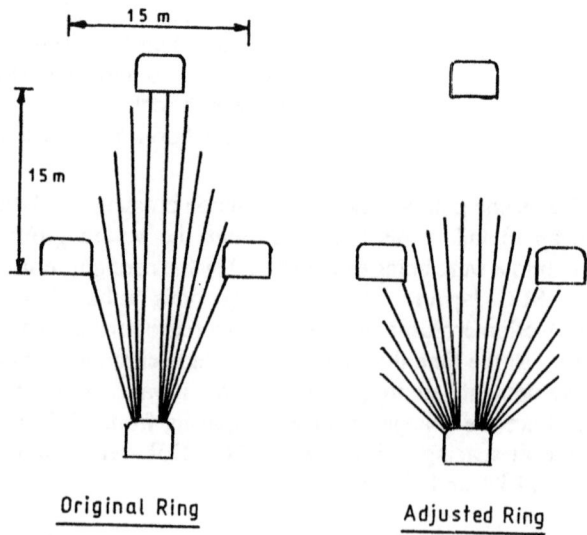

Original Ring   Adjusted Ring

Fig 9   Strike section showing change in ring pattern.

MINING BELOW THE SILL AND THE RETURN TO THE CONVENTIONAL MINING METHOD

The decision on returning to the conventional mining method was made before production from SLC commenced because of the necessary lead time on development. The conventional method still offers some advantages compared with experiences gained in SLC production so far. Besides the mine personnel being used to it, ore extraction and nickel recovery are high. The method offers flexibility as many drawpoints can be available for draw at the same time and hole closure is not as great a problem because of the large blasts which can be taken. Both broken ore and drilled-unblasted reserve can therefore be built up. Drill drives can be mined smaller presenting fewer support problems. The SLC layout has long hauling distances and it has been necessary to increase the mine's fleet of scooptrams while mining in the SLC area in order to maintain production levels. This has increased costs.

In order to avoid a production hiatus and to quickly build up broken ore reserves, the two conventional stopes below the sill will be brought into production as a double block ahead of SLC 2 (Fig 5). Once this block is commissioned, the SLC 2 will commence retreating towards the footwall.

Experience in the SLC 1 has demonstrated that because of poor ground conditions close to the conventional stopes, a slot for the SLC independent of the conventional stopes speeds up the opening of the SLC area. The slot for the SLC 2 will therefore be independent of the conventional double block. At the same time that SLC 2 comes into productiom, conventional stopes down dip can be brought into production, and the slot for the sill can be cut.

CONCLUSION

The decision to use a sublevel caving method immediately above and below the sill is considered to have been appropriate. The extraction of 89% achieved to date on the 450 level is satisfactory. However the rate of extraction has so far been well below expectations because of various problems.

The main problems affecting SLC production have been training of the labour force, support, hole closure, cutbacks, difficulties with slot cutting and ventilation. The effect of these problems has been that only a low number of crosscuts have been available for production at the same time.

It is clear now that insufficient attention was paid to rock mechanics at the initial planning stages. The problems of ground conditions have now been largely overcome by improved support methods and by modifying the SLC layout itself. It is felt that with these changes, production rates will improve.

The cutting of a slot through the sill should ensure that the sill caves as mining progresses below it. What is still not clear is the rate of caving of this sill and the effect on dilution this will present. The rate of caving of the sill will be monitored by use of extensometers. Waste dilution will be minimised by maintaining an even draw across the stope. Even with the experience of sublevel caving gained so far it is considered beneficial to return to the conventional sublevel stoping methods. The introduction of the 570/582 double block ahead of the SLC 2 will avoid a production hiatus and enable the bringing in of further stopes down-dip at the same time as the SLC 2.

ACKNOWLEDGEMENTS

Over the past few years several mining engineers and geologists from A.A.C.S. and BNC have studied the implications of the presence of the gabbro sill at Shangani. Most of the ideas summarised in this paper are originally theirs. This fact is gratefully acknowledged. We wish to thank the management of BNC and A.A.C.S. for permission to publish the paper.

REFERENCES

1. Butcher D.W. and Mills R.G. (1983). The change over from open pit to underground mining at Shangani and the development of a suitable underground stoping method. A.A.C.S. Internal Report.

2. Viljeon M.J, Bernasconi A, Van Coller N, and Viljeon R.P. (1976). The Geology of the Shangani Nickel Deposit. Econ. Geol. 71, 76-95.

3. Laubscher D.A. and Taylor H.W. (1976). The importance of a Geomechanics clssification of jointed rock masses in mininh operations. Smyposium for Rocks Engineering, Johannesburg. 119-135.

4. Marano G (1989) Rock Mechanics assessment of mining below the sill at Shangani. A.A.C.S. Internal Report.

5. Marano G (1990) The Geotechnical characteristics of the Gabbro sill at Shangani Mine. A.A.C.S. Internal Report.

6. Steffen O.K.H., Stacey T.R., Hume H, and Houghton D.A. (1979) Report on the effects of underground mining at Shangani Mine. S.R.K. Internal Report.

7. Just G.D. (1972) Sublevel caving mining design principles. Trans. Instn. Min. Metall. (Sect. A Min. Industry), 81. A214-A219.

8. Playle R. (1990) Shangani Ring Drilling above the sill (Internal Report A.E.C.I.).

Advances in cost-effective support technology for rock reinforcing in some deforming ground condition

A. D. Wilson
Geological Consultant, African Associated Mines (Pvt) Ltd, Bulawayo, Zimbabwe

SYNOPSIS

The paper covers the various types of rock reinforcing as used by Shabanie and Mashaba Mines (Pvt) Ltd., under block cave, prebreak cave and sub-level cave mining conditions.

The development of a patented integral steel strapping system is discussed and the various applications, under which some of the designs have been used successfully, are noted.

The system is based on the installed steel support components rapidly becoming dynamic by making use of the steel's tension capacity under deforming conditions.

Comparisons are made with conventional " W " straps in terms of application potential and the use of shotcrete.

Cost comparisons are made for the reinforcing methods discussed and comments made on installation capabilities under current levels of mechanisation within Shabanie and Mashaba mines.

It should be noted that TENDON STRAPS and any like facsimiles thereof are **subject to several international PATENT PENDING applications and, consequently, authority must be obtained from Shabanie and Mashaba Mines (Pvt) Ltd., for manufacture or use of these units unless they are purchased from a manufacturer duly authorised and licensed by Shabanie and Mashaba Mines (Pvt) Ltd.**

BACKGROUND

Shabanie and Mashaba Mines (Pvt) Ltd., operates Shabanie and King mines which are large chrysotile asbestos producers and both rely heavily on geomechanics classification to indicate support requirements [1,2].

The massive mining methods employed are block caving, prebreak caving, false-footwall and sub-level caving as described in detail elsewhere [3-5].

The support methods used on both mines, prior to incorporation of tendon straps, were generally centred around various combinations of active rock reinforcement from rockbolts (and cablebolts) and surface restraint from mesh-reinforced shotcrete or TH arches backed with concrete. "Post-bolting" of shotcrete with rings of rockbolts was introduced to pin the layer to the rock and maximise the confinement. However, failures were still evident in areas where major deformation occurred as a result of stress concentrations. Typical failures of the support were:-

- unravelling of material between rockbolts;

- "swelling" failure of shotcrete resulting in a tension-fractured layer with much reduced confining capacity (**Plate 1**);

- buckling of TH arches and cracking/rupture of the backing concrete;

- floor-lift related to high abutment stresses and/or low rock strengths.

The failure mechanisms indicated the need for more rigorous initial confinement of the rock mass. The answer appeared to lie in generating a significant tensile force between rockbolt heads so that any movement of rockblocks at or near the surface would be contained. Various straps made from plate were used over a wide range of applications and found to have merit.

However, major drawbacks included:-

- the large surfaces were prone to severe blast and concussion damage;

- it was not possible to successfully enclose these straps in shotcrete or concrete because the flat surfaces quickly debonded; exposing the plate to further damage;

- the plate thickness selected (3mm) resulted in the rockbolts "pulling through" the steel. Fairly rapid corrosion was also a problem in some high risk areas (which are generally wet and blocky).

- the plate straps were expensive and their continued availability uncertain because of production and import constraints.

Plate 1;   Typical shotcrete "swelling" failure

DEVELOPMENT

Recognising the various drawbacks above, a fabricated steel strap was developed to emulate the plate strap but designed to have better tear-out characteristics. It was also configured such that it could be used alone or integrated directly within shotcrete. The materials from which the new "Tendon strap" was made are all manufactured within Zimbabwe and their availability is relatively predictable and not linked to imports.

Description:

The typical Tendon strap consists of three general component types, each of which can vary in dimension depending on the application required;-

1)   A series of parallel, solid steel bar Tendons which exert the confinement;

2)   A series of flat steel plates with suitable holes in them; the plates being welded at regular intervals to the Tendons. These plates act as the foci of confining forces which are then transmitted to the anchoring rockbolts/ cables.

A system of interlocking of plates at overlap positions has been developed for "special cases" but is not considered a necessary "standard".

3)   A series of steel spacers which aid in the distribution of confining forces between Tendons and do not allow the latter to be pushed apart under point loads. These spacers are typically NOT welded to the Tendons but tightly wrapped or crimped to minimise movement.

A typical "standard" Tendons strap, as used in Zimbabwe by Shabanie and Mashaba mines as well as Anglo American's Shangani and Epoch mines, consists of 4 Tendons of 10mm diameter (rebar or mild-steel round-bar), 3 plates, spaced at 950mm hole centres and 4 spacers distributed equidistantly between all plates (**Fig. 1**).

Design principles:

Tendon straps are designed as a surface restraint system regularly pinned to the rock surface by anchors, which are either in the form of bolts or cable. Deformation or unravelling of the rock mass causes tension in the Tendons which then actively confine the movement of rock blocks with the direct aid of the anchored rockbolts or cables. The Tendons act in unison by allowing a degree of yield; the rock mass is thereby assisted to support itself.

A major consideration in the Tendon strap design is the ability to shotcrete through/ over the Tendons thereby giving a well-distributed load-bearing surface which has far greater reinforcing potential than, say, diamond (chain)-mesh or weld-mesh reinforced shotcrete. A variation of design, which has very little flat steel surface vulnerable to debonding, is being tested. (**Fig 2.**)

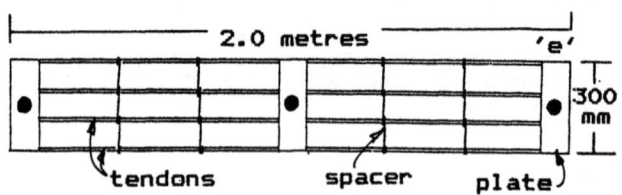

Fig. 1 Typical Tendon strap layout.

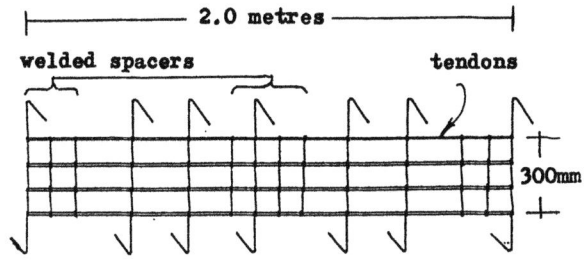

Fig. 2 Shotcrete Tendon straps.

Range of uses:

- by themselves applied directly against the rock surface to act against key-blocks; or
- in combination with diamond-mesh or weld mesh to cater for blocky material; or
- installed on top of shotcrete to give added confinement; or
- installed within concrete flooring to give protection against floor-lift.

All applications are governed by the geomechanics rock mass rating (**RMR**). This in situ rating must be subsequently adjusted for mining stress considerations to give the final mining rock mass rating (**MRMR** [6]).This indicates the amount and type of support required. If the RMR is 3b or better, the rock can be reinforced effectively with Tendon straps.If the RMR is below 3b, Tendon straps can be used to reinforce the shotcrete or concrete lining. The key geomechanics indicators are intact rock strength, joint orientation/condition and frequency [2].

Variations:

The configuration of Tendons (and the other items) depends upon the anticipated tensional load or confinement they will have to exert. This must be matched to the capacities of rockbolts/anchors used. The plate thicknesses can also vary; they are generally reinforced so that the tear-out force matches the combined tensile capacity of the tendons. Example options are detailed below:-

	TABLE 1		Reinforced Plate	
Anchor diameter	Tendon No. x	Diam. (mm)	thkns. (mm)	x e (mm)
16mm rebar	4 x	10	8	x 100
20mm rebar	4 x	12	10	x 150
25mm rebar	4 x	14	12	x 180
15.5mm cable	4 x	14	12	x 180

(Refer **Fig 1**)

Rust deterrent:

Part of the design specifications required the Tendon straps to be dipped in an oxide primer. To date, there have been no failures as a result of corrosion. In extreme cases of rust potential, or long-life requirement, the units can be resin dipped, powder coated or galvanized. However, as none of these techniques is perfect, a sealing layer of shotcrete is generally preferred.

Strap anchors: Rockbolts or tensioned cable.

These are critical items within the Tendon Strapping system as most of the dynamic tensional/shear forces are focused on them.
It is, therefore, most important that they are:-

- Long enough to be well anchored in more stable ground;
- Ultimately fully grouted in the hole (after pre-tensioning if required);
- Designed such that their ultimate tensile strength is at least as great as the tensile strength of one Tendon strap.

Friction tubes:

These are used to initially pull the Tendons tight up against the rock mass surface and make the straps more immediately "active". **The tubes are only required if the final rockbolts cannot be tensioned.** They also give some added shear strength to the rockbolt.

The use of friction tubes allows some controlled de-bonding of the rockbolts and, hence, there is a distributed yield in the first 0.5m of bolt anchor.

The use of friction tubes (or "Split-Sets") without final rockbolts or cable anchors installed within them is not recommended because the tensile forces set up in the straps readily guillotine the thin tubes.

Alternative pinning options would be:-

- suitable length, solid-steel end-anchored bolts (which are later full-column grouted) **or;**
- continuously threaded bolts. The torquing up of the nuts effectively pulls the straps tight against the rock or shotcrete **or;**
- tensioned cable bolts are an effective deep-reinforcing alternative and this technique is being pursued for future support recommendations in certain areas.

INSTALLATION TECHNIQUES:
SUMMARY OF STEPS.

There are four main options which vary according to the type of rockbolts or anchors used by the individual mines. All the Zimbabwean mines which use the Tendon strapping system currently employ Options 1 and 4 below.

Option 1: Use of friction tubes (or SHORT 0.5m Split-Set) in separate cycle to grouting of non-tensioned rockbolts Fig.3.

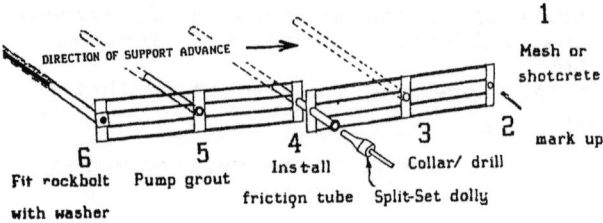

Fig. 3 Installing with friction tubes; overlapped end-plates.

Steps
1   Decide whether shotcrete is required or if only diamond mesh or weld mesh is needed to give the initial confinement to small rock blocks between straps (dependant on adjusted geomechanics classification and interpreted ground conditions) 6.

2   Mark up the Tendon strap locations; their radial spacing being determined from structural and stress considerations. Typical options in current use are 1.2m or 0.6m; start 0.3m on either side of the drive centreline and working progressively down the sidewalls. Tendon straps are overlapped at their ends, thereby sharing a common friction tube (or Split-Set) **(Plate 2)**.
3   Collar the hole and drill to final length through strap plates (assists alignment); , azimuths generally perpendicular to sidewall or adjusted to account for joint orientations.

NOTE:
The rock hole diameter is dependent on rockbolt manufacturer's recommendation for Split Sets. If friction tubes are to be used to initially pull the straps up tight, the hole must be just slightly larger than the outside diameter of tubes to be used. For example, a 40mm hole needed for a 38mm OD friction tube; the hole length at this diameter needs to be longer than the tube to ensure flush fitting. The remainder of hole can be of a decreased diameter sufficient to allow for internal bolt plus grout.

4   Insert friction tube; push into hole with "Split-set" pusher dolly on a jackhammer.   This gets the strap tight up against the sidewall. (Be careful not to distort the tube as this will prevent the next steps).

Plate 2. Installing common friction tube at overlap position.

5   Fill hole with grout through tube by either:-

a)   Pumping with liquid grout (applicator tube must start at the deep-end of the hole and be forced out by grout pressure; grout viscosity must be such that it will not run out of "up" holes);   OR

b)   Shoot sufficient cementitious capsules into hole after activation per manufacturer's instructions (diameter of capsules to suit chosen tube internal diameters);   OR

c)   Use suitable resin cartridges if rockbolts are going to be spun in.

6   Insert the final rockbolt anchor plus washer and hammer tight onto friction tube. Grout must be pushed out of the hole as the rockbolt is finally located **(Plate 3)**.

Plate 3. Grouting of rockbolts; showing 2 bolts awaiting final setting.

NOTE:

The installation of a "solid" rockbolt inside the initial tube is essential to give sufficient tensional and shear resistance to deformation transmitted by the strap in tension.

Follow the standards.

If supervision is not good, the practical time lapse between installing friction tubes/Split Sets and the grouted rockbolts can be significant.

This can be dangerous because:-

- There is a false sense of security with only friction tubes (or Split Sets) because the system "looks good".

- The friction tubes are readily distorted by either rock-block movements or the effects of lateral tension in the straps, especially in an "active" stress environment.

- In-hole rubble generated by on-going deformation can block the free passage of grout and rockbolt, leaving an unsatisfactory system.

Therefore, steps 5 and 6 must be done within 24 hours of step 4 otherwise there could be a chance of movement cutting off the hole.

A benefit of having the combined effect of the friction tube and rockbolt in a grouted medium is the immediate restraint and enhanced shear resistance in the first 0.5 metre from the excavation surface.

Option 2: Using only end-anchored tensioned rockbolts.

STEPS 1 to 3 here are the same as in Option 1 above, the rock hole diameter depends on supplier's recommendation for the particular end-anchor bolt used.

4   Insert the rockbolt and tension as per design specifications to ensure Tendon strap pulled tight up against sidewall.

5   Follow up with full-column grouting of rockbolt to ensure lasting confinement (some bolts can be grouted prior to tensioning).

Option 3: Using continuously threaded bolts.

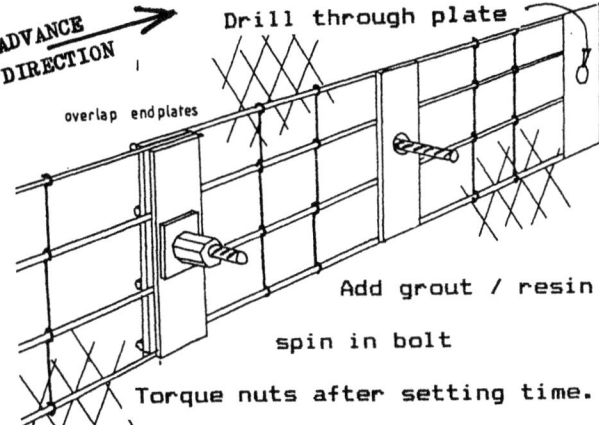

Fig. 4   Installing with continuously threaded rockbolts.

Steps 1 to 3 here are the same as in Option 1 above.

4   **Fill hole with grout (or sufficient cementitious/resin capsules), install bolt, allowing adequate length for strap/washer/nut to protrude out of hole.** The straps can be hung on the bolts and the nuts 'just on' so that small alignment adjustments can take place prior to final setting.

5   **After suitable setting time, torque up the nuts** so that the strap is pulled tight up against sidewall.

NOTE:
If pre-tensioning of bolts is required, use quick setting resin at the end of the hole with slow set in the remainder. The bolt can then be spun in and the nuts torqued immediately to give a quick and final installation.

Option 4: Using cable bolts.

The steps here are essentially the same as for Option 3 above except that sufficient cable must protrude to allow for the tensioning cylinder plus barrel/wedge clamp.

The Tendon strap design must allow for and match the increased tensional capacities of the cable used especially if significant deformation is expected. (Table 1).

Current applications use two configurations of cable/strap/bolts in the reinforcing of turnouts and the support of wide-span service installations:-

a)   **Using cables for the strap overlaps** to give better depth of anchorage as well as more shear resistance in this 'continuous situation (**Fig.5**). More than two straps can share one cable in this situation which is typical at wide turnout spans and service excavations.

284

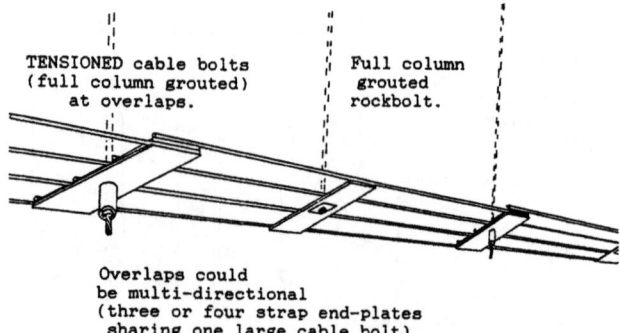

TENSIONED cable bolts
(full column grouted)
at overlaps.

Full column
grouted
rockbolt.

Overlaps could
be multi-directional
(three or four strap end-plates
sharing one large cable bolt)

Fig. 5 Cables at strap overlaps.

b) **Using cables to secure the centre-plate** and rockbolts in the end-plates of "special cases" where overlapping is not required. (**Fig.6**). This is typically used at wide turnout spans.

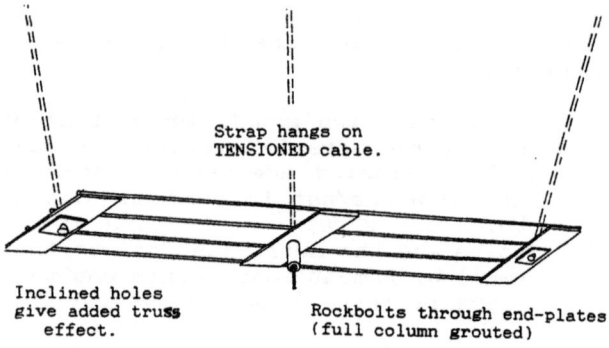

Strap hangs on
TENSIONED cable.

Inclined holes
give added truss
effect.

Rockbolts through end-plates
(full column grouted)

Fig. 6 Centre-cables with rockbolts.

NOTE: In all 4 Options above, a final layer of shotcrete can be applied to increase the active confinement of the system as well as giving a smoother surface less prone to damage from LHDs.

CURRENT APPLICATIONS OF TENDON STRAPS.

1 Containment of Spalling from either high horizontal or vertical stresses.

When Tendon straps are used in overlapping horizontal bands, either alone or in conjunction with shotcrete, diamond mesh or weld mesh, they effectively contain spalling or scaling surfaces (**Fig. 7**). **The rockbolt or cable-bolt lengths must cater for anchorage in the more stable ground** and these anchors should ideally be tensioned and fully grouted.

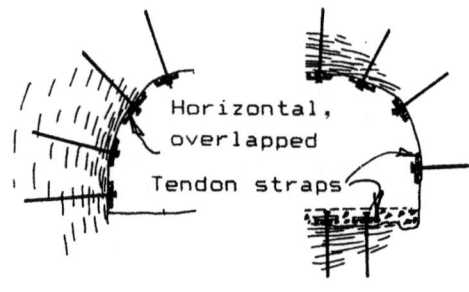

Horizontal,
overlapped
Tendon straps

Fig. 7 Spalling control.

2) Confinement of Sloughing or Blocky ground (Fig. 8).

When straps are used alone, or in conjunction with diamond mesh or on top of shotcrete, plastic deformation or gravitational unravelling of blocky ground can be effectively contained. This is achieved by the tensional confinement exerted from horizontally installed overlapping Tendon straps. **Rockbolt lengths must ensure good anchorage and their orientations should cater for predominant joint sets (depends on environment).** If there is likely to be significant plastic flow, such as in highly sheared zones where the rock bolt anchors may not be secure, the use of longer cable-bolts should be considered for depth of anchorage. However, if these shear zones are narrow, Tendon straps overlapped and rockbolt-anchored in "competent" ground are effective. The restraining surface prevents the propagation of frittering failure.

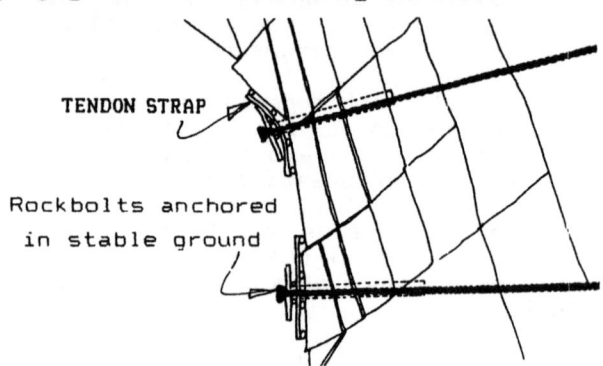

TENDON STRAP

Rockbolts anchored
in stable ground

Fig. 8 Confining blocky/sloughing ground.

3) Reinforcing Confinement of Grizzly Drawpoint turnouts or Drawpoint Brows (Fig.9).

Used as a rock-reinforcing system and then covered with shotcrete, Tendon straps in drawpoint situations **effectively maintain crown pillar integrity by keeping the apexes secure and containing relaxational failure of the brows.** A typical grizzly construction system would comprise of:-

a) **Shotcrete layer applied** (± 50mm); optional, not required if " insitu classification rating" is 3A or better;

b) **Install horizontal overlapped Tendon straps** from the brow line to wrap around the corners into the drive; typically 0.6m radial spacing;

then EITHER:-

c) Shotcrete through/over the TENDON STRAPS so that there is + 100mm of wear available; or

d) **Shutter up the brow and pump/pour concrete** (vibrated) as normal; allow for a wearing thickness.

285

e) **Repair blast-damaged/wear areas immediately** upon steel exposure.

The benefit in this system is that the straps effectively reinforce the shotcrete and the need for expensive (imported) yielding steel arches is minimised unless the ground conditions are very poor. Tendon straps can also effectively be used as an integral part of the concrete construction of grizzly, ore-pass and floor reinforcment in these areas of high development density.

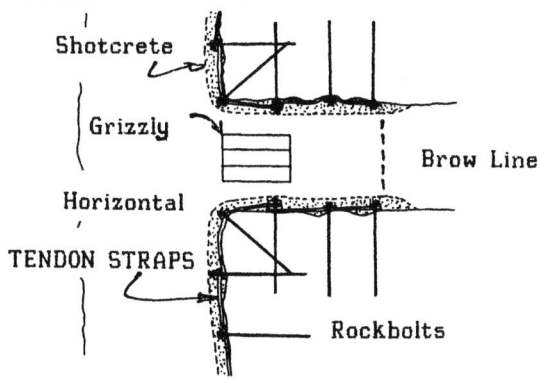

Fig. 9 Drawpoint turnouts and brows.

4) Reinforcement/Confining of Turnout Apexes (Bullnoses); (Fig. 10).

Unravelling of acute apexes in LHD layouts is a common problem which effectively increases the turnout area resulting in greater unsupported spans. **These problems can be largely solved if Tendon straps are used alone or in conjunction with mesh and/or shotcrete to confine the deformation in the narrow "pillar" zone.** In extreme cases (depending on stresses, joint orientation, degree of overbreak, etc.,) multiple lengths of Koepe rope or cable-bolt strand can be used around the apexes; with the rope ends grouted deep in the stable ground [7]. In these cases, a series of vertically installed Tendon straps down the "point" of the apex and along sidewalls effectively distributes the confining effect of the Koepe rope. The exposed rope is then either shotcreted over or protected by initially running the rope through suitably bent piping. This system can be added to by building up the apex with shuttered reinforced concrete which has good hangingwall continuity and is itself strapped back to stable ground.

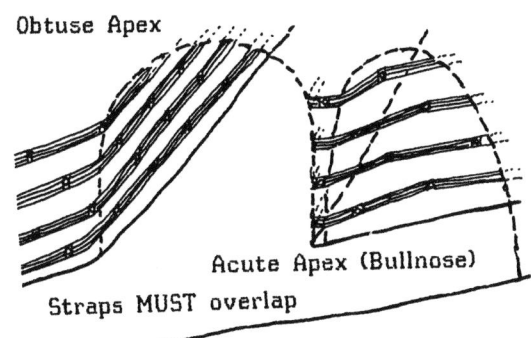

Fig. 10 Confining apexes

5) Confinement of Pillar Deterioration (Fig. 11).

Straps can be used in the confining of deteriorating narrow pillars or to contain differential spalling due to varying lithology in pillars. **Even though the pillar may effectively "fail", the confinement of the "failed" mass maintains some load-bearing capacity.** In these cases, the rockbolts/cable anchors should extend through the pillar from surface to surface and be fully grouted. Other applications of Tendon straps in seam mining include support of gullies, permanent in-seam serviceway pillars, crossovers and in early support at the face where blocky ground may be intersected (dykes, etc.).

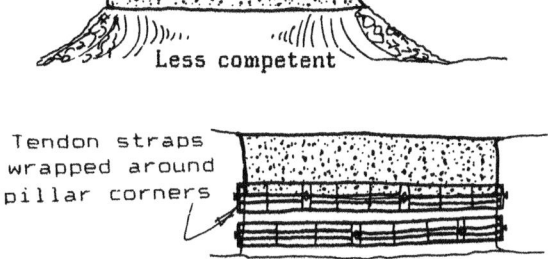

Fig.11 Confining pillar deterioration.

6) Confinement of Floor Lift (Fig. 12).

Tendon straps can be of great confining benefit in areas of high induced stresses resulting in floor-lift damage. The key to success here (as in all cases) is to **install the support early (preferably soon after development because it is difficult to rip up tracks/floors, etc., later).** The dominant direction of straps should be such that they are parallel to the longest exposed length of the expected rock blocks. An overlapped chequer-board pattern, where common rockbolts secure overlaps and crossovers, is the most effective. The straps should be secured allowing 150mm space between strap and rock. The longest practical rockbolts should also be used or tensioned cablebolts in extreme cases. Concrete is then poured and vibrated to get good all-round load distributing contact.

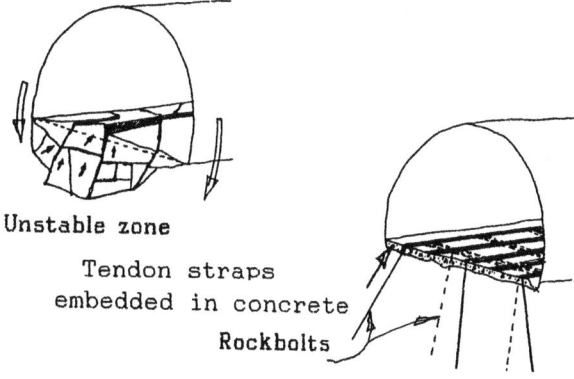

Fig. 12 Controlling of floor-lift.

7) Stabilisation of large spans at crossovers, turnouts and Service Installations (Fig. 13).

When used in a chequered pattern and secured with long rockbolts or suitable cablebolts at overlaps, Tendon straps aid the stability of large spans by confining the movement of rock blocks thereby minimising high arch type failures. This system is particularly effective across zones of different competency. Analysis of joint orientations and conditions are essential here to optimise the rockbolt or cable anchor configurations and lengths. Further surface restraint from pre or post-shotcreting ensures an integral confining support system.

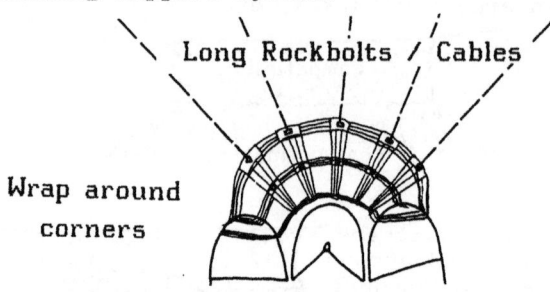

Fig. 13 Stabilising large spans/ turnouts.

8) Prevention of scaling from Ventilation passes and reinforcing of shafts (Fig.14).

Where ventilation passes are intersected by potentially unstable blocky or sheared zones, the use of overlapped Tendon straps installed across the zone and anchored into "solid" ground, reduces scaling and minimises the need for access/lashing in live MRA systems. The restraint can be augmented by weld-mesh or, preferably, shotcrete to create a more active liner. This system is useful where there could be shear movements along the zone (say in a footwall shear); the Tendon straps will deform but still retain confinement on the entrapped rock mass. The Tendon strap system can be used in shafts, serviceways and, theoretically, in orepasses or sand and stone passes. However, unless the latter passes are finally tubbed with concrete and repaired as the concrete wears, the steelwork may deteriorate from abrasion and impact damage.

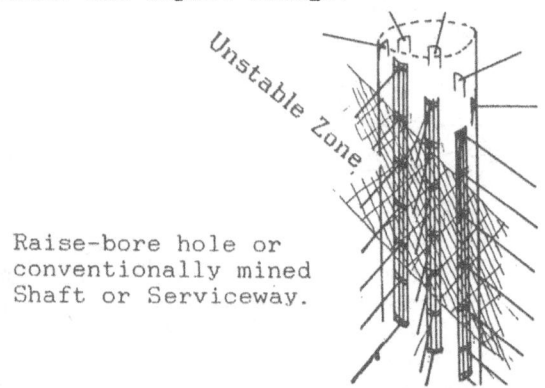

Fig. 14 Support within Service Shafts.

OTHER POTENTIAL APPLICATIONS

- Pit slope stability of blocky/problem areas in association with mesh.

- Tunnel support in conjunction with shotcrete.

- Application in coal mines for roof support, across blocky dykes, strike gulleys and centre gulleys as well as in pillar confinement.

- In conjunction with cable bolts in open stope stability; installed from hangingwall drives prior to stoping.

- In conjunction with tensioned cablebolting in stabilising slope faces in civil engineering applications.

COMPARISON WITH OTHER SUPPORT COMPONENTS: EXPERIMENTAL RESULTS

The relative confining potential of samples of 4 different support components were tested using a horizontally mounted Instron 250Kn capacity servo-hydraulic actuator ram system at the University of Zimbabwe's Civil Engineering Department (Plate 4).
The components chosen and their relative configurations are listed in Table II.

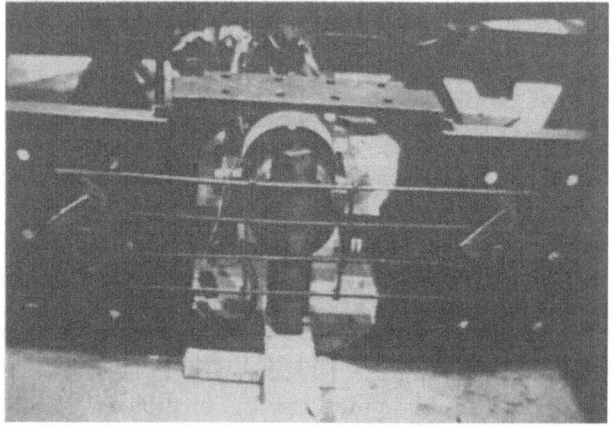

Plate 4. Set-up of Instron ram system.

Samples used: TABLE II

Sample	Configuration	Length width (centres)
Weld Mesh	75mm x 75mm x 3.5mm Diam.	950mm x 500mm
Diamond Mesh	75mm x 75mm x 4mm Diam.	950mm x 500mm
"W" strap	330mm x 3mm	950mm x 330mm
Tendon strap	4 x 10mm Diam.bars	950mm x 300mm

Machine set-up.

All samples were individually fitted to the rigid frame of the machine by 25mm mild-steel bolts and nuts. The ram system then impinged on the centre of the test length, simulating the effect of a key-block(s) being forced against the support unit between anchor points. The degree of deformation (mm) caused by the ram was simultaneously plotted against the confining force (Kn) exerted against the ram by the test pieces.

Comparative confinement potential

The results from the 4 types listed in **Table II** are graphically represented in **Fig.15** and described briefly below.

-Weld mesh.

The welds began to fail with very little deformation (50mm) and applied virtually no confining force (0.3Kn - 1.1Kn). The failure is typical of that seen where the welding is the weakest link in the system. This data was not even plotted.

-Diamond mesh.

Insignificant amounts of confinement were registered up to 300mm deformation. Thereafter, an increasing confinement was evident until strands began to fail at 510mm deformation. The maximum confinement achieved was 37Kn.

A reason for the large deformation prior to exerting confinement is explained by the sample changing its configuration. As the ram actuator travelled, the mesh strip pulled together as a narrow band and only began loading when the strands were almost straightened. In practice, this will not occur because of spaced multiple anchor points over the mesh. However, the maximum load data is probably realistic for that area of mesh and confirms the high degree of "swell" of broken material that can be contained by this mesh underground.
This system is very passive and does not prevent the generation of broken material.

-"W" strap.

There was a constantly increasing load with deformation up to 100 mm; thereafter the load increased more rapidly to 40Kn. A slight drop in load was noticed which coincided with a "shift" in the plate on the frame; the load climbed to 60Kn and dropped again. The plot was then very erratic and it became more evident that the bolt was pulling through the plate. The maximum confining load registered, before the bolt finally pulled through the end of the plate, was 69Kn. The results confirm underground observations that the confinement potential is limited by the plate thickness; numerous cases of "pulled-through" straps are seen in deforming ground conditions **(Plate 5)**.

-Tendon strap.

The confinement up to 100mm deformation was less than that for the "W" strap because the two ´\/´ troughs in the "W" strap give a minor beam effect. However, the confining capacity of the tendon strap accelerated more rapidly thereafter. A relatively steady deformation under load situation pertained right up to 147Kn when the one tendon failed in a conventional "necking" pattern. The three remaining still carried a residual load but the tests were considered terminated after the first failure.

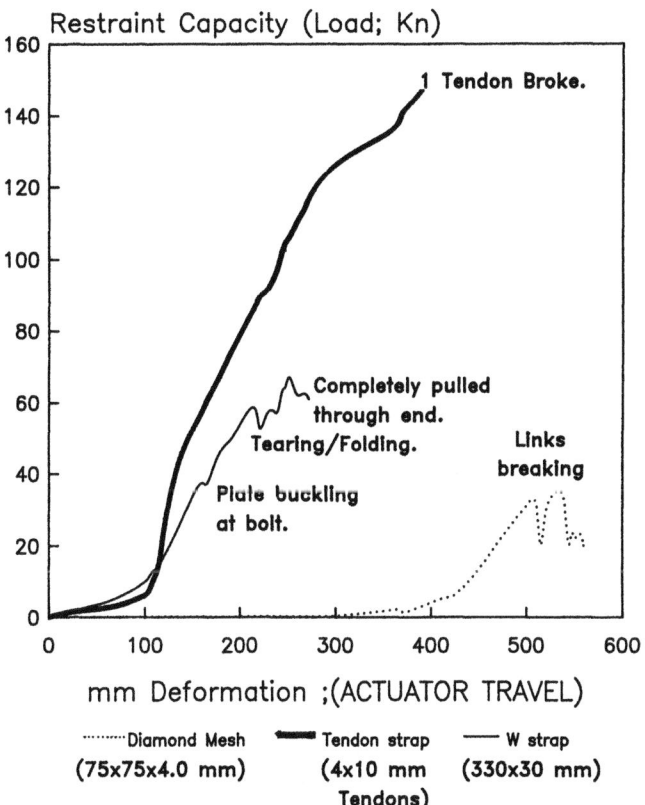

Restraint Capacity (Load; Kn)

mm Deformation ;(ACTUATOR TRAVEL)

········Diamond Mesh	▬▬ Tendon strap	── W strap
(75x75x4.0 mm)	(4x10 mm Tendons)	(330x30 mm)

Fig. 15 Comparison of confinement.

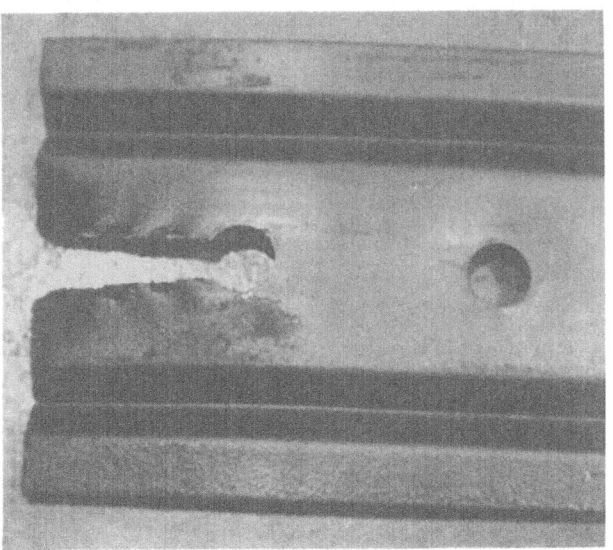

Plate 5. Typical tearing failure of"W" strap 3mm thick.

Hence, on the basis of underground experience and confirmatory laboratory-scale testwork, the confining potential exhibited by the ´standard´ Tendon strap was far in excess of that shown by the sample "W" straps or 75mm x 75mm x 4mm diameter diamond mesh.

Practical implications

From the results indicated in **Fig.15**, the Tendon straps can withstand a large imposed deformation but there is obviously a finite limit that can be tolerated before the units either begin to fail or shear the anchor bolt. In practice, no shearing of 16mm rebar anchor bolts has been seen at either Shabanie, King or reported from Shangani or Epoch mines. In the ground conditions which pertain at Shabanie and King, the deformation is held well within these limits by the straps; the latter "tighten-up" the system which would have deteriorated further in previous non-strap based support options.

Another factor which assists the overall situation is the yield exhibited by the rockbolt anchors. The bolts debond close to the collar and this allows the entire contained surface to move towards the void; this uses the bolts to their tensile capacity and reduces the shear effects on the heads.

INTEGRATION WITH SHOTCRETE.

Apart from being able to install Tendon straps on top of shotcrete to act as external reinforcing, a more integral system of shotcreting over tendon straps can often be more practical. This also negates any damage from LHD´s and assists in high corrosion-risk areas. Both Shabanie and King mines employ shotcrete over Tendon straps in specified areas and minimal problems of shadow/voids are seen. The only areas of minor void formation are at wide overlaps of mesh; even this is not considered significant in the overall confinement context. An example of satisfactory flow of shotcrete through mesh plus straps is shown in **Plate 6** where the arrow indicates advance of the shotcrete layer.

Plate 6. Shotcrete over mesh and Tendon straps; minimal voids generated.

It is significant to note that the tendon straps do not necessarily prevent cracking of shotcrete but effectively minimise the deformation that occurs subsequently. Mesh or steel fibre reinforced shotcrete fails in tension with very little applied deformation and, as borne out by laboratory work [8], the peak load capacities are quickly reached. Thereafter the continued application of load results in rapid deterioration from failure of the enclosed mesh. This is where the application of external tendon straps to shotcrete is beneficial because the higher capacity of the straps takes over and holds the cracked shotcrete together.

CHOICE OF SUPPORT TECHNIQUE: MRMR vs COST.

The degree of deformation seen in development is directly related to the insitu rock mass rating (RMR) and anticipated mining rock mass rating (MRMR) [6]. The adjustments of RMR to MRMR are essential to ensure adequate support pressures in the production life of development concerned.

The application of Tendon straps enhances the reinforceability of relatively competent RMR class 2B to 3B ground and also assists the tensile strength of a shotcrete layer in poorer class 4A to 5A ground. With a decrease in MRMR, plastic deformation becomes more dominant and pinned, reinforced linings are necessary. The effective deep anchorage of bolts/cables securing heavier Tendon straps over mesh-reinforced shotcrete is considered a viable alternative to massive reinforced concrete linings in some deforming conditions.

At Shabanie mine, but more particularly at King mine, there has been a reduction in use of steel arches set in massive concrete in favour of tendon-strapped reinforced shotcrete.

Tendon straps combined with galvanised mesh have been used very successfully at Shabanie for reinforcing and stabilising areas that would previously have been routinely shotcreted had the equipment and labour resources been available.

Drawpoint brow areas have also been successfully supported with Tendon straps and shotcrete in ground that was 3A-3B RMR; this would have previously been TH arched with concrete. These changes to enhanced rock reinforcement have resulted in more cost-effective support with an increase in ease and speed of installation.

The various support components used within the Group are listed in **Table III** together with their cost of installation. The distribution and combination of the components, as related to RMR and MRMR and recommended by the Rock Mechanics Department, are shown in **Table IV**.

Rope lacing ´r´ and massive reinforced concrete linings ´L´ are not part of the recommendations but are included sequentially for comparison. The cost per linear metre for each of the combinations is shown in **Table V**.

Table III.

SUPPORT CODE	Description of components	Estimated Costs; (30/11/90) Z$/linear metre ‡‡
a	Local rockbolts to pin joint intersections.	95
b	3.0m long rebar bolts plus washers @ 1.0m x 0.6m	165.4
c1	b + Tendon straps (+ 0.5m friction tubes) @ 1.2m	256.2
c2	b + Tendon straps (+ 0.5m friction tubes) @ 0.6m	512.4
d1	b + mesh reinforced shotcrete (150mm thick; 30Mpa)	636
d2	d1 + post-bolted with 3.0m bolts @ 1.0m x 0.6m spacing	801.4
x	75x75x4mm diameter galvanised diamond (chain) mesh	160
e1	d1 + Tendon straps @ 1.2m external or shotcreted in.	892.2
e2	d1 + Tendon straps @ 0.6m external or shotcreted in.	1148.4
f	d1 + Heavier Tendon straps @ 0.6m + cable bolting.	1303.5
h	Spiling bolts; 4m long @ 0.3m spacing	137.3
L	Reinforced massive concrete lining	4200
p	b + Yielding steel arches set in concrete/shotcrete.	3715.4
r	Rockbolts @1m with tensioned rope lacing over mesh.	675.4
t	Avoid development if possible; otherwise = L.	

(‡‡ Assume a 3.2m x 3.2m end; STORES plus LABOUR cost ONLY.)

Table IV.

SUPPORT TECHNIQUES: GEOMECHANICS RELATED.
(Modified after D.H.Laubscher[6])

MRMR ¦	RMR- 1A/B	2A	2B	3A	3B	4A	4B	5A	5B
1A/B									
2A									
2B	a								
3A	b	a	a	a					
3B	b	b/ c1	b/ c1	b/ c1	c1/ c2				
4A	c1+x	c1+x /c2	c1+x /d1	c1+x /d2	c2+x /d2	c2+x			
4B	c2+x /r	c2+x /r	c2+x /e1	e1 /e2	e2 /f	e2 /f	f /p		
5A			e1 /e2	e2 /f	f /p	h+f /p	h+f /L	h+f /L	t
5B						h+f /p	h+f /L	t	t

Table V.

COST (z$) OF SUPPORT TECHNIQUES (AS AT 30/11/90)
(Insitu rock mass rating)

MRMR ¦	RMR- 1A/B	2A	2B	3A	3B	4A	4B	5A	5B
1A/B									
2A									
2B	95								
3A	165	95	95	95					
3B	165 256	165 256	165 256	165 512	256				
4A	416 512	416 636	416 801	416 801	672	672			
4B	672 675	672 675	672 892	892 1148	1148 1303	1148 1303	1440 3715		
5A			892 1148	1148 1303	1303 3715	1440 3715	1440 4200	1440 4200	?
5B						1440 3715	1440 4200	?	?

INSTALLATION CAPACITIES.

Shabanie and King mines do not use drilling jumbos or mobile shotcrete or scissor-jack rigs, therefore the degree of support mechanisation is limited. Hence the installation capacities quoted **below** relate to the use of GD bolt drilling, Picolla/Minicrete shotcrete machines and Spedel grout pumping in LHD layout situations. Materials handling to site is effected by transporter LHDs or via sand and stone passes for shotcrete materials.

One of the advantages of the Tendon strapping system is that a large area of immediately effective rock reinforcing can be installed relatively quickly and with minimal equipment and labour.

This has also proved to be an effective repair technique at all mines where currently used. Accelerating major deformation was effectively terminated in the highly stressed slot/turnouts ahead of an advancing cave in block WC/308 at King. Many kilometres of progressively unravelling development have been successfully stabilised at Shabanie. In all cases, the speed and effectiveness of installation have been the key benefits.

The average installation capacities at Shabanie and King mines, and from which the cost estimates were based, are summarised in **Table VI** below.

Table VI.

INSTALLATION CAPACITY: (Numbers of items that can be installed COMPLETE per 8-hr SHIFT).

SUPPORT ITEMS	Installed No.	units	1-10 Men	11 + Men	items per man-shift
Rockbolts 1.8m (initial support)	60	ea	6	1	8.6
Rockbolts 3.0m (initial support)	40	ea	6		6.7
Spiling bolts; 4.0m x 0.3m spacing.	25	ea	6		4.2
Reinforced Shotcrete; Linear metre.	1	lin m	10		0.1
Tendon straps @ 1.2m +F/T +3m bolt	12	set	5		2.4
Tendon straps @ 0.6m +F/T +3m bolt	12	set	5		2.4
Mesh+Tendon straps @1.2m+F/T+3mR/b	10	set	7		1.4
Mesh+Tendon straps @0.6m+F/T+3mR/b	10	set	7		1.4
TH arches (0.5m spacing; = 1 lin.m)	2	set	12	1	0.2
with floor channels;+					
'U' pins for TH arches;+					
bird mesh shuttering;+					
shotcrete cover.					
floor bolts	30	ea	6		5.0
floor straps +3m bolts	10	set	7		1.4
LHD bricks	17 /metre	272	ea	8	34.0

F/T = Friction tube;   (ASSUMES DEVELOPMENT = 3.2m x 3.2m)
R/B = Rockbolt

SUMMARY

Tendon straps are a support system which have been shown to be effective in numerous mining stability situations and also have potential in other application areas as yet untested.

In the significant deformation situations at both Shabanie and King mines, the confinement given by Tendon straps has been sufficient to terminate the damage and the rock mass appears to stabilise and reconsolidate under the dynamic surface restraint.

The ease and speed of Tendon strap installation is of great benefit and, **depending on the rock MRMR, their use as rock reinforcement goes a long way to substitute for or can significantly enhance the performance of shotcrete linings.**

Tendon straps can be installed with various industry-standard, solid or strand steel rock anchors, to produce a cost-effective and safe support system.

ACKNOWLEDGEMENTS

The author gratefully acknowledges the following: Mr. C.B. Parshotham for the underground experimentation monitoring; Dr. M. Mansell from the University of Zimbabwe Civil Engineering Department for laboratory assistance, and the Directors of African Associated Mines (Pvt) Ltd. for permission to publish this paper.

REFERENCES

1. **Laubscher, D.H.** Design aspects and effectiveness of support systems in different mining conditions. Trans Instn. Min. Metall .(Sect.A), vol.93, Apr.1984

2. **Wilson, A.D.** Geomechanics Classification; Practical applications, Assoc.of Mine Managers of Zimbabwe (1985).

3. **Fomison, D.W.** Mining Operations at Shabanie and Mashaba Mines. Proc. Conf. Mining and Metallurgical Operations in Zimbabwe, Vol. II, 1982, pp236-264.

4. **Marano, G. and Everitt, A.P.** Selection of mining method and equipment for Block 58, Shabanie Mine, Zimbabwe, Proc. Conf. African Mining. 1987, pp 229-238.

5. **Buchanan, G.F.,Whewell, B.W.,DeKock, H.C.G., Parshotham, C.B.** The false footwall mechanised caving method, King section, Gaths mine, Zimbabwe. Proc.Conf. African Mining. 1987. pp 53-62.

6. **Laubscher, D.H.** A geomechanics system for the rating of rock mass in mine design. J. S. Afr.Inst. Min. Metall., vol.90, No.10, Oct.1990. pp.257-273

7. **Brumleve, C.B.** Rock reinforcement of a caving block in variable ground conditions, King Mine, Zimbabwe. Proc.Conf. African Mining, 1987,pp.31-46.

8. **Kirsten, H.A.D. and Labrum, P.R.** The equivalence of fibre and mesh reinforcement in shotcrete used in tunnel-support systems. J. S. Afr. Inst. Min. Metall., vol. 90, No.7. Jul. 1990. pp.153-171.

Artisanal mining in the Lake Victoria goldfields, Tanzania

J. H. Bills
M. P. Martineau
J. G. Park
SAMAX Ltd, London, England

Summary

Tanzania currently produces an estimated 8-15 tonnes of gold per annum from artisanal operations. Gold rushes reminiscent of those recorded from the earliest development of the USA or Australian goldfields are a regular occurrence. Totally new discoveries are being made yearly, at least one of which is already the basis for a major mine evaluation. Mining communities of up to 25,000 persons can be established within the space of a few weeks and show a high degree of internal organisation. After a decade of informal operation during which recorded production and tax receipts have been minimal, a gold buying programme has been instituted and gold mining licence agreements have been signed with commercial companies. Economic, safety and environmental concerns provide a pressing need to ease this transformation to commercial mining. Current examples of artisanal operations are described as a historical record of a transient phase similar to that preceding commercial development in Western Australia or Western USA.

General

According to published and other estimates Tanzania produces between eight and fifteen tonnes of gold per year; production on this scale has probably existed for the last twelve years. With the exception of the Buck Reef Mine operated by the State Mining Corporation, which produces less than 100 kilos of gold per annum, all production is small scale from artisanal workings.

Until early 1990, when the Bank of Tanzania initiated a gold buying programme, essentially all production was traded unofficially and most smuggled from the country. The annual loss of foreign currency earnings has been estimated by the Tanzanian Business Times (7 December 1990) to be 172 million US dollars or about half of Tanzania's current annual foreign exchange earnings and almost triple the value of coffee exports, the largest official foreign exchange earner. The gold mining has supported a large and well organised, but unofficial, economy not unlike that found in the gold rush camps as recorded photographically and in written records of Western Australia. Directly and indirectly small-scale gold mining in Tanzania provides employment for perhaps 300,000 persons and has operated at about 300 sites. A major portion of the foreign currency earnings are recycled by the buyers through the importation of consumer goods.

Recently gold production as observed by SAMAX geologists working in the goldfields appears to be in decline. They believe that this arises from a number of factors notably the exhaustion of the shallow, softer, higher grade ores from prior discoveries, from water problems and lack of air in deeper workings, from a move towards gemstone fields and from a decrease in the rate of new discoveries.

The intense phase of the gold rushes in Australia typically lasted less than ten years with individual fields booming for only a few months to perhaps three years before the miners drifted to newer and reportedly bonanza discoveries. The same pattern appears to be followed in Tanzania.

The Start of the Boom

Commercial mining, which was never on a major scale, declined rapidly before and immediately following independence and ceased entirely in 1972. Diamond mining also started to decline and by 1976 a large number of people with some skills in loaming, in gold prospecting and in underground mining methods were unemployed. In 1976 the first major new discovery and boom occurred at Bulyanhulu in the Southern part of the Lake Victoria Goldfields. This was a totally new discovery unrecorded by the excellent geological maps of the area or by previous loaming by the Geological Survey, and occurring in a deeply laterized scrubland plain. By any standards it was a major deposit as subsequent commercial exploration has demonstrated a resource exceeding three million tonnes assaying 14gm gold per tonne open along strike and at depth in a single vein over 1.6km long and up to 4 metres wide. Within a year 50,000 people were reported to be working the reef or providing support services to the miners. Production is unknown but at its peak and based on current rates per man was probably in the range of five to ten kilos of gold per day. The cycle at Bulyanhulu was never completed because the site was cleared by the State and passed to the State Mining parastatal STAMICO in 1978 and hence to International mining interests. Nonetheless illegal mining continues to the present day surreptitiously, at night and on an unquantifiable scale.

The Present Situation

Fourteen years later totally new discoveries on the laterized plains continue to be made through the simple expedients of

dollying quartz and loaming. SAMAX is aware of six such new discoveries made during 1990 but the real number is probably far higher. At least three of these stimulated population explosions in excess of 10,000 people.

The history of one such 1990 discovery, Mwagi Magi, illustrates just such a boom or "Kuhira". Gold was discovered on 9th May 1990 by a cattle herder who observed gold bearing quartz float three kilometres from a road and 30km from the nearest old mine. On 25 May a claim, nominally 1,500ft by 600ft, was pegged giving legal title to a syndicate of small miners to organise the mining and to establish the village. Only Two other claims were subsequently pegged, yet by early July, the village had grown to a town supporting seven to ten thousand people and was serviced by regular landrover taxis and by regular buses. Queues of unofficial buyers in new station wagons clogged the road and production from perhaps a two hectare area was reported to equal four to five kilos per day extracted from gold bearing quartz clasts occurring at the base of the 3-6m thick laterite profile. A clean, thriving regularly laid out townsite of wattle and grass huts had been created with all services available. The first major collapse of undermined laterite had already occurred resulting in the first deaths. Shafts were already being sunk to explore the quartz veins in the soft volcanic and ultramafic host rocks. Squareset shafts were being skilfully and continuously timbered to a depth of 20 metres from locally cut round hardwood with resulting nearby deforestation.

By mid July however, the population had begun decline and was then perhaps 5,000; the basal gravels were reaching exhaustion; and most production was from 1-2 centimetre wide quartz veins in bedrock. Many people had migrated to a new discovery, Lugunga, 5km to the North of Mwagi Magi; but the decline was principally a result of the exhaustion of easily mined shallow quartz rubble and mining difficulties in the crumbly rock around the deep shafts. By mid-November there were 107 active shafts into bedrock and the population had shrunk still further to a hard core of 3-4,000 miners. Production had reportedly declined to a kilo a week. On past evidence mining activity will continue to decline as water logging and poor ventilation in the shafts makes production more and more difficult.

Completion of the life cycle may be deduced from the history of the Nyakafura deposit, situated about 4km to the west of Mwagi Magi and which was discovered in 1980. Nyakafura had, at its peak, a population of up to 25,000 miners and supporters working on a 150 to 200 metre long 1 metre wide quartz vein and associated rubble beds. Local enrichment of free gold at the water table produced bonanza hand-sorted grades often exceeding 300g/t. Although nominally abandoned after two years because of uncontrollable ingress of water at a 40m depth it is probable that in fact the grade declined to uneconomic levels below the enrichment zone. This interpretation is supported by the history of Nzega Ndogo, 130km South East of Nyakafura a similar vein discovery with high primary grades which occurs within a SAMAX prospecting licence. Here production continues after three years to a depth of over 60 metres below the water table solely on the basis of baling without either mechanised equipment or pumps.

Although vein deposits are the most common basis for the gold rushes, production also comes from stockwork prospects. At SAMAX's Mwaluzilwa project a 3kms long, 80m wide shear zone hosts a network of en-echelon, vertical, oxidized sulphidic shear and sub-horizontal ladder veins in intermediate volcanoclastics. High grading of centimetre to metre wide lenses has continued with a workforce of about 1000 people for a period of over three years and shows no signs of abatement as new surface discoveries continue to be made along strike and within the zone.

Throughout the gold fields production methods are simple and labour intensive; only rarely are mechanized equipment such as compressors and jack-hammers or water pumps seen on the mining areas. These are invariably in a poor state of repair and lack spares. Explosives are even rarer.

Claim Organisation and Operation

Gold claims are pegged, and after a lengthy procedure and in accordance with the Mining Act, registered in the name of a single owner or partnership.

Occasionally a committee of a village will appoint an agent to apply for claims on their behalf. The claims are administered by a committee and secretary who maintain detailed records of visitors, people living and mining on the claim and gold production (official or unofficial). Monthly returns listing labour, costs, employees and production are required to be made but in these remote areas and given the problems of control, returns until recently, rarely reached officials. In consequence claims tended to be renewed without adequate documentation.

Individual pit owners sponsor teams of 4-6 miners to carry out shaft sinking and extraction of gold ore. Shafts are sunk at a rate of 2-3 metres per day. Waste rock and ore is hoisted to the surface in sacks or woven bark baskets on wooden windlasses or 'rolas' fashioned from locally cut green trees, using home made plaited sisal or bark ropes.

Mining is entirely by hand using outworn pickaxes, sledge hammers, chisels and shovels. Quartz reef is broken in the shaft and further cobbed and hand sorted at pit head. At some localities firesetting is used routinely to break quartz reefs. Undercutting at the base of the laterite profile to extract quartz rubble is a regular cause of collapse.

The grade or ore is measured in fractions of a 'tola' (11.65g) and in chapas (2g) per "Karai" (a metal dish with a capacity of 10-15 kilos of broken stones). The cost of processing are such that ore taken for crushing must be expected to average half a chapa per karai which roughly translates to 40g/t. Dollying and panning of a small handful of picked ore at the shaft head is used to monitor this. In practice SAMAX sampling and observation indicates that many goldfields operate at an overall loss - a few winners and many losers - and that on average the treated grade may be closer to 25g/t. Mining costs are such that in soft rock high grade stringers not much more than a centimetre wide may be selectively mined to a depth of 30m. In consequence it is on occasion possible for the artisanal miners to highgrade deposits that would be too low in overall grade for commercial companies to consider, whilst a vein grading a uniform fifteen grams - if such exists - could not be treated. Reports of average vein grades in excess of 90g/t, as at Matinge in the southwest of the goldfield, must be treated with caution as such grades are usually the result of extensive cobbing and hand sorting.

Claim holders normally sublease the actual workings to "pit owners". The "pit owners" control a few metres of strike length and sponsor teams of four to six miners. The "pit owners" may either work pits at the claimholder's expense in return for "shift" of their own (which may be one to three days per week) or they work at their own expense and for their own benefit and grant to the claimholder the right to work their pit on a regular basis (typically one shift per week). The miners themselves receive a cut of the ore mined. Negotiated rates vary widely between deposits, and probably within them, but on one rich field where the claim owner subleased the "claimholder" received 30% of the mined crude ore - selected from a group of sacks - whilst the pit owner and miners received 70%. The bagged crude ore, at this point, may either be auctioned to "buyers of stone" or treated individually by the "pit owner" and the "claim holder".

Equipment to crush, grind and recover gold are typically rented, and labour hired, to carry out the comminutions and gold recovery process. At the crushing stage the gold ore is broken by hammer or quartz cobble on a flat quartz block called a 'jiwe'; the rock chips being contained within a plaited rope bandana or 'ngata'. This is typically work for the old men. The minus 1cm product is passed to pounders for secondary crushing.

Pounding is one of the highest paid tasks on the goldfields and takes place with much rhythmic chanting in circles of four to fifteen workers. Within a major goldfield up to twenty such groups may be active producing a wave of sympathetic vibration throughout the camp. This secondary crushing is an interactive process: a 2-4 kilo charge of crushed product is pounded in hardwood mortars using motor-truck half shafts (called "Kinu no axel"). Oversize is recycled through the use of a 1mm sieve ("cheko-cheo") comprising a punched tin sheet typically operated by children. One labourer is expected to crush a karai of stone in a twelve hour shift and for this will receive about 500 shillings ($2.50).

The recovery of free gold from the sand or 'mchanga' is achieved first by sluicing and finally by amalgamating with mercury. Free gold is trapped on old sacking by washing the sand over a single stage, two metre long, wooden sluice or 'kabuta'; the sand is introduced on to the sluice through a wooden feed box with crude punch plate as its base situated at the top of the sluice. The sacking is next washed in a karai and panned to produce a heavy mineral concentrate. Mercury is then added to amalgamate with the free gold particles, excess mercury recovered by squeezing the amalgam in a piece of cloth and finally fumed over an open fire in a twist of silver cigarette paper within a circle of eager onlookers before being secreted away for final sale.

Tailings are often sold for regrinding and a second washing. Regrinding is commonly womens' work and almost everywhere this takes place indoors. The tailings sand, referred to as 'marudio' (to repeat) is reportedly washed as many as ten times. Sieved tailings are ground like flour on worn granite slabs using quartz or granite 'kusaga' (grinding stones). These gold recovery methods are crude and inefficient. Even after repeated recycling recoveries seldom exceed 75%.

Dealers - notoriously shy over the observance of their activities - congregate around the margins of the workings with scales and a practised eye for fraud, and the true density of gold and the typical fineness of each field. From the data available gold purchases average 63 to 80% purity according to field.

Camp Organisation

A claimholder has no permanent control over his workers who drift at will from field to field according to production rates and rumours of new discoveries and greater riches. In the initial development of a gold camp the miners build makeshift huts from wattle and grass thatch and a claimholder seeking to attract a stable labour force may erect rows of 'single mens' quarters. Entrepreneurs are quick to follow the miners and within weeks of startup camps show a clear segmentation ranging from "high street" stalls of smoked fish and ugali, tailors, second hand clothes dealers and bottle shops, through "hotels" to the more obvious brothels. The camp will also be serviced by bicycle taxis or four wheel drive vehicles, depending on access and richness, and will be supplied with water by the 'wachotmaji' who transport up to 160 litres of water a trip in plastic containers on bicycles from a far away as 15km. In addition to the claim holders committee CCM party officials invariably have a presence on the gold camp and law and order is maintained by the 'Sungi-sungu', an unofficial 'police force' who dispense summary punishments of beatings and 'running out town' to petty criminals and drunkards. Much of the crushing, grinding and washing of gold ore is done around the edge of the gold camp; only at the height of the dry season does washing move away to water sources. The smithies too are located at the edge of the camp, daily reforging tools in simple charcoal hearths blown with rotary bellows driven by the ubiquitous hand turned bicycle wheel and string belt.

Economics

Although fortunes are undoubtedly occasionally made, as in most historic gold rushes, the lot of the miner is not a happy one. Prior to the introduction of a gold buying programme by the Bank of Tanzania the dealer price for gold was about 1,600 Tsh per gram ($8). Competition for this foreign currency earner has raised the official bank price to 2,950 Tsh ($15) per gram and a dealer price which fluctuates between 2,500 and 3,500 Tsh per

gm ($13-18). Allowing for the average 73% gold content reported on the local radio and in the press, and using the official exchange rate, this would equate to price of $640 per ounce (Dec 1990). At the parallel market rate however this would equate closely to the free world price.

The Economy of Water

Few goldfields have access to a clean supply of water and many have no local access to water at all. Water for processing and personal use is brought in by bicycle at a retail costs of about 3Tsh per kilometre per 20 litre canister. It is thus one of the highest cost items in the operation; for instance at Mwaluzilwa this equates to 3Tsh per litre and at Matinge to 8Tsh per litre.

Health and Safety

The Bank of Tanzania is reported in the press to expect to purchase three tonnes of gold and sell three tonnes of mercury per annum to enable its recovery. Illegal dealers probably purchase about six tonnes of gold per year and may supply about two tonnes of mercury; all is evaporated. As yet there are no records of mercury levels in Lake Victoria fish or the expected adverse effects of breathing in mercury fumes.

A large body of anecdotal evidence from minesites and local health workers however indicates widespread pulmonary problems (silicosis) from the crushing of quartz by pounders and regrinders. Stories of the collapse of rock pounders within three months from poor diet and silicosis, are hopefully exaggerated.

Human excrement litters every area of the workings and this together with the limited water supplies from open water holes poses a constant and major danger of disease. Death from major collapses of the workings are widely reported. In the villages they talk of individual incidents killing up to 90, however there is no independent confirmation of such incidents. Deeper workings lack ventilation; following local custom miners string bark thongs to trees in a cone to "attract oxygen" to the shafts and on a more scientific basis lower chickens in baskets to test air before descending; bunches of fresh plucked grass are taken down to "purify the air". Nonetheless death from "gas", from collapse, and disease is common, unreported and unremarked.

Warnings of ukimwe (AIDS) abound and condoms are available from mine secretaries but in parallel with the spread through truckers plying the roads, the transitory lifestyle of a mining community and the migratory habits of the workers must carry its own long term cost.

Transition to Commercial Mining

The disparity between the objectives of a claimholder scraping a living by the highgrading of a shallow deposit and the Commercial Company seeking to bulk mine a large but low grade global deposit of which the high grade veins constitute only a minor (but essential) part has led to extreme difficulty in negotiation. This has not been aided by the clauses in the mining code forbidding the purchase of claims by non-Tanzanians. It is not the purpose of this paper to address whether the National Interest is served better through Commercial Mining with its overall higher recovery of gold and provision of foreign exchange or by indigenous mining with its higher local employment. What is evident is that the two modes are mutually exclusive within a single goldfield. The supply of mechanical equipment and explosives to small scale miners can be shown in several fields to exacerbate the safety and health problems; of itself this permits deeper mining but does not reduce waste. Transfer of title to a Commercial Company sterilizes a working area for several years during exploration and feasibility studies.

During 1990 the Government of Tanzania has introduced an Investment Code, a "one stop" Investment Promotion Centre", a gold buying programme yielding access to foreign exchange, a

tightening up and supervision of the application of Mining Act regulations and a programme of removal of illegal and unregistered miners. These actions are likely to alleviate many of the problems facing Commercial Companies, foreign or local, in taking prospects through the transition from "small miner" to commercial operation.

This paper is essentially historical, designed to provide a written record of what has probably been a transitory phase in the evolution of Tanzania's mining industry. The practices appear to be directly comparable to those of gold rushes around the world and no different in kind from those found in the most advanced mining countries, within or only slightly beyond living memory. Should the Tanzanian "boom" lead to a similar development of commercial mining, then the new discoveries of a generation of Tanzanian "frontiersmen" will amply repay the pains of their first development.

Metallurgy 4

Metallurgical circuit development and optimisation subsequent to commissioning of the tailings leach plant stage III at Nchanga Division of Zambia Consolidated Copper Mines Ltd, Zambia

S. Chowdhury
J. A. Mawer
P. Mukuka
Zambia Consolidated Copper Mines Ltd, Nchanga Division, Chingola, Zambia
D. V. Stone
Techpro Mining and Metallurgy, Ashford, Kent, England

SYNOPSIS

The stage III expansion to the Tailings Leach Plant (TLP) at Nchanga Division of Zambia Consolidated Copper Mines Limited (ZCCM), Zambia, was designed to treat 50,000 tpd solids consisting of a mixture of current arisings and reclaimed copper tailings. The plant was commissioned in 1986.

As a result of a continuing post-commissioning development programme, a number of areas within the metallurgical circuit were identified as constraints on plant throughput and performance.

Investigations were carried out with the objectives of improving metal recovery as well as optimising the consumption of solvent extraction reagents, acid and lime. Specific areas which were investigated included the preparation of slurry feed to the plant, modifications to the leach circuit, improvements to the solvent extraction plant and recovery of metal from plant residues.

Subsequent to these initial investigations and a review of the various metallurgical circuit options, commitment was made to a number of projects to enhance the performance of TLP Stage III, as well as other upstream operations. These projects are discussed in this paper.

By desliming a portion of the feed to the Concentrator to eliminate the very fine, submicron particles, the physical properties of the feed to the plant will be improved. The leaching circuit will be modified to incorporate a single stage leach, with improved control of acid addition. This allows the introduction of a raffinate leaching circuit, utilising acid which was previously being discarded. The dilute copper bearing leach liquor, generated by the raffinate leach circuit will be separately treated through a one-stage solvent extraction circuit. This change will be made possible by the conversion of the existing solvent extraction streams from an in series three extraction and two strip stage operation, to a series/parallel mode.

Due to the scale of the operations at TLP Stage III, these projects, which incur relatively low capital investment, show potential for substantial increases in revenue.

INTRODUCTION

Operations of Zambia Consolidated Copper Mines Limited (ZCCM) centre around five major production divisions located on the Zambian Copperbelt. Aspects of tailings treatment, at the major mineral beneficiation and hydrometallurgical complex located at the Nchanga Division's Chingola plant, form the basis of this paper.

Nchanga Tailings Leach Plant (TLP) Stage II, treating 900,000 tonnes per month consisting of current tailings, reclaimed tailings, low grade concentrates and leach residue, was commissioned in 1974[1] (Holmes et al). Commissioning of the Stage III expansion treating 1,500,000 tonnes per month of reclaimed and current tailings commenced in 1986[2] (Hampsheir).

During the start up of the Stage III plant, a number of areas of the flowsheet were identified as having potential for improvement. The paper by Mwenechanya et al[3], covered initial circuit modifications carried out by ZCCM after commissioning of the plant.

During 1988/89 ZCCM, in conjunction with Techpro Mining and Metallurgy (Techpro), a UK based Metallurgical and Mining Consultancy and Engineering Company, carried out a number of investigations and studies. These looked at additional areas of the metallurgical circuit where further improvements were desired. The outcome of these investigations was that ZCCM approved, for implementation, a number of TLP associated projects at Nchanga Division.

PLANT DESCRIPTION

Concentrator

A combination of open pit and underground mining methods are utilised at Nchanga Division, generating a complex run of mine ore consisting of oxide and sulphide copper and cobalt mineralisation.

Crushing and milling operations are conducted at two plants located adjacent to the major open pit and underground mines respectively.

The feed to the flotation plant consists of a combination of milled products and fine slimes fractions. Dependent on ore type, this feed is treated through either a cobalt circuit or a differential copper circuit. A general

overview of the circuit is given on the Concentrator block flow diagram in Fig.1. A more detailed description is provided in the article by Brooks and Fleming[4].

Products from the flotation plant consist of a cobalt concentrate, a copper smelter concentrate (predominantly acid insoluble), a high grade leach concentrate (mainly oxide mineralisation) and a low grade sulphide concentrate which is roasted prior to leaching at the High Grade Leach Plant.

The current final tailings, together with tailings material reclaimed from disused disposal dams, form the feed to the TLP Stage III Plant.

Tailings Leach Plant (TLP) Stage III

A detailed description of the TLP Stage III circuit has been given by Mwenechanya et al[3], however a summary plant description is provided below and is shown in Fig 2, the TLP Stage III block flow diagram.

Under normal operating conditions, current and reclaimed tailings, forming the main feed to TLP Stage III, are thickened separately, although the facility does exist to transfer material between the two preleach thickeners, which together have a total design capacity of 50,000 tpd of solids.

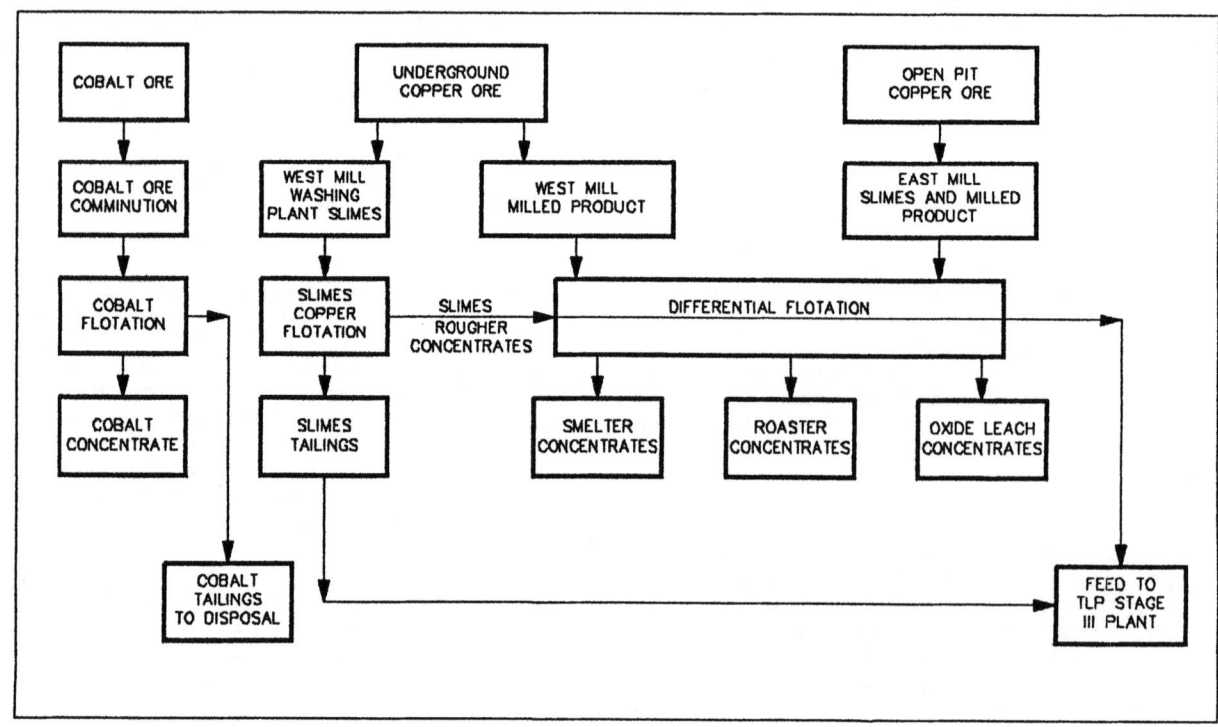

FIG.1 – NCHANGA CONCENTRATOR BLOCK FLOW DIAGRAM

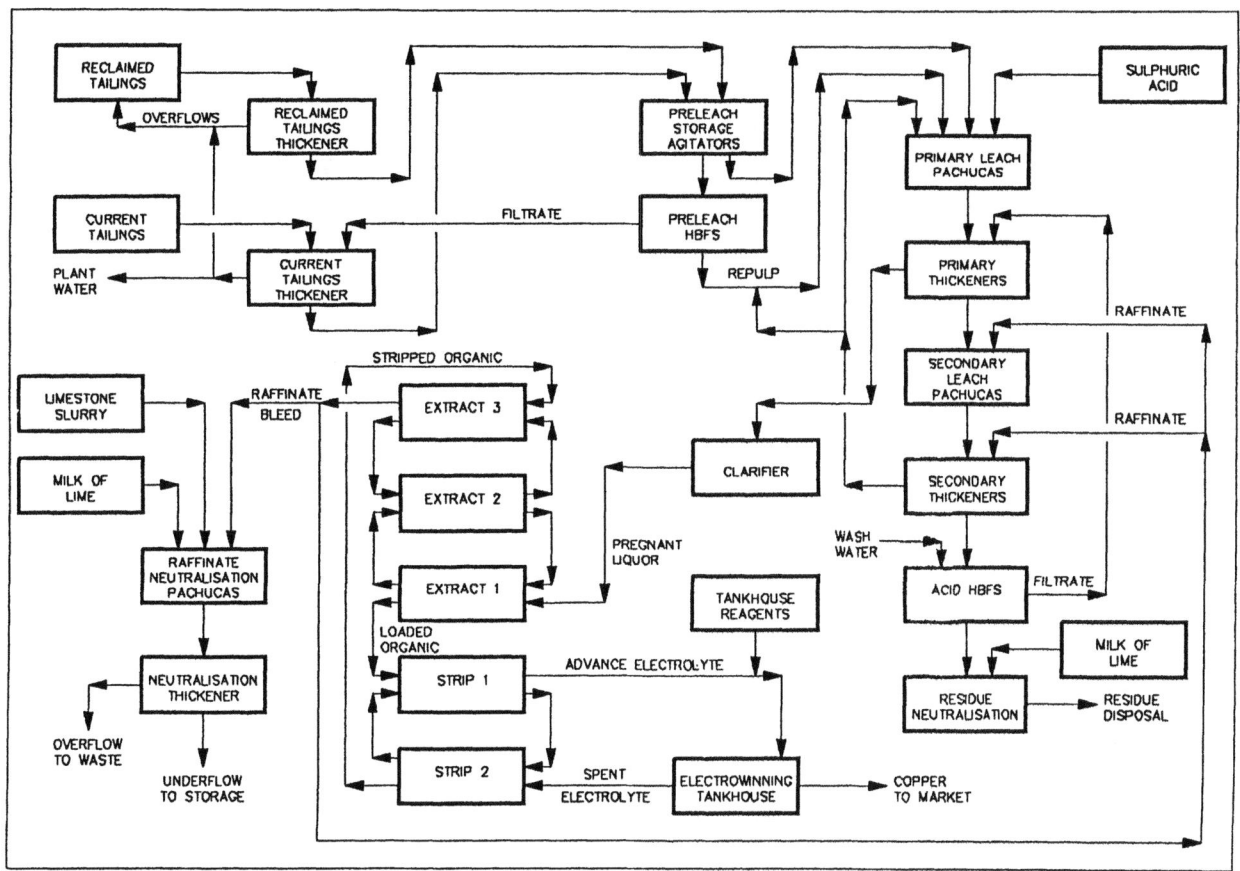

FIG.2 — TLP STAGE III BLOCK FLOW DIAGRAM

Preleach slurry is then further dewatered, by means of horizontal belt filters (HBF's), prior to contacting with leach liquor. Filtrate from the preleach HBF's is returned to the preleach thickeners, the overflows from which return to the plant water system.

Leaching operations are carried out in air agitated leach pachucas and to a lesser extent in the wash thickeners. The slurry pH is controlled by sulphuric acid addition to the primary leach pachucas with secondary leaching being carried out between the primary and secondary thickeners.

Washing of residue is conducted by means of acid HBF'S with filtrate returning to the leach circuit and filter cake being repulped with either neutralised raffinate or water.

Final plant residue is neutralised with milk of lime prior to disposal in a valley type tailings dam.

Pregnant liquor from the primary thickener overflow is clarified then fed to the solvent extraction circuit, where copper values are extracted through four parallel streams, each consisting of three extraction and two stripping stages in series.

Raffinate bleed, at a flowrate of approximately 1000m³/hr, is neutralised with a mixture of limerock slurry and milk of lime.

Thickened precipitates are collected into a mined out open pit with the neutral thickener overflow being used for acid HBF cake repulping or discarded.

FLOWSHEET DEVELOPMENT

A programme of circuit development by Nchanga Division, combined with testwork investigations carried out by ZCCM Technical Services Unit, has continued since plant commissioning. These investigations resulted in a number of supplementary projects being initiated under Techpro project management.

Some of the priority projects were constructed and commissioned in 1989/90.

Projects proposed have necessitated both the installation of more efficient equipment for particular operating duties and also modifications to improve utilisation of existing equipment.

These projects are discussed below:

Reclamation of Tailings Dam Material

Throughputs have been restricted in this area of operations due to screening constraints and also as a result of a high incidence of coarse material and vegetation arising in the reclaim preleach thickener.

It is planned to incorporate a more effective screening facility upstream of the main transfer pumpstation, in order to remove trash oversize material from the circuit.

Removal of Ultra-Fine Fraction From Current Concentrator Tailings

Various testwork programmes and studies have been undertaken by the ZCCM Technical Services Unit, to ascertain the suitability of a fine particle removal facility at the Concentrator and the most advantageous location for such a unit.

The conclusion reached from the testwork was that the submicron fraction should be removed from the washing plant slimes, arising from underground copper ore, prior to flotation. The initial conceptual design of the plant was carried out by ZCCM, with detailed design and construction management by Techpro.

Equipment installed includes a scalping trommel screen, two banks of cyclones, associated services and pumping systems.

The design criteria for the new plant are summarised in Table I, together with the indicated benefit in overall metal recovery.

It is anticipated that a number of additional benefits will result from installation of this plant, as indicated below:

- Improved preleach thickener performance at TLP III, due to removal of slow settling ultrafine fraction, resulting in higher reclamation rates, improved thickener utilisation and hence higher overall plant throughputs.

- Increased preleach HBF dewatering capability at TLP III and a resultant reduction in raffinate bleed required to maintain a solution balance. This leads to

	Design Parameters	Copper Balance tonnes/month
1. Plant Feed		
Total throughput, tpd solids	2000	--
Slurry density, kg/m³	1187	--
Percentage Passing 79.5 Micron, %	85	--
2. Cyclone Performance		
Mass split to overflow, wt%	10-15	
Cyclone d50, micron	6	
Copper loss to overflow, tpm	--	(88)
3. Flotation Performance		
Additional Acid Insoluble recovery, tpm	--	39
4. Preleach Thickening		
Copper gained by increased reclamation, tpm	--	300
Copper gained by improved transfer of reclaimed tailings to current tailings thickener, tpm	--	195
Percentage reduction in current tailings thickener area required per unit throughput, %	25	
5. Preleach Filtration		
Percentage reduction in preleach filter form time, %	20	--
Copper gained due to reduction in raffinate bleed, tpm	--	21
6. Raffinate neutralisation		
Retained sulphuric acid, tpm	846	
Reduction in lime consumption, tpm	423	
Copper balance, tpm		467

Table I - Slimes Treatment Plant Design Criteria

- reductions in copper and acid losses to final residues and hence reduced lime consumptions.

- Testwork results have indicated that recovery of acid insoluble copper, from the concentrator slimes, is improved on removal of the ultra fine fraction.

The improved operational aspects have to be offset by the loss of copper to the cyclone overflow, this is however outweighed by the benefits above.

Leach Circuit Modifications

Reference has previously been made to the loss of copper and especially acid, resulting from the requirement to bleed raffinate from the leach circuit to balance solution flows.

The limited ZCCM group production of sulphuric acid necessitated a review of measures to reduce consumption and fully utilise available acid. This applies especially to the TLP III plant as it is the highest consumer of sulphuric acid on the Copperbelt.

Following this review, a number of modifications were evaluated for the leach circuit at TLP III, incorporating various configurations of new and existing thickening and filtration equipment, necessary to utilise the raffinate bleed to achieve improved results.

A summary of the findings is given in Table II.

During 1989, TLP III operated for part of the year with an acid constraint. Investigations by Nchanga Division and Technical Services Unit led to the conclusion that the acid required to leach the last say, 5% of "readily leachable" copper, was in the order of 10 tonnes of acid per tonne of copper, as compared to 2.5 tonnes of acid for the material as a whole. The acid situation was such that a decision was taken by Nchanga Division, to shut down secondary leaching and thickening, to reduce consumption by gangue and accept a lower leach efficiency. However, by paying particular attention to the control of primary leach, these losses were reduced to an almost negligible level and the plant has been operating in this mode since that time.

This decision released leach pachucas and thickeners from the plant and enabled the circuit shown in Fig. 3 to be introduced, without the necessity of providing major new capital equipment.

It can be seen from this block flow diagram that the most significant circuit change will be the use of raffinate bleed from the main leach circuit for a raffinate leaching subcircuit.

The net result of these modifications will be a potential increase in copper production of about 15 000 tonnes per annum, assuming additional tonnage is treated through raffinate leach. Alternatively, it is anticipated that acid losses will reduce by approximately 100 tonnes per day, with a resultant reduction of lime consumption by 60 tonnes per day, if part of the TLP III design tonnage is routed through raffinate leach.

	Single Stage Thickener Recovery	Two Stage Thickener Recovery	Simple Filtration Recovery	Filtration Recovery with Thickener
Leach Efficiency, %	85	85	85	85
Wash Efficiency to SX, %	75.7	96.6	97.1	99.0
SX Efficiency, %	95	99	95	99
Acid Soluble Copper Recovery, %	61.1	82.2	78.8	84.3
Incremental Copper, tpd	31	42	40	43
Comparative Estimated Capital Cost	1.00	1.42	4.17	4.33

Table II - Comparison of Raffinate Leaching Circuit Configurations

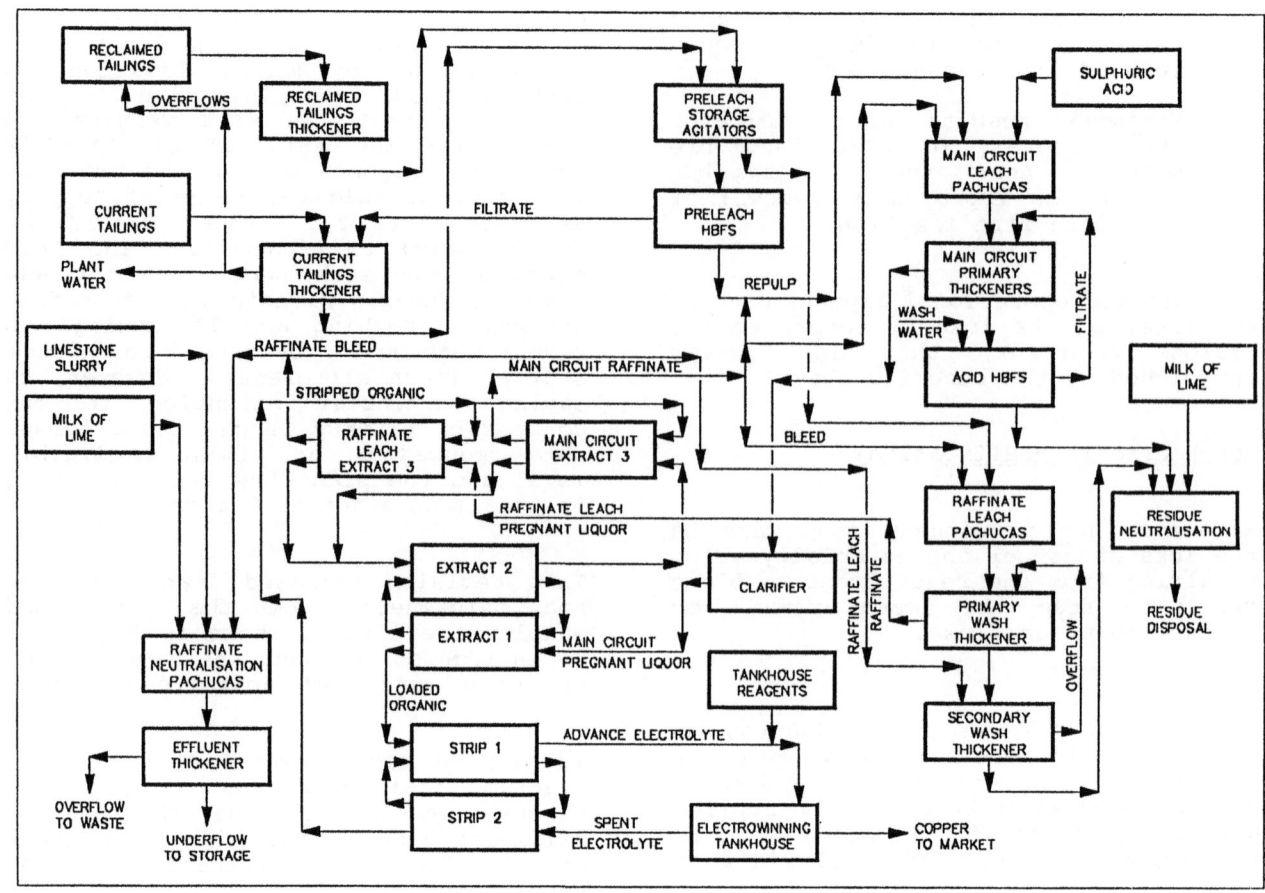

FIG.3 — TLP Ⅲ WITH RAFFINATE LEACHING BLOCK FLOW DIAGRAM

Presently, tonnage throughput of this new circuit is limited, by preleach thickening capacity, to 50 000 tpd which is the same as for TLP III, hence if this tonnage is being treated, the benefit gained by operating raffinate leaching is in acid conservation. To realise additional copper production over and above what the main leach circuit is designed for, will require additional preleach thickening capacity. This option will be considered, if the rate of reclamation of old tailings can be substantially increased, with improvement in reclamation capacity, by installing more reclamation modules and improved screening facility on the Tailings Dams.

Another positive aspect of operating with this dual circuit is that, if there was a constraint in treating material through the main circuit, say in filtration as a result of more filters being out of circuit than scheduled, then there would indeed be an increase in copper production directly attributable to raffinate leaching.

The possible benefits are therefore largely a matter of choice, to suit prevailing conditions.

The new circuit makes maximum use of existing equipment. As a consequence scarce foreign exchange requirements are minimised.

Solvent Extraction Plant

One of the requirements, of the introduction of the raffinate leaching circuit, has been the provision of a separate solvent extraction facility for the raffinate leach solution. This has been achieved by conversion of two conventional series configuration streams into the series/parallel configuration, as shown in Fig 3 and Fig 4.

Leaching, washing and solvent extraction plant design criteria, for the existing and proposed circuit configurations, are given in Table III.

PROJECT IMPLEMENTATION

Subsequent to definition of process requirements by ZCCM, Techpro were responsible for providing detailed engineering and project management services, in order to implement the above projects.

RAFFINATE LEACHING DESIGN CIRCUIT

1.	PLANT FEED	STAGE III DESIGN	MAIN CIRCUIT	RAFFINATE LEACH CIRCUIT	
	Total throughput (tpd)	50,000	41,000	9,000	
	Acid soluble copper (%)	0.76	0.8	0.55	(taken preferentially from reclaimed material which has known gangue acid consumption)
2.	LEACHING AND WASHING				
	Leach residence (hours)	3.9 (including	2.0	2.5	
	Leach efficiency (%)	86 secondary	85	85	
	Wash efficiency (%)	96 leach)	96	96	
3.	SOLVENT EXTRACTION				
	Pregnant liquor flow (m^3ph)	3200	2700	1556	
	Pregnant liquor copper ($kgpm^3$)	4.1	4.2	1.1	
	Raffinate copper ($kgpm^3$)	0.1	0.05	0.01	
4.	RAFFINATE BLEED				
	Flow ($m^3.p.h$)	914	778	778	
	Sulphuric Acid (t.p.d) to neutralisation	135	Nil	40	
5.	COPPER PRODUCTION (t.p.a)	112,000	98,000	15,000	

Table III - Stage III and Raffinate Leaching Operating Comparison

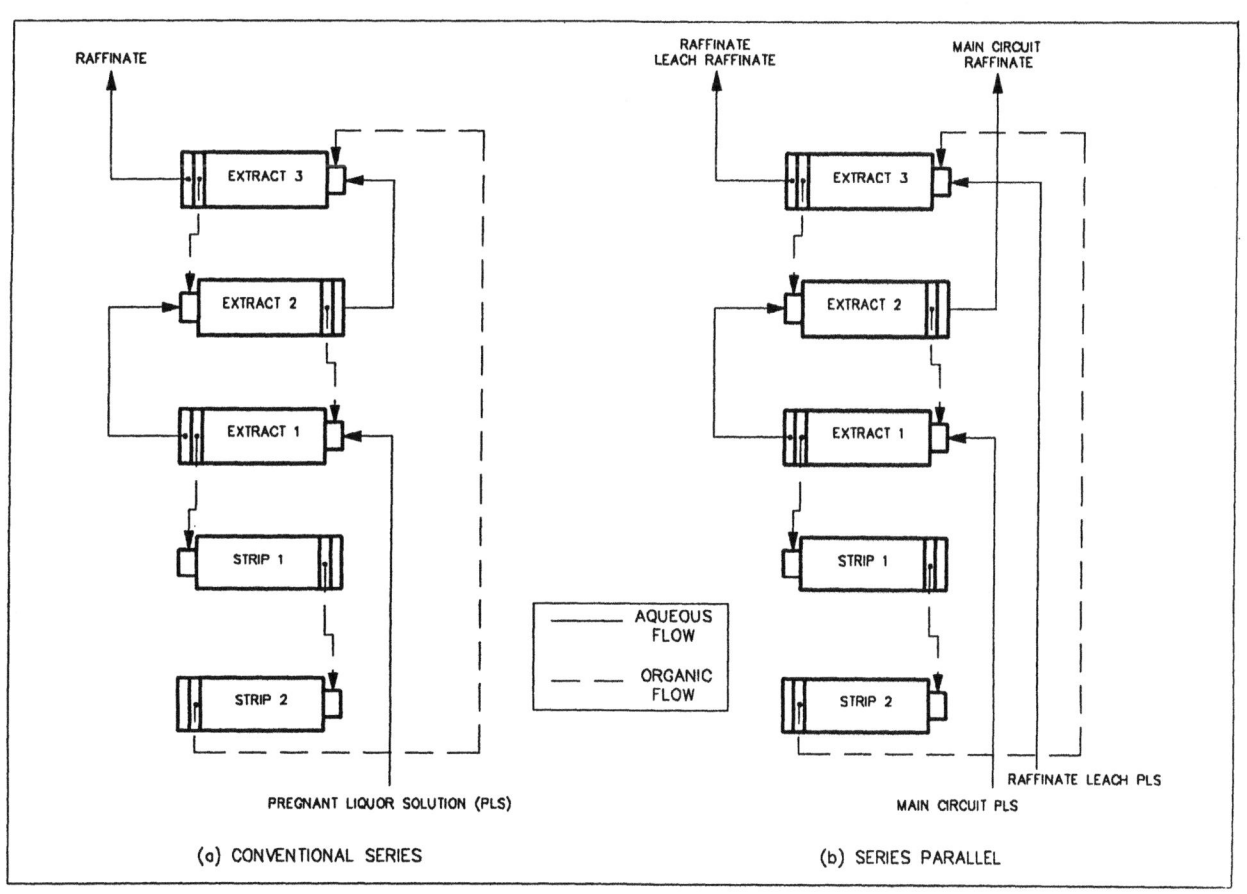

FIG.4 - CONVENTIONAL AND SERIES PARALLEL CONFIGURATIONS

Tailings Reclamation

Detailed engineering has been completed by Techpro, procurement progressed during 1990 with commissioning in March 1991.

Fine Particle Removal at Concentrator

Detailed engineering of this facility was completed by Techpro in March 1990 as described below.

West Mill washing plant slimes are routed, via existing storage tankage, to the new separation plant. The slurry passes through a scalping trommel screen with 2 mm aperture and is then pumped to either one of two banks containing 128 cyclones.

Each of these separation cyclones is fitted with a 11.0 mm diameter vortex finder and a 6.5 mm diameter underflow spigot.

The overflow material is collected and pumped, directly via a 1.5 km pipeline, to the residue neutralisation pachucas at the TLP Stage III. Production underflow is diluted with water, to the required density, prior to being pumped to the existing slimes flotation circuit at the Nchanga Concentrator West Mill.

The plant was constructed, under Techpro management and was incorporated into the existing circuit without production losses, the plant was commissioned in November 1990.

Raffinate Leaching

ZCCM chose the two stage thickening recovery route out of the options investigated for the Raffinate Leach circuit, mainly in view of the high potential incremental copper production indicated and relatively low capital (especially foreign exchange) requirement.

In view of the benefits envisaged, ZCCM decided that the project should proceed on a 'fast-track' basis and Techpro were appointed to manage the project in March 1990.

One of the prime objectives of the plant design was to retain circuit flexibility, such that the new circuit could be commissioned without interrupting the existing operations of TLP Stage III.

Implementation of the project construction and commissioning was also governed by the requirements of a high degree of interaction with the existing installations. Evidence of this being that it was necessary to install thirty six terminal points, these being connections between the main circuit and the Raffinate Leach circuit.

New pregnant leach solution and solvent extraction plant raffinate pumping systems were required and, in order to accommodate large variations in flowrates between the main circuit and Raffinate Leach Circuit, major modifications were carried out on thickener underflow pipework systems.

Slurry and solution pipework was rubber lined and 6 km of pipework were installed.

The project was initially planned to take 12 months from start to finish. The actual project implementation was that the new installation was constructed within this time period and commissioning of the new circuit commenced at the start of February 1991.

FUTURE PROJECTS

One of the main objectives of the Nchanga Division Metallurgical department in conjunction with the Technical Services Unit is the continued search for improved efficiency and increased production. Longer term feasibility studies and appraisals are currently being undertaken by Nchanga Division, to assess the economics of additional projects to enhance the copper production and reduce operating costs at Nchanga Division.

Consequently, a number of additional projects are in the conceptual design, feasibility or detailed design phase with ongoing input from both ZCCM Nchanga Division and Techpro.

Solvent Extraction Plant

Further improvement in operational control of this plant is being sought. An ongoing project is proceeding, with the objective of replacing the existing mixer/drive units with improved design. This will require further series/parallel conversions to maintain circuit flexibility during installation of the new drives. Reduction of solvent loss is also a target for the TLP Stage

III management and provides significant potential for cost savings.

Recovery of Sulphide Values Reporting to Plant Residue

Techpro have already undertaken a number of conceptual engineering studies, in order to assess a variety of options specified by Nchanga Division for the recovery of Sulphide (Acid Insoluble) Copper from the TLP Stage III residue. Recovery methods utilising separation and flotation circuits have been reviewed.

It is believed that this material has been liberated from the tailings, as a result of the leach process unlocking these values. The existing circuit has no facility to recover the sulphides, which consequently report to final residue.

Pilot scale metallurgical testwork, using conventional flotation cells, has shown that a total copper recovery of 10% at a grade of 20% is possible.

This will realise an additional 8 000 tpa of copper for the Division. The circuit would consist of desliming followed by two stages of flotation of the coarse fraction. Subsequent treatment would be to blend with smelter feed or roast and leach in existing facilities.

Also under investigation is the use of column flotation, which could result in a treatable product without the necessity to deslime the material prior to flotation.

CONCLUSION

Completion of the current programme of projects will improve production capabilities of both the Concentrator and Tailings Leach Plant and reduce operating costs related to acid and lime consumption.

The importance of taking a broad overview of interactive metallurgical circuits of this nature has been demonstrated.

It is also clear that it is necessary to continually assess the potential within operating plants for development and enhancement to optimise recoveries and reduce production costs.

Further improvements and developments are planned and will continue to enhance the operation of this major tailings leach facility.

ACKNOWLEDGEMENT

Thanks are due to the management of Zambia Consolidated Copper Mines Limited and Techpro Mining and Metallurgy, for permission to publish this paper.

REFERENCES

1. J.A. Holmes, L.N. Stewart, A.D. Deuchar, J.D. Parker, Int. Symp on Ext Metallurgy of Copper, J.C. Yannopoulos and J.C. Agarwal (Eds) Vol II, 905, AIME, New York 1976.

2. P.R. Hampsheir, Proceedings of the Institution of Mechanical Engineering, 200, 131, London (1976)

3. S.K. Mwenechanya, J.A. Mawer, S.Chowdhury, Proceedings of International Conference 'Copper 87', Santiago, Chile Vol 3, 255.

4. M.J. Brooks, I.T.R. Fleming, Mining Magazine, July 1989, 34-39.

New developments in X-ray fluorescence analysis of metallic and mining samples

S. Uhlig
Siemens AG, Dept. AUT V371, Karlsruhe, Germany

SYNOPSIS

Traditionally X-ray fluorescence analysis (XRF) is used for both qualitative element identification and quantitative determination. In recent years there has been an increase in the number of XRF users requiring software packages to run semi-quantitative analyses. The general requirement is to analyse a great number of elements within a short time and to obtain more or less reliable values on the composition of a totally unknown sample. Such semi-quantitative analytical programs are now available. What can the XRF user in the mining industry expect from such a program? What does "semi-quantitative" analysis mean? My contributution to this conference will give some answers to these questions based on the experiences in our application laboratories.

INTRODUCTION

In most fields of mining, it is a quite normal requirement for a very large number of samples to be analysed very quickly. Elemental analysis in geochemical exploration, process and quality control in mineral dressing or metallurgy are typical applications. X-ray fluorescence spectrometry (XRF) is a non-destructive analytical method for qualitative, semiquantitative, and quantitative determination of a sample's elemental composition. It can be used to identify all elements from boron to uranium, and to determine concentrations of these elements in metals, pressed powder samples, glass beads and fluids. Depending on the specific application, concentrations from 1 ppm up to 100 % can be determined.

X-RAY FLUORESCENCE ANALYSIS (XRF)

The fundamentals of XRF analysis will only be summarized. The sample to be analysed is irradiated by high-intensity X-rays. Modern laboratory X-ray fluorescence equipment uses an X-ray end-window tube as the source of primary radiation. The inner electrons of the elements in the sample will be activated by the X-ray beam in such a way that element-specific X-ray fluorescence radiation is emitted. This (secondary) radiation consisting of several wavelengths is dispersed into individual spectral lines by reflection from an analyzer crystal. The intensities of the fluorescence radiations are measured and compared with intensities of known standard samples. Based on the measured intensities, concentrations of elements or compounds are calculated using the latest analytical software packages. Interelement influences (matrix effects) including absorption and enhancement (secondary emission) are corrected applying easy-to-use software packages for evaluation of "fundamental" parameters such as mass absorption coefficients and fluorescence yields. The advantages of X-ray fluorescence analysis include multi-element capability, high precision, and short measuring times

SAMPLE PREPARATION

For the normal routine analysis of trace and major elements in geological samples (rocks, soils, etc.), sample preparation is quite simple and rapid: crushing, grinding (<< 0.05 mm) and pressing of powdered material to pellets for XRF analysis. Depending on the analytical demand, fused beads may be preferable for the analysis. Solid samples, for example metals, can be analysed directly after polishing the measuring surface. For analysing liquid samples, liquid cups or filter preparations can be used. With special enrichment methods, trace element analyses in the ppb range (e.g. metals in water samples) are also possible. For controlling the chemical composition of mineral concentrates, an easy and fast preparation method for XRF is to analyse loose powder samples in liquid cups.

X-RAY SPECTROMETER SYSTEM

The Siemens SRS 303 is a microprocessor-controlled sequential X-ray fluorescence spectrometer with an integrated 3.0 kW X-ray generator and X-ray tube with rhodium anode for universal XRF analysis (Fig. 1).

The use of large number sample changers (SRS 303 AS with magazine for 72 samples) with fully automatic hardware and software control, facilitates continuous analysis during night and week-end operation. A routine analysis program combines minimum operator input with maximum flexibility, e.g. totally different measurement programs for each sample.

The spectrometer is linked to a microprocessor-based computer for automatic operation and to support powerful and user friendly software routines for sample analysis. Fundamental parameter software packages for calculation of interelement influences (matrix effects) facilitate and optimize the evaluation of concentrations. Applying semi-quantitative analysis, short overview determinations of more than 60 elements in samples of totally unknown composition can be realized within some minutes.

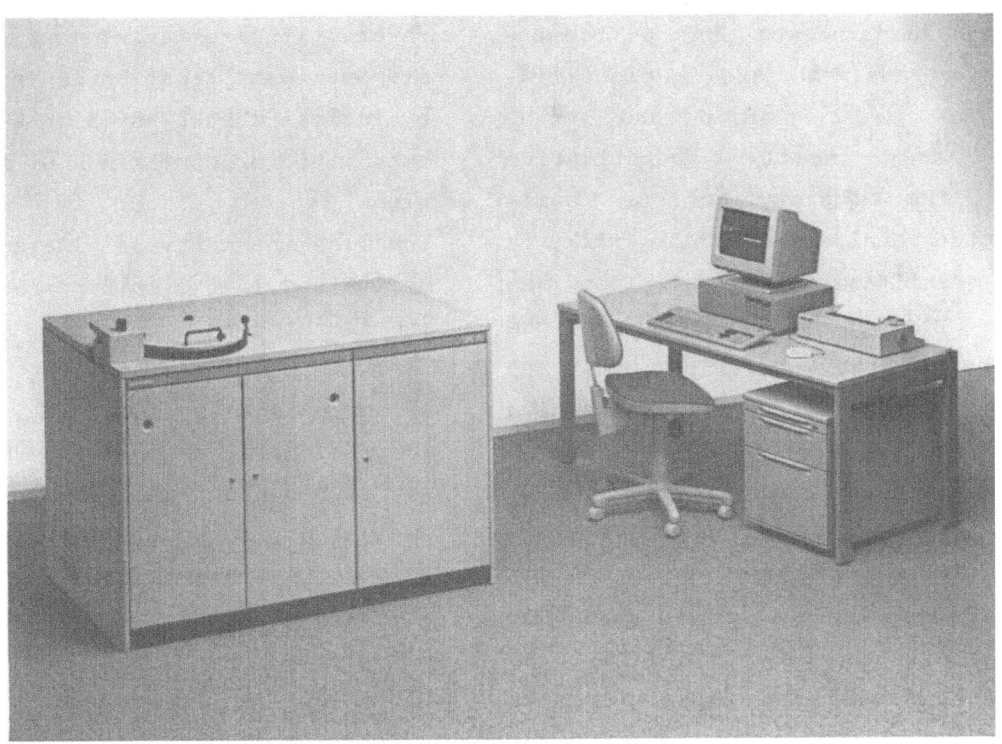

Fig. 1: Sequential X-ray Spectrometer SRS 303

The spectrometer software provides for the generation of custom reports, special calculations (e.g. "moduli") and storage of data. Exploration statistic software packages allow rapid evaluation, interpretation and graphic documentation of the analytic data and prompt planning and action. The XRF analytical software is open for easy integration with customized software packages and computer networks without influencing the reliability of the standard software product. Analyses results can be stored as standard LOTUS 1-2-3 worksheet and ASCII data files to be accessed by commercially available programs.

SEMI-QUANTITATIVE XRF ANALYSIS

Introduction

A precise quantitative analysis by XRF needs careful consideration of sample preparation, instrument conditions, and reference samples. In the mining industry there is very often a demand to analyse totally unknown samples with a minimum of preparation, e.g. classifying raw materials (minerals, scrap etc.) or multi-element analysis of exploration samples. For this purpose, manufacturers of wavelength-dispersive X-ray spectrometers now offer special program packages for semi-quantitative analyses.

The usual requirements for a "semi"-quantitative analysis are a classification into major, minor and trace elements and "nearly" quantitative results. The quality of a semi-quantitative analysis depends (like a "fully" quantitative analysis) on the counting statistics. Therefore the reduction of the measuring time to achieve a very fast semi-quantitative analysis will be limited by the user's demand on the analytical quality.

Concentration calculations are based on fluorescence intensities from a layer at the surface of a sample, whose thickness may vary, depending on the element and basic material, from several centimeters down to a few layers of atoms. Thus, the limit of accuracy in semi-quantitative analysis depends also on the surface quality of the sample.

Method

The SIEMENS semi-quantitative program SSQ is based on the measurement of scans, i. e. several angle ranges are continuously scanned using different optimized measurement conditions (excitation, crystal, collimator, detectors). Compared with the line and background measurement technique, this method offers a series of advantages:
- The graphical display of the scans with marked elements enables easy and clear interpretation of the results (Fig. 2).
- The automatic background determination from the course of the scans is resistant to interfering lines.
- As the scans generally contain

several lines of an element, the element identification can be secured by logical checks, such as K-alpha/K-beta ratio or comparison of K- and L-lines.
- The line intensity is always determined in the current line maximum and therefore a line shift, e. g. by chemical shift of the light elements, is automatically adjusted.
- In problematic cases or if there are any doubts about the automatic evaluation, scans are available for interactive interpretation using the standard software.

Semi-quantitative analysis, on the other hand, generally handles completely unknown samples (i. e., at the beginning of a measurement, it is not always clear whether an element is only a trace element or achieves concentrations of up to 100 %). If the high sensitivity of a modern X-ray spectrometer is used, higher concentrations inevitably lead to a detector overflow. If line or measurement conditions are selected for concentrations up to 100 %, the capacity of the spectrometer is lost in the ppm ranges.

SSQ offers a modern and intelligent solution for this particular problem; the spectra are always registered with optimum sensitivity. To accomplish this, detector overflow is automatically recognized, and the corresponding lines are measured with reduced tube current. These conditions are automatically taken in account when the intensities are corrected (Fig. 3).

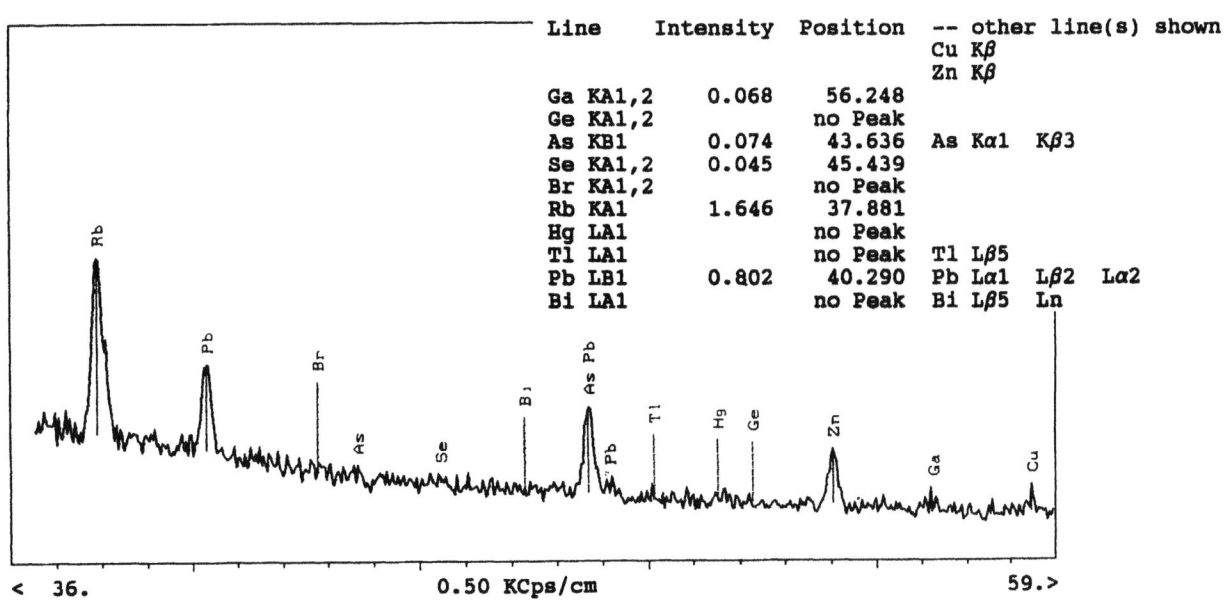

Line	Intensity	Position	-- other line(s) shown
			Cu $K\beta$
			Zn $K\beta$
Ga KA1,2	0.068	56.248	
Ge KA1,2		no Peak	
As KB1	0.074	43.636	As $K\alpha1$ $K\beta3$
Se KA1,2	0.045	45.439	
Br KA1,2		no Peak	
Rb KA1	1.646	37.881	
Hg LA1		no Peak	
Tl LA1		no Peak	Tl $L\beta5$
Pb LB1	0.802	40.290	Pb $L\alpha1$ $L\beta2$ $L\alpha2$
Bi LA1		no Peak	Bi $L\beta5$ Ln

Fig. 2: Easy and clear interpretation of results with optimized scan display

showing automatically identified elements, and peak positions of not

detected elements

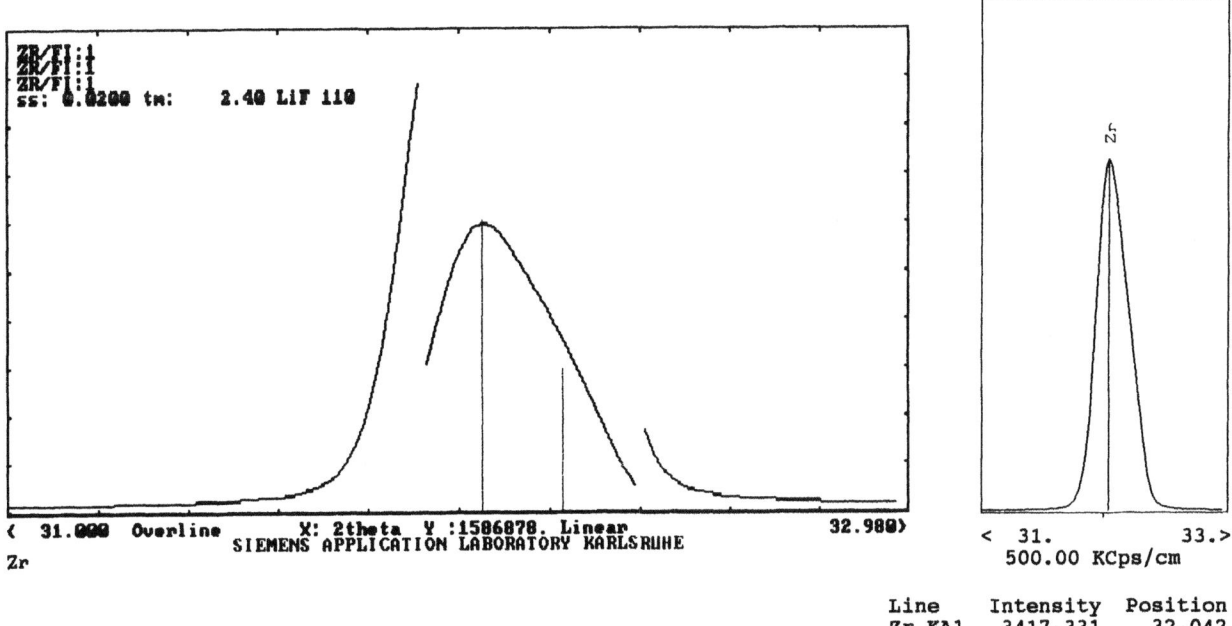

Line	Intensity	Position
Zr KA1	3417.331	32.042

Fig. 3: Scan of Zr KA1,2 line in a high Zr sample, detector overflow was

automatically recognized and the tube current automatically reduced;

and display of stored Zr KA1,2 spectra with corrected intensities

Evaluation

The basis for all further evaluations is the determination of the net intensities of selected fluorescence lines. This requires a reliable calculation of the background and the elemental line. Our semi-quantitative program SSQ automatically determines the background under the spectral line from the course of the scan and is therefore resistant to disturbing elemental lines . Despite the relatively low presence of X-ray overlap in the spectra, the influence of measured intensities by line overlaps cannot be avoided. SSQ minimizes the problem by using crystals and collimators with high resolution and by selecting lines with a very low influence. The remaining overlaps are automatically corrected using the overlap coefficients.

X-ray fluorescence analysis is principally a relative method, as the spectrometer only provides intensities of fluorescence lines. Therefore standard samples with exactly defined and certificated concentration values are required in order to fix the relation between measured intensity and concentration in the sample. Since spectrometers of the same type differ only in their sensitivity (pulse rates per %), the standard samples can be measured and evaluated at the manufacturer's site - keeping in mind the required precision of semi-quantitative analyses. Transfer of these values to the user's specific X-ray system and the compensation of the system's drift can be done by using the reference samples which are provided

with the software package for semi-quantitative analyses.

For correction of interelement effects, the quantitative analysis generally uses methods with fixed correction coefficients, which are either defined empirically or from fundamental parameters (empirical and "theoretical" alphas). As these correction coefficients are valid only for a single very narrow concentration range for each element in the samples, a fundamental parameter method providing individual calculation if the concentrations vary widely, as in semi-quantitative analysis. Therefore, a program package for semi-quantitative XRF analysis should include several possibilities to correct the interelement effects.

CONCLUSIONS

X-ray fluorescence analysis is a very flexible analytical method for the full range of element determinations in geochemical exploration. XRF analysis is not only an adequate analytical method in central laboratories, but also in forward field laboratories. Inexpensive sample preparation and low costs per sample/element determination will compensate for the capital investment in an X-ray spectrometer system - together with the advantages of a clean, fast and operator friendly analytical system.

X-ray fluorescence spectrometry in the mining industry

L. Akino
E. Chikukwa
BSR Ltd, Bindura, Zimbabwe

ABSTRACT

A change from conventional wet chemical analysis to the non-destructive x-ray fluorescence spectrometry method in the nickel mining and processing industry is described, highlighting speed, cost and scope advantages. Problems encountered with respect to reference standards acquisition, calibration and manpower training are also highlighted. The social implications of the change have been assessed and controlled.

INTRODUCTION

With the exception of a few laboratories in the iron and steel and ferroalloy industries, mining laboratories generally rely on wet analysis and flame atomic absorption spectrometry. Besides high operating costs in terms of reagents and glassware, wet analysis is also labour intensive and requires a relatively high degree of technical skills of the operatives. This introduces a problem of data subjectivity due to an over-reliance on analyst skill making control of data quality difficult.

For processing plants like those of Bindura Nickel Corporation, the long turn-round times associated with wet analysis makes the data generated only of historical importance and do not serve the desired purpose of control. Where it does, time constraints still persist.

With increasing demands for volume output, accuracy and cost-effectiveness, the BSR laboratory has switched to x-ray fluorescence spectrometry.

GENERAL

Operations

The BSR laboratory is a referral laboratory for the Bindura Nickel Corporation Limited. The laboratory provides services to six operations daily for process control purposes. These include two mines and two concentrators (Trojan and Madziwa), a smelter and a refinery. In addition, general analytical services are provided to the geological sector for mining development and prospecting.

On average, 2000 analyses are reported daily for process control and for others, the turn-round time is three days. About 60% of samples received are solids and these take up 95% of total man-hours.

Typical annual operating costs for 1990 included the following:

Labour	450 473
Reagents	95 000
Acids	120 184
General equipment, filter paper and accessories	211 921
Total	Z$877 578

JUSTIFICATION

Due to near obsolescence of virtually all instruments at BSR laboratory, there arose a need to re-equip and three options were open to the laboratory.

a)   Replace all instruments as they were.
b)   Acquire an x-ray fluorescence spectrometer as the laboratory workhorse.
c)   Acquire an inductively coupled plasma spectrometer for multi-element analysis.

Table 1 below is a financial comparison of the three options, showing operational costs savings of $140 000 on materials, $254 800 on labour per year (at 1989 rates) and $300 000 on additional instrumentation on choosing the XRF options.

DYNAMIC AND ELEMENTAL RANGE

Application of the x-ray fluorescence system ranges from the parts per million (ppm) level through minor to major element levels. This adequately covers our analytical ranges which are summarised in Table II.

The elemental range covered is more than adequate for the present BSR needs giving ample analytical flexibility.

SPEED OF ANALYSIS

Real analytical time is typically 10 seconds for simultaneous analysis of six elements with 20 to 30 seconds for subsequent elements. However, all plant routine requirements are covered by a 30 second analysis. This contrasts sharply with the previous overnight analyses required for gravimentric nickel and copper determinations. A sample required several analysts each doing specific elements. Typical time comparisons are given in Table III.

Such speed advantages coupled with automation facilities leave more labour available for other activities. Furthermore, the speed is such that the plant load does not fully commit the analytical time available, hence a capacity to offer external services has been created with reduced labour. This can be a lucrative revenue generation facility once adequate programming has been achieved.

EASE OF OPERATION

Once XRF spectrometry is set up, accuracy is centrally controlled by the spectroscopist removing the inherent problems of data subjectivity due to an over-reliance on analyst skill and state of mind. For running the system, virtually no technical skills are required as the operation is simple.

IMPLEMENTATION

Having these potential advantages, the challenge was now to transform them to real benefits.

STANDARDS PREPARATION

Like all other spectroscopic techniques, x-ray fluorescence spectrometry depends heavily on both primary calibration and setting-up standards. However, unlike the steel and cement industries where such standars are available commercially, the nickel industry has no such facilities. The very few nickel-containing standards do not suit BSR's purposes, and so an in-house standards preparation programme was initiated.

This extensive exercise required external assistance (mainly the University of Zimbabwe) and this fully stretched the laboratory facilities since routine services still had to be provided without compromise. Insistence on accuracy and suitability of the materials as setting up standards put a heavy burden on the preparative team.

Table II illustrates the programmes covered in this exercise.

Programme priorities were rated on the basis of:

i) Routine work-load, hence labour and materials savings.
ii) Urgency to processing plants.
iii) Ease of analysis and availability of analytical facilities.

The few programmes which were serviceable on installation, afforded enough labour to hasten the outstanding preparatory work. This also made the 'redundant' labour useful in providing the vital back-up services to the instrument during the initial running period.

OPERATING STAFF

Training was provided for the operators with the University of Zimbabwe's assistance in the principles, concepts and operations of an x-ray fluorescence spectrometry based system. This is still a major challenge as most of the staff has been in-house trained on atomic absorption spectrometry, with x-ray fluorescence spectrometry being a new 'unheard of' technique.

Training has therefore required the various stages of selling to the staff the need for change, the change itself the new staff roles and then the technical details. It is expected to take at least 2 years to have all operators confident with the technology.

Training has also been provided for the system managers with a view to extending the service to trace analysis as practised at the University of Cape Town.

MAINTENANCE STAFF

Besides setting up a local agent for the maintenance, the supplier has offered full servicing and maintenance training for the plant technicians and engineers once they have had 10 - 12 months running experience on the instrument.

Installation was conducted with BSR engineers under instruction from the installation engineer as this was part of the training.

It is expected that within a few years (2 - 4 years) maintenance will be fully in-house with factory referrals made only in extreme circumstances.

SOCIAL IMPACT

In a country like Zimbabwe, technology advancement is slow in most areas. When the need to incorporate new technology is justified and can be financed, it is unfortunate that social considerations can upset such developments.

The introduction of x-ray fluorescence spectrometry brought much apprehension amongst the laboratory staff as fears of job security became very real. Retrenchment was thought to be possible due to technology substitution of labour(17 jobs in this instance) with redundancy of individuals due to academic limitations. With concerted efforts, job security was assured and redundancy fears cleared.

However, some doubts still exists, as workers who do not fully understand the extent to which this technical development will go, still feel insecure, fearing that extended developments will ultimately lead to redundancy.

Workers continue to be reassured of employment. Redeployment of staff is half way through. This has brought some confidence in them as it has been voluntary and with some now asking for transfers.

ACKNOWLEDGEMENTS

The authors thank the management of Anglo American Corporation Services Limited and Bindura Nickel Corporation Limited for permission to publish this paper.

Table 1 : <u>Financial Comparisons of Options on Laboratory Re-instrumentation Programme</u>

	Replacement as is		XRF Option		ICP Option	
Capital Outlay	Potentiometric Cobalt Titrator and electrodes	$ 42 000				$ 42 000
	2 x Copper electroplating banks (each with 6 positions) and					
	platinum-ware	65 000				65 000
	Leco CS 444	250 000				250 000
	3 x Electronic Balances	60 000				60 000
	Double beam UV / Vis	110 000		$ 110 000		
	Fusion Furnace	35 000		35 000		35 000
	Complication Furnace	15 000		15 000		15 000
	XRF Instrument			574 556		
	Linisher and Pelletiser			70 000	ICP instrument	527 548
		$ 667 000		$ 804 556		$ 994 548
Operating Costs	1. Reagents	80 000		24 000		80 000
	2. Acids	68 000		7 000		68 000
	3. Ancillary equipment	80 000		80 000		100 000
	4. Instrument Spares/repair and service	205 000		60 000		255 000
	5. Industrial gases	11 000		3 000		80 000
	6. Glassware	60 000		12 000		60 000
	7. Filter papers	90 000		20 000		90 000
	8. Stationery	5 000		5 000		5 000
	Total	$ 599 000		$ 211 000		$ 738 000
Labour Compliment	1. Analysts	15		11		15
	2. Lab Assistants	15		8		15
	3. Labourers	14		8		14
		44		27		44

Table 2 : <u>Analytical Programmes for X-Ray Fluorescence</u>

Sample	RANGE %							
	Ni	Cu	Co	Fe	S	SiO$_2$	Cao	MgO
Leach Alloy	55 - 75	18 - 30	0,20 - 1,0	0,25 - 0,80	4,5 - 7,0			
Madziwa Mines : Heads								
Tails	0,50 - 1,50	0,20 - 0,30		5 - 7	0,50 - 3,0			
Trojan Mines : Heads								
Tails	0,010 - 0,90	0,01 - 0,05	0,20 - 0,50	5 - 7	1 - 2			
Furnace Cons	8,0 - 12,0	2,6 - 6,0	0,20 - 0,46	20 - 30	15 - 25	10 - 25	0,6 - 2,0	6 - 15
Concentrates - Trojan, Epoch & Shangani	7 - 20	0,5 - 0,40	0,2 - 2,0					
Furnace and Converter slags	0,11 - 0,80	0,6 - 0,30	0,10 - 1,0	35 - 70	0,5 - 1,0	10 - 40		
Cobalt Cake	3 - 10	0,2 - 0,5	25 - 60			2 - 6		
Leach Residue, Fe Sludge & BCL Sludge	4 - 25	3 - 55	0,2 - 3,0	0,2 - 3,0	10 - 25	1 - 2		
Furnace Matte	15 - 25	5 - 10	0,9 - 2,0	35 - 45	20 - 30			

Table 3 : <u>Analytical Time Comparisons</u>

Sample	Analysis	Previous	Labour	XRF	Labour
1 Trojan Mine Cor & Bulk	Ni	3 hrs	1 LA	40 sec	1A
2 Trojan Grabs	Ni, SNi	3 hrs	1 LA	40 sec	1A
3 Madziwa Chips & Core	Ni	3 hrs	1 LA	40 sec	1A
4 Madziwa Grabs & Belts	Ni, SNi	3 hrs	1 LA	40 sec	1A
5 Trojan Heads & Tails	Ni, SNi, S	3 hrs	1 LA	40 sec	1A
6 Trojan Conc	Ni, S, Fe, SiO_2, CaO, MgO	6 hrs	1A	40 sec	1A
7 Trojan Bank Cons	Ni	3 hrs	1 LA	40 sec	1A
8 Trojan Bank Tails	Ni, SNi	3 hrs	1 LA	40 sec	1A
9 Madziwa Heads & Tails	Ni, SNi, S	3 hrs	1 LA	40 sec	1A
10 Madziwa Final Cons	Ni, Cu	3 hrs	1 LA	40 sec	1A
11 Madziwa Bank Cons	Ni, Cu	3 hrs	1 LA	40 sec	1A
12 Madziwa Bank Heads & Tails	Ni	3 hrs	1 LA	40 sec	1A
13 Concentrate	Ni, Cu, Co	24 hrs	1A	10 sec	1A
14 Blend Feed	Ni, Cu	3 hrs	1 LA	10 sec	1A
15 Blend Residue	Ni, Cu, Co	3 hrs	1 LA	10 sec	1A
16 Blend Sludge	Ni, Cu, Co	3 hrs	1 LA	10 sec	1A
17 Dried Blend	Ni, Cu, Co	3 hrs	1 LA	10 sec	1A
18 Furnace Conc	Ni, Cu, Co, Fe, S, SiO_2, CaO, MgO	6 hrs	1 LA	10 sec	1A
19 Furnace Slag	Ni, Cu, Co, Fe, S, SiO_2, CaO, MgO	6 hrs	1 LA	10 sec	1A
20 Granulated Slag	Ni, Cu, Co	3 hrs	1 LA	10 sec	1A
21 Furnace Matte	Ni, Cu, Co, Fe, S	3 hrs	1 LA	10 sec	1A
22 Conv. Slag	Ni, Co, Cu, FeO, S, SiO_2	6 hrs	1A	10 sec	1A
23 Leach Alloy	Ni, Cu, Co, Fe, S	6 hrs	2A	10 sec	1A
24 Leach Residue	Ni, Cu, Co, Fe, S, SiO_2	6 hrs	2A	10 sec	1A
25 Cyclone Overflow	Ni, Cu, Co, Fe, S	6 hrs	2A	10 sec	1A

KEY :

1 LA	=	1 Laboratory Assistant
1A	=	1 Analyst
SNi	=	Sulphide Nickel
LA	=	Laboratory Assistant

Geology

Groundwater flow model for Konkola underground copper mine, Zambia

S. C. Mulenga
Konkola Mine, Zambia (currently Department of Geology, Imperial College of Science, Technology and Medicine, London, England)
M. H. de Freitas
Department of Geology, Imperial College of Science, Technology and Medicine, London, England

SYNOPSIS

Konkola Underground Copper Mine is located in the Copperbelt Province of Zambia and has the largest known ore reserves of all the eight operating mines owned by Zambia Consolidated Copper Mines Limited. Konkola has long been recognised as one of the wettest, if not the wettest mine in the world. An average of 360,000m^3/d is pumped from the mine, giving a ratio of 113:1 of water to ore hoisted. The large volumes of water encountered and expected during mining constitute a major cost in mine planning and development.

Mining has now reached a depth where management decisions concerning future production require as accurate an assessment as possible of groundwater inflows and a numerical model of groundwater flow to the mine has been developed, as outlined by Mulenga and de Freitas (1): this model is currently undergoing refinement. Many problems were faced in designing the model and this paper describes the decisions that had to be taken in order to build the model for simulating groundwater flow to and through the mine and predicting water level drawdown, and mine discharge.

The paper describes four aspects of this work:

1. The decisions required to represent the geology of the mine in the model.

2. The decisions required for the input parameters to be used.

3. The calibration of the model.

4. Preliminary results from the model. Relevant aspects of field geology, mining engineering and associated hydrogeological studies are described, their inter-relationship explained and the uncertainties associated with them demonstrated.

INTRODUCTION

Konkola Underground Copper Mine, formerly known as Bancroft, is situated at the northern end of the Zambian copperbelt, for details, see Garlick (2). The stratified copper deposit is sandwiched between two major aquifers. Consequently, the unsupported open-stoping mining method used at the mine requires dewatering of the aquifers during mine development, so that collapse subsequent to ore extraction occurs in de-watered strata. Large inflows of groundwater encountered and expected during mining are a constant operational problem, Whyte and Lyall (3). Currently, an average of 360,000m^3/d is pumped from the mine to surface. The water pumped to ore hoisted ratio is 113:1.

The associated costs of mine drainage are very substantial. Pumping alone, excluding the cost of mining and setting up pumping complexes, accounts for about 10% to 15% of the total mine operational costs (1984-1990 figures), and the management required for future mine development needs to know whether this figure will increase or decrease with an increase in the depth of mining. In order to base these management decisions on a firm scientific and technical basis, it has been necessary to test the adequacy of the hydrogeological data for making such predictions and, for this, a numerical model was required.

REPRESENTATION OF GEOLOGY

Basic geology

The mine is located in the nose of the Kirilabombwe anticline and wedged between two major faults; the Lubengele and the Luansobe as shown in Figures 1a and 1b. The ground between these faults is broken and blocky and, in some ways, can be likened to a box of sugar cubes. This means that flow through the rock mass in the hangingwall and in

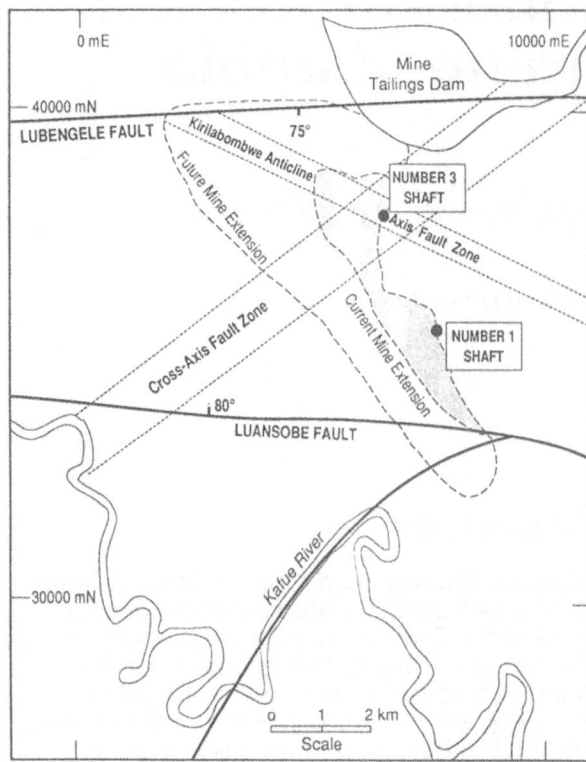

Fig. 1a. Konkola Mine showing the major aspects of structural geology and the limits of mining.

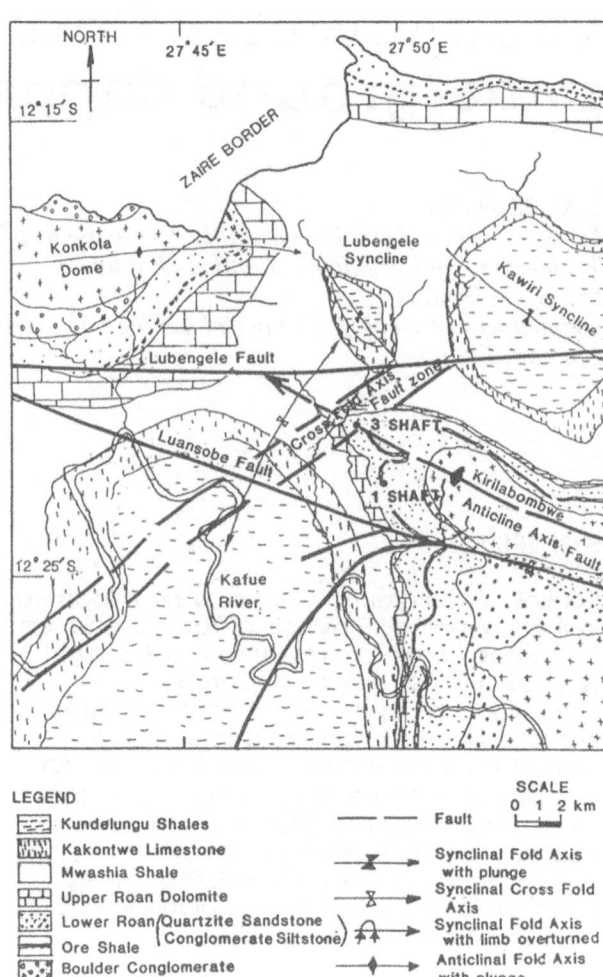

LEGEND

Kundelungu Shales
Kakontwe Limestone
Mwashia Shale
Upper Roan Dolomite
Lower Roan/Quartzite Sandstone (Conglomerate Siltstone)
Ore Shale
Boulder Conglomerate
Basement Complex (Granite, Gneiss, Schist)

Fault
Synclinal Fold Axis with plunge
Synclinal Cross Fold Axis
Synclinal Fold Axis with limb overturned
Anticlinal Fold Axis with plunge

SCALE
0 1 2 km

Fig. 1b. Geological map of the Konkola mine area.

the footwall close to the ore body is dominated by the transmission of water through fissures, even though the rock material contains pores (in general porosity does not exceed 20%) and can consequently transmit groundwater by porous flow. Figure 2 illustrates the extent to which surface hydrology above the mine reflects the jointing and general pattern of fracture in the ground.

The Ore Shale formation is sandwiched between two major aquifers, those of the Hangingwall and those of the footwall. The hangingwall aquifer is composed mainly of dolomites, limestones and interbedded dolomitic siltstones and shales. The footwall aquifer consists of feldspathic sandstones, conglomerates and quartzites which overlie the granites and gneisses of the Basement Complex (Figure 3).

The geology shown in Figures 1a, 1b and 3 was represented in the numerical model by seven horizontal layers from surface to 1320m level based on mine level maps. The mine levels represented were as follows: 450m (1480ft), 565m (1850ft), 670m (2200ft), 810m (2650ft), 960m (3150ft) and 1320m (4000ft), each having the ability to be given different values for their hydraulic conductivities, the conductivity of each layer being assumed to decrease with depth to reflect the effect of increase in the confining stress. Superimposed upon this

general structure were zones of much higher conductivity that represented the Luansobe and Lubengele faults and the zones of internal deformation associated with the Axis fault zone and the Cross-Axis fault zone. Boundary conditions for the model were as follows. The Basement Complex was taken as defining the lower impermeable boundary at the bottom of the model and extended upwards to ground surface towards the east where it outcrops. The northern boundary of the model was defined by the Lubengele fault and mine tailings dam, and the western boundary by the intersection of the Lubengele and Luansobe faults. At this point of intersection, the Kafue river is also encountered and its extension south and east was used as the southern boundary of the model for the reasons shown in Figure 2. The Kafue river and its tributaries, and the tailings dam, were designated constant-head areas. Everywhere else the boundaries were defined as variable head. These boundaries defined a very large volume of ground within which the mine exists and represents the limits of what is called the Far Field model. The area of the

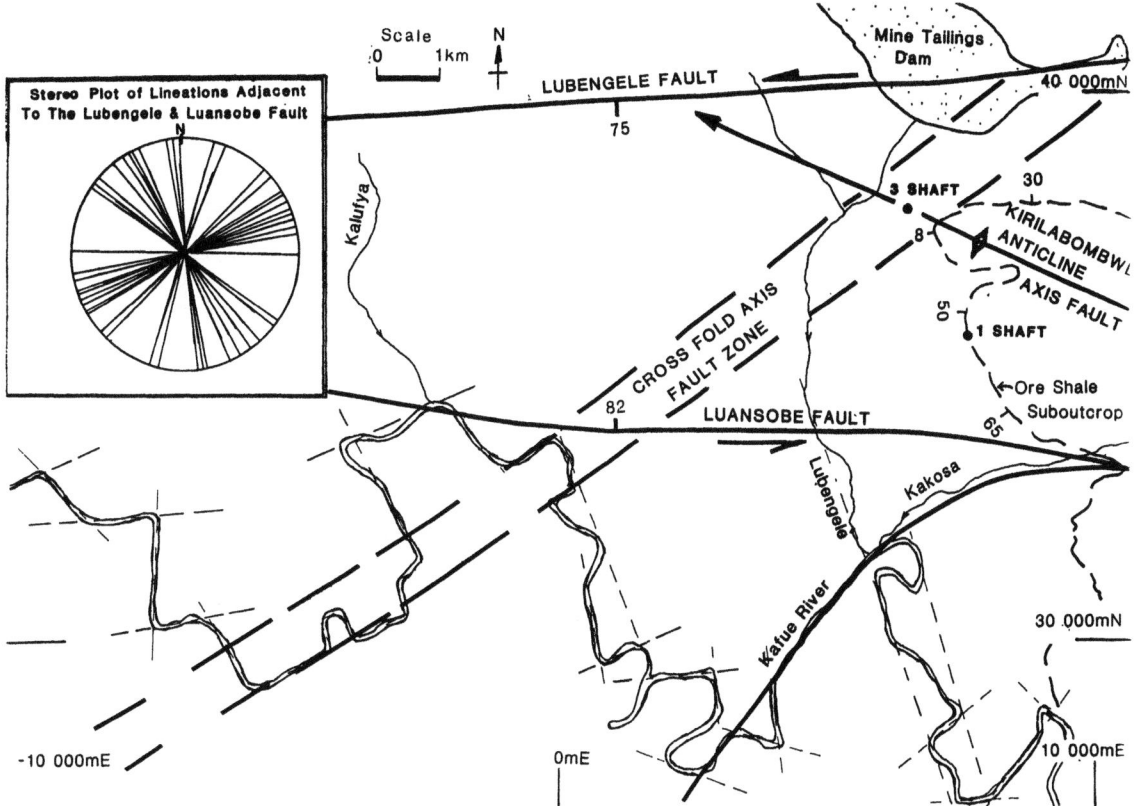

Fig. 2.   Relationship between the major elements of surface hydrology, especially the Kafue River, and fracture patterns.

mine and the western limit of future mining define a more detailed model called the Near Field model whose boundaries are the Lubengele fault and the tailings dam in the north, the limit of mining to the west, the Luansobe fault and sections of the Kafue river which lie along the Luansobe fault to the south, and the outcrop of the Basement Complex to the east. These boundaries are illustrated in Figure 4.

Being in a mine, almost all the data available on ground conditions had been gathered with mineral exploration and exploitation in mind, and very little in the way of systematically gathered information for hydrogeological purposes had been obtained other than the measurement of water levels in open holes that crossed the hangingwall aquifer. Consequently, the density of input data to represent geology varied significantly between the Far Field and Near Field models, and this is reflected by the element size used in the Far and Near Fields, see Figure 4.

Hydrology and hydrogeology

The Kafue river controls the drainage system of the Konkola area. It flows over the hangingwall aquifer and along the faults in the vicinity of the mine. The tailings dam is located on the Lubengele fault which is close to and dips

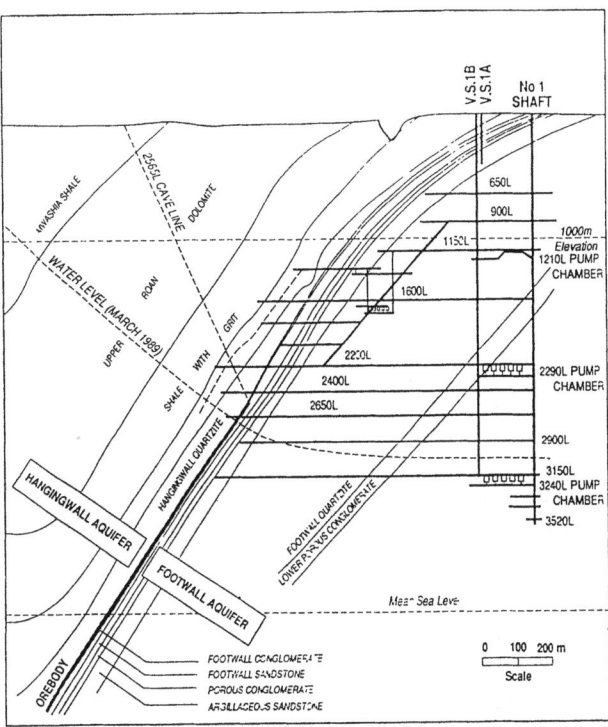

Fig. 3.   Generalised geological section at No. 1 shaft showing the ore body, aquifers and water levels.

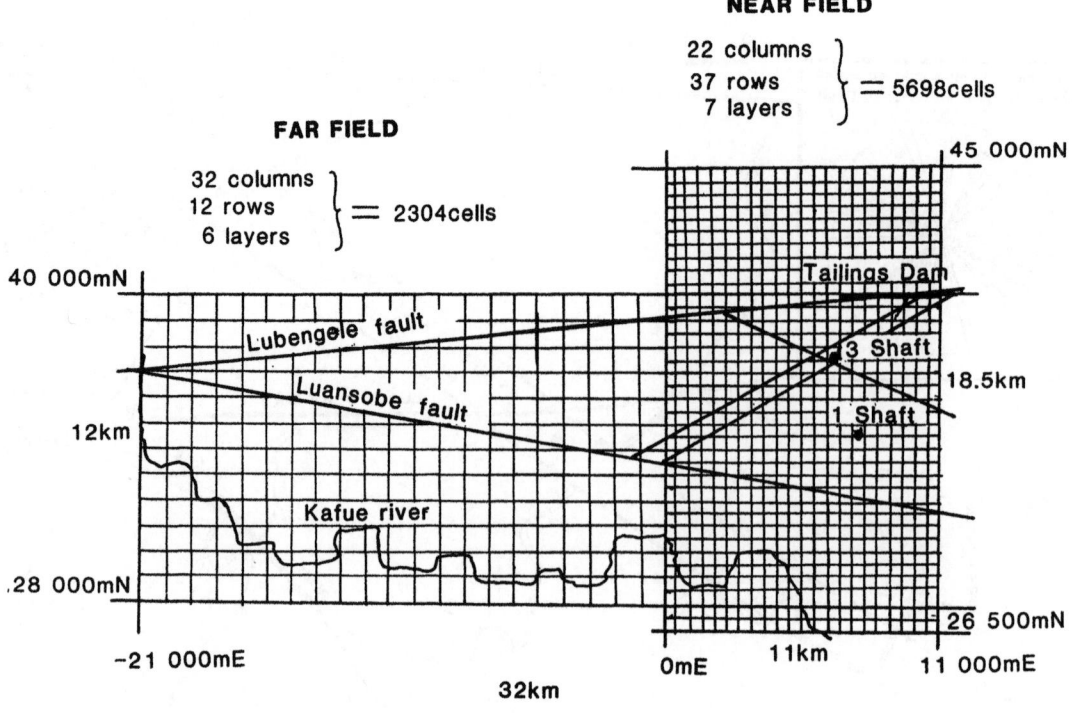

Fig. 4.   Boundaries of the Far Field and Near Field model.

towards Number 3 Shaft. The dam itself is founded directly on the superficial deposits which cover bedrock as a thin veneer. These include the alluvium of a former river which occupies the bottom of a shallow valley across which the dam is built and in which the tailings reservoir is impounded. Although attempts have been made to seal the base of the reservoir using tailings, there is no guarantee that a hydrological seal exists between the reservoir and the drift upon which it rests nor is there any guarantee that leakage beneath the tailings dam is prevented. Thus recharge from the river and the dam must be considered as potential sources of water that can enter the mine mainly through the faults and associated fracture zones. For these reasons, the total heads for the Far Field boundaries were as follows: the Lubengele fault and Kafue river were constant head recharge boundaries equal to ground

level; the eastern boundary of the Basement Complex was a no-flow boundary of variable head with time; ground level has the potential for accepting recharge from effective rainfall.

INPUT PARAMETERS AND CHOICE OF MODEL

The simulation of any groundwater problem involves the solution of the fluid mass balance equation over a specified volume of ground subject to the internal parameters of hydraulic conductivity and storage, and to boundary conditions, Wang and Anderson (4), Fawcett et al (5), Singh and Atkins (6) and Freeze and Cherry (7).

The potential flow which describes the change in flux in response to a change in potential is given by the Laplace Equation, which for

(a) Steady State Flow

$$\Delta Q = \left[K_x \frac{\delta^2 h}{\delta x^2} + K_y \frac{\delta^2 h}{\delta y^2} + K_z \frac{\delta^2 h}{\delta z^2} \right] dx.dy.dz = 0$$

and for

(b) Transient State Flow

$$\Delta Q = \Delta^2 h = \frac{S}{Kb} \frac{\delta h}{\delta t} \neq 0$$

where S = coefficient of aquifer storage
   b = saturated aquifer thickness.
   k = hydraulic conductivity
   h = total head
   t = time

In most cases, the values for (k) were very poorly known and values for (S) are almost non-existent: the difficulties with defining the input parameters are described below.

Input parameters

Historically, systematic records of underground dewatering boreholes, discharges, pressure heads and surface borehole water levels have been maintained. There was no data on storage coefficient and transmissivity. Values of hydraulic conductivity were only poorly known. With these restrictions the approach adopted was to input the measured values of discharge and head in the model, and use the model to obtain matching values of hydraulic conductivity, which could then be used for the initial simulations.

From the knowledge of the mine geology, it was clear that substantial differences in values of hydraulic parameters would exist, especially in areas of faulting and in fracture zones. Groundwater modelling in a mine poses other problems too. Mining activity disturbs the ground and, unlike the situation with most models for water supply, the model for the mine must not only be provided with data on these hydraulic parameters as they exist in undisturbed ground but also with data on how these values can change as the effects of open-stoping are felt in the hangingwall aquifer, and the effects of stress relief are felt in the footwall aquifer. It is therefore very difficult to obtain representative values of storage which permit a comparison to be made between drawdown and discharge from the limited data available. Thus the input data had no reliable values of storage which could be used, and this is a difficulty that has to be resolved by further field observations.

The model

Measurements of water level and mine discharge have shown repeatedly that a transient system exists. Although it is obvious that the Konkola mine groundwater flow regime is non-steady in the long term, over the short term, i.e. one or two months, the situation can be approximated to steady state and, in view of the paucity of information on storage at present, it was decided to start using the model to simulate steady state flow. A decision then had to be made concerning the type of model to be used. Water clearly flows mainly through the fractures of the rock in this mine, although some porous flow must also exist, however, the distribution, number, orientation and conductivity of these fractures is beyond prediction with the quality of data at hand. Thus it has been decided to model the ground as if it were a continuum rather than to attempt a representation of fracture flow. Bearing in mind the large area of the mine and its great depth, the relative scales of the real situation and its model implies that representing the ground as a continuum will not significantly affect the quality of the predictions obtained.

These considerations led to the selection of MODFLOW, the three-dimensional finite difference groundwater flow model developed by McDonald and Harbaugh of the United States of America Geological Survey (8). Simulation of the mine groundwater flow was achieved by modelling the problem on the two separate scales described earlier, a Far Field and Near Field.

The Far Field model was governed by the natural hydrogeological boundaries which surround the mine and the dimensions of the ground thus represented are 32km by 12km and 1.32km deep. The Far Field model has six horizontal layers, each divided into 1km square blocks, and thus an overall total of 2304 model blocks.

The Near Field model is used to simulate in more detail the water levels within the mine area and to provide predictions of mine discharge and drawdown. It covers the area of past, present and future mining activity. Boundary conditions could be obtained from a combination of results from the Far Field model and observed groundwater levels. The volume of ground represented is 11km by 18.5km and 1.32km deep. It is made up of seven layers which correspond to the mine levels. Each layer is divided into 500m square blocks. The Near Field model has a total of 5698 model blocks.

MODEL CALIBRATION

The reliability of a model can be gauged by the accuracy with which it can reproduce past events and the Konkola model is presently undergoing these trials. In order to reproduce the observed groundwater behaviour, the model was calibrated by inputting data for June 1970 to March 1988. The data comprised total volume of mine pumping, dewatering borehole discharges and pressure heads, values of hydraulic conductivities obtained from borehole pressure build-up tests and flow tracers, and the Kafue river and tailings dam ground level elevations.

The Far Field boundaries were used to generate the total heads for the Near Field boundaries.

In the Near Field model, the boundary conditions were set at values obtained from the Far Field simulation coupled with observed water levels from borehole data. 1970 records were used as a starting point. Simulation of faults and fracture systems was achieved by introducing high hydraulic conductivity zones. Constant-head values were set at various block nodes relevant to the Kafue river and the tailings dam. The recharge into the system from the river and dam were simulated using the MODRIV module, and mine drainage was simulated using the MODWEL module.

In these initial runs the variation in the value for hydraulic conductivity used was of the order of thirty between the ground in the fracture zones and the ground outside these zones, the range being from 0.03m/d to 1.0m/d. These values appear at first sight to be somewhat low for the type of problem under consideration; this is because the volume of fractures through which the water is transmitted is substantially smaller than the volume of rock in which they are situated, and because a "continuum" model is being used in which parameter values must apply to entire model blocks. Therefore, although the individual fractures have high conductivities when these are averaged over the volumes occupied by a model block, the resulting value is substantially smaller. It should also be noted that the only parameters which could be calibrated were water level maps and mine discharge. So the calibration runs were directed to reproducing drawdown for a given total discharge using values of hydraulic conductivity that seemed appropriate from the initial simulations. Table 1 shows the level of agreement obtained between the model simulation and the real situation.

Grid Location	Discharge (m3/d)	
Layer	Measured	Simulated
3	92	168
3	216	153
3	11893	9395
3	262	237
4	38336	21587
4	30393	21213
4	18445	14536
5	27805	21633
5	23194	21458
5	46511	41119
5	1823	2113
5	5000	5185
6	18000	21381
6	15063	16569
6	14440	16428
6	2120	4642
6	5000	5070
6	1000	2890
6	25600	25545
6	8766	11149
7	27772	30205
7	38275	39007
7	3158	3765
Total mine discharge	363164	335448

Table 1. Comparison between measured discharge at dewatering locations in the mine and the discharge predicted by simulation at grid locations where these dewatering points were sited.

Results

The model has been able to reproduce the shape of groundwater level map and of cone of depression as shown by comparing Figure 5(a) with 5(b) constructed using borehole water levels around the mine. The importance of the Luansobe and Lubengele faults, and their associated fracture zones, as the main channel ways of groundwater flow, and of the Kafue river and tailings dam as the major surface water recharge sources, is well demonstrated because the shape and magnitude of the cone of depression could not be reproduced by the model until these sources of recharge had been introduced: see Figure 6.

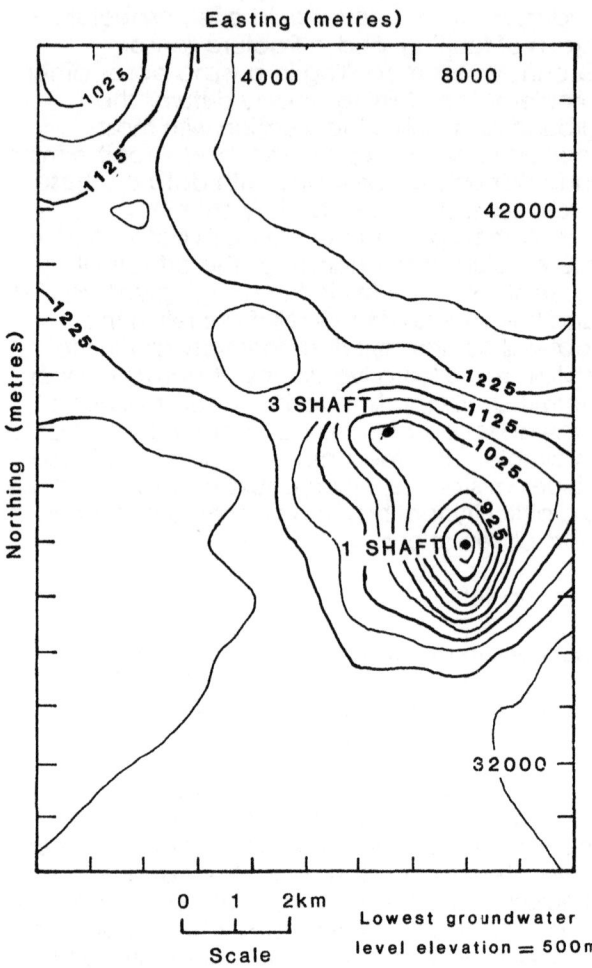

Fig. 5a. Simulated groundwater contour map for March 1988.

In the discharge simulation there was a wide variation in degrees of accuracy attained at individual block nodes. The discrepancy between simulated and measured discharge ranged from -44% to +189%. In the water level simulation, there was also a variation in degrees of accuracy obtained between measured and simulated levels. The measured lowest groundwater level was 428m whilst that simulated was 485m. This clearly shows that representative field data is essential if meaningful simulations are to be achieved at a scale equivalent to the ground represented by either a single model block or small groups of model blocks.

Model uncertainties

Although the model is proving a great help in building a better understanding of the Konkola mine groundwater problem, it cannot yet be relied upon to make accurate simulations. This is because at the time of writing this paper model refinement is far from complete. The data set is made up from a few field values of hydraulic

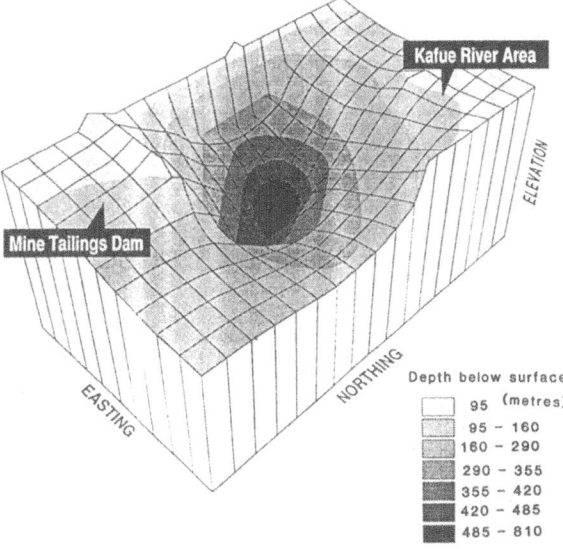

Fig. 6.   Three-dimensional representation of draw-down around and above the Konkola Mine as predicted by numerical modelling.

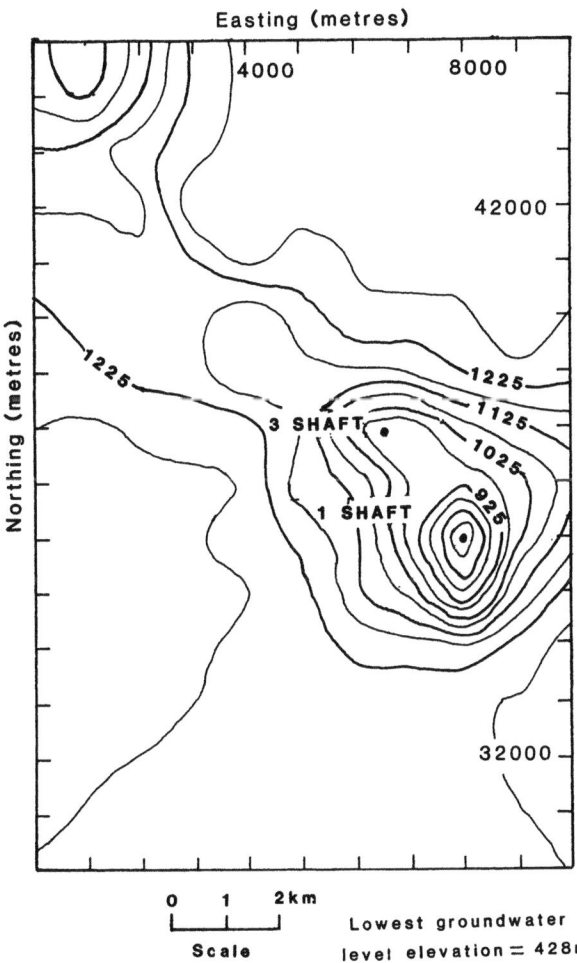

Easting (metres)

Fig. 5b.   Map of actual groundwater contours for March 1988.

parameters that were augmented by extrapolated values and hence the model can only give a broad picture. Simulations have shown that boundaries have an over-riding influence on the hydrogeology of the mine and very little is known about them. Hydraulic properties such as hydraulic conductivity, transmissivity and storage are absolutely crucial, but there are no adequate field values for these parameters to cover the whole area.

Current research has shown that recharge from the Kafue river and tailings dam is significant and yet the relative proportions of their recharge to the total recharge are presently unknown (1). To remove these uncertainties, field investigations to obtain representative values of hydraulic parameters would have to be carried out. The cost of this activity would be offset by the savings made in eventually reducing the cost of pumping. Thus the hydrogeological programme envisaged is one which primarily aims to find a permanent solution for reducing water inflow into the mine, by investigating various groundwater control measures. Removing uncertainties so that the model can make accurate predictions of discharge and drawdown is merely a means of achieving this goal and not an end in itself.

FINANCIAL DECISIONS

Pumping of water to surface alone, i.e. excluding the cost of excavating and setting-up pumping complexes, accounts for about 10% to 15% of the annual total mine cost (Konkola Mine Annual Reports. This cost is bound to significantly rise as the mine expands. In the twenty-year mine plan, production is scheduled to more than double the current output. To achieve this target, substantial lowering of groundwater levels will be required. In the hangingwall aquifer, drawdowns of at least 20 metres per year at Number 1 Shaft and 15 metres per year at Number 3 Shaft will be necessary. Pumping costs could increase from their current value of 10% to 15% to approach 20% of the annual total mine operational cost.

Verification of geological parameters, especially fault zones and aquifer boundaries, is estimated to cost about 0.07% of the annual pumping cost. Hydrogeological investigations directed to securing values of transmissivity and storage within the key areas of the mine are estimated to cost approximately 7% of the annual pumping cost. Investigations required to define the rate of flow through the mine and the hydraulic connections between the mine and its surrounding hydrogeological boundaries are estimated as costing about 8% of the annual pumping cost. On-going monitoring of instrumented sites to obtain values of head and appropriate values of transmissivity and storage which operate during mining are estimated as costing in the region of 0.2% of the annual pumping cost.

From this preliminary schedule, it is evident that an investment equal to approximately 1.9% of current annual mine operational cost is required to provide the factual data needed to assess groundwater control in the mine where pumping alone accounts for 10% to 15% of current operational costs. Since the anticipated working life of Konkola Mine extends beyond 20 years, this figure should be viewed as a good investment in providing a scientific basis upon which the formulation of a permanent and cost effective groundwater management solution could be based.

CONCLUSIONS

1. The model has already greatly helped to enhance the current understanding of the Konkola mine groundwater flow problem and highlighted the problems involved in understanding groundwater flow in a mining environment.

2. A knowledge of the basic geology of the mine and of the region in which the mine is located has proved to be of great value in creating the numerical model, especially when hydrogeological data is missing and values have to be assumed.

3. The insertion of zones of high conductivity has a profound effect upon the response of the model and their incorporation into the model was the most important of the decisions that were made in order to match water levels to mine discharge.

4. Calibration runs that link hydraulic gradients to mine discharge provide an indication of the values for hydraulic conductivity but do not reveal the flows in any particular fissure or fissure system. The model will therefore never be a substitute for ground investigation and monitoring.

5. It is quite clear that a numerical model for a mine, once created, has to be supported by a parallel programme of field studies during the life of the mine in order that the time dependent values for hydraulic conductivity, storage and boundary conditions may be updated as the relaxation of ground around the mine evolves.

ACKNOWLEDGEMENTS

The authors wish to thank the management of Zambia Consolidated Copper Mines Limited for the research sponsorship and permission to publish this paper. Special appreciation is expressed to Adrian Butler, Paul Johnston and Dr. Howard Wheater of the Department of Civil Engineering, Imperial College, for their invaluable assistance with the construction and initial calibration of the model.

REFERENCES

1. Mulenga, S. and de Freitas, M.H. (1990). Preliminary results of current investigations in the groundwater inflow problem at Konkola Underground Copper Mine - Zambia. In: International Journal of Mine Water (in press).

2. Garlick, W.G. (1961). Structural evolution of the copperbelt. In: The Geology of the Northern Rhodesian Copperbelt, edited by F. Mendelsohn, London: Macdonald.

3. Whyte, W.J. and Lyall, R.A. (1969). Control of groundwater at Bancroft Mines Limited, Zambia. Ninth Commonwealth Mining and Metallurgical Congress, Mining and Petroleum Geology Section, paper 16, 32pp. Institution of Mining and Metallurgy, London.

4. Wang, H.F. and Anderson, M.P. (1982). Introduction to groundwater modelling - finite difference and finite element methods. W.H. Freeman and Company, San Francisco, USA, 237pp.

5. Fawcett, R.J., Hibberd, R.N. and Singh, R.N. (1984). An appraisal of mathematical models to predict water inflows into underground coal workings. In: International Journal of Mine Water, Vol.3, No.2, pp33-54.

6. Singh, R.N. and Atkins, A.S. (1984). Application of analytical solutions to simulate some mine inflow problems in underground coal mining. In: International Journal of Mine Water, Vol. 3, No. 4, pp1-27.

7. Freeze, A.R. and Cherry, A.J. (1979). Groundwater. Prentice Hall Inc., Englewood Cliffs, New Jersey, 604pp.

8. McDonald, M.G. and Harbaugh, A.W. (1988). A modular three-dimensional finite difference groundwater flow model. In: Techniques of Water-Resources Investigations of the United States Geological Survey, Book 6, Chapter A1.

9. Konkola Mine Annual Report (1984-1990). In: Zambia Consolidated Copper Mines Limited Annual Reports 1984-1990.

Controls on Archaean gold mineralization in the Mashava area, Zimbabwe

T. G. Blenkinsop
Department of Geology, University of Zimbabwe, Harare, Zimbabwe

SUMMARY

The Mashava area (formerly Mashaba) has produced just less than 7000 kg of gold from late in the last century to 1964. Gold has been obtained from every major lithology in the area, including the oldest Tokwe gneisses, Sebakwian greenstones and banded iron formations (BIF); the Mushandike granite; the Mashaba Igneous Complex; Bulawayan greenstones, metasediments and BIF; and Shamvaian greenstones and metasediments. Total gold production and production per square kilometre (concentration) was much greater in Sebakwian BIF than other lithologies. The distribution of gold production was almost exclusively concentrated in the west of the Mashava-Masvingo greenstone belt. South of Mashava, production defined a broad NW-SE zone that is truncated against a major structure called the Jenya-Mushandike Dislocation Zone (JMDZ). South of Masvingo, there was a NW-SE zone of production along the southwest margin of the greenstone belt, a more prominent NE-SW zone in the centre of the belt, and a zone near the northern edge of the greenstone belt, parallel to several splays of the JMDZ. There were no large workings along the JMDZ itself. Most deposits were controlled on a smaller scale by fractures, deformation zones, and folds. Auriferous fluids moved through broad zones in the crust, and gold was deposited in smaller structures which allowed sealing and pore fluid pressure cycling. There may have been two main periods of mineralization, firstly at some time between the Sebakwian and Bulawayan, and secondly at the end of the Shamvaian, when lode deposits formed in all lithologies. No single source rock can be identified.

INTRODUCTION

Most gold in Zimbabwe has been produced from greenstone belts in the Archaean craton, which contain three types of deposit: lode (veins and deformation zones), iron formation, and volcaniclastic[1]. Many of these deposits have characteristics of both lode and iron formation types, and together they have accounted for 95% of total gold production from the country[1]. Several aspects of mineralization in these deposits are not well understood. There is a long-standing controversy about the relative importance of structural and lithological influences, and the scales on which they affect mineralization. Another problem is the age of mineralization, both in absolute terms and relative to local geological events, and a third problem is the source of gold.

This study examines lode and iron formation gold in the Masvingo-Mashava greenstone belt and adjacent granites and gneisses (Figure 1). Although the relatively small total of 7000 kg gold has been produced from an area of 1386 km^2, it is an appropriate area to study problems of lode and iron formation deposits because it includes a range of Archaean lithologies

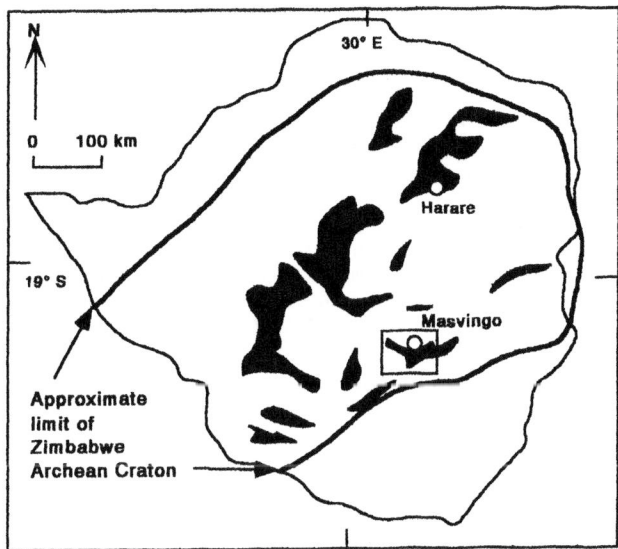

Figure 1. Main greenstone Belts and the Zimbabwe Archaean Craton within Zimbabwe (after Wilson[6]) The box outlines the area shown in Figures 2 and 3.

with ages from the earliest gneisses (age 3.5 Ga) to the latest Archaean granites (2.6 Ga). The geological history of the area is known in some detail[2,3,4], and a wealth of data on gold production is available from the records of over 300 claims and workings in the area, all from lode or iron formation deposits. Records have been examined from the last years of the last century up to 1964, the period for which they are readily available[3,4]. Extra gold was obtained by reprocessing sands at some mines: this was not included in the analysis of production, grades and concentrations from individual mines, in order to standardize these figures for production from ore rocks only, avoiding possible bias towards those mines where reprocessing occurred.

The aims of this study are to use these data to evaluate the roles and scales of influence of lithology and structures on gold mineralization in lode and iron formation deposits, to relate the mineralization to the local geological history and to the evolution of the craton, and to make some deductions about the source of gold. Some important conclusions can be drawn by integrating an analysis of distribution and production of gold mines in the area with a study of local and regional geology. The method and general conclusions may be significant for gold exploration in other greenstone belts.

330

LITHOLOGY AND STRATIGRAPHY OF THE MASHABA AREA

Lithostratigraphic groups used here (Sebakwian, Bulawayan and Shamvaian) follow the terminology in Wilson[2]. The oldest rocks are gneisses belonging to the Tokwe segment[5], with an approximate age of 3.5 Ga, infolded with mafic and ultramafic metavolcanics and banded iron formations (BIF) assigned to Sebakwian Group[3]. These outcrop in the southwest, west and north of the Masvingo-Mashava greenstone belt (Figure 2). They are clearly intruded by the Mushandike granite, which occupies most of the area between Masvingo and Mashava (Figure 2), and has been dated at 2.9 Ga[6]. A large ultramafic body called the Mashaba Igneous Complex (MIC), consisting of dunite (altered to serpentinites), harzburgite, pyroxenite, minor gabbro, and dolerite dykes, lies mainly east of Mashava. Arms extend from the central region to the northwest, to the northeast, to the east along the northern margin of the greenstone belt, and along the southwest boundary of the greenstone belt. The MIC is thought to represent the remains of a magma chamber for the overlying Bulawayan volcanics[5].

The majority of the greenstones in the Mashava-Masvingo belt are Bulawayan Group basic metavolcanics, with a presumed age of ~2.7 Ga. Other lithologies in this group include conglomerates, grits and quartzites at low grades of metamorphism, BIF, and schists. Bulawayan BIF can be seen lying unconformably on the Mushandike granite.

Younger sedimentary rocks of the Shamvaian Group are separated from Bulawayan rocks by a major unconformity. They can be divided into a lower part including conglomerates, grits, pelitic sediments and BIF, and an upper part of limestone, BIF and pelitic sediments, all at low grades of metamorphism, and separated by another unconformity or tectonic contact. They form the main part of the greenstone belt south and east of Mashava (Figure 2).

All the rocks described above were intruded in places by massive granites that are part of the Chilimanzi suite, age ~2.6 Ga[5] (in the north and southwest of Figure 2).

LITHOLOGICAL CONTROLS ON MINERALIZATION

Every Archaean lithology in the area has produced significant amounts of gold with the sole exception of the Chilimanzi suite granites, which have no recorded production. Lithologies have been grouped on the basis of Wilson's mapping[3]. For example, Bulawayan rocks are subdivided into greenstones (metavolcanics), BIF (including some quartzites) and schists (including all metasediments). The largest number of mines (50) occurred in Bulawayan greenstones, followed by Tokwe gneisses (27) (Figure 3), but total production has been much greater from Sebakwian BIF (3132 kg), followed by Tokwe gneisses (1714 kg) and then Bulawayan greenstones (609 kg) (Figure 4). The average grade obtained from ore rocks was restricted to the narrow range of 5 to 14 g/t, with the highest figure (14g/t) coming from the lower part of the Shamvaian Group, followed by Tokwe gneisses (10g/t) and then Bulawayan schists and grits (8 g/t) (Figure 5).

More revealing statistics are obtained by normalising the number and production of mines in each lithology by its outcrop area. Both the number of mines per square kilometre and production per square kilometre were an order of magnitude greater in Sebakwian BIF than any other lithology (2.87 mines/km² and 599 kg/km² respectively). The next

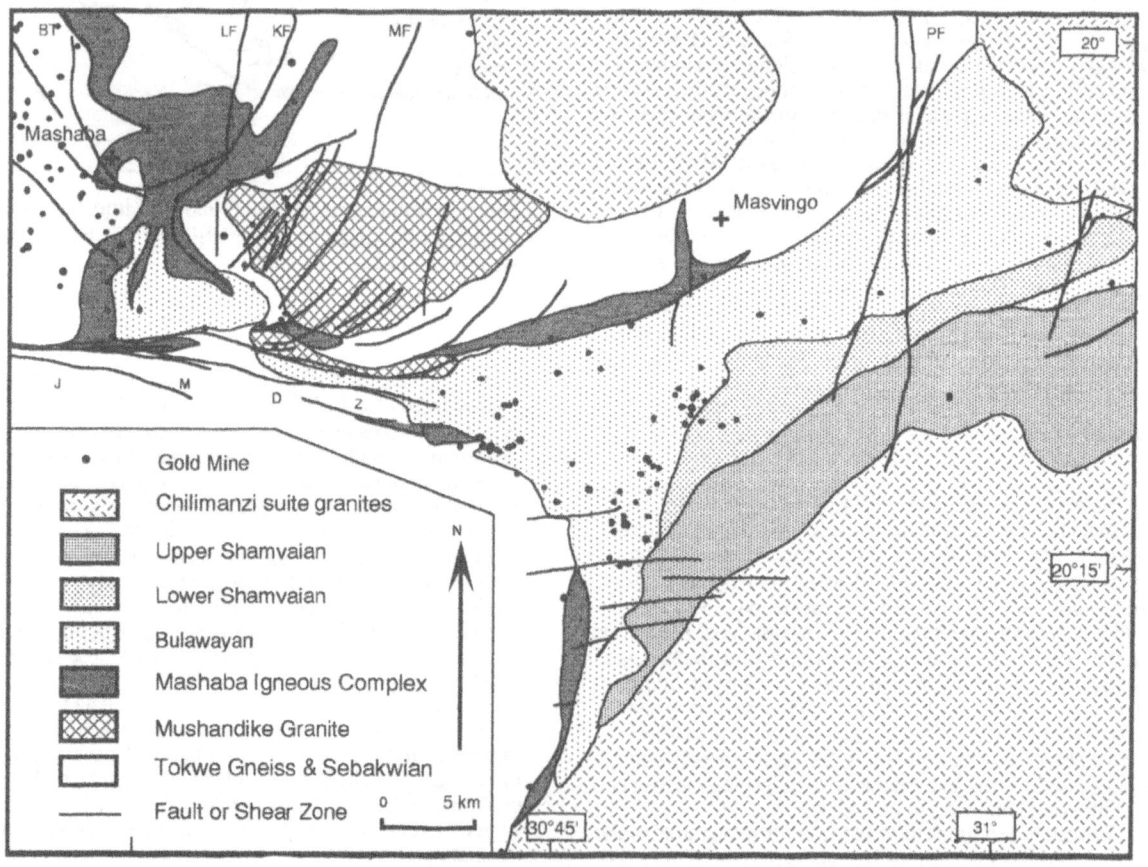

Figure 2. Geology of the west part of the study area (after Wilson [3,4]) JMDZ - Jenya Mushandike Dislocation Zone, BT - Basal Thrust of the Mashaba Igneous Complex, LF - Lochinvar fault, KF - Kenilworth fault, MF - Mushandike fault, PF - Popoteke fault.

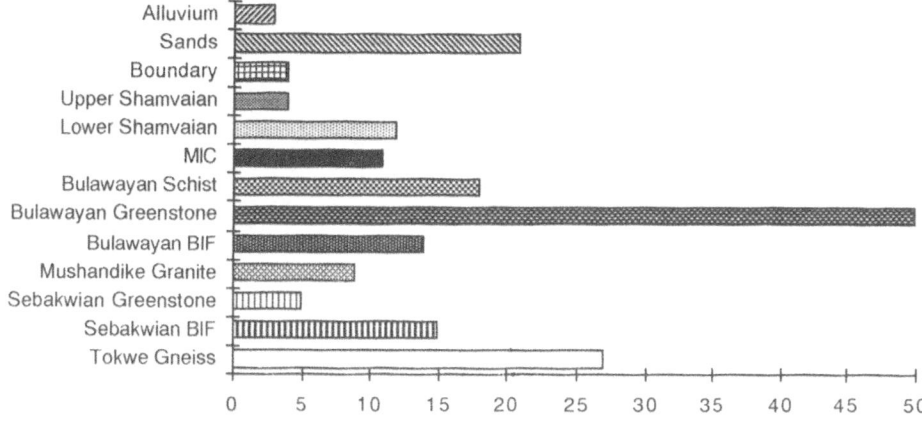

Figure 3. Histogram of number of gold mines grouped by lithology.

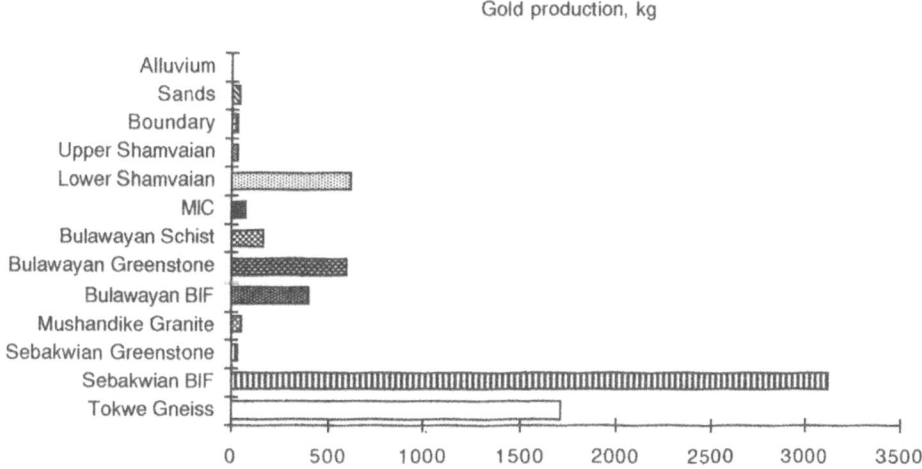

Figure 4. Histogram of production of gold mines grouped by lithology.

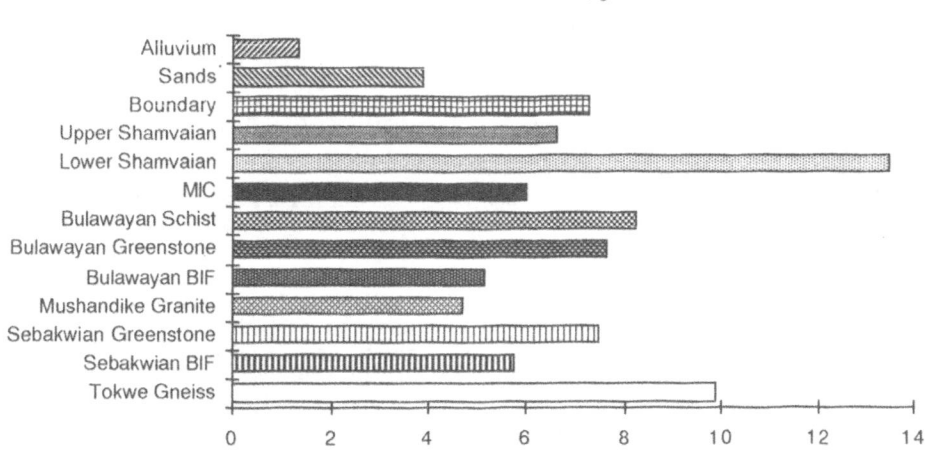

Figure 5. Histogram of grade of gold mines in g/t grouped by lithology.

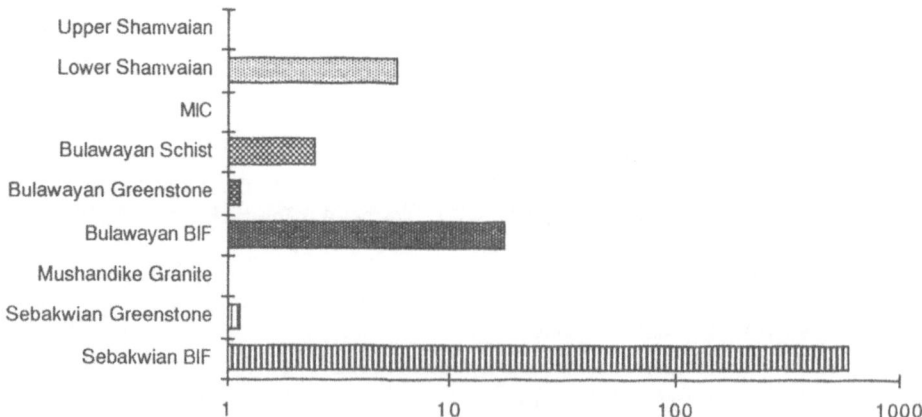

Figure 6. Histogram of concentration of gold in kg/km^2 grouped by lithology.

TABLE 1. Gold production from BIF, BIF area, BIF gold concentration, and BIF proportion

	Total Production	Area	Concentration	Proportion of BIF
	kg	km^2	kg/km^2	%
Upper Shamvaian	26	17.5	1.5	5
Lower Shamvaian	0	1.9	0	2
Bulawayan	410	24	17.4	4
Sebakwian	3132	5.2	599	14

highest concentration was in Bulawayan BIF (17.4 kg/km^2) followed by lower Shamvaian sediments (5.83 kg/km^2) (Figure 6).

Separate analysis of iron formation gold further emphasises the dominance of the Sebakwian, which had the largest total BIF production (Table 1). This is not simply due to variation of BIF outcrop area, which is lower in the Sebakwian than the Bulawayan. The concentration of gold in Sebakwian BIF was therefore much higher (by a factor of 34) than in Bulawayan BIF (Table 1). A bias may possibly be introduced into these figures because Sebakwian production was dominated by a single mine (Empress Mine, 2611 kg). However, even after subtracting this production from the total, more gold has been produced from the Sebakwian than the Bulawayan, and the concentration of gold was more than five times greater in the Sebakwian. No gold was produced from BIF in the Lower Shamvaian, and only a small amount from BIF in the Upper Shamvaian. An overall decrease in gold concentration from the Sebakwian to the Shamvaian is also reported by Foster[7], who pointed out the length of BIF compared to total area of outcrop decreased from the Sebakwian to the Bulawayan and Shamvaian in data for the whole craton. This study shows a large decrease in the proportion of outcrop area occupied by BIF from the Sebakwian to the Bulawayan and to the Lower Shamvaian, but there is an increase in the proportion of BIF in the Upper Shamvaian (Table 1). Concentration of gold in BIF correlates well with the proportion of BIF in the Mashava area, and in the whole craton.

Lithological influences in gold deposition are primarily evident from the range of four orders of magnitude in concentration of gold (Figure 6). BIF, especially in the Sebakwian, clearly exerted a major control on gold concentration. Although many BIF deposits show small-scale structural influences, their overall stratabound geometry and their very high concentration of gold point towards a syngenetic component of mineralization. The correlation between gold concentration in BIF and proportion of BIF

relative to other lithologies shown in Table 1, and noted for the whole craton[7], suggests that gold mineralization and formation of BIF were linked. A combination of both syngenetic and epigenetic processes was suggested by Foster[7] in his study of gold deposits over the whole craton, but Groves et al.[8] have proposed that sulphidisation of fluids by iron in oxide facies-BIF alone may explain gold concentration in Zimbabwe BIF.

Foster[7] showed that the abundance of gold mineralization in BIF was affected by the lithologies associated with BIF, with mafic and ultramafic rocks correlating with higher concentrations. This effect is not apparent in Sebakwian and Bulawayan BIF gold concentrations, since both are associated with similar mafic and ultramafic rocks, but it may be significant that the Shamvaian BIF, with a lower concentration, is interleaved with pelitic schists. The most simple explanation for the influence of BIF on gold concentration is that a component of BIF gold mineralization, particularly in the Sebakwian, is syngenetic, but field evidence is equivocal since many deposits also show evidence of some structural control.

STRUCTURES IN THE MASHAVA AREA

The major structure of the area is the Jenya-Mushandike Dislocation Zone (JMDZ), which comprises the Jenya fault system west and south of Mashava, and the Mushandike shear zone south of the Mushandike granite[5] (Figure 2). The JMDZ is a major crustal lineament extending for over 100 km in the south part of the Zimbabwe craton, with a long and complex history. Early penetrative deformation along the JMDZ may displace Sebakwian and Tokwe gneisses by 10 to 20 km in a sinistral sense. A later phase of deformation involved penetrative dextral strike-slip movement, and movement on a network of smaller shear zones in the Mushandike granite[9], which join the JMDZ. These shear zones have sub-horizontal stretching lineations, and record both sinistral and dextral shear. They may have formed as a conjugate network, which was rotated by dextral simple shear in the Mushandike granite. The latest phase of deformation was discrete faulting on the Jenya fault, shown by 1-2 km of dextral offset on the Great and East dykes, west of the study area.

The base of the northwest arm of the MIC is a thrust which defines another major regional structure that extends over 45 km to the northwest. The MIC has been thrust towards the southwest along its basal thrust[3]. There are several NW-SE striking faults sub-parallel to the basal thrust to the west of Mashava, which have sinistral strike separations of a few kilometres.

Smaller structures within the study area include a number of NNE-trending faults that cut the Mushandike granite and the MIC, such as the Lochinvar, Kenilworth and Mushandike faults (Figure 2). These have dextral strike separations of a few hundred metres to several kilometres, and do not appear to affect the Chilimanzi granite. The north-striking Popoteke fault cuts the Masvingo-Mashava greenstone belt east of Masvingo, and has 2 km of sinistral strike separation (Figure 2). A number of small east -west faults cut the southern part of the greenstone belt.

Folding can be seen in Sebakwian, Bulawayan and Shamvaian quartzites, BIF, conglomerates and limestones. Tight to isoclinal folds in the Sebakwian with northerly plunges can be distinguished from later Bulawayan folding of poorly defined geometry, which pre-dated the MIC. However the main episode of folding was post-Shamvaian, which created the major tight to isoclinal folds with moderately plunging, northeast-trending fold axes[3].

STRUCTURAL CONTROLS ON MINERALIZATION

Gold has been produced almost entirely from the western end of the study area. There was only a single small mine in the whole of the narrow eastern part of the greenstone belt outside the boxed area of Figure 1. Within Figure 2, gold mines were conspicuously concentrated in two areas: around Mashava in the east, both within the greenstone belt and older gneisses; and mainly in the greenstone belt south of Masvingo. This concentration is emphasised by the pattern of total production, which was dominated by the outputs of the three largest mines, (Empress Mine, 2611 kg, Texas Mine, 839 kg and Cambrian Mine, 751 kg) in the area just south and east of Mashava (E,T,anc C in Figure 7), and by the production of the Coronation and Koodoo mines (503 kg and 101 kg) south of Mashava (N and K in Figure 7).

The distribution of mines around Mashava defines a broad linear zone that trends NW-SE towards the JMDZ where it is terminated (Figure 7). This zone is parallel to the basal thrust of the MIC, and several other faults. The Mushwe claims were worked from quartz veins marking the basal thrust to the MIC, which has also been the site of asbestos formation. Wilson and Nutt [10] have attributed both gold mineralization and asbetsos formation to a common fluid flux associated with the basal thrust, during the "main deformation" event of Wilson[5], in which crustal blocks moved towards the southwest. South of Masvingo, there was a prominent NE-SW zone of production in the centre of the greenstone belt, and a NW-SE zone along the southwest margin of the belt that terminates against the JMDZ. These zones cut across units and delimit gold distribution on the largest scale. They may overlie important lines of crustal weakness that allowed transport of auriferous fluids, and that were restricted to the west of the greenstone belt.

Only four mines lay directly on traces of the JMDZ, with the small total production of 22 kg. Lack of significant gold production directly located on large scale structures has been noted in several places[11,12], for example in the Superior Province, Canada, where it is well known that gold is found in

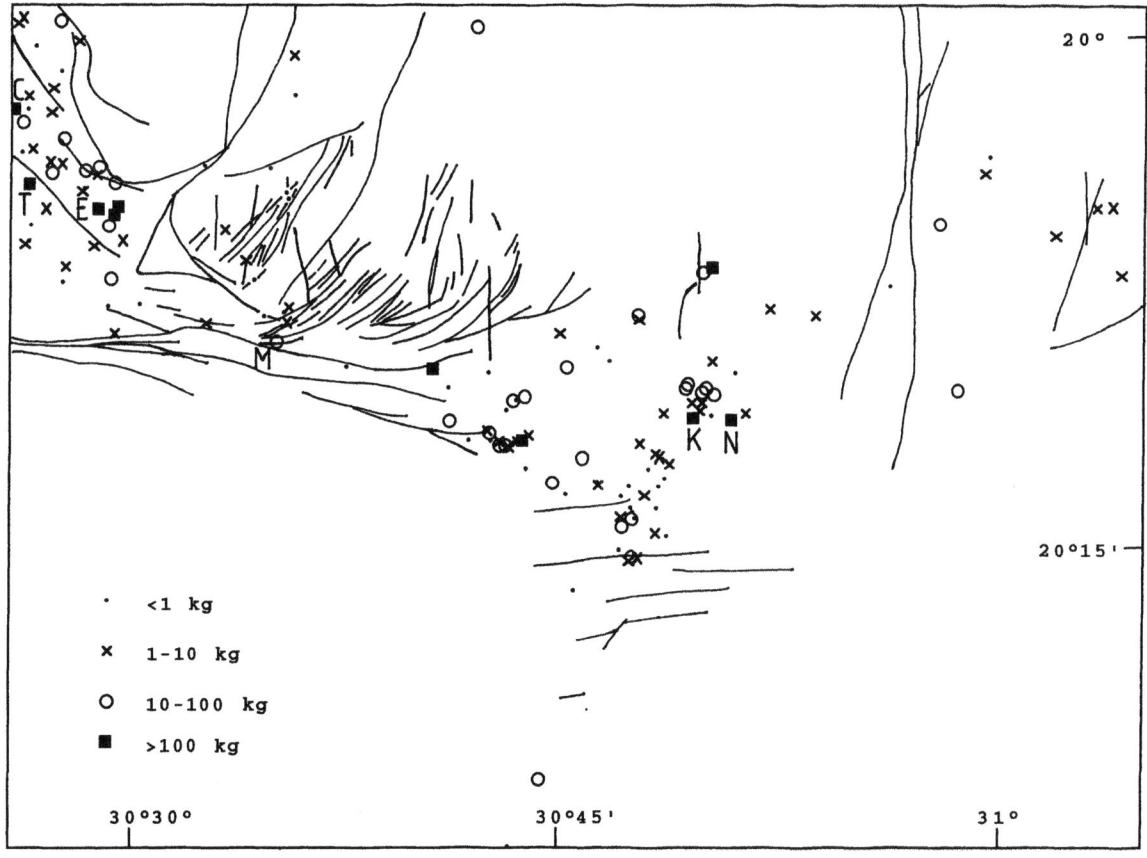

Figure 7. Location and output of gold mines in the west part of the study area. Major structures are shown by lines. C- Cambrian Mine, E- Empress mine, K- Koodoo mine, M- Margrate Neilsen mine, N- Coronation mine, T- Texas mine.

smaller, subsidiary "splay-structures" adjacent to the major "breaks". This seems to be the case for the JMDZ, which did not have large mines directly on the structure itself. However, there is a clear relationship between the eastern end of the JMDZ, and a zone of production trending ENE towards Masvingo near the north edge of the greenstone belt (Figures 2 and 7). This zone is parallel to several splays of the JMDZ where it splits at its eastern end. The Margrate Neilsen Mine (production 53 kg, M in Figure 7) shows very clearly how mineralization is located in smaller structures adjacent and related to the JMDZ. The mine exploited mineralization within micaceous quartz schists in a dextral strike-slip shear zone a few metres wide. Fabrics and rock types are similar to those on the adjacent part of the JMDZ, which is a few hundred metres further south. This shear zone joins the JMDZ, and was active together with the JMDZ in the same dextral shearing event.

Structural influences on gold also operated at a smaller scale, in common with many other gold deposits within Zimbabwe[10,13]. Many deposits occur in localised deformation zones or veins[3,4] a few of which are large enough to be marked at the scale of Figure 7. This is clearly demonstrated in eight deposits in and around the western end of the Mushandike granite, which all occur in north to northeast striking shear zones (not all shown on Figure 7), which are part of the network that links with the JMDZ. Yet it is noticeable that very similar shear zones in and near the granite to the east have not yielded gold, possibly because they do not overlie the crustal conduits that created the NE-SE zone of production crossing the western part of the granite. Descriptions of BIF deposits show some of them were strongly influenced by small-scale structures including fractures, fold hinges and cataclastic zones[3,4], so that a clear distinction cannot be made between lode and iron formation types of deposit in many cases. The Empress Mine is a good example[3]. The size and shape of the ore bodies is determined by the intersection of E-W fractures with the BIF. Mineralization also occurs in the footwall of a quartzite, which may have acted as an impermeable barrier to mineralizing fluids, except where it was breached by E-W fractures, giving rise to a rich zone of mineralization in the quartzite.

These observations support the hypothesis that gold deposition may have been caused by episodic fluctuations in fluid pressure, which requires a structural setting where fluid pathways can be sealed and periodically ruptured, possibly by seismic processes[14,15]. This may not have been possible along the largest structures such as the JMDZ, which remained permeable because they are long and continuous[16], but could have occurred where smaller barriers exist such as faults and impermeable strata, as in the Empress example. Tectonic/fluid interaction may therefore have been a fundamental control on gold deposition in the study area.

AGE OF MINERALIZATION

The maximum age of a gold deposit is the age of the rock in which it occurs. On the basis of existing age data[5]. maximum ages can be assigned to gold deposits in the following units: Sebakwian (3.5 Ga), Mushandike granite (2.85 Ga), Bulawayan and MIC (2.7 Ga), Lower Shamvaian (2.65 Ga), Upper Shamvaian (2.63 Ga). A histogram showing the maximum age of gold in the study area is shown in Figure 8. A minimum age of ~2.6 Ga is likely for all gold because of the complete absence of gold deposits in the Chilimanzi suite granites, which also cut across some deposits. The maximum possible accumulation of gold by any age can be found by subtracting gold younger than this age from the total amount of gold: this is shown in Figure 9 for both total gold and gold in BIF. The stepped shape of both curves permits a variable rate of total gold accumulation and rate of gold accumulation in BIF through the Archaean. A maximum of about two-thirds of the total could have accumulated from the Sebakwian to the Bulawayan, and almost no more before the Shamvaian. It is also possible that the total amount of gold was deposited between the Shamvaian and the Chilimanzi granite at 2.6 Ga. The shape of the curves for total and BIF gold is remarkably similar.

The occurrence of gold in the relatively late dextral shear zones of the Mushandike granite, and within structures that can be related to post-Shamvaian deformation, shows that some mineralization over a large part of the study area was post - Shamvaian. Wilson suggested that the major portion of mineralization occurred at this time[3]. This has interesting similarities to the emerging consensus that gold was deposited late in the evolution of greenstone belts[17], (at least in greenschist facies domains), and to the results of direct isotopic dating, which show that gold was younger than any geological events recognised in the belts[18]. However, significant gold accumulation is allowed between 3.5 and 2.85 Ga by the data in Figure 9. Gold deposition in this earlier period is strongly suggested by the order of magnitude decrease in gold concentration from BIF in the Sebakwian to Bulawayan BIF, and the correlation between gold concentration in BIF and proportion of BIF noted above. Gold concentrations are very similar in Sebakwian and Bulawayan greenstones, indicating that they may have been mineralised in a single, later event. These observations are most consistent with two main periods of mineralization: an early episode when gold was concentrated in Sebakwian BIF, and a late episode at the end of the Shamvaian, when lode deposits were formed in all lithologies. This history would also lead to the similarity between the shape of the curves for total and BIF maximum accumulations shown in Figure 9.

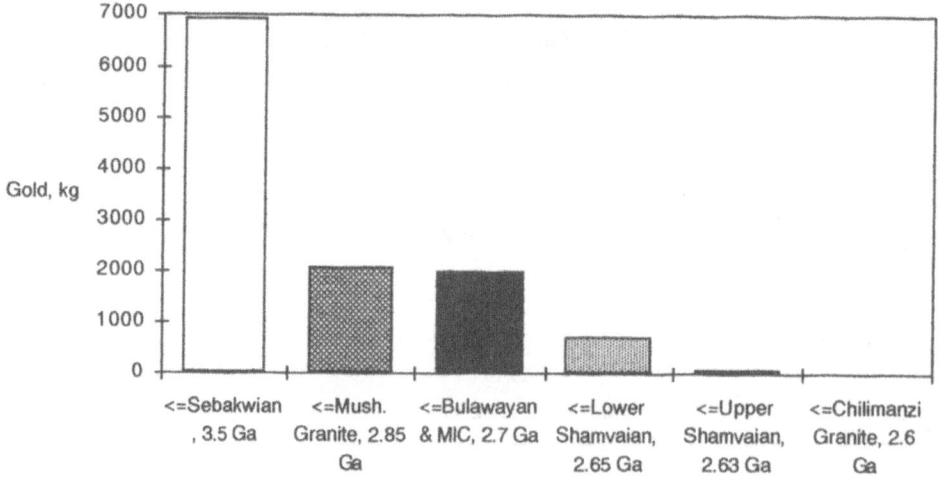

Figure 8. Maximum age of gold mineralization, given by the maximum age of the rock in which the gold occurs.

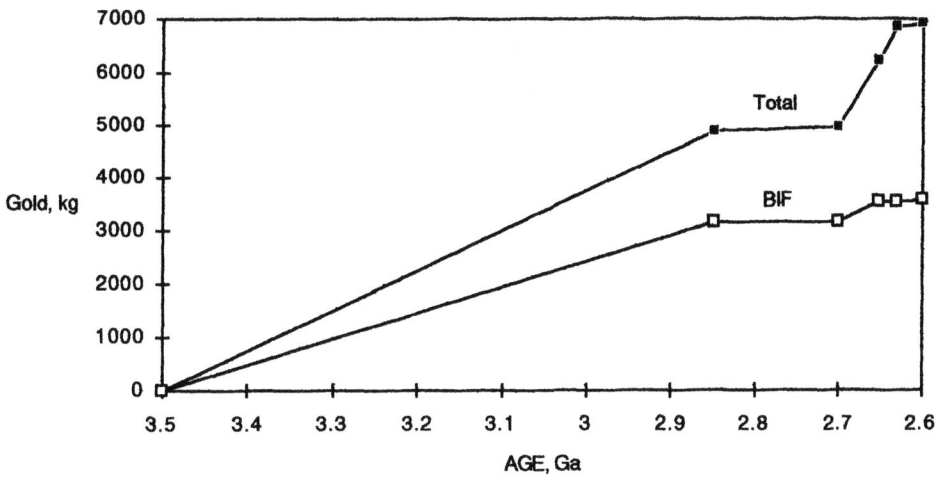

Figure 9. Maximum possible gold accumulation by age, calculated at each age by subtracting all gold younger than this age from the total amount of gold.

SOURCE OF GOLD

Limited conclusions can be drawn from this study about sources of gold. Identification of gold sources is complicated since several stages of concentration may have occurred, each starting from a different source. Unlike the results of Foster[7], there is a complete lack of correlation between concentration of gold excluding BIF deposits and concentration within BIF of similar age in the study area. This implies that BIF was not a source of gold in other deposits at any stage. Two other possible source rocks for Zimbabwe gold deposits have been suggested[7]: sulphides deposited in the inter-pillow areas of pillow-lavas, leached from the tholeiitic lavas themselves; and immiscible sulphide melts that scavenged gold from komatiitic magmas immediately prior to eruption. The latter explanation was proposed for deposits within the Mashaba Ultramafic Suite, which includes the MIC. However, production of significant gold from every major lithology, and lack of gold concentration focussed on the outcrop of a single lithology, suggest either that there is no single gold-enriched source rock, or that gold remobilisation has been very extensive.

CONCLUSIONS

Gold production in the Mashava area was concentrated in Sebakwian and Bulawayan banded iron formations (BIF), and in Upper Shamvaian grits and conglomerates. Lithological control on the sites of gold deposition operated on an intermediate scale by concentrating gold in these units. The correlation between gold concentration in BIF and proportion of BIF relative to other lithologies suggests that some BIF mineralization could have been syngenetic, but small-scale structures including fractures, fold hinges and cataclastic zones also played a role in creating many BIF deposits.

Gold production was almost entirely restricted to the west end of the Masvingo-Mashava greenstone belt, where it occurred along a broad NW-SE zone around Mashava, which is parallel to the basal thrust of the Mashava Igneous Complex and several other NE-trending faults. NE-SW and NW-SE zones of production occurred south of Mashava. The zones cut cross the major lithostratigraphic units of the area, and may have overlain crustal weaknesses through which fluids could pass. The NW-SE zone around Mashava is terminated by the Jenya Mushandike Dislocation Zone (JMDZ), which has not produced significant gold directly from the structure itself. However, there was a zone of production parallel to splays of

the JMDZ at its eastern end. Fractures and deformation zones control most deposits on a smaller scale.

Mineralization may have occurred in two main periods, firstly at some time between 3.5 and 2.85 Ga, and secondly at the end of the Shamvaian just before 2.6 Ga. An early period of mineralization in Sebakwian BIF is suggested by the exceptionally high gold concentration, which may be partly syngenetic. Most gold in other lithologies may have been introduced in the later period, during the final stages of Shamvaian deformation, including dextral shear along the JMDZ, before the intrusion of the Chilimanzi suite granites. A late pulse of mineralization is also seen in several other greenstone belts around the world.

No single source rock for the gold can be identified. The auriferous fluid flux occurred over a large area and was therefore possibly derived from deeper crustal levels, from where it was focussed upwards along broad zones and deposited in smaller structures, which were periodically sealed, allowing fluid pressure cycling. Fluctuations in fluid pressure and gold deposition did not occur directly in the JMDZ. Control of gold deposition by fluid pressure may be common in other greenstone belts in Zimbabwe, where structural controls are already recognised, which implies that more detailed models of the interaction between structure, fluid pressure, and mineralization may be important for exploration.

REFERENCES

1. Foster, R.P. Archaean gold mineralization in Zimbabwe: Implications for metallogenesis and exploration. Bicentennial Gold 88, Melbourne, May 1988, p. 62-72.
2. Wilson, J.F. A Preliminary Reappraisal of the Rhodesian Basement Complex. Special Publication of the geological Society of South Africa, 5, 1-23.
3. Wilson, J.F. The Geology of the Country around Mashaba. Bulletin of the Rhodesian Geological Survey, 68, 1968, pp. 239.
4. Wilson, J.F. The Geology of the country around Fort Victoria. Bulletin of the Rhodesian Geological Survey, 58, 1964, pp. 147.
5. Wilson, J.F. A craton and its cracks: some of the behaviour of the Zimbabwe block from the late Archaean to the Mesozoic in response to horizontal movements, and the significance of some of its mafic dyke fracture patterns. Journal of African Earth Sciences, 10, 1990, 483-501.
6. Moorbath, S., Taylor, P.N., Orpen, J.L., Treloar, P., and Wilson, J.F. First direct radiometric dating of Archaean stromatolitic limestone. Nature, 326, 1987, 865-867.

7. Foster, R.P. Major controls of Archaean gold mineralization in Zimbabwe. Transactions of the Geological Society of South Africa, 88, 1985, p. 109-133.

8. Groves, D.I., Phillips, G.N., Ho, S.E., Henderson, C.A., Clark, M.E., and Woad, G.M. Controls on distribution of Archaean hydrothermal gold deposits in Western Australia. Gold'82: The Geology, Geochemistry and Genesis of Gold Deposits, edited by R.P. Foster, Geological Society of Zimbabwe Special Publication 1, 1984, 689-712.

9. Blenkinsop, T.G., Dhilwayo, J., and Muranda, S.C. Intracratonic shearing on shear zones of the Mushandike granite, Zimbabwe. Proceedings of the Second Symposium of Science and Technology, Research Council of Zimbabwe, in press.

10. Wilson, J.F. and Nutt, T.H.C. The nature and occurrence of mineralization in the early precambrian crust of Zimbabwe. In: Precambrian Continental Crust and its Economic Resources, Edited by Naqvi, S.M., Elsevier, Amsterdam. in press.

11. Dube, B. Contrasting structural styles of gold-only deposits and occurrences in the Canadian Appalachians and their relationship to major fault zones: the example of western Newfoundland. Abstract, NUNA research conference on Greenstone Gold and Crustal Evolution, Geological Association of Canada/Society of Economic Geologists, Val d'Or, Quebec, Canada, 1990, p. 35.

12. Heather, K.B., Fyon, J.A., Muir, T.L., and Troop, D.G. Some empirical characteristics of Archaean gold deposits: Implications for Depositional and genetic models. Abstract, NUNA research conference on Greenstone Gold and Crustal Evolution, Geological Association of Canada/Society of Economic Geologists, Val d'Or, Quebec, Canada, 1990, p. 47.

13. Nutt, T.H.C., McCourt, S., and Vearncombe, J.R. Structure of some gold and antimony-gold deposits from the Kaapvaal and Zimbabwe cratons. Publication No. 12, Geology department & University Extension, The University of Western Australia, p. 63-80.

14. Sibson, R.H. Earthquake rupturing as a mineralising agent in hydrothermal systems. Geology, 15, 1988, p. 701-704.

15. Sibson, R.H., Robert, F.R., and Poulsen, K.H. High angle reverse faults, fluid-pressure cycling, and mesothermal gold-quartz deposits. Geology, 16, 1988, 551-555.

16. Cox, S.F. Fluid pressure regimes and fluid dynamics during deformation of low-grade metamorphic terranes - implications for the genesis of mesothermal gold deposits. Abstract, NUNA research conference on Greenstone Gold and Crustal Evolution, Geological Association of Canada/Society of Economic Geologists, Val d'Or, Quebec, Canada, 1990, p. 30.

17. Robert, F., Phillips, G.N., and Kesler. S.E. Introduction. Abstract volume, NUNA research conference on Greenstone Gold and Crustal Evolution, Geological Association of Canada/Society of Economic Geologists, Val d'Or, Quebec, Canada, 1990, p. 6 - 10.

18. Clark, M.E., Krogh, T.E. and Archibald, D.A. U-Pb and Rutile Ages and 40Ar/39Ar Biotite ages for the Victory Mine, Kambalda, Western Australia: constraints on the age and P-T-Time conditions of mineralization. Abstract, NUNA research conference on Greenstone Gold and Crustal Evolution, Geological Association of Canada/Society of Economic Geologists, Val d'Or, Quebec, Canada, 1990, p. 23.

Geological setting of gold deposits in the Mutare Greenstone Belt, Zimbabwe

K. G. Chenjerai
Geological Survey Department, Harare, Zimbabwe

ABSTRACT

The Mutare Greenstone Belt is defined here as stretching from the Odzi River in the west right to the Mozambique border in the east.

The Mutare Greenstone Belt is an east-west trending synclinorium of ultramafic, mafic and banded-iron formations corresponding to the Bulawayan Group. Overlying this succession and occupying the core of this structure are metasediments of the Mbeza System (Shamvaian). Both the northern and southern margins are intruded by granites. The northern limb of the synclinirium is intruded by quartz-dolorites and felsites.

The axis of the regional folding trends east-west along the centre of the syncline. Fractures and shear zones parallel to the regional fold axis host more than 90% of the known gold occurrences. The gold deposits occur as quartz veins filing fractures or as impregnations in shear zones. A few gold deposits are hosted in shear zones that are oblique to the regional deformation and cut across different lithologies.

All the Archean rock types within this greenstone belt are known to host gold mineralisation. More than 50% of the known gold occurrences are in the ultramafic and mafic metavolcanics. About 25% are hosted in felsic intrusives. The metasediments host about 15% and the banded-iron formations account for about 8%.

On a regional scale, there appears to be no direct lithological control on the gold distribution, however, a strong structural control is apparent. On the small scale of the individual deposit, the gold mineralisation seems to have been affected by a complex interplay between structure and lithology.

INTRODUCTION

For the purpose of this study, the Mutare Greenstone Belt is defined as stretching from Odzi River in the west to the Mozambique border in the east (Fig 1).

Fig. 1. General geology of the Mutare greenstone belt.

The historical production records have been taken from Geological Survey Bulletins 32[1] and 45[2]. Some of the mines stopped producing prior or shortly after the publication of the two Bulletins. However, a few mines have persistently produced and these will be dealt with briefly later on in the discussion.

The objectives of this paper are to look at the geological setting of gold deposits in the Mutare Greenstone Belt and to study the possible controls of the gold mineralisation, an area under current research by the author.

GENERAL GEOLOGY

The Mutare Greenstone Belt is an east-west striking synclinorium of ultramafic and mafic lavas with intercalations of banded iron-formations[1]. This succession corresponds to the C2, 7Ga Upper Bulawayan[3]. The core of this structure is occupied by predominantly metasediments of the Mbeza System which corresponds to the Shamvaian Group[3]. Both the north and south margins of the synclinorium are embayed by intrusive granites and in the northern limb the greenstones are intruded by the Penhalonga quartz-diorite.

1. Bulawayan Group

Underlying the Mbeza metasediments are the ultramafic, mafic and iron-formation rocks which have been downfolded into a synclinal structure approximately 2 000 m thick[4]. The ultramafic lavas show a wide range of lithological variations from a very foliated talcose schist to a massive fairly jointed rock, devoid of foliation. The spinifex textures and columnar structures strongly indicate that the ultramafic rocks are komatitic lavas[5].

The mafic volcanics generally lack foliation and vary from lapilli to pillow lavas, mainly of tholeiitic composition.

The banded iron formations occur as discontinuous units within and at the contact of the ultramafic and mafic lavas. Bedding is poorly developed and intercalations of quartzite and argillite are common[4]. The baded ironstones are ferruginous quartzites composed of alternating bands of whitish cherty quartz and black iron ore magnetite[6]. The banding, where developed, ranges from a mm to a cm scale and disseminated sulphides, notably pyrite, are common.

2. Shamvaian Group

Uncomformably overlying the predominantly volcanic rocks of the Upper Bulawayan and filling the core of the synclinorium is a succession of metasediments most of which were derived by erosion of the older greenstones[1]. At the base is a well developed basal conglomerate composed of angular to subrounded pebbles of actinolite schist, banded iron-formation, metabasalt, serpentinite and quartz within a greywacke matrix. Overlying the basal conglomerate are varying thicknesses of grit, arkose, greywacke, phyllites and slates.

3. Felsite

The felsite appears to be limited to the volcanics of the Bulawayan Group. The felsite north of Penhalonga Valley occurs as a more or less continuous lenses. Invariably associated with the felsite are quartz porphyries which appear to grade into and out of the very fine grained, commonly cryptocnystalline quartzo-feldspathic felsites. Rounded to sub-angular quartz phenocrysts are prominent in the very fine grained felsic matrix of the quartz porphyries. At the Redwing and Old West mines, the felsite has a sill-like geometry.

4. Penhalonga Quartz-Diorite

The Penhalonga quartz-diorite is a two phase diorite with an older fine grained dark greenish-grey massive rock and a younger massive, medium to coarse grained dark grey rock[1]. The Penhalonga quartz-diorite is similar to the diorite outcropping in the area around the Old Mutare Mission.

5. Granite

North of the Penhalonga Valley, the granites range from massive to foliated with the foliation increasing towards the contact with the greenstone belt. South of the greenstone belt, the granite(s) have an irregular contact with the greenstones and range from massive to gneissic.

STRUCTURE

The Mutare Greenstone is a tightly folded synform[1] with the Mbeza metasediments occupying the central portion of the belt. The synclinal form of the greenstone belt is notably asymmetrical, with the southern limb dipping nearly vertical to a major synclinal axis which plunges to the east at an almost imperceptible angle[7]. The northen limb dips southwards at a shallow angle to a minor syncline, the axis of which plunges eastwards at approximately 25° from the horizontal.

Current research by the author in the Penhalonga Valley has shown that the Penhalonga Valley possibly occupies a syncline. The banded iron-formation separating the Mbeza Valley from the Penhalonga Valley is folded and this marks an anticlinal axis between the two synclines.

The Mbeza syncline is a symmetrical fold which terminates approximately five kilometres to the east of the Mozambique border where[7] the axial plunge becomes westerly[7]. The eastward plunging Penhalonga syncline becomes the major fold in Mozambique.

In the Penhalonga Valley, the dorminant east-west striking foliation is present in all lithogies ranging from a fracture cleavage in the metabasalt to a slaty cleavage in the ultramafic rocks. The foliation which is parallel to the lithological contacts, dips steeply either north or south (Fig 2).

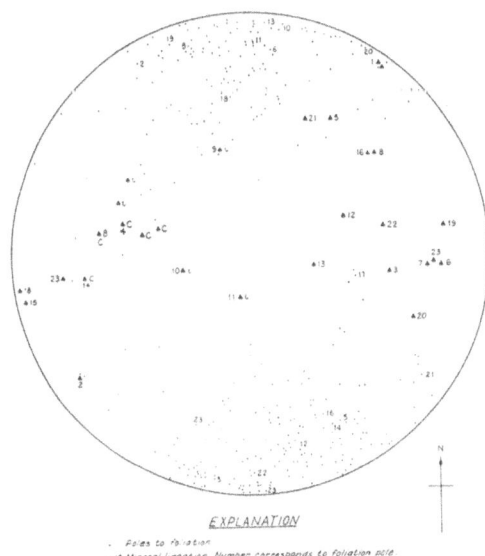

Fig. 2. Poles to foliation.

The mineral lineation defined mostly by actinolite, plunges shollowly towards the east (Fig 2).

Lapilli structures in the metabasalt and pebbles of the basal conglomerate define a stretching lineation that plunges shallowly to the east. The sub-vertical foliation and the sub-horizontal mineral lineation as well as stretching lineation are indicative of

strike slip movement.

GOLD MINERALISATION

The two-fold classification of stratabound and non-stratabound deposits[9], is used in the following discussion :

Stratabound	: Iron-formations Volcaniclastic-hosted
Non-stratabound	: Veins Mineralised shear zones

No volcaniclastic-hosted gold deposits have been recognized in the Mutare Greenstone Belt.

The distribution of gold deposits in the Mutare Greenstone Belt is shown in Figure 3. The individual gold deposits generally strike east-west, parallel to the foliation in the country rocks. The amount of dip varies from medium to high angle either towards north or south. Some gold deposits such as the Rezende shear are at an angle to the general east-west direction.

Nine mines have recorded a total production of more than 300 kg of gold each and these are shown in Figure 3.

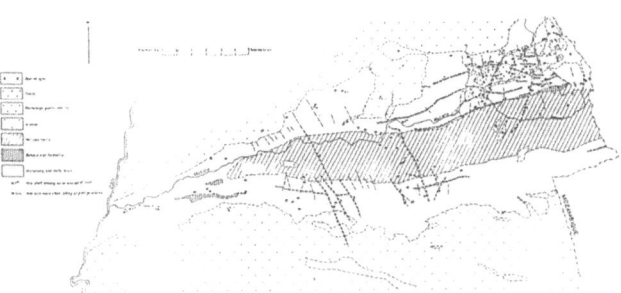

Fig. 3. Distribution of gold deposits in the Mutare greenstone belt.

1. Gold in Ultramafic and Mafic Lavas

In terms of aerial distribution (Fig 1), the ultramafic and mafic lavas are the commonest lithologies in the Mutare Greenstone Belt. The proportion of gold mines in different host lithologies is shown in Figure 4 and 52% of the known gold deposits are hosted in the komatiitic and tholeiitic lavas. 76% of the total gold production has come from deposits in mafic volcanics with the ultramafic contributing 16,7%. The total of 78,7% for the whole craton compares very well with the 76% from the Mutare Greenstone Belt.

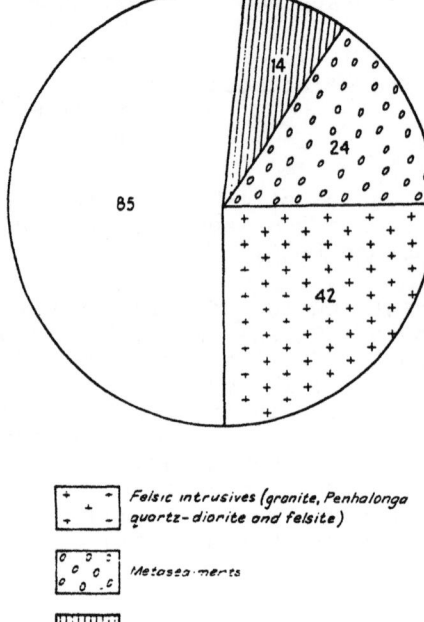

Figure 4. Number of mines in different host lithologies.

Legend:
+ Felsic intrusives (granite, Penhalonga quartz-diorite and felsite)
o Metasediments
Banded iron formation
Ultramafic and mafic lavas

Three styles of gold deposits in the ultramafic and mafic lavas have been noted[2].

(i) Quartz veins within the gold belt with wall rocks of schist. The veins contain galena and pyrite and, less commonly, pyrrhotite, chalcopyrite, crocoite and bismuth minerals. The Rezende shear falls in this category.

(ii) Shear zones in which the schists are mineralised along shear zones, quartz being rare or absent. Characteristically such shear zones carry gold and a great deal of arsenic which is oxidised near the surface to scorodite. The Champion deposit is a typical example, and

(iii) Veins occurring on the main contact of the greenstone and the granite. These appear to be unimportant.

Four mines have recorded a total production of more than 300 kg each and two of these are described in detail below.

a. Champion Mine : the quartz vein is one of a number of shear zone-hosted veins branching off the main east-west striking Speedo shear[11]. Having a mean width of just over a metre, the vein dips 70° - 80° towards the north and is encased in a narrow zone of mineralised and silicified schist[12]. Lensoid quartz stringers appear to be fairly persistent in the body. The sulphide mineral is arsenopyrite accompanied by chalcopyrite, pyrite, pyrhotite and galena. The Champion Mine has produced 329 kg of gold.

b. Rezende Mine : the Rezende shear dips at an average 70° towards north-west. The lode is a body of lensoid quartz stringers varying in width from less than a metre to more than 2 metres. The quartz varies from white to dark grey. Crack-seal textures are very common giving the lode a banded appearance. Pyrite is the main sulphide mineral with pyrrhotite, chalcopyrite and scheelite being accessory. The Rezende shear cuts across both the basaltic greenstones and the Penhalonga quartz-diorite and is terminated in the north-west by a fault. The Rezende Mine has recorded a total production of 26 110 kg of gold and is currently being reclaimed.

2. Gold in Felsite, Penhalonga Quartz-Diorite and Granitic Rocks

Gold deposits in granitic rocks are insignificant. The major gold deposits are in the felsite and Penhalonga quartz-diorite with 25% of the known gold deposits being hosted by these rocks. The quartz veins are commonly encased in a quartz-sericite schist. The veins are generally narrow lenses ranging from less than 10 cm to almost 100 cm wide. Pinching and swelling of the veins both down dip and along strike is a common characteristic. The quartz is often laminated representing a typical crack-seal texture. The mineralisation of these veins is fairly simple, usually consisting of free gold, pyrite and galena.

In terms of gold production, these deposits have contributed 9% of the total gold production (Fig 5).[12] For the whole of the Zimbabwean craton, similar deposits have accounted for 13,3% a figure that compares fairly well with the 9% calculated for the Mutare Greenstone Belt.

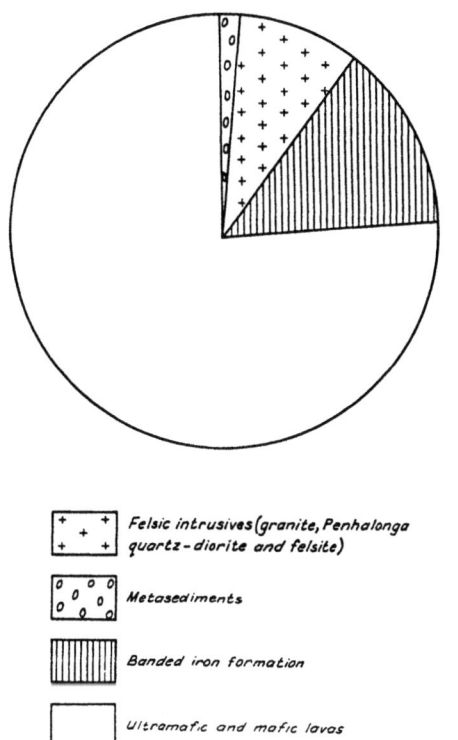

Felsic intrusives (granite, Penhalonga quartz-diorite and felsite)

Metasediments

Banded iron formation

Ultramafic and mafic lavas

Figure 5. Percentage gold production.

The four mines that have produced more than 3000 kg each are briefly described below.

a. The Kent Mine occurs within a quartz-sericite schist defining a shear zone in the Penhalonga quartz-diorite. The shear strikes east and dips 50° towards the north. Coarse crystals of pyrite and pyrrhotite in the orebody have been noted. The mine produced a total of 944 kg of gold.

b. King's Daughter : two generally narrow quartz veins striking east-west and dipping 20° - 30° towards the south. The host rock is quartz porphyry and felsite. Galena, arsenopyrite, pyrite and chalcopyrite are the main sulphide minerals with accessory pyrrhotite[4]. Good values appeared to correlate with an increase in the galena content of the veins[14]. The mine produced 1 008 kg of gold.

c. Old West : three styles of sulphide mineralisation at the Old West Mine have been noted[15].

(i)   Mineralisation within the felsite; the felsite hosts quartz veins ranging in width from less than 10 cm to about 30 cm and dipping 30° - 40° towards the east. The quartz is translucent white to grey and contains pyrite, galena, sphalerite, pyrrhotite and gold.

(ii) Along the felsite/quartz-diorite contact is a zone of disseminated sulphide enrichment. This zone extends 3 to 5 cm into the quartz-diorite, and

(iii) Massive sulphide veins occur both in the felsite and quartz-diorite. Where the veins occur in the felsite they tend to be planar bodies, 3 to 10 cm wide, and consist mainly of pyrite and galena with lesser chalcopyrite, spalerite, pyrrhotite and gold. There are no major textural and mineralogical variations when the veins pass into the quartz-diorite.

The Old West Mine is still producing more than 400 kg of gold annually.

d. Redwing Mine : the Redwing orebodies[16] are very similar to the Old West orebodies[15]. In the west, the Kent Dyke has displaced the Redwing felsite.

The Redwing Mine is still producing more than 500 kg of gold per year.

3. Gold in Metasediments

15% of the known gold deposits in the Mutare Greenstone Belt are hosted in Mbeza metasediments (Fig 4). The individual gold deposits are essentially lensoid quartz stringers filling fractures and shear zones in the metasediments.

These deposits have contributed 1% of the total gold production, with most of the mines producing a total of less than 15 kg each (Fig 5). A figure of 6,2% total gold production from sedimentary rocks of the whole Zimbabwean craton has been calculated[10], a figure that does not compare well with the 1% from the Mutare Greenstone belt. However, on the whole, the two figures generally indicate the small gold output from metasedimentary host rocks.

4. Gold in Banded Iron Formations

As shown in Figure 4, 8% of the known gold deposits in the Mutare Greenstone Belt, are hosted in banded iron formations. These stratabound deposits have contributed 14% of the total gold production (Fig 5), a figure that compares fairly well with the calculated 12,8% for the whole of Zimbabwe[10].

BIF-hosted mineralisation is essentially narrow quartz vein stringers e.g. the Home Mine[1]. At the Arthur Mine, the mineralisation is hosted partly in brecciated banded iron formations and partly in the Shamvaian metasediments[1]. The vein is white quartz carrying a little galena[1]. The Pension orebody is a essentially brecciated banded iron formation which has been thickened by faulting and folding and has been almost completely mineralised with pyrite.

POSSIBLE CONTROLS OF GOLD MINERALISATION

The structural control of gold mineralisation on an orebody scale in the Odzi Greenstone Belt has been noted[2]. At Redwing Mine, north-south caused shortening brittle fracture within the felsite and the formation of numerous easterly-dipping parallel, low angle fracture planes[17]. Mineralisation along these planes gave rise to a series of parallel veins in the felsite. Continuing deformation folded the felsite and veins into anticlines and synclines of small altitude which plunge at a shallow angle to the east.

For the rest of the greenstone belt, as is shown in Figure 3, a structural control on the gold deposits is apparent. More than 90% of the known gold deposits strike east-west, parallel to the regional lithological contacts and foliation. At the Kenya Mine, the orebody is a mineralised shear zone parallel to the foliation and dipping south. Quartz vein impregnations parallel to the foliation are common occurrence e.g. the Arram, Ester and Leap Year Mines. The mechanisms of this structural control and the partitioning of strain in the greenstone belt are under current research by the author.

The banded iron formations appear to have responded to the deformation by brecciation as noted at the Arthur and Pension Mines. At the Penhalonga Mine, a mineralised shear zone transgress from the talc schists into banded iron formation, parallel to the banding.

All Archean rock types within the greenstone belts of the Superior Province, Canada, are known to host gold mineralisation[18]. Individual gold deposits can be contained in a number of lithologies, a fact that has also been noted in the Mutare Greenstone Belt, particularly at the Old West, Redwing and Rezende deposits. This implies structural rather than lithological controls are dominant.

CONCLUSIONS AND THEIR RELEVANCE TO GOLD EXPLORATION

It is apparent that, on a regional scale, there is structural control of the gold mineralisation in the Mutare Greenstone Belt. However, at an orebody scale, the control of the location, dimension and orientation of the oreshoots appears to have been affected by a complex interplay between structure and lithology e.g. the Redwing orebody (author currently investigating this).

Second order shear zones tend to carry gold mineralisation as compared to first shear zones in Western Australia[18]. Within the Superior Province, gold mineralisation is hosted in small scale structures within larger deformation zones[18]. Though not enough research has yet been done in Mutare Greenstone Belt, the south-westerly extension (Mutare-Masvingo) has been shown to be a large major transcrustal strike slip shear zone with a sinistral sense of displacement[19].

In summary, identification of the Mutare-Masvingo transcrustal shear zone in the Mutare Greenstone belt may make it possible to predict subsisiary structures likely to host gold deposits.

343

ACKNOWLEDGEMENTS

The impetus for this paper was provided by Dr. T. Blenkinsop of the University of Zimbabwe. His encouragement and help goes beyond many thanks.

I would also like to thank all colleagues for their constructive criticism and useful discussion.

REFERENCES

1 Phaup, A.E. (1937) The Geology of the Umtali Gold Belt, S. Rhod. Geol Surv Bull 32.

2 Swift, W.H. (1956) The Geology of the Odzi Gold Belt, S. Rhod. Geol Surv Bull 45.

3 Wilson, J.F. (1979) A preliminary re-appraisal of the Rhodesian basement complex Spec. Publ. Geol. Soc. S. Afr., 5, 1 - 23.

4 Roberts, A.E. (1979) A gold bearing felsite horizon near Umtali-Rhodesia. Spec. Publ. Geol. Soc. S. Afr., 5.

5 Orpen, J.L. (1982) Final Report on E.P.O. 584 Geological Report.

6 Chenjerai, K.G. (1989) The Geology of the Penhalonga, Mutare District. Unpublished. Zim. Geol. Surv. Technical Files.

7 Roberts, A (1974) Final Report on E.P.O. 400.

8 Eisenlohr, B.N., Groves, D.I. and Partington, G.A. (1989) Crustal scale shear zones and their significance to Archaean gold mineralisation in Western Australia. Mineral Deposit 24 pp 1 - 8.

9 Foster, R.P. and Wilson, J.F. (1984) Geological setting of Archaean gold deposits in Zimbabwe, 521-551. In Foster, R.P., Ed., GOLD '82: The Geology, Geochemistry and Genesis of Gold Deposits, A.A. Balkema, Rotterdam, 753 pp.

10 Foster, R.P. (1985) Major Controls of Archaean Gold mineralisation in Zimbabwe. Trans. Geol. Soc. S. Afr., 88, 109 - 133.

11 Kalbskopf, S. (1982) Champion Mine, Odzi District. Unpublished. Zim. Geol. Surv. Tech Files.

12 Chenjerai, K.G. (1989) A Progress Report on the Champion Mine, Champion Farm (State Land), Mutare District. Unpublished. Zim. Geol. Surv. Tech Files.

13 Maufe, H.B. (undated) Rough notes re: Kent Mine dump. S. Rhod. Geol. Surv. Tech Files.

14 Stagman, J.G. (1959) Report on the King's Daughter Mine. Unpubl. S. Rhod. Geol. Surv. Tech Files.

15 Fabiani, W.M.B. (1980) The Geology of a felsite intrusion in the Penhalonga Valley - with special reference to the Old West Mine. Unpublished. Spec. Hons. project. University of Zimbabwe.

16 Hatherly, R.S. (1974) The Geology of the Redwing Mine Penhalonga. Unpubl. Spec. Hons. Project. University of Zimbabwe.

17 Harrison, N.M. (1979) The Geology of the Redwing Gold Mine, Penhalonga, Umtali District, Rhodesia. Spec. Publ. Geol. Soc. S. Afr., 5.

18 Colvine, A.C., Fyon, J.A., Heather, K.B., Marmont, S., Smith, P.M. and Troop, D.G. (1988) Archaean Lode Gold Deposits in Ontario. Ontario Geol. Surv. Miscellaneous Paper 139.

19 Vearncombe, J.R., Barley, M.E., Eisenlohr, B.N., Groves, D.I., Houston, S.M., Skwarnecki, M.S., Grigson, M.W. and Partington, G.A. (1988) Structural control on Mesothermal Gold Mineralisation : Examples from the Archaean Terrains of Southern Africa and Western Australia. Econ. Geol. Mon. 6 p 124 - 134.

General 2

Regional strategies for the minerals industries of the states of the SADCC

Paul Jourdan
Institute of Mining Research, Harare, Zimbabwe

SYNOPSIS

The SADCC has been in existence for eleven years but little progress has been made in regional coordination of the minerals sector even though this sector is the most important exporter of the region. There is however scope for minerals development at the regional level.

THE SADCC

The SADCC initiative was formally established in April 1980 at the Lusaka Summit in Zambia. The SADCC initially comprised nine southern African states namely, Angola, Botswana, Lesotho, Malawi, Mozambique, Swaziland, Tanzania, Zambia and Zimbabwe. Namibia joined as the tenth member in 1990. The members aimed to develop the region and to reduce their dependence on South Africa for transport, imports and the supply of electric power through "collective self-reliance". By 1989, 81% of the 6.3 billion US dollars for regional projects, of which half was secured, was for the rehabilitation of the regional communications system, particularly the railways.

The principal role of the SADCC minerals sector is that of a **foreign exchange earner** and in this regard it generates more than any other sector. The region derives over sixty percent of its foreign exchange from mineral exports worth, in 1988, 5.7 GUSD. Over the last decade minerals have on average constituted 64% of total export receipts, including oil exports, and 40% excluding oil. In 1988 the principal mineral exporters were, by value, Angola 37%, Botswana 25%, Zambia 15%, Zimbabwe 11% and Namibia 11% of total SADCC mineral exports. Thus these five states constituted 99% of all the mineral exports of the region.

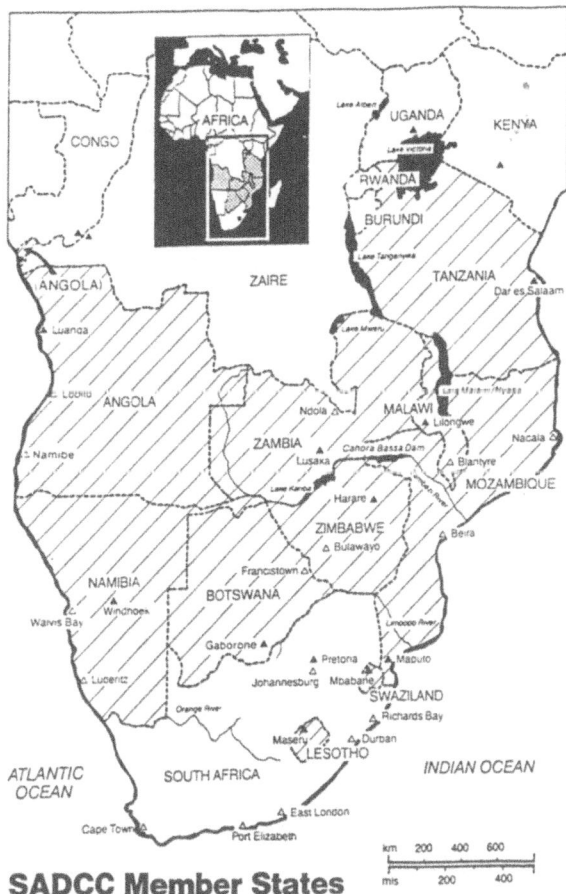

SADCC Member States

The SADCC region is rich in minerals but less than 10% of the mineral production is consumed within the region. The principal mineral exports are:

Angola:	Oil and diamonds,
Botswana:	Diamonds and base-metals,
Namibia:	Diamonds, uranium and base-metals,
Zambia:	Copper and cobalt,
Zimbabwe:	Gold and base-metals.

In 1988 oil, diamonds and copper made up over 80% of the total value of regional mineral output.

```
SADCC: PRODUCTION OF PRINCIPAL MINERALS

Mineral  1970 1975 1980 1985 1988  %     %World MUSD  Producer3
Units=kt                           Change 1988¹ 1988² 1988  %
=============================================================
Oil Mt     5.1  8.8  6.8 11.6 24.6 385%   .8%  2490 Ang 100%
Diamond⁴   5.5  6.7  8.4 14.5 17.4 216%  18.7% 1400 Bot  87%
Copper     743  734  694  570  505 -32%   5.7% 1122 Zam  84%
Nickel     8.6 15.6 30.5 29.4 34.0 295%   4.0%  326 Bot  66%
Uranium      0    0 4.77 3.39 3.80 nap    8.8%  265 Nam 100%
Fe-Chrome  163  182  211  195  225  38%   7.2%  235 Zim 100%
Gold t    14.0 11.5 11.8 15.3 14.9   6%    .8%  221 Zim  96%
Steel      349  467  689  643  600  72%    .1%  100 Zim  99%
Coal      4619 4972 4674 4173 6422  39%    .1%   74 Zim  79%
Cobalt    2.05 1.95 3.65 4.73 5.45 166%  15.0%   73 Zam  92%
Asbestos   221  299  284  199  209  -5%   4.8%   61 Zim  89%
Zinc      99.6 84.6 51.8 53.4 57.5 -42%    .8%   51 Nam  65%
Silver t  92.7 82.1  158  127  159  72%   1.1%   40 Nam  68%
Lead      97.4 63.4 52.7 47.4 50.8 -48%   1.5%   32 Nam  87%
Chromite   504  876  553  526  561  11%   4.6%   25 Zim 100%
Tin       2.15 1.76 2.02 2.71 2.06  -4%   1.0%   14 Nam  58%
Iron Ore  9456 9090 1621 1100 1021 -89%    .1%   14 Zim 100%
=============================================================
¹SADCC output as a % of World production in 1988, ²value of
SADCC output in 1988 in MUSD, ³principal SADCC producer & %
of SADCC production (volume), ⁴million carats.
Source: IMR SADCC Databank 1989.
```

A significant proportion of mining in the region has some degree of state control. The Zambian state owns 60% of the parastatal copper/cobalt mining company ZCCM and all petroleum production is partly owned by the Angolan parastatal Sonangol in partnership with Cabinda Gulf Oil Company of the USA, (Cabgoc) now owned by Chevron, Elf Aquataine (French), Petrofina (Belgian) and Texaco (USA). The states of the region owned 45% of the value of the major minerals produced in 1988, excluding oil. Including oil, the state share was 47%. State ownership by mineral varies considerably, from zero in the case of chromite mining, to 79% for regional coal extraction.

The rapid decline in mineral prices from 1980 meant that by 1986 most of the major mining houses were in a state of severe crisis. These included Botswana RST (nickel/copper), with a crippling 714 MUSD debt, Bindura Nickel (BNC), Zimalloys, the Zambian copper giant, ZCCM, MCM and African Associated Mines (asbestos). Due to relatively good diamond and gold prices, Debswana and the Zimbabwean gold mining companies were financially more stable.

Although the mining industries of the region face similar problems, the legislative structure that the governments have set up to overcome them varies considerably from state to state especially as regards their attitude to the mining trans-national corporations (TNC's). The colonial mining laws were by and large extremely favourable to the TNC's

particularly with regard to mineral rights, repatriation of profits, labour rights and the training of indigenous personnel.

Largely as a reaction to this perceived pro-TNC bias by the colonial administrations, on gaining independence many of the new governments introduced a new legislative regime which generally vested all mineral rights with the state, introduced heavier taxation, compulsory training of locals, extensive mining safety regulations and diminished repatriation of profits. From greater legislative control, the new regimes turned to acquiring increased state participation in the local subsidiaries of the mining TNC's as the commodities "boom" continued into the early seventies, in order to gain direct control of their mining industries and to retain as much of the surplus value as possible within the country concerned. The ex-Portuguese colonies (Angola and Mozambique), which only gained independence in 1975, from the outset embarked on a policy of state control of their mining industries, but this was also in part as a result of the settler exodus from these countries and the resultant abandonment of mining operations.

The major effect of the strict legislative regime and the increasing state participation in mining industry ownership was to cause a rapid decline in new foreign investment in mineral exploration and mining. With the exception of diamonds in Botswana and oil in Angola, there has been no major foreign investment in the mining industries of the SADCC since the early seventies. With the onset of the Global Crisis and the resultant decline in base metal prices in the early eighties, the states of the region started to look to the development of new mineral resources as the value of their "traditional" minerals declined in real terms. In order to attract investors, especially for small and medium scale operations, the governments embarked on a new phase of mining and investment laws revision.

The lack of investment has been particularly severe in exploration. Although several SADCC States have good gold mining potential, particularly Zimbabwe and Tanzania, they have almost entirely missed out on the spectacular gold boom of the eighties. Investment went instead to developed countries like Australia, Canada and the USA.

In 1981 Zimbabwe presented the SADCC Council of Ministers with a report on "Regional Cooperation in the Mining Industry", which outlined the importance of mining in the regional economies, emphasised the importance of southern African minerals as raw materials for the industrialised countries.

The clearest policy statement for the SADCC mining sector comes from the January 1985 Lusaka workshop:

"The ultimate objective of the (mining sector development) programme is to achieve coordinated investment and production in the SADCC region, through region-wide planning and promotion of both new projects and reactivation of existing capacity in mining, and mining linked activities."

Six years later tnere still was no evidence of any coordination through the SADCC Mining Coordinating Unit of either mining investment nor mineral production - only a seemingly endless stream of studies financed by the industrialised countries. What went wrong?

The reasons why coordination of the mining sector took so long to get underway and why so little has been achieved are twofold: Firstly the step-wise project approach of the SADCC without trade integration is extremely difficult to implement for productive projects beyond infrastructure and regional services. In this regard it is not only mining that has failed, there are no SADCC production facilities in any sectors. But this still does not explain why the mining sector has achieved so little in the regional coordination of non-productive mining aspects such as regional research and training strategies or facilities and the rationalisation of mineral processing facilities. Seconly, the weakness of the Mining Sector Coordinating Unit in Lusaka has contributed to the lack of progress.

The SADCC Mining Programme consists of 54 projects at an estimated cost of 70 MUSD of which 35 MUSD (50%), mainly foreign, has been secured. Several regional studies have been completed on the local (regional) manufacture of mining machinery and spare parts, the local production of mining chemicals and explosives, a study on regional iron and steel production (jointly with the SADCC Industry Sector), a regional mining manpower survey, a study on small scale mining, a study on mining data management, a report on fertilizer minerals, study on industrial minerals development, a market study for semi-finished copper products, a market survey of possible products from Sua Pan brines in Botswana, a feasibility study on the establishment of a refractory industry, a hydrogeological investigation of the SADCC region, a prefeasibility study on the establishment of an alumina/aluminium industry, a study on the processing of

lime in the sub-region and a study on regional coal exploitation. In addition, a regional study on methods of promoting investment in small and medium scale mining will be carried out in 1990/91.

What has always been admired as a strength of the SADCC, the decentralised structure with country sectoral responsibility, has in the case of mining proved to be a weakness as there is no effective means for the Secretariat in Gaborone to keep tabs on the efficacy of the sectoral co-ordination units, nor is there an adequate mechanism for the member countries to assess the units and to take remedial action.

What then are the options for strategies for the regional planning of the minerals sector? The answer to this question is probably best approached by presenting a series of scenarios ranging from a minimum situation that assumes that the level of regional integration will remain limited, to a maximalist situation that would allow the full realisation of the regional market.

1. The Minimum Strategy: The SADCC continues as is:

Although it will be difficult to realise the full fruits of regional integration under this scenario, there are nevertheless several areas where the regional mining and mineral processing industry might benefit. The first of these is the area of infrastructure and services. The minerals sector has already been the principal beneficiary of the substantial improvement in transport links particularly for export and there is still room for improvement in this area, particularly the rehabilitation of the Benguela line and the Limpopo line. There is also room for further improvement in the minerals handling capacity of the ports of Beira, Maputo and Dar es Salaam. These kinds of projects have been shown to be possible under the current SADCC structure and will probably receive funding as they facilitate the export of minerals to the donor countries. However, the rehabilitation of the transport system would also make intra-regional trade in minerals, mineral-based products and mineral inputs possible, except for the new SADCC member, Namibia, which has no rail link to the rest of the region. The, now shelved, project for a trans-Kalahari railway linking Namibia to Botswana for the export of the huge coal reserves in Botswana and, possibly, those of eastern Namibia, needs to be reassessed in the light of Namibia's independence and increasing oil and coal prices. A project of this type, if viable, would also most probably attract funding from potential First World customers.

The same applies to the regional electricity grid. Energy infrastructural projects aimed at interconnecting the SADCC region have proved to be successful under the SADCC project system and mining, as the region's largest consumer of electricity (34% in 1988), will benefit from the more stable and possibly cheaper supply. The project to connect Cabora Bassa to the Zimbabwean grid will be of particular importance to the development of ferrochrome smelting in that country.

Similar to infrastructure, projects for the establishment of research and development facilities to serve the region have been possible under the current SADCC system, particularly in agriculture. This has not occurred in mining, but it has been argued that this has been mainly due to the weakness of the SADCC Mining Coordinating Unit in Lusaka, rather than an inherent problem of the SADCC. With current and planned increased donor support to the Unit it is likely that more will happen in this regard. There are clearly numerous areas where the regional mining and mineral processing industry could benefit from applied research and there is also no shortage of locations in the region to base such facilities.

The SADCC studies into the regional situation have shown that there exists a high degree of complementarity in areas of manufacturing of supplies for the mining industry and in certain downstream processing, which cannot be realised without regional trade and payment mechanisms. The current situation of soft currencies and acute forex shortages further inhibits the setting up of regional facilities as there is no means of guaranteeing the potential regional market.

Without trade integration there are still numerous areas where the mining sector could benefit from the regional context such as the rationalisation of mining manpower training, particularly if a funding mechanism was created, possibly in the form of a regional scholarships clearing house. Limited regional cooperation in minerals marketing whereby larger minerals producers marketed the product of smaller producers could take place under the aegis of the Unit and a regional minerals marketing monitoring unit could be created, but a regional minerals marketing authority would not be possible under the current regional structure.

The SADCC Mining Coordination Unit could initiate or operate several other regional strategies that would be viable under the current SADCC project system including the establishment of a regional minerals, mining and geology databank, the encouragement of regional geology, mining

and mineral processing professional organisations, the organisation of regional conferences and workshops on minerals related issues to enhance the technical level of mining professionals, the facilitation of joint exploration programmes on geological features shared by two or more member countries.

A system whereby all the SADCC members undertake to purchase from a regional production facility should be pursued, but it would be difficult to manage as all the members would have to agree to any change in production range or volume and it would be extremely difficult to get members to honour their commitment if they could purchase the item cheaper on the world market. Nevertheless, for limited crucial areas with a large influence on the rest of the economy, namely fertilisers and iron and steel, regional rationalisation of existing facilities and the establishment of new facilities for the regional market, may be possible under an enforceable commitment to purchase system. In this regard it is recommended that a fertiliser board and an iron and steel board be created as these two areas straddle several of the current SADCC sectors (mining, industry and agriculture) and are vital to the development process.

A possible mechanism to overcome the payment problems associated with the establishment of a regional mineral-based manufacturing facility (such as a fertiliser plant) would be to devise a system whereby the forex content of the product is determined (with recourse to an independent referee) and payment is made in two parts: hard currency for the forex part (the facility would need to pay off its forex capital costs and limited forex running costs) and regional credits for the rest (the credit would allow companies in the country with the regional facility to import items from the rest of the region). In this way the forex cost of the product would almost certainly be cheaper than the full forex price of extra-regional sources. In addition the forex component will steadily decrease as the forex capital costs are paid off.

In conclusion, given the current level of regional integration in the SADCC, strategies for the mining sector will be principally limited to the improvement of the regional infrastructure and the creation of regional research and information facilities. The bulk of the regional potential for mineral based industries and mining input supply industries cannot easily be realised under the present SADCC system.

2. The Maximum Strategy: The SADCC moves towards a customs union with intra-regional currency convertibility.

The classical theory of customs unions emphasises the aspects of trade creation and trade diversion. Trade creation takes place when production (exports) from one of the partners displaces that of another, higher-cost, member. Trade diversion takes place when imports from a lower cost external (to the union) producer are replaced by a higher-cost union producer so the union theoretically ends up paying more, but the forex saving could compensate the extra cost. Thus if a union is on average trade creating rather than diverting it is considered beneficial and will increase "welfare". The main problem with these concepts, as far as they go, is that they were principally developed for the First World context assuming full employment and currency convertibility. This situation is not apparent in the SADCC where industrial capacity stands idle due to a shortage of raw and intermediate products (particularly mineral or mineral based), not because of a shortage of (local) funds to buy the inputs, but because there is no forex available. In this way many downstream industries are also affected because the local source of a vital input cannot produce.

The trade creation/diversion type of analysis also fails to appreciate the dynamic aspect of new industries being established that are only possible in the larger economies of scale, behind the joint outer tariff wall. In the context of an extremely underdeveloped region such as the SADCC, where many production facilities do not yet exist, this aspect is of even greater importance than groupings where facilities already exist (but would be rationalised). However, trade creation has two edges: On the one hand, by replacing the high-cost producer, the product becomes available to the union at a lower price, but, on the other hand, this entails the closure of the high-cost plant in one of the member countries (though, in the SADCC this would not occur much as often there are no facilities). This would be part of the polarisation effect where most facilities producing for the union will be drawn to the member with the already existing industrial base, whether this was an expansion of an existing industry (which might replace a competitor in another member, if one exists) or the establishment of a new industry for the regional market.

A customs union will increase the financial and technological economies of scale making regional (union) companies viable instead of having to rely solely on foreign TNCs for investment. Indigenous intra-regional companies (IRCs) are more likely to have policies that are in line with regional development policies than TNCs and, more importantly, they would not be repatriating their surplus outside the region.

The most problematic issue that would face a SADCC customs union would be that of industrial polarisation. Without corrective measures, regional industrial development will centre on the already most developed member, Zimbabwe. Therefore an industrial location policy would be absolutely essential from the outset with industries being allocated to the partners on a more equitable basis, based on their raw materials.

For instance, regional non-ferrous metallurgy could be based on the existing facilities in Zambia (Zamefa and Kabwe) while basic iron and steel production and engineering could be based in Zimbabwe (Zisco, Lancashire Steel, Zemco, Issels, Crasters, Connolys, etc...) with mini-mills throughout the region (Cifel, Scaw, AlAf, Siderurgia Nacional, etc...). Fertiliser plants could be established in Tanzania (Kilamco and Minjinji) and/or Mozambique (Pande and Evate).

As Zimbabwe already possesses many mineral-based industries and mining input industries (albeit often on a small, high-cost, import substitution scale) a regional strategy would require the closure or relocation of some of these industries, such as high-cost fertiliser production (Sable and Zimphos) and non-ferrous metallurgy (such as Almin, Cafca and Radiator and Tinning).

However, on the other hand, the lost employment and revenue could be compensated for by the expansion of its ferrous and engineering industries for the regional market and more efficient agriculture due to cheaper fertilisers. This would require a high level of regional planning and consensus which would need the creation of a permanent regional planning body to investigate the optimal distribution of regional production facilities with the highest degree of equity. It would also have to predict the relative gains and losses for each country for each project and could take corrective action in the further distribution of industrial plants. The polarisation effect could also be compensated for by giving the less developed members more than their share of revenues (such as in the SACU) and/or by setting aside a proportion of regional revenues for a development bank to fund projects in relatively depressed areas of the region and in other sectors.

Forex constraints, for both current consumables and new projects, are most probably the main problem facing the mining industry of the SADCC except for Botswana and Namibia. The shortage of foreign exchange to purchase essential mining inputs and spares, and for new equipment, has caused many of the mines in the region to operate below capacity, particularly in Zambia and Zimbabwe.

Very few SADCC members would be likely to accept this maximum strategy of a customs union with compensatory mechanisms and intra-SADCC currency convertibility due to current commitments (the SACU countries), feared loss of sovereignty over revenue collection or perceived loss from their present position. At an intuitive level, from the actions and statements of governments regarding the SADCC over the last ten years, it appears that Mozambique, Angola and, to a lesser extent, Tanzania, would favour a movement towards full economic integration; Swaziland would be opposed, as it seems to see more benefits from the SACU; as would Lesotho, mainly due to its geographic position; Malawi's position is unclear, but it is unlikely to oppose South Africa at present; Botswana might go along if there was a industry location system that compensated for the loss of SACU advantages; Zambia could also probably be persuaded if there was an industrial location system to protect and expand its industries, many of which also exist in Zimbabwe; geographically Namibia would have difficulty in joining and might therefore remain in the SACU, at least until such time as it has a rail connection to the rest of the SADCC; which leaves Zimbabwe as the crucial player.

Although the regional planning of the minerals sector of the SADCC may be in the interests of the SADCC "nations" not all the governments are likely to pursue this strategy as far as the creation of a customs and/or currency union. This is due to the objective reasons, stated above, and subjective reasons, such as a lack of appreciation of long-term gains by local elites; the distrust of their neighbours' ability to produce a product of adequate quality and quantity, and for timely delivery; distrust of their fiscal and exchange rate policies that would affect them in the event of trade integration and intra-regional currency convertibility; and, of great importance, their perceived loss of status. An important aspect of most of the SADCC government elites is that they are alienated from the production process (which is in the hands of settlers/TNCs) and thus political power is their only method of access to the national surplus. However, a union that included a cash payment to the local government (such as the SACU revenues) could in fact increase the amount of surplus

that they had access to. This cash payment aspect could have been an important factor in the survival of the SACU up till now.

The possible demise of the apartheid system in South Africa will remove a major unifying factor in the SADCC and could thus further weaken the region's commitment to integration and collective self-reliance.

3. A Post-Apartheid SADCC: The extension of the SACU to include SADCC.

The recent unfolding events in South Africa have raised the possibility of the establishment of a post-apartheid government in the not too distant future, which has in turn raised the question of a post-apartheid South Africa's and SACU's relationship to the SADCC and vice versa.

The obvious major change would be that South Africa would replace Zimbabwe as the dominant regional exporter and South African capital, particularly mining companies, would seek to re-penetrate the region, particularly countries with unexploited mineral potential such as Angola, Tanzania and Mozambique. There would be very little change as regards the SACU except for the outstanding renegotiation for the payment lag and the raising of the stabilisation factor. However, South African exporters trying to penetrate the SADCC region would come up against the same problems as experienced by Zimbabwe: namely that, although the countries are desperate for their products, they are too forex poor to buy them and will thus to a large degree continue to import from the West through aid credits and commodity import programmes.

The effect on the regional minerals sector of the creation of a sub-continental customs union would be tremendous as South Africa is itself the home of some of the world's largest mining TNCs which have the financial, technical and managerial capability to carry out virtually any mineral exploration or development project. Almost all mining inputs are already manufactured in South Africa and therefore these industries in Zambia and Zimbabwe would most probably have to be protected and/or rationalised. Rather one factory producing low-cost crushers for the whole region than ten little high-cost facilities for each local market.

The SADCC region has energy reserves that would interest South Africa. Current oil production from Angola exceeds sub-continental consumption. Due to South Africa's heavy reliance on high-sulphur coal for energy generation, in turn due to its oil vulnerability, it is currently facing a major ecological crisis (acid rain). There is huge untapped hydropower potential, particularly in Angola, which could be linked into the South African grid as was done for Cabora Bassa in Mozambique. Natural gas resources in the SADCC (Namibia, Angola, Tanzania and Mozambique) could be used to produce nitrogen fertilisers for the South African market. In this regard AAC has recently shown interest in the development of the Mozambican Pande deposit. The other essential "mineral" that South Africa lacks is water. The Lesotho highlands water scheme could ultimately export 70 cubic metres per second to South Africa by the year 2025 and provisional plans have been drawn up for the import of water from the Zambezi River.

To conclude, the formation of a post-apartheid sub-continental economic grouping would make even more economic sense in the long-run than the SADCC, but would have to resolve an even greater industrial polarisation problem. Yet the disproportionate strength of the South African economy and its convertible currency based on gold could also be a positive factor in that union would offer immediate advantages to the war torn and partially destroyed economies of Mozambique and Angola and might act as a binding factor so long as the regional location of industry was satisfactorily addressed.

References

1. Anglin, D.G. *Economic liberation and regional cooperation in southern Africa: SADCC and PTA*, in <u>International Organisation</u>,vol.37, no.4, 1983.
2. Balkay, B. *Mining and mineral resource-based industries in the SADCC group of countries*, mimeo, UNIDO, Vienna, 1984.
3. Balkay, B. *The SADCC Group: Scope for effective sovereignty in the raw materials sector,* mimeo, Institute for World Economics, Budapest, 1985.
4. Boyd, J.B. *A subsystemic analysis of the Southern African Development Coordination Conference,* in <u>African Studies Review</u>, vol.28, no.4, 1985.
5. Brown, R. & Faber, M. *Some policy and legal issues affecting mining legislation and agreements in African Commonwealth countries,* Commonwealth Secretariat, London, 1977.
6. Chr. Michelsen Institute. *SADCC intra-regional trade study,* CMI, Bergen, 1986.
7. Cobbe, J.H. *Integration among unequals: The Southern African Customs Union and Development,* in <u>World Development</u>, vol.8, no.4, 1980.
8. Hanlon, J. *SADCC in the 1990's: development on the front line,* Economist Intelligence Unit, London, 1989.
9. Haughton, S.H. *Geological History of Southern Africa,* Geological Society of South Africa, Johannesburg, 1969.
10. Jourdan, P. *Mining in the SADCC* in <u>Raw Materials Report,</u> vol.3, no.3, 1985.
11. Jourdan, P. *Mining in the SADCC: A regional approach to minerals development,* mimeo, paper presented at the AGID <u>Geosciences in Development</u> conference in Nottingham, 1988.
12. Jourdan, P. *Problems and prospects of mining in the SADCC region,* mimeo, paper presented at the SAPES/ICDA conference on <u>SADCC: Problems and prospects of political and economic cooperation,</u> Gaborone, 1989.
13. Jourdan, P., Lebrun, M., Chitambo, A., Wapakwenda, S. & Sweta, W. *Mining in the SADCC: Progress and Prospects,* in <u>Raw Materials Report</u>, vol.6, no.1, 1988.
14. Kalyala, D. & Mudenda, G. *The effects of the world economic recession on the mining sector in the SADCC region,* in Amin, Chitala & Mandaza (eds) <u>SADCC: Prospects for Disengagement and development in southern Africa</u>, Zed Books, London, 1987.
15. Maasdorp, G. *The Southern African Customs Union - An assessment,* in <u>Journal of Contemporary African Studies</u>, vol.2, no.1, 1982.
16. Pelletier, R. *Mineral resources of south-central Africa,* OUP, Cape Town, 1964.

Development of a minerals information and evaluation system

J. G. Voss
E. Alaphia Wright
I. Koppe
Department of Mining Engineering, University of Zimbabwe, Harare, Zimbabwe

SYNOPSIS

This paper is a 'progress report' on the project, "The Mineral Resources Data Bank", launched in mid 1988 and based in the Department of Mining Engineering, University of Zimbabwe. The basic concepts forming the framework of the system being built are explained. Reasons for the hardware and software selected are presented and a prototype, built to test systems functionality, described. Important aspects learnt from the development so far, include the viability of building such a system and the confirmation of the urgent need for the completion of the project.

INTRODUCTION

The exploitation of mineral resources in Zimbabwe is of major importance for the country's economy. For the mineral industry to maintain its position and contribute to long term development, it is necessary to (1) explore and evaluate new resources, (2) develop mineral resources for exploitation, (3) avoid wasteful exploitation of deposits already in production, while paying attention to the protection of the environment and (4) develop the industry in its technology and economics.

In order to support activities aimed at achieving the above objectives, the documentation of relevant information on mineral resources, the rapid access to, and efficient means of handling such information, becomes paramount.

At present, data on mineral resources and information on their exploitation are collected and stored by several Government departments and other institutions. These bodies include (1) the Geological Survey Department, (2) the Department of Mining Engineering in the Ministry of Mines, (3) the various offices of the commissioner of Mines and (4) the Institute of Mining Research.

It has been recognised that the collection, documentation and evaluation of information on mineral resources can effectively be realised through the basis of a computerised Mineral Resources Data Bank. A major requirement for such a data bank, is that it should serve not merely as a documentation facility, but also as a support instrument for relevant evaluation practices in the Minerals Industry.

The project, "The Mineral Resources Data Bank Project" was therefore launched in mid 1988 to undertake development work for such a data bank system. The project is based at the Department of Mining Engineering, University of Zimbabwe.

This presentation discusses some of the details considered in developing the data bank. To start with, the basic concepts are explained followed by a description of the approaches employed. The hardware and software requirements are examined and the details of a prototype, developed to test the basic concepts, are described. Finally, important lessons learnt from the research work so far are discussed.

BASIC CONCEPTS

On the whole, computerised mineral data banks are now routinely in use in several countries worldwide. Some of these cover only specific minerals, while others cover several minerals and whole regions. Three examples of such systems are (1) "AMIS", a mineral resources data base system operated by the Bureau of Mines, of the Government of the United States of America[1,2], (2) "GEOFIZ", operated by the Federal Bureau for Geological Sciences and

Mineral Resources in Hanover, Federal Republc of Germany, and (3) the SADCC Mining Data Base, Institute of Mining Research in Harare [3].

By and large, these systems are basically sophisticated filing systems with report generation facilities. In terms of evaluation, such systems seldom go beyond the practice of computing summary statistics on such details as: market share of a given mining group, production levels of given minerals in given years, and so on.

The Mineral Resources Data Bank (MRDB) system was concieved as a 'Working Data Bank'. In such a system data is stored in a series of levels; from global details such as occurences in the country right down to the original exploration data for given deposits. The storage levels are to be linked together in such a way that it is possible to follow cases of interest right down to the reserve estimation and mine planning stage.

Fig. 1 shows a function diagram for the basic concepts of the system. The implications were that such a system could only be effected on a computing system with adequate capacity. The 'workstation' was therefore selected as the minimum computer configuration to be considered.

Quite early in the execution of the project, it was decided that the system can best be built on the basis of commercially available software. The complete system would then be made up of two major groups of modules. Group 1 would contain the data management modules and group 2 will be made up of the planning and evaluation system. A major requirement then would be the building of the necessary interfaces between the two groups of software modules.

APPROACHES EMPLOYED

Procedures

As a project of a similar nature has not been executed before, the work was started more or less from scratch.
First of all, a working group was formed consisting of representatives from several institutions dealing with mineral resources information. The institutions represented in the working group include The Departments of Mining Engineering and Geology of The University, The Geological Survey Department of the Ministry of Mines and the Mining Development Unit , also of the Ministry of Mines.

The working group was tasked with (1) the identification of end users and their special requirements, (2)charting the cooperation formalities for collection of mineral resources information, (3) defining access facilities and (4)working out measures to ensure confidentiality of the data to be handled. In addition, the building of a prototype to test the validity of the proposed concepts for the MRDB system was set down as a priority.

Description of The Prototype

The database management part of the prototype was developed and implemented within a dBase III plus environment. The function diagram in Fig. 1 shows the relationship between the system and the user. Fig.2 shows a flow chart of the data management module(s) for the proposed full sized system. This framework was also implemented in the prototype.

The prototype consists of a main program 'MRDB' and two supporting sub programs 'MANIPUL' and 'DATADIS'. MANIPUL is employed as a processing module which handles raw input and writes up data in relevant formats for storage on disks and tapes. The stored data can then be processed using DATADIS for data display and the production of relevant reports. Both sub programs also employ a procedure named 'CURSOR', which facilitates a comfortable users' interface with the system.

Four main groups of information are handled by the system. These are refered to as geographical, general, geological and production information. Fig. 3 gives a summary of the information concerned.

In the prototype, data for given deposits can also be written out to disks, from both MANIPUL and DATADIS, for further processing through the evaluation modules. For evaluation and planning purposes, given occurences are identified directly from the data management modules of the system (Figs. 4 & 5).

The above are in addition to the usual data entry and query facilities offered by dBase III plus. Fig. 6 shows a typical response to queries concerning production and mineral types. The entries have been ficticiously selected for reasons of confidentiality.

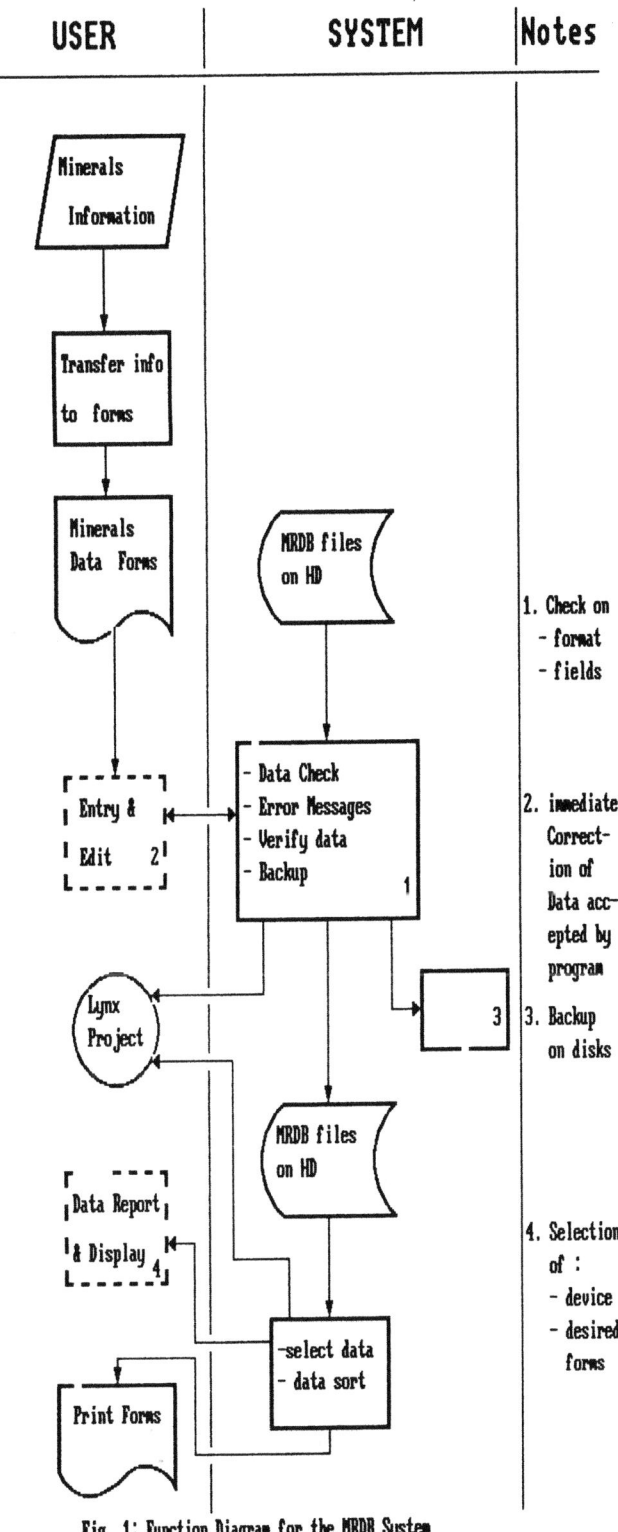

Fig. 1: Function Diagram for the MRDB System

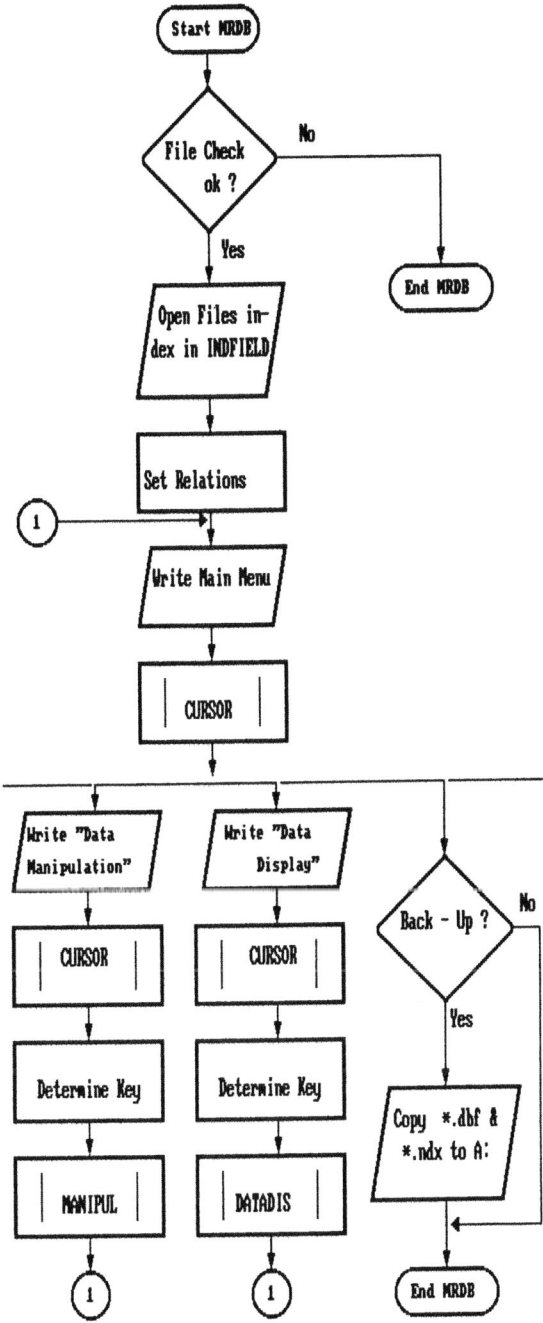

Fig. 2: Flowchart for the Main Program "M R D B"

HARDWARE & SOFTWARE CONSIDERATIONS

Hardware

By the very nature of mineral deposits, the amount of data from exploration can easily become quite large. Fig 7 shows the record size (in Bytes) for the prototype implementation. With a record size of 737 Bytes, it is evident that not many records can be held in memory when using a standard microcomputer. It has therefore been decided that the required computer resources for implementing a full sized working system, will have to be in the workstation class at the minimum.

Fig. 8 shows the initial configuration of the computers on which the system is being built, as installed in the Department of Mining Engineering, University of Zimbabwe. The hardware had since been upgraded to bring the CPU's up to 16 MB Ram each with a total of just under 1200 MB hard disk capacity.

358

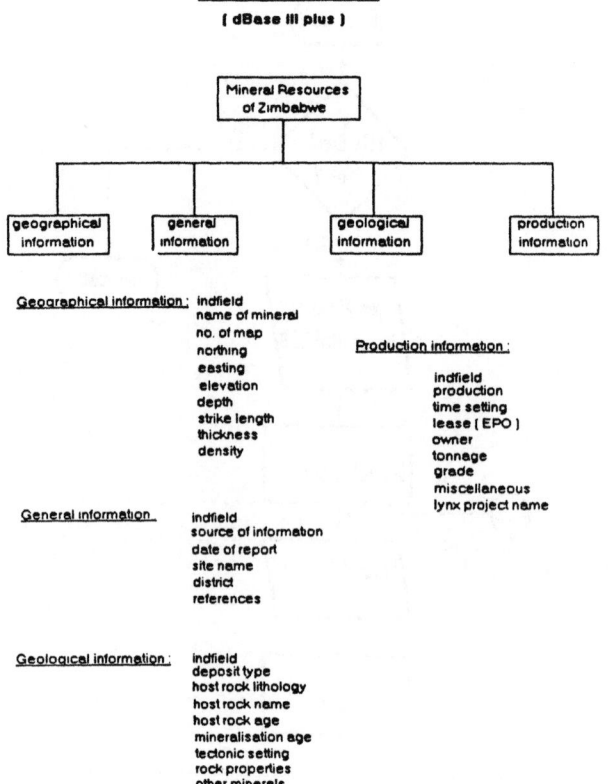

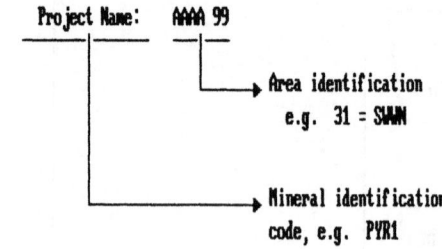

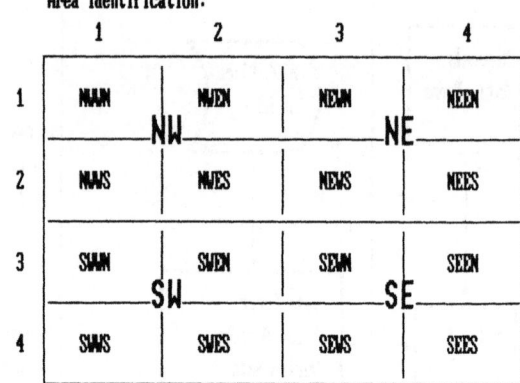

Fig. 3: Basic Information groups handled by MRDB

The department is currently (November 1990) in the process of exchanging the mother boards for the two HP series 9000 model 350 for those of the more powerful model 375. Peripherals for the main system include an A0 digitiser, (CALCOMP 9100), used for map data entry; several A3 digitising tablets, an A1 plotter (HP DraftPro), and several A3 Plotters (HP 7475 A).

The prototype of the MRDB system has been implemented on a 640K Ram, 40MB hard disk microcomputer. The capacity of the microcomputer, an HP Vectra PC, was quite adaquate for building and testing the functionality of the prototype MRDB.

Software

An overall requirement of the system is that it has to offer multi-user facilities. This led to the choice of a UNIX based Operating System. The proposed full sized MRDB system will run under the HP-UX operating system as supplied for the HP 9000 series computers.

With respect to the mine planning software, extensive studies were undertaken to consider available mining packages. It was decided to select software offered by Lynx Geosystems of Vancouver Canada. Two supporting reasons for this choice include: (1)the Lynx

Example:
Total Number of Areas : 16
Estimated number of minerals/area: 100
Total Number of Projects: 1600

Fig.4: Scheme for Project Identification for MRDB Evaluation Module

Lynx Project Management

according to MRDB Management

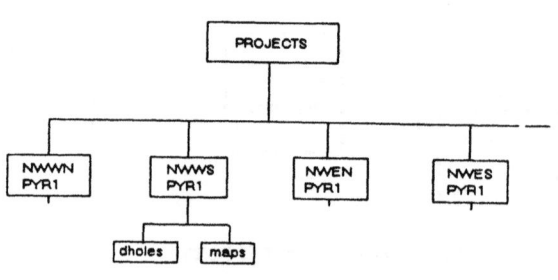

Fig. 5: Overall project Management Scheme for the Evaluation Module

Software had a version running under HP-UX, and (2)Lynx had been able to arrange for local maintenance support to be provided in Harare.

13.09.89

MRDB - Production and Lynx

MINERAL	SITE NAME	PROD	TIME	SET	TONNAGE	GRADE	LYNX PROJECT
AG	norton	y	s 24.08.60		5000	3.00	AG001101
AG	ytong	n	.	.	10000	12.00	AG004401
ASB	tengenenge		.	.	30000	30.00	ASB02201
AU	nyanyadzi	n	.	.	6000	4.00	AU002302
AU	biriiri	n	.	.	1000	2.50	AU002301
CO	copper	?	.	.	500000	10.00	CO002301
DIA	secret	?	.	.	50000	9.00	DIA03201
EVA	gona rhe zhou		.	.	18000	18.00	EVA04401
FE	luck	?	.	.	1000000	11.00	FE004401
GEM	limework		.	.	2000	7.00	GEM04301
GRF	nkulu		.	.	550000	20.00	GRF01101
MN	malawi		.	.	1000	12.00	MN002301
MN	shebeen		.	.	20000	7.00	MN002303
MN	uz		.	.	8000	4.00	MN002304
NI	loop	?	.	.	10500	5.00	NI002401
NI	ik	y	s 30.07.77		3000	4.00	NI004301
PYR	xtree		.	.	150000	12.00	PYR04201

Fig. 6: Example Results to a Query on Production according to Sites

MRDB FILES UNDER DBASE III PLUS

	HEADING	FIELD SIZE	TOTAL
GEOGRAPH.DBF :	Indfield	6	
	Name of mineral	4	
	No. of map	20	
	Northing	10	
	Easting	10	
	Elevation	7	
	Depth	7	
	Strike length	7	
	Thickness	6	
	Density	6	83 Byte
GENERAL.DBF :	Indfield	6	
	Source of inform.	1	
	Date of report	8	
	Site name	40	
	District	40	
	References	80	175 Byte
GEOLOG.DBF :	Indfield	6	
	Deposit type	24	
	Host rock lith.	40	
	Host rock name	25	
	Host rock age	15	
	Mineralis. age	15	
	Tectonic setting	68	
	Rock properties	68	
	Other minerals	60	321 Byte
PRODUC.DBF :	Indfield	6	
	Production	1	
	Time setting	12	
	Lease (EPO)	16	
	Owner	60	
	Tonnage	10	
	Grade	5	
	Miscellaneous	40	
	Lynx project name	8	158 Byte

Total Bytes per Record :   737

Fig. 7: Record Size (in Bytes) for Prototype MRDB implementation

DEPARTMENT OF MINING ENGINEERING AND METALLURGY
UNIVERSITY OF ZIMBABWE

HEWLETT PACKARD 9000   SERIES 300   SYSTEMS CONFIGURATION

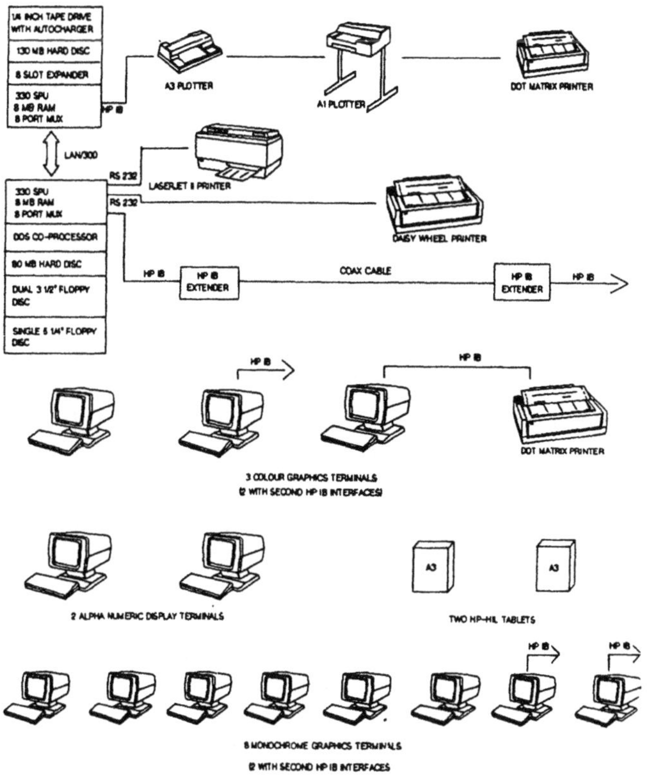

Fig. 8: Initial Computer Configuration for the System on which MRDB is being built

Unlike the choice of dBase III plus for building the prototype MRDB, the selection of the data management modules for the full sized system has not been so easy. To date (November 1990), several possible softwares are being examined. These include: (1) ORACLE, a relational Database product of Oracle Corporation of Belmont, Califonia, USA, (2) ARC-INFO , A Geographic Information Systems package, also from the USA, and (3) some of the ERDAS application programs of ERDAS Inc of Atlanta, Georgia, USA.

The necessary interfaces between the data management part of the full sized MRDB and the Lynx mine modelling software is being written in-house.

DISCUSSIONS & CONCLUSIONS

The successful construction and use of the prototype MRDB has served to validate the basic concepts. In addition, invaluable experience concerning some major aspects, to be considered in building the full sized system, have also been accumulated while working with the prototype.

Surprisingly, the need for such a minerals information and evaluation system appears to have been originally underestimated. With increased activities in minerals evaluation there is now an urgent need to complete the full sized system as soon as possible. Also, the amount of effort required for developing the system have similarly been underestimated. To speed up the development, a Senior Research Fellow was employed from the start of November 1990 to work full time on the project.

Further, the advantages of cooperation in a project such as this surfaced time and time again during the work on the prototype. Such cooperation is being continued for the construction of the full sized system.

The Zimbabwean mining industry appreciates the advantages to be derived from the use of mining software for mine modelling purposes[4]. On the completion and commissioning of the full sized MRDB, it is envisaged that various institutions involved with mineral resources will be able to benefit from the added advantages that would come with the use of such a system. For a start, it would then be possible to make full use of readily accessible information for the various analyses required to support effective decision making concerning mineral resources.

Two important details remain to be finalised. These are (1) the details covering confidentiality of the data to be collected and (2) the general procedures for the administration of the full sized MRDB, when once it has been completed and put into operation.

References

1. Stone, J.P. & Kass, L.M. (1982), "The Bureau of Mines Automated Minerals Information System- AMIS, in Proceedings of the 17th APCOM Symposium, Society of Mining Engineers of AIME, New York, pp 740-753.

2. Kass, L.M., (1984), "Development of Microcomputer workstations for the Automated Minerals Information System (AMIS)", in Proceedings of the 18th APCOM Symposium, IMM London, pp 631-643.

3. Jourdan, P. (1989), "The SADCC Mining Database", Annual Report of the Institute of Mining Research, Harare, pp 92-100.

4. Wright, E.A.,(1990), "Mining Software in Zimbabwe - An Evaluation of Users' Interaction with Existing Packages", 2nd Symposium of the Research Council of Zimbabwe, Harare, 10 Pages.

Minerals and the environment in Ghana

J. Stocks
Jay Mineral Services Ltd, Truro, England
J. G. A. Renner
Renner & Associates, Accra, Ghana
P. C. Acquah
Minerals Commission, Accra, Ghana

INTRODUCTION

The Ghanaian economy is principally dependent for foreign exchange on cocoa, timber and mineral products, especially gold. 35,000 people are employed in the mining sector, including 11,000 in small scale production. Mining produces about 20% of Ghana's foreign exchange, a proportion which is rising.

Major economic changes in the country were initiated in April 1983 when the Government of Ghana introduced a new reform package known as the Economic Recovery Programme. This aimed to revitalise productivity, improve the financial position of the public sector and encourage the expansion of private investment. The Programme was favourably received by international agencies and Ghana has benefitted from a substantial inflow of loans and development funds. A more flexible foreign exchange rate and a new investment and mining code have strengthened the incentive offered to foreign capital. As a result of these changes the mining sector is very active with over 70 prospecting and reconnaissance licences in the hands of local and foreign companies and a wide range of projects at various stages of development. Ashanti Goldfields Corporation (Ghana) Ltd, the country's leading gold producer, has embarked on a major expansion programme and new gold output is coming on stream from Teberebie Goldfields Ltd, and Canadian Bogosu Resources Ltd amongst others.

ENVIRONMENTAL POLICY

Historically the pressures on the mining sector have been economic to generate employment, earn foreign exchange and boost local and national economies. Although a national Environmental Protection Council (E.P.C.) has been in existence for some time, this has in the past been underfunded and lacking effective teeth. The result is some historic environmental damage from mining activity. Pollution of surface water courses from tailings and other sources has occurred, a factor of concern in a country where many people outside urban centres are as yet unconnected to piped water. There has been some defoliation associated with roasting of sulphide ores and some dereliction.

The mining sector has not been alone in causing environmental problems and in March 1988 the Government of Ghana gave the lead role to the E.P.C. to prepare an Environmental Action Plan which would help make Ghana's economic development strategy more environmentally sustainable. Six committees were formed as an aid to developing a National Environmental Policy. The topics covered by these were:-

a) Mining, Industry and Hazardous Chemicals.
b) Land Use Management.
c) Marine and Coastal Systems.
d) Forestry and Wildlife.
e) Water Management.
f) Human Settlements.

The Committee on Mining, Industry and Hazardous Chemicals began work in July 1988 with terms of reference from the E.P.C. which included:-

1) Potential negative impacts of surface mining;
2) Health and safety hazards to miners;
3) Environmental and public health concerns related to gold ore processing;
4) Issues of solid waste management, effects of effluents on water quality and air pollution;
5) Need for systematic impact assessment procedures.

The Committee consulted widely and reported in November 1988. A questionnaire for mining companies and other relevant organisations was circulated, available data collected, and experience in other countries was carefully evaluated. It was determined that existing legislation is generally broad enough to cover environmental control but there is an absence of specific regulations. The allocation of responsibility was identified as a particularly important topic.

The E.P.C. called a seminar of all experts in late 1988 after the individual committee reports had been received. This seminar identified the lack of provision for environmental education as a serious omission and a consultant was retained to make recommendations on this topic and to act as

361

a general editor for the six reports. He was asked to assign priorities to the various projects put forward by the committees.

A National Conference was then called, to which the edited report was presented. This Conference was open to the general public and delegates were also sent from District Assemblies to ensure that the views of the population at large were taken into account. Following the Conference a group of consultants from the World Bank and the University of Ghana was asked to undertake an economic appraisal of the various projects proposed and to prepare documentation for identified projects.

The edited report, in the form of a draft Environmental Action Plan (E.A.P.), was thoroughly reviewed to identify any gaps in a workshop which took place from 29th October to 2nd November 1990. Finalisation of the document is almost complete, ready for submission to Government by the end of 1990.

In parallel with this activity, the E.P.C. has prepared Draft Guidelines for Environmental Impact Assessment. The proposed E.I.A. system consists of three stages:-

i)   Initial screening of project.
ii)   Initial environmental evaluation.
iii)   Full assessment.

A booklet is in preparation to help project planners and investors, as well as decision makers and regulatory agencies, to integrate environmental considerations into development projects at an early stage in the planning process. The objective of the E.I.A. is to bring into focus any possible negative impacts on the environment at such an early stage in the planning process that these impacts may either be mitigated or avoided. The E.I.A. also permits reconciliation of ecological needs and concerns with economic considerations, and promotes the effective participation of concerned groups in the development process. Overall the E.I.A. seeks to compare all feasible alternatives, and determines which one represents an optimum mix of environmental and economic costs and benefits.

Drilling, mining, blasting, quarrying and sand winning are listed as projects requiring E.I.A. Proposed administrative procedures include the formation of Technical Committees by Regulatory Agencies. Their duties would include vetting

proposals and recommending whether full E.I.A. is required. If so the Company would be asked to undertake an E.I.A. and send it to the Technical Committee for comment and recommendations to the general public by means of national advertising and consultation with the District Assembly and local traditional leaders etc. The Technical Committee would then prepare a further review taking into account the views of the general public and this would be submitted through the

Regulatory Agency to the responsible Ministry for decision.

THE MINING SECTOR

The Minerals Commission under the instructions of the Ministry of Lands and Natural Resources, in December 1989 went out to tender on a study on the effect of mining on the environment. This study, sponsored by the World Bank, aims to formulate guidelines to minimise any adverse effects to increased mining activity on the environment. The study will address soil, water, land use, crops, other vegetation and air. The consultant will also prepare a socio-economic assessment of the impact of new mines on the local population including job creation, changes in settlement patterns and health effects.

The study commenced on 5th November 1990 with field visits by selected consultants and Government officials. The field visits are expected to cover eleven operating mines in the country which variously win gold, diamonds, manganese and bauxite. During the study, the consultants will gather some baseline information, prepare an assessment of the impact of new mines - particularly surface mines - and related investments on Ghana's environment. It is expected that the environmental study will culminate in the formulation of guidelines for draft regulations for the construction and operation of both underground and surface mines in the country. The consultants are due to report by April/May 1991.

Once the National Environmental Action Plan is in place, there will be a clear requirement for E.I.A. for new mining projects. In the interim the Minerals Commission is already anticipating new legislation by requiring E.I.A. from all new applicants for mining leases. It is also a policy of some international agencies, such as the I.F.C., to require E.I.A. on projects seeking funding. An example of this is presented below.

THE SANSU PROJECT - A CASE STUDY

Ashanti Goldfields Corporation has been mining gold in the Obuasi District of Ghana since 1897. The Company is 55% owned by the Government of Ghana and 45% by Lonrho PLC, and currently employs over 10,000 Ghanaians. Historically virtually all production has come from underground mining of high grade quartz and sulphide orebodies. Recent prospecting has disclosed a halo of low grade gold values surrounding the high grade zones. These provide a low grade bulk mining target, particularly close to surface where the sulphides have decomposed to oxides.

The Sansu Project embraces outcrops along a 6.5km strike length. 25 deposits of significance have been delineated. It is proposed to recover gold from the better grade ore in an Oxide Plant, with lower grade material fed to a heap leach operation. Initial reserves total some 15 million tonnes but ongoing exploration may add substantial additional reserves. Mining will be by standard open pit benching with limited blasting because of the weathered nature of much of the ore. Heap leach production is scheduled to peak at about 650,000 tonnes per annum and Oxide Plant capacity will be just over 1 million tonnes per annum. It was estimated that the project would create almost 1,000 new jobs.

As part of its funding arrangements with the I.F.C., Ashanti Goldfields agreed to retain independent consultants to undertake a full E.I.A. of the project. A baseline study collected data on:- air quality; surface and groundwater; flora and fauna; topography, soil quality and land use; and socio-economic factors. Important factors to emerge from the baseline study are briefly described below.

Air quality in the area was good, with windblown dust and smoke from woodfuel the only detectable pollutants. In the vicinity of Obuasi township and existing mining operations, local rivers are considerably polluted by industrial effluents and domestic wastes. Both surface and groundwater in the Sansu area are of good quality, but with rather high arsenic levels caused by widespread local mineralisation.

The area around Obuasi has witnessed extensive mining activity for almost 100 years. As a result of this and the associated high population levels the whole district is ecologically impoverished.

Almost nothing remains of the original tropical rain forest and the area is now characterised by extensive forest plantations, small subsistence farms, human settlements and industrial buildings. Both the population and variety of fauna have been affected by habitat destruction, hunting and farming. As far as could be ascertained, no threatened or endangered species are living in or close to the project area.

Topography is hilly and soil analysis indicates that just over half of the area is potentially good for agriculture and forestry. Some 40% of the land is currently forest plantation, with 40% lying fallow and remaining land taken up by small farms and villages.

The township of Obuasi has grown from 12 houses in 1897 to its present population of 74,000. The company operates a 150 bed hospital available to all local residents. Malaria accounts for 60% of illnesses in 14,000 outpatients and 1,000 inpatients treated per month. Settlements closest to the project are the villages of Sansu, Alata and Bidiem with populations of 1,115, 46 and 113 respectively. Villagers here are typical of many small settlements in the Ashanti region. Facilities are poor, with no connected electricity and reliance upon surface water sources. There is a junior school at Sansu, but the nearest health care and secondary school facilities are at Obuasi, 15km by road.

An environmental impact assessment of the project was undertaken, and reached the following principal conclusions.

Airborne dust would be the only air pollutant and this would arise from a variety of activities including drilling, blasting, loading, hauling and crushing of rock. The area's high rainfall throughout most of the year was expected to minimise dust generation except during the dry season from November to February. The company's proposals to include a dust extraction unit at the crusher and regularly spray roads by water bowser in dry weather were accepted as adequate subject to monitoring. In the absence of local standards, it was recommended that World Bank Guideline Standards of $100\mu g/m^3$ annual geometric mean should be adopted, and additional dust suppression measures implemented if monitoring were to show non-compliance or inadequacy of original measures.

Water abstraction from the Oda River was predicted to peak at $5,000m^3$ per day and hence impact on water resources was accepted to be minimal. The project was designed for total recirculation of process water and incorporated extensive measures to prevent cyanide contamination of surface and groundwaters. A comprehensive monitoring programme was recommended to ensure early detection of accidental leaks and spillages and facilitate quick remedial measures. Project design allowed for occasional neutralisation and discharge of cyanide solution during periods of unusually heavy and prolonged rainfall. Settlement ponds and silt traps were recommended at strategic locations to prevent runoff carrying a heavy load of suspended solids.

In view of the impoverished nature of the area and the lack of endangered species, impact on flora and fauna were not viewed as significant.

The project would directly affect 1,000ha of land. Some of this would be removed permanently from the useful land bank. The Consultants recommended early implementation of revegetation trials on tailings to assist satisfactory reclamation at the end of the project.

The main negative socio-economic impacts of the project would be destruction of some small farms and the need to resettle the villages of Alata and Bidiem. A scheme of compensation for farms was agreed between the company and village leaders, and compensation for displaced villagers would be according to statutory levels established by the Government's Land Valuation Board. On the positive side, known reserves were expected to yield a total revenue of US$786 million, of which the Government of Ghana would receive US$260 million in royalties, taxes and dividends. Allowing for a multiplier factors, the project would create about 1,000 new jobs many of which would be suitable for villagers. It was clear from consultations with local villagers and their leaders that the project was welcomed because of the perceived benefits of employments and improvement of local infrastructure.

The consultants overall conclusions were:-

"The Sansu Project as proposed will have little adverse impact on air quality, flora and fauna, and land use. There is considerable potential

for adverse impact on surface and groundwater but the proposed control measures together with an extensive monitoring programme should be adequate to prevent significant adverse environmental effect. Reclamation should be technically possible on all disturbed land areas, although land use potential will be reduced over a total of about 1,000 acres (400ha).

Having regard to all the above, we consider the adverse environmental impact of the project will be relatively small, but that the socio-economic impact will be of great importance both locally and nationally and that this consideration outweighs all others."

CONCLUSIONS

During the past 3 years, Ghana has been proceeding in an orderly and logical fashion towards a National Environmental Policy in a manner that could be a model for other developing countries.

Within the near future the Policy will be in place, supported by the necessary legislation and regulations, and with the Institutions to ensure compliance. The implementation of this clear and coherent framework will benefit all parties. The people of Ghana will be assured that economic growth takes place only in an environmentally sustainable fashion and that the price of development is not irreparable damage to the natural environment. The developer will be able to plan with a clear understanding of legislation and regulatory requirements and with the knowledge that his project will then be in harmony with the goals of the host country.